Edited by
Edward P. Riley,
Sterling Clarren,
Joanne Weinberg,
and Egon Jonsson

Fetal Alcohol Spectrum Disorder

Titles of the series
"Health Care and Disease Management"

Martin, W., Suchowersky, O., Kovacs Burns, K., Jonsson, E. (eds.)

Parkinson Disease

A Health Policy Perspective

2010
ISBN: 978-3-527-32779-9

Rapoport, J., Jacobs, P., Jonsson, E. (eds.)

Cost Containment and Efficiency in National Health Systems

A Global Comparison

2009
ISBN: 978-3-527-32110-0

Lu, M., Jonsson, E. (eds.)

Financing Health Care

New Ideas for a Changing Society

2008
ISBN: 978-3-527-32027-1

Rashiq, S., Schopflocher, D., Taenzer, P., Jonsson, E. (eds.)

Chronic Pain

A Health Policy Perspective

2008
ISBN: 978-3-527-32382-1

Forthcoming

Clarren, S., Salmon, A., Jonsson, E. (Eds.)

Prevention of Fetal Alcohol Spectrum Disorder FASD: Who Is Responsible?

2011
978-3-527-32997-7

Related Titles

Miller, N.S., Gold, M.S. (Eds.)

Addictive Disorders in Medical Populations

2010
ISBN: 978-0-470-74033-0

Mitcheson, L., Maslin, J., Meynen, T., Morrison, T., Hill, R., Wanigaratne, S., Padesky, C.A. (Foreword by)

Applied Cognitive and Behavioural Approaches to the Treatment of Addiction: A Practical Treatment Guide

2010
ISBN: 978-0-470-51062-9

*Edited by Edward P. Riley, Sterling Clarren,
Joanne Weinberg, and Egon Jonsson*

Fetal Alcohol Spectrum Disorder

Management and Policy Perspectives of FASD

WILEY-BLACKWELL

The Editors

Prof. Dr. Edward P. Riley
San Diego State University
Center for Behavioral Teratology
6363, Alvarado Ct. #209
San Diego, CA 92120
USA

Prof. Dr. Sterling Clarren
University of British Columbia
Faculty of Medicine – Pediatrics
L408 – Oak Street 4480
Vancouver, BC V6H 3V4
Canada

Dr. Joanne Weinberg
University of British Columbia
Department of Cellular &
Physioloical Scinces
Health Sciences Mall 2350
Vancouver, BC V6T 1Z3
Canada

Prof. Dr. Egon Jonsson
University of Alberta
University of Calgary
Department of Public
Health Science
Institute of Health Economics
10405 Jasper Ave
Edmonton, Alberta T5J 3N4
Canada

Series Editor

Prof. Dr. Egon Jonsson
University of Alberta
University of Calgary
Department of Public
Health Science
Institute of Health Economics
10405 Jasper Ave
Edmonton, Alberta T5J 3N4
Canada

Cover
Photo: PhotoDisc/Getty

Limit of Liability/Disclaimer of Warranty: While the publisher and author have used their best efforts in preparing this book, they make no representations or warranties with respect to the accuracy or completeness of the contents of this book and specifically disclaim any implied warranties of merchantability or fitness for a particular purpose. No warranty can be created or extended by sales representatives or written sales materials. The Advice and strategies contained herein may not be suitable for your situation. You should consult with a professional where appropriate. Neither the publisher nor authors shall be liable for any loss of profit or any other commercial damages, including but not limited to special, incidental, consequential, or other damages.

Library of Congress Card No.: applied for

British Library Cataloguing-in-Publication Data
A catalogue record for this book is available from the British Library.

Bibliographic information published by the Deutsche Nationalbibliothek
The Deutsche Nationalbibliothek lists this publication in the Deutsche Nationalbibliografie; detailed bibliographic data are available on the Internet at <http://dnb.d-nb.de>.

© 2011 Wiley-VCH Verlag & Co. KGaA, Boschstr. 12, 69469 Weinheim, Germany

Wiley-Blackwell is an imprint of John Wiley & Sons, formed by the merger of Wiley's global Scientific, Technical, and Medical business with Blackwell Publishing.

All rights reserved (including those of translation into other languages). No part of this book may be reproduced in any form – by photoprinting, microfilm, or any other means – nor transmitted or translated into a machine language without written permission from the publishers. Registered names, trademarks, etc. used in this book, even when not specifically marked as such, are not to be considered unprotected by law.

Typesetting   Toppan Best-set Premedia Ltd, Hong Kong
Printing and Binding   Strauss GmbH, Moerlenbach Germany
Cover Design   Adam-Design, Weinheim

Printed in the Federal Republic of Germany
Printed on acid-free paper

ISBN: 978-3-527-32839-0

ISSN: 1864-9947

Contents

Preface *XVII*
List of Contributors *XIX*

1 **Prenatal Alcohol Exposure, FAS, and FASD: An Introduction** *1*
Tanya T. Nguyen, Jennifer Coppens, and Edward P. Riley
1.1   Introduction *1*
1.2   History *2*
1.3   Diagnosing the Effects of Prenatal Alcohol Exposure *3*
1.3.1 Fetal Alcohol Syndrome *3*
1.3.2 Fetal Alcohol Spectrum Disorder(s) *4*
1.4   Risk factors influencing FAS and FASD Conditions *4*
1.5   Prevalence and Impact of FAS and FASD *5*
1.6   Prevention *7*
1.7   Interventions *8*
      Acknowledgments *10*
      References *10*

Part One   Incidence, Prevalence, and Economic Aspects of FASD *15*

2 **Researching the Prevalence and Characteristics of FASD in International Settings** *17*
Philip A. May
2.1   Introduction *17*
2.2   Maternal Risk Factors and FASD *17*
2.3   Determining the Prevalence of FASD: How the Methods Have Influenced the Rates *20*
2.4   The Prevalence of FASD from In-School Studies *21*
2.5   Summary Rates of FASD and Their Meaning *22*
      References *24*

Fetal Alcohol Spectrum Disorder–Management and Policy Perspectives of FASD. Edited by Edward P. Riley, Sterling Clarren, Joanne Weinberg, and Egon Jonsson
Copyright © 2011 WILEY-VCH Verlag GmbH & Co. KGaA, Weinheim
ISBN: 978-3-527-32839-0

3		**Frequency of FASD in Canada, and What This Means for Prevention Efforts** *27*
		Suzanne C. Tough and Monica Jack
3.1		Introduction *27*
3.2		Challenges to Obtaining Accurate Incidence and Prevalence Rates *27*
3.3		Incidence of FASD *29*
3.3.1		National Rates in Canada (see Box 3.1) *29*
3.3.2		Provincial Rates in Canada *30*
3.3.2.1		British Columbia *30*
3.3.2.2		Alberta *30*
3.3.2.3		Saskatchewan *30*
3.3.2.4		Manitoba *31*
3.3.2.5		Other Provinces *31*
3.4		Prevalence of FASD *31*
3.4.1		Child Welfare Systems *31*
3.4.2		Corrections Systems *33*
3.4.3		Aboriginal Communities *34*
3.5		Rate of Exposure to Risk *35*
3.6		Gaps in the Data *37*
3.7		Policy Considerations *37*
3.7.1		Establish Baseline Rates of FAS/FASD and Track Them Over Time *37*
3.7.2		Continue with Intervention Efforts *38*
3.7.3		Assess and Intervene in Areas with Higher FAS/FASD Frequency *38*
3.7.4		Intervene to Prevent FAS/FASD Where Risk is Higher *39*
3.7.5		Work Towards Developing a Consistent Message *40*
3.7.6		Key Players *40*
3.8		Conclusions *41*
		Acknowledgments *41*
		References *41*
4		**Costs of FASD** *45*
		Nguyen Xuan Thanh, Egon Jonsson, Liz Dennett, and Philip Jacobs
4.1		Introduction *45*
4.2		Methods *46*
4.2.1		Literature Search *46*
4.2.2		Inclusion Criteria *46*
4.2.3		Cost Adjustment *47*
4.3		Results *47*
4.3.1		Search Results *47*
4.3.2		Summary of Studies Included in the Review *47*
4.3.3		Summary of Methods Used in the Reviewed Studies *49*
4.3.4		Summary of Results of the Reviewed Studies *51*

4.3.4.1	Annual Cost of FAS/FASD for the US, Canada, and the Province of Alberta	51
4.3.4.2	Annual Cost per Case	55
4.3.4.3	Lifetime Cost per Case	56
4.4	Discussion	57
4.5	Conclusion	58
4.6	Appendices to Chapter 4	59
4.6.1	Appendix 1: Search Strategy	59
4.6.2	Appendix 2: Summary of Included Studies	64
4.6.3	Appendix 3: Excluded studies that consider the costs of FAS/FASD	68
	References	68

Part Two Causes and Diagnosing of FASD 71

5 Direct and Indirect Mechanisms of Alcohol Teratogenesis: Implications for Understanding Alterations in Brain and Behavior in FASD 73
Kristina A. Uban, Tamara Bodnar, Kelly Butts, Joanna H. Sliwowska, Wendy Comeau, and Joanne Weinberg

5.1	Introduction	73
5.1.1	Mechanisms of Alcohol's Teratogenic Effects	73
5.1.2	Direct Mechanisms of Alcohol's Actions on the Fetus	74
5.1.3	Indirect Mechanisms of Alcohol's Actions on the Fetus	75
5.1.3.1	Alcohol Effects on Prostaglandins	76
5.1.3.2	Alcohol-Induced Disruption of Cell–Cell Interactions or Cell Adhesion	76
5.1.3.3	Alcohol and Oxidative Stress	76
5.1.3.4	Disruption of Endocrine Balance	77
5.1.4	Neurobiology of Stress	78
5.1.5	FASD and Stress Responsiveness	81
5.2	Fetal Programming: Programming of the HPA Axis by PAE	82
5.3	Altered Epigenetic Regulation of Gene Expression: A Possible Mechanism Underlying Fetal Programming of the HPA Axis and Altered Neuroendocrine-Immune Interactions	84
5.4	Prenatal Alcohol Exposure: Early Experience, Stress Responsiveness, and Vulnerability to Depression	87
5.4.1	Interactions Between Central Monoaminergic Neurotransmitters and the HPA Axis	88
5.4.2	FASD, Stress, and Depression	89
5.4.3	Prevention and Treatment of Depression in FASD Populations	90
5.5	FASD and Substance Abuse	91
5.5.1	Neurobiology of Addiction	91
5.5.2	Stress and Substance Use	92
5.6	Summary and Policy Considerations	93
	Acknowledgments	94

Glossary   94
Abbreviations   97
References   98

6   Genetic Factors Contributing to FASD   109
Albert E. Chudley
6.1   Introduction   109
6.2   The Evidence   110
6.3   Genetic Factors in Alcohol Metabolism   111
6.4   Some Genetic Factors   115
6.5   Epigenetics, the Environment and Nutrition   116
6.6   Conclusions, and Some Policy Recommendations   118
6.7   Glossary   119
References   123

7   Diagnosis of FASD: An Overview   127
Gail Andrew
7.1   History of Diagnosing FASD   128
7.2   How Does Prenatal Alcohol Exposure Cause Damage?   130
7.3   Screening for FASD   131
7.4   The Diagnostic Process   133
7.5   FASD Across the Lifespan   136
7.5.1   Diagnosis in the Neonatal Period and Early Infancy   137
7.5.2   Diagnosing in Toddlerhood   138
7.5.3   Diagnosing in School Age   138
7.5.4   Diagnosing in Adolescence and Adulthood   140
7.6   Implications of a Diagnosis of FASD   140
7.7   Conclusion and Future Directions   141
7.8   Policy Considerations   142
References   142

Part Three   Prevention Policies and Programs   149

8   FASD: A Preconception Prevention Initiative   151
Lola Baydala, Stephanie Worrell, and Fay Fletcher
8.1   Introduction   151
8.2   Prevention Strategies   151
8.2.1   The National Registry of Evidence-Based Programs and Practice   152
8.2.2   LifeSkills Training   152
8.2.3   The Alexis Working Committee   153
8.2.4   The Adaptations Committee   154
8.2.5   Community Member Participation   155
8.3   Research Relationships   155
8.3.1   Capacity Building   156
8.4   The CIHR Guidelines for Research Involving Aboriginal People   156

8.5	Summary *157*	
	Acknowledgments *158*	
	References *158*	
9	**Bringing a Women's Health Perspective to FASD Prevention** *161*	
	Nancy Poole	
9.1	Introduction *161*	
9.2	Applying Gender-Based Analysis to FASD Prevention *162*	
9.3	Developing a Framework for Women-Centered Prevention Practice *163*	
9.3.1	Women-Centered Care *165*	
9.3.2	Harm-Reduction Orientation *165*	
9.3.3	Collaborative Care *166*	
9.4	Evidencing the Framework *166*	
9.4.1	Research on Women-Centered, Trauma-Informed Care *167*	
9.4.2	Research on Harm-Reduction Practice *168*	
9.4.3	Research on Collaborative Care *169*	
9.5	Conclusions *170*	
	References *171*	
10	**Next Steps in FASD Primary Prevention** *175*	
	Robin Thurmeier, Sameer Deshpande, Anne Lavack, Noreen Agrey, and Magdalena Cismaru	
10.1	Introduction *175*	
10.2	Current State of FASD Primary Prevention in North-Western Canada *176*	
10.2.1	The Born Free Campaign *178*	
10.2.2	The Mother Kangaroo Campaign *180*	
10.2.3	The With Child/Without Alcohol Campaign *181*	
10.2.4	Summary of Results *181*	
10.3	Campaign Evaluation: What Is It and Why It Is Important *182*	
10.4	Incorporating Social Marketing Strategies *184*	
10.5	Creating Behavioral Change: Protection Motivation Theory *185*	
10.6	Future Considerations for Health Promoters and Policy Makers *188*	
	References *189*	
11	**Preventing FASD: The Parent–Child Assistance Program (PCAP) Intervention with High-Risk Mothers** *193*	
	Therese M. Grant	
11.1	Introduction *193*	
11.2	FASD Prevention *193*	
11.3	Background *194*	
11.4	The PCAP Intervention *195*	
11.4.1	Relational Theory *195*	
11.4.2	Stages-of-Change *196*	

11.4.3	Harm Reduction	197
11.5	PCAP: A Two-Pronged Intervention	197
11.5.1	Between the Client and the Case Manager	197
11.5.1.1	Establishing Trust	197
11.5.1.2	Working with the Family	197
11.5.1.3	Role-Modeling	198
11.5.2	Between the Client and the Community Service Providers	198
11.6	Preventing Alcohol- and/or Drug-Exposed Births	199
11.6.1	Substance Abuse Treatment	199
11.6.2	Family Planning	199
11.7	PCAP Outcomes	200
11.8	PCAP Cost Effectiveness	201
11.9	PCAP Intervention with Women who Themselves Have FASD	201
11.10	Policy Recommendations: Collaborative Approaches for Preventing Alcohol-Exposed Pregnancies	202
	References	204

12 FASD in the Perspective of Primary Healthcare 207
June Bergman

12.1	Primary Care Approaches to FASD	208
12.2	Barriers to Screening	210
12.3	Impact of Healthcare Reform	211
	Reference	212

Part Four FASD and the Legal System 213

13 The Manitoba FASD Youth Justice Program: Addressing Criminal Justice Issues 215
Mary Kate Harvie, Sally E.A. Longstaffe, and Albert E. Chudley

13.1	Introduction	215
13.2	The Legislative Context	216
13.3	The Information Gap	217
13.4	The Manitoba FASD Youth Justice Program	220
13.5	Screening	221
13.6	The Preassessment Period	222
13.7	Medical Assessment	222
13.8	The Doctor's Report and Its Use	223
13.9	Sentencing Conferences	224
13.10	The Sentencing Process	225
13.11	The Statistical Outcomes	225
13.12	One-Day Snap-Shot of Age of Majority Youth (28 February 2010)	226
13.13	Other Initiatives	226
13.13.1	"This is Me"	226

13.13.2	This is Me Life Books	227
13.13.3	The Icons Project	227
13.13.4	Youth Accommodation Counsel	229
13.14	Strengths and Challenges	229
	References	231

14 Understanding FASD: Disability and Social Supports for Adult Offenders 233

E. Sharon Brintnell, Patricia G. Bailey, Anjili Sawhney, Laura Kreftin

14.1	Fetal Alcohol Spectrum Disorder (FASD) is a Disability	234
14.1.1	Primary Disabilities Associated with FASD	234
14.1.2	Secondary Disabilities Associated with FASD	235
14.1.3	The Social Determinants of Health and FASD	236
14.1.4	Human Rights and FASD	237
14.1.5	Incarceration and FASD	237
14.2	Correctional Environment in Canada for Adults with FASD	238
14.2.1	Treatment Programs	240
14.2.2	Recidivism and Alternative Sentencing	240
14.2.3	Release Planning	242
14.2.4	Correctional System Needs	243
14.3	Interventions and Social Supports for Adults with FASD after Release	244
14.3.1	Client-Centered Lifelong Multisectoral Supports	245
14.3.2	Employment and Housing	246
14.3.3	Training and Programs	247
14.3.4	External Executive Function Support	248
14.3.5	FASD Costs	249
14.3.6	Developmental Disability Assistance	250
14.4	Policy Considerations for Adults with FASD	251
	References	253

15 Policy Development in FASD for Individuals and Families Across the Lifespan 259

Dorothy Badry and Aileen Wight Felske

15.1	Introduction	259
15.2	Birth	261
15.3	Childhood	261
15.4	Adolescence/Teenage Years	264
15.5	Adulthood	265
15.6	A Disability Paradigm for FASD	267
15.7	Cultural Fairness	268
15.8	Life Trajectory Policy Model	269
15.9	Conclusions	270
	References	271

16 The Impact of FASD: Children with FASD Involved with the Manitoba Child Welfare System *275*
Linda Burnside, Don Fuchs, Shelagh Marchenski, Andria Mudry, Linda De Riviere, Marni Brownell, and Matthew Dahl
16.1 Introduction *275*
16.2 Study One: Children in Care with Disabilities *277*
16.3 Study Two: The Trajectory of Care for Children with FASD *278*
16.4 Study Three: Youth with FASD Leaving Care *280*
16.5 Study Four: The Cost of Child Welfare Care for Children with FASD *282*
16.6 Study Five: Economic Impact of FASD for Children in Care *284*
16.7 Conclusions *292*
References *293*

17 British Columbia's Key Worker and Parent Support Program: Evaluation Highlights and Implications for Practice and Policy *297*
Deborah Rutman, Carol Hubberstey, and Sharon Hume
17.1 Introduction *297*
17.2 Background *297*
17.3 Program Model and Components *298*
17.4 Literature *299*
17.5 Evaluation Methods *300*
17.6 Formative Evaluation Findings *301*
17.6.1 Activities and Role of the Key Worker *303*
17.6.2 Regional and Provincial Supports for Key Workers *305*
17.6.3 Parents', Caregivers' and Community Partners' Perceptions of the Program *307*
17.6.4 Program Challenges *308*
17.7 Summative Evaluation Findings *309*
17.7.1 Increased Knowledge about FASD *309*
17.7.2 Shifts in Parenting Strategies and Responses *310*
17.7.3 Feeling Supported *310*
17.7.4 Increased Access to Services and Resources *311*
17.8 Discussion *311*
17.8.1 Promising Practices *312*
17.9 Policy Considerations *313*
17.10 Conclusions *314*
References *315*

18 FASD and Education Policy: Issues and Directions *317*
Elizabeth Bredberg
18.1 Introduction *317*
18.2 Where Do Students with FASD Fit Into the Education System? *318*
18.2.1 Eligibility *319*
18.2.2 Exclusion and Discipline *319*

18.2.3	Accountability: Curriculum and Learning Outcomes	320
18.3	Students with FASD within Special Education Systems	320
18.4	Education Professionals and FASD	321
18.5	Inter-Agency and Community Supports for Students with FASD	322
18.6	Policy indications	322
18.7	Conclusions	325
	References	325

19 Shifting Responsibility from the Individual to the Community 327
Audrey McFarlane

19.1	Introduction	327
19.2	Why Do We Need to Make a Shift?	327
19.3	Examples of Individual's Situations	328
19.4	One Model of Community of Care	330
19.4.1	Diagnosis	330
19.4.2	Prevention	332
19.4.3	Intervention	333
19.4.4	Outcomes	333
19.5	History	334
19.6	Future	335
19.7	Policy Considerations	336
19.8	Conclusions	337
	Acknowledgments	338
	References	338

20 A Social Work Perspective on Policies to Prevent Alcohol Consumption during Pregnancy 339
Mary Diana (Vandenbrink) Berube
References 348
Appendix to Chapter 20 350

21 A Cross-Ministry Approach to FASD Across the Lifespan in Alberta 353
Denise Milne, Tim Moorhouse, Kesa Shikaze, and Cross-Ministry Members

21.1	Introduction	353
21.2	The Impact of FASD	354
21.3	Overview of Strategies	356
21.4	FASD Service Network Program	356
21.5	Ministry Initiatives Based on the Strategic Plan	357
21.5.1	Awareness and Prevention	357
21.5.1.1	Parent–Child Assistance Program	357
21.5.1.2	Generating Awareness/Skills Development in Justice	358
21.5.2	Assessment and Diagnosis	358
21.5.2.1	Development of an Assessment and Diagnosis Model for Aboriginal and Remote Communities	358

21.5.2.2	Adult Assessment and Diagnosis Demonstration Project	*359*
21.5.2.3	Development of FASD Clinical Capacity	*359*
21.5.3	Supports for Individuals and Caregivers	*360*
21.5.3.1	Employment Supports for People Affected by FASD	*360*
21.5.3.2	Employment Supports and Services	*361*
21.5.3.3	FASD: Supporting Adults Gain and Maintain Employment	*361*
21.5.3.4	AVENTA Addiction Treatment for Women Demonstration Project	*361*
21.5.3.5	Kaleidoscope Demonstration Project	*361*
21.5.3.6	FASD Community Outreach Program Demonstration Project	*361*
21.5.3.7	Step-by-Step Demonstration Project	*362*
21.5.3.8	Well Communities–Well Families Demonstration Project	*362*
21.5.3.9	Service Coordination and Mentorship	*362*
21.5.3.10	FASD Videoconference Learning Series	*362*
21.5.3.11	Supports through Justice	*363*
21.5.3.12	First Nations and Inuit Supports	*363*
21.5.3.13	The WRaP (Wellness, Resiliency and Partnerships) Coaching Demonstration Project	*363*
21.5.4	Training and Education	*364*
21.5.4.1	Development of e-Learning Modules	*364*
21.5.4.2	Promising Practices, Promising Futures: Alberta FASD Conference 2009 and 2010	*364*
21.5.4.3	IHE Consensus Development Conference on FASD: Across the Lifespan	*365*
21.5.4.4	FASD Education and Training	*365*
21.5.4.5	Leading Practices Workshops	*365*
21.5.4.6	Building an Educated Workforce	*365*
21.5.5	Strategic Planning	*366*
21.5.6	Research and Evaluation	*366*
21.5.6.1	Corrections and Connections to Community	*366*
21.5.6.2	FASD Community of Practice Research	*367*
21.5.6.3	Research Project on School Experiences of Children with FASD	*367*
21.5.7	Stakeholder Engagement	*367*
	References	*368*

22 Critical Considerations for Intervention Planning for Children with FASD *369*
John D. McLennan

22.1	Introduction	*369*
22.2	The Development of a Rational Service System for At-Risk Children	*369*
22.3	Factors Supporting the Development of Separate Specialized Services for Subgroups of At-Risk Children	*370*

22.3.1	What is the Prevalence of the Special Subpopulation?	370
22.3.2	What is the Prevalence of Specific Difficulties/Needs within the Special Subpopulation?	371
22.3.3	What Is the Attributable Risk of the Particular Disorder for These Specific Difficulties?	371
22.3.4	What Is the Effectiveness of Interventions for the Subpopulation?	372
22.3.5	Is There Evidence for Unique Benefits to Support Separating Out Services for the Subpopulation?	372
22.3.6	Are There Risks in Delivering Services Separately for the Subpopulation?	372
22.4	Should Separate Specialized Services Be Developed for Children with FASD?	372
22.4.1	What Is the Prevalence of FASD?	372
22.4.2	What Is the Prevalence of Specific Difficulties/Needs within a Population of Children with FASD?	373
22.4.3	What Is the Attributable risk of FASD for these Specific Difficulties?	375
22.4.4	What Is the Effectiveness of Interventions for FASD?	378
22.4.5	Is There Evidence for Unique Benefits to Support Separating-Out Services for Children with FASD?	379
22.4.6	Are There Risks in Delivering Services Separately for Children with FASD?	381
22.5	Policy Considerations: Strengthening the Service System for a Broader Range of Children At-Risk	383
	Acknowledgments	383
	References	383

Part Five Research Needed on FASD 387

23	**FASD Research in Primary, Secondary, and Tertiary Prevention: Building the Next Generation of Health and Social Policy Responses** 389	
	Amy Salmon and Sterling Clarren	
23.1	Introduction	389
23.2	Mapping Prevention: What Research is Needed Now, and Why?	390
23.2.1	Primary Prevention: Social Support and Determinants of Women's Health	390
23.2.2	Accurate Diagnosis of FASD: Preventing Secondary Disabilities and Reaching out to Mothers (and Potential Mothers)	393
23.3	Conclusions: Drawing a Road-Map for Integrated, Supportive, and Effective Care	396
	References	397

24 Focusing Research Efforts: What Further Research into FASD is Needed? *399*
Sara Jo Nixon, Robert A. Prather, and Rebecca J. Gilbertson
24.1 Introduction *399*
24.2 FASD and Heterogeneity: An Encouraging Outcome *399*
24.3 Models: Moving Beyond Description *402*
24.4 Applying Neuroscience: Beyond the Mother? *403*
24.5 Summary *405*
References *406*

Part Six Personal Views from People Living with FASD *411*

25 Living with FASD *413*
Myles Himmelreich

26 Charlene's Journey *419*
Charlene Organ

Appendix: FASD Consensus Statement of the Jury *433*
Acknowledgments *433*
Process *433*
Conference Questions *434*
Introduction *434*
Question 1 *434*
Question 2 *435*
Question 3 *436*
Question 4 *438*
Question 5 *439*
Question 6 *441*
Conclusion *442*
Jury Members *443*
Conference Speakers and Topics *443*
Planning Committee *446*
Scientific Committee *447*
Communications Committee *448*
Disclosure Statement *448*
Institute of Health Economics *448*
IHE Board of Directors *448*
Chair *448*
Government *449*
Academia *449*
Industry *449*
Other *449*
CEO *449*
FASD Research and Resources *450*

Index *451*

Preface

In October 2009, the Institute of Health Economics (IHE) staged a consensus development conference to address key questions about the prevention, diagnosis, and treatment of Fetal Alcohol Spectrum Disorder (FASD). Experts in the field presented scientific evidence to a "jury" about prevention and the social determinants that may induce drinking during pregnancy, the importance of diagnosis, the impact of FASD across a person's lifespan, and the community supports needed for those living with FASD, as well as their families. After two days of hearings, the jury developed a statement which answered eight relevant and common questions about FASD, including suggested policy changes for enhanced prevention, and for improving the lives of people with FASD and their families. That statement is available in the Appendix of this book.

During the planning of that conference, IHE invited the experts – who included researchers, clinicians, economists, epidemiologists, social workers, and judicial workers – to expand on their speeches and write chapters for a book that would aim at a worldwide health policy-making audience. An overwhelming majority of the speakers were interested, and this book is the result of their hard work.

In addition to policy makers, this book is for anyone interested in FASD, including those with the condition, family members and other caregivers, researchers, clinicians and others in healthcare and social services, and the justice sector. The chapters describe the impact of FASD on the individual, their families and society, and the many complex issues involved in the condition's prevention, diagnosis, and treatment. The book ends with personal accounts of life with FASD, written by Myles Himmelreich and Charlene Organ, that not only powerfully illustrate the challenges created by having FASD, but also serve as a reminder that FASD does not – and should never – define a person.

We would like to acknowledge the Government of Alberta FASD Cross-Ministry Committee, Canada Northwest FASD Partnership, Health Canada and the Public Health Agency of Canada for their financial support of the consensus development conference and the production of material to this book. We would also like to thank Minister Janis Tarchuk, who was Minister of Children and Youth Services at the time of the conference and Deputy Minister, Fay Orr, for their strong support. Special recognition should be given to Ms Denise Milne, who represented the Cross-Ministry Committee and assisted greatly in all aspects of the conference

Fetal Alcohol Spectrum Disorder–Management and Policy Perspectives of FASD. Edited by Edward P. Riley,
Sterling Clarren, Joanne Weinberg, and Egon Jonsson
Copyright © 2011 WILEY-VCH Verlag GmbH & Co. KGaA, Weinheim
ISBN: 978-3-527-32839-0

preparations. We are very grateful to the Honorable Anne McLellan for serving as Chair of the Jury for the conference, to Dr Gail Andrew, who acted as the Scientific Chair, and Ms Nancy Reynolds who acted as moderator. Special mention should be made to the Honorable Iris Evans, Minister of International and Intergovernmental Relations for Alberta, who has been a tireless supporter of initiatives to improve the lives of those affected by FASD. We would also like to give special thanks to Gail Littlejohn for her support in the editing process of this publication.

On behalf of the Institute of Health Economics (IHE) (www.ihe.ca)

Egon Jonsson
John Sproule
Liz Dennett

List of Contributors

Noreen Agrey
Saskatchewan Prevention Institute
1319 Colony Street
Saskatoon, Saskatchewan
Canada S7N 2Z1

Gail Andrew
Alberta Health Services
Glenrose Rehabilitation Hospital
10230 111 Ave
Edmonton, Alberta
Canada AB T5G 0B7

Dorothy Badry
University of Calgary
Faculty of Social Work
2500 University Drive NW
Calgary, Alberta
Canada T2N 1N4

Patricia G. Bailey
University of Alberta
Occupational Performance
Analysis Unit
1-78 Corbett Hall
Edmonton, Alberta
Canada T6G 2G4

Lola Baydala
University of Alberta
Department of Pediatrics
Misericordia Child Health Clinic
16930-87 Avenue
Edmonton, Alberta
Canada T5R 4H5

June Bergman
University of Calgary
Department of Family Medicine
Faculty of Medicine and Dentistry
#1707, 1632-14th Avenue NW
Calgary, Alberta
Canada T2N 1M7

Mary Diana (Vandenbrink) Berube
Alberta Children and Youth Services
Ministry Support Services
12th Floor, Sterling Place,
9940-106 Street
Edmonton, Alberta
Canada T5K 2N2

Yagesh Bhambhani
University of Alberta
Occupational Performance Analysis
Unit
1-78 Corbett Hall
Edmonton, Alberta
Canada T6G 2G4

Fetal Alcohol Spectrum Disorder–Management and Policy Perspectives of FASD. Edited by Edward P. Riley,
Sterling Clarren, Joanne Weinberg, and Egon Jonsson
Copyright © 2011 WILEY-VCH Verlag GmbH & Co. KGaA, Weinheim
ISBN: 978-3-527-32839-0

List of Contributors

Tamara Bodnar
University of British Columbia
Department of Cellular &
Physiological Sciences
2350 Health Sciences Mall
Vancouver, BC
Canada BC V6T 1Z3

Elizabeth Bredberg
Bredberg Research and Consulting in
Education (BRACE)
2620 W 37th Ave
Vancouver, British Columbia
Canada V6N 2T4

E. Sharon Brintnell
University of Alberta
Occupational Performance Analysis
Unit
1-78 Corbett Hall
Edmonton, Alberta
Canada T6G 2G4

Marni Brownell
University of Manitoba
Manitoba Centre for Health Policy
Community Health Sciences
Faculty of Medicine
408-727 McDermot Avenue
Winnipeg, Manitoba
Canada R3E 3P5

Linda Burnside
Manitoba Family Services and
Housing
Disability Programs and Employment
& Income Assistance
305–114 Garry Street
Winnipeg, Manitoba
Canada R3C 4V7

Kelly Butts
University of British Columbia
UBC Institute of Mental Health
Department of Psychiatry
5950 University Blvd
Vancouver, BC
Canada BC V6T 1Z3

Albert E. Chudley
University of Manitoba
Program in Genetics and Metabolism
Professor, Department of Pediatrics
and Child Health
Department of Biochemistry and
Medical Genetics
FE 229-840 Sherbrook Street
Winnipeg, Manitoba
Canada MB R3A 1R9

Magdalena Cismaru
University of Regina
Faculty of Business Administration
3737 Wascana Parkway
Regina, Saskatchewan
Canada S4S 0A2

Sterling Clarren
Canada Northwest FASD Research
Network
L408-4480 Oak Street
Vancouver, BC
Canada V6H 3V4

Wendy Comeau
University of British Columbia
Department of Cellular &
Physiological Sciences
2350 Health Sciences Mall
Vancouver, BC
Canada BC V6T 1Z3

Jennifer Coppens
University of Alberta
Doctor of Medicine Program
Faculty of Medicine & Dentistry
11025 Jasper Avenue #608
Edmonton, Alberta
Canada T5K 0K7

Matthew Dahl
University of Manitoba
Manitoba Centre for Health Policy
Community Health Sciences
Faculty of Medicine
408-727 McDermot Avenue
Winnipeg, Manitoba
Canada R3E 3P5

Linda De Riviere
University of Winnipeg
515 Portage Avenue
Winnipeg, Manitoba
Canada R3B 2E9

Liz Dennett
Institute of Health Economics
#1200, 10405 Jasper Ave
Edmonton, Alberta
Canada T5J 3N4

Sameer Deshpande
University of Lethbridge
Center for Socially Responsible Marketing
D548, 4401 University Drive
Lethbridge, Alberta
Canada T1K3M4

Aileen Wight Felske
Mount Royal University
Faculty of Health and Community Studies
4825 Mount Royal Gate SW
Calgary, Alberta
Canada T3E 6K6

Fay Fletcher
University of Alberta
Faculty of Extension
2-254 Enterprise Square
10230-Jasper Avenue
Edmonton, Alberta
Canada T5J 4P6

Don Fuchs
University of Manitoba
Faculty of Social Work
Winnipeg, Manitoba
Canada R3T 2N2

Rebecca J. Gilbertson
Comprehensive Biobehavioral Core,
Clinical and Translational Science
Institute and Department of
Psychiatry
P.O. Box 100256
Gainesville, FL 32610-0256
USA

Myles Himmelreich
2534a 15 Avenue SE
Calgary, Alberta
Canada T2A 0L5

Charlene Organ
106 27 132 Avenue NW
Edmonton Alberta
Canada T5E 0Z4

Therese M. Grant
University of Washington School of Medicine
Department of Psychiatry and Behavioral Sciences
Fetal Alcohol and Drug Unit
180 Nickerson Street, Suite 309
Seattle, WA 98109-1631
USA

List of Contributors

Mary Kate Harvie
Provincial Court of Manitoba
5th Floor, 408 York Ave
Winnipeg, Manitoba
Canada MB R3C 0P9

Carol Hubberstey
Nota Bene Consulting Group
2776 Dewdney Avenue
Victoria, British Columbia
Canada V8R 3M4

Sharon Hume
Nota Bene Consulting Group
2708 Dunlevy Street
Victoria, British Columbia
Canada V8R 5Z4

Monica Jack
University of Calgary
Department of Peadiatrics
Alberta Health Services, Public Health
Innovation and Decision Support
c/o 2888 Shaganappi Trail NW
Calgary, Alberta
Canada T3B 6A8

Philip Jacobs
Institute of Health Economics
#1200, 10405 Jasper Ave
Edmonton, Alberta
Canada T5J 3N4

Egon Jonsson
Institute of Health Economics
#1200, 10405 Jasper Ave
Edmonton, Alberta
Canada T5J 3N4

Laura Kreftin
University of Alberta
Occupational Performance Analysis
Unit
1-78 Corbett Hall
Edmonton, Alberta
Canada T6G 2G4

Anne Lavack
University of Regina
Faculty of Business Administration
3737 Wascana Parkway
Regina, Saskatchewan
Canada S4S 0A2

Sally E.A. Longstaffe
University of Manitoba
Manitoba FASD Centre
Manitoba FASD Network
Children's Hospital Room CK 265,
840 Sherbrook Steet
Winnipeg, Manitoba
Canada R3A 1S1

Shelagh Marchenski
University of Manitoba
Faculty of Social Work
Winnipeg, Manitoba
Canada R3T 2N2

Philip A. May
University of New Mexico
Center on Alcoholism
Substance Abuse
and Addictions (CASAA)
2650 Yale SE
Albuquerque, New Mexico 87108
USA

Audrey McFarlane
Lakeland Centre for Fetal Alcohol
Spectrum Disorder
Box 479
Cold Lake, Alberta
Canada T9M 1P3

John D. McLennan
University of Calgary
Department of Community Health
Sciences
TRW Building, 3rd Floor, 3280
Hospital Drive NW
Calgary, Alberta
Canada T2N 4Z6

Denise Milne
Alberta Children and Youth Services
Community Partnerships
10th Floor, Sterling Place,
9940-106 Street
Edmonton, Alberta
Canada T5K 2N2

Tim Moorhouse
Alberta Children and Youth Services
Research Innovation
6th Floor, Sterling Place,
9940-106 Street
Edmonton, Alberta
Canada T5K 2N2

Andria Mudry
University of Manitoba
Faculty of Social Work
Winnipeg, Manitoba
Canada R3T 2N2

Tanya T. Nguyen
San Diego State University/University
of California
San Diego Joint Doctoral Program in
Clinical Psychology
6363 Alvarado Ct #103
San Diego, CA 92120
USA
San Diego State University
Department of Psychology and the
Center for Behavioral Teratology
6363 Alvarado Ct #209
San Diego, CA 92120
USA

Sara Jo Nixon
University of Florida
Comprehensive Biobehavioral Core
Clinical and Translational Science
Institute
P.O. Box 100256
Gainesville, FL 32610-0256
USA

Nancy Poole
BC Centre of Excellence for Women's
Health
4500 Oak Street E311, box 48
Vancouver, British Columbia
Canada V6H 3N1

Robert A. Prather
University of Florida
Department of Psychiatry
P.O. Box 100256
Gainesville, FL 32610-0256
USA

Edward P. Riley
San Diego State University
Department of Psychology and the
Center for Behavioral Teratology
6363 Alvarado Ct #209
San Diego, CA 92120
USA

Deborah Rutman
University of Victoria
Nota Bene Consulting Group and
School of Social Work
1434 Vining Street
Victoria, British Columbia
Canada V8R 1P8

Amy Salmon
Canada Northwest FASD Research
Network
L408-4480 Oak Street
Vancouver, BC
Canada V6H 3V4

Anjili Sawhney
University of Alberta
Occupational Performance Analysis
Unit
1-78 Corbett Hall
Edmonton, Alberta
Canada T6G 2G4

Kesa Shikaze
Health and Wellness
23rd Floor, Telus Plaza NT, 10025
Jasper Avenue
Edmonton, Alberta
Canada T5J 1S6

Joanna H. Sliwowska
Poznan University of Life Sciences
Institute of Zoology
Department of Histology and
Embryology
ul. Wojska Polskiego 71c
60-625 Poznań
Poland

Nguyen Xuan Thanh
Institute of Health Economics
#1200, 10405 Jasper Ave
Edmonton, Alberta
Canada T5J 3N4

Robin Thurmeier
University of Regina
Faculty of Business Administration
3737 Wascana Parkway
Regina, Saskatchewan
Canada S4S 0A2

Suzanne C. Tough
University of Calgary
Departments of Pediatrics and
Community Health Services
Alberta Centre for Child, Family and
Community Research
Child Development Centre
c/o 2888 Shaganappi Trail NW
Calgary, Alberta
Canada T3B 6A8

Kristina A. Uban
University of British Columbia
Department of Psychology
2136 West Mall
Vancouver
Canada BC V6T 1Z4

Joanne Weinberg
University of British Columbia
Department of Cellular &
Physiological Sciences
2350 Health Sciences Mall
Vancouver, BC
Canada BC V6T 1Z3

Stephanie Worrell
University of Alberta
Department of Pediatrics
Misericordia Child Health Clinic
16930-87 Avenue
Edmonton, Alberta
Canada T5R 4H5

1
Prenatal Alcohol Exposure, FAS, and FASD: An Introduction
Tanya T. Nguyen, Jennifer Coppens, and Edward P. Riley

1.1
Introduction

Prenatal Alcohol Exposure (PAE) can result in a wide range of physical, psychological, behavioral, and social problems that affect the individuals, their families, and their communities. Indeed, PAE is a major public health issue placing undue burden on all aspects of society. Among the most severe outcomes of PAE is the Fetal Alcohol Syndrome (FAS), which is characterized by growth deficits, facial anomalies, and neurobehavioral problems. However, FAS is not the only detrimental outcome of heavy gestational alcohol exposure, and the majority of individuals affected by such exposure do not meet the diagnostic criteria of FAS. Currently, PAE is increasingly understood as the cause of a continuum of effects across many domains. Fetal Alcohol Spectrum Disorder (FASD) is a nondiagnostic term used to identify the wide array of outcomes resulting from prenatal exposure to alcohol. These outcomes range from isolated organ damage or subtle developmental disabilities to stillbirths and FAS. Perhaps the most pervasive outcome following prenatal alcohol exposure is what is now commonly referred to as an Alcohol-Related Neurodevelopmental Disorder (ARND). While individuals with ARND may exhibit many of the alcohol-related brain and behavioral abnormalities of FAS, they may not display the characteristic facial dysmorphia required for an FAS diagnosis. Although cases of FASD are often not as easily recognized as FAS, they can be just as serious. Unfortunately, missed diagnoses of FASD can have devastating consequences, placing heavy emotional, financial and social stresses on the individual and all parties involved (Riley and McGee, 2005).

Although the relationship between alcohol consumption during pregnancy and abnormal fetal development has been alluded to throughout history (Warren and Hewitt, 2009), FAS went unrecognized until the late 1960s and early 1970s (Lemoine et al., 1968; Jones and Smith, 1973; Jones et al., 1973). Since those initial defining case studies, the scientific literature on the effects of PAE on the developing fetus has grown rapidly. A simple search of pubmed.gov (U.S. National Library of Medicine) using "fetal alcohol syndrome" as a search term turned up almost 3500 citations. This research has improved our understanding of the relationship

Fetal Alcohol Spectrum Disorder–Management and Policy Perspectives of FASD. Edited by Edward P. Riley, Sterling Clarren, Joanne Weinberg, and Egon Jonsson
Copyright © 2011 WILEY-VCH Verlag GmbH & Co. KGaA, Weinheim
ISBN: 978-3-527-32839-0

between alcohol exposure and developmental deficits, and has resulted in an increased social awareness of the risks of drinking during pregnancy, prevention efforts to reduce these risks, and development of intervention programs to help promote positive outcomes for individuals with FASD. However, despite our current knowledge and the progress that has been made, many challenges remain in understanding how alcohol exerts its effects, in developing efficacious and effective prevention and intervention programs, and how best to improve the daily functioning of these individuals.

1.2 History

It has been suggested that the adverse effects of alcohol on the developing fetus have been recognized for centuries. Some of the earliest references date back to Greek and Roman mythology and Judeo-Christian tradition, such as the ancient Carthaginian custom that forbade bridal couples from drinking wine on their wedding night, and the belief that alcohol consumption at the time of procreation leads to the birth of defective children (Jones and Smith, 1973). Passages in Robert Burton's *The Anatomy of Melancholy* allegedly quote Aristotle describing an association between alcoholic mothers and disabled children in *Problemata*: "… foolish, drunken and harebrained women [for the] most part bring forth children like unto themselves, morose and languid" (Burton, 1621). However, there remains much controversy regarding the validity of these claims and sources. Although many authors have assumed Burton to be quoting Aristotle's words verbatim, there is no evidence of any such statement in *Problemata*, nor in any of Aristotle's other works (Abel, 1999). Others have claimed that the Carthaginians did not truly understand that drinking during pregnancy caused problems; rather, they believed that intoxication at the *exact moment of conception* led to the birth of a deformed offspring (Calhoun and Warren, 2007).

More recent and credible historical reports, however, have documented alcohol's teratogenic effect. During the 1700s, a group of English physicians described children born to alcoholic mothers as "weak, feeble, and distempered" (Royal College of Physicians of London, 1726). A deputy medical officer of the Convict Prison in Parkhurst, England, noticed that imprisoned pregnant alcoholic women had high rates of miscarriage, and that those offspring which survived displayed distinctive patterns of birth defects (Sullivan, 1899). From these observations, Sullivan concluded that alcohol had a direct effect on the developing embryo.

Despite these observations and early animal studies supporting an association between gestational alcohol exposure and adverse outcomes (e.g., Stockard, 1910), the first clinical accounts of alcohol's teratogenic effects were not published until the late 1960s. In 1968, Lemoine and colleagues published their report entitled "Outcome of children of alcoholic mothers" (Lemoine *et al.*, 1968), which established a connection between maternal alcohol consumption during pregnancy and abnormal fetal development, describing common problems of children born to mothers who drank heavily during pregnancy. Unfortunately, the authors did not present any diagnostic criteria to facilitate the recognition of fetal alcohol effects in future cases (Hoyme et al., 2005), and the paper was published in French, which limited its wide availability. As a result, FAS remained unrecognized for five more years until Jones and colleagues reported a series of case studies which documented consistent patterns of physical and developmental abnormalities in infants and children exposed to alcohol *in utero* (Jones and Smith, 1973; Jones and Smith, 1975; Jones et al., 1973). These authors coined the term "fetal alcohol syndrome," and laid the foundation for the diagnosis of this disorder.

1.3
Diagnosing the Effects of Prenatal Alcohol Exposure

1.3.1
Fetal Alcohol Syndrome

There are several suggested diagnostic schemas for FAS (e.g., Bertrand et al., 2004; Chudley et al., 2005; Hoyme, 2005) and, while there are minor differences between them, all require anomalies in three distinct areas: (i) prenatal and postnatal growth deficits; (ii) facial dysmorphology; and (iii) central nervous system (CNS) dysfunction. Typically, growth retardation is defined as evidence of prenatal or postnatal weight or height at or below the 10th percentile, after correcting for age, gender, race, and other appropriate variables. The Canadian guidelines also recommend evidence of a disproportionately low weight-to-height ratio at or below the 10th percentile. Most guidelines recommend three essential dysmorphic features—a smooth philtrum, a thin upper vermillion border, and small palpebral fissures—although the revised Institute of Medicine (IOM) guideline requires only two of the three characteristics (Hoyme et al., 2005). Finally, a diagnosis of FAS requires evidence of CNS abnormality. Within this criterion, the diagnostic schemas differ more substantially. For example, the revised IOM guideline only requires evidence of structural brain abnormalities, such as diminished head circumference at or below the 10th percentile. The CDC criteria are more extensive, outlining structural, neurological, and functional CNS dysfunction. Structural anomalies may be evidenced by the two criteria delineated in the IOM guidelines, as well as brain abnormalities observed with neuroimaging techniques. Seizures or other signs of neurological damage not attributable to postnatal insult may qualify as evidence of neurological problems. Lastly, functional abnormalities are defined as a global cognitive deficit (such as a decreased IQ), or deficits in three different functional CNS domains, which include cognition, behavior, executive functioning, and motor functioning. The Canadian guidelines outline eight domains that must be assessed: hard and soft neurologic signs; brain structure; cognition; communication; academic achievement; memory; executive functioning and abstract reasoning; and attention deficit/hyperactivity. Diagnosis requires evidence of impairment in three of these domains.

1.3.2
Fetal Alcohol Spectrum Disorder(s)

It is now recognized that there is a spectrum of deficits arising from PAE; FASD is the umbrella term used to describe this broad range of outcomes. Since the term FASD is not diagnostic, some of the guidelines (Chudley et al., 2005; Hoyme et al., 2005) use the terms ARND or ABRD (alcohol-related birth defect) to describe these FASDs. ARBD is a term which refers to individuals with a confirmed history of PAE and who display congenital birth defects, such as physical malformations or organ abnormalities. The ARND classification refers to individuals with a confirmed history of PAE who have behavioral and cognitive deficits related to CNS dysfunction. For example, an association between maternal alcohol use and sudden infant death syndrome (SIDS) has been suggested (e.g., Burd and Wilson, 2004). This would make SIDS an FASD in those cases where PAE was suspected, if other causes could be ruled out. Similarly, an increased risk of congenital heart defects has been associated with prenatal alcohol exposure; thus, such heart defects might be considered an FASD/ARBD if the mother drank heavily during pregnancy. Behavioral problems in children exposed to alcohol *in utero*, but who do not meet the diagnostic criteria of FAS, are perhaps the most commonly cited type of FASD/ARND.

1.4
Risk factors influencing FAS and FASD Conditions

The variation in the range of phenotypes of individuals with PAE suggests that alcohol's teratogenic effects can be moderated or exacerbated by other variables. Not every woman who drinks heavily during pregnancy will give birth to a child with an FASD (Warren and Foudin, 2001), and not all children with an FASD have the same deficits (Bertrand et al., 2004). In fact, there have been reports of discordance among twin pairs in regards to FAS (Warren and Li, 2005; Streissguth and Dehaene, 1993). Numerous biological and environmental factors have been shown to influence the effects of alcohol on the developing fetus, with the most obvious and important factors being those related to the nature of the PAE. The amount of alcohol consumed is highly correlated with the severity of outcome; typically, a higher level of alcohol consumption, along with longer duration of exposure, will generally lead to more adverse effects (Bonthius and West, 1988; Maier, Chen, and West, 1996). However, a linear relationship between dosage and severity may not always be expected. Studies in both animals and humans have revealed that the pattern of alcohol consumption may moderate dose effects. A binge-like exposure results in more severe neuropathology and behavioral alterations than does chronic exposure (Bonthius, Goodlett, and West, 1988), and those women who binge drink are at a higher risk of having a child with neurobehavioral deficits than those who drink chronically during pregnancy (Maier and West, 2001). In fact, Jacobson et al. (1998) have proposed that describing consumption by the average number of drinks per occasion is more useful in predicting outcome than the average number of drinks per week. A high peak blood alcohol concentration induced during binge episodes appears to be a significant risk factor for prenatal injury (Streissguth et al., 1993; Warren and Foudin, 2001).

Phenotype can also differ as a function of the developmental timing of alcohol exposure. For example, exposure during different critical periods of development will strongly influence not only the specific systems affected but also the severity of the deficit, as different organ systems develop at different rates and times during gestation. The clearest example of this "critical period" relates to the facial features required for a diagnosis of FAS. Studies in mice have shown the dysmorphic facies to be a result of alcohol exposure during a limited period of gestation, the human equivalent of which would be gestation weeks 3 and 4 (Sulik, 2005). Alcohol exposure during the first trimester interferes with the proliferation, migration, and differentiation of precursor cells in the cerebral cortex (Cook, Keiner, and Yen, 1990; Miller, 1993; Miller, 1996); however, exposures at other times might have other effects, such as altered synapse formation or changes in myelinization. Alcohol exposure during the third trimester interferes with the development of specific brain structures, including the hippocampus, cerebellum, and prefrontal cortex (Livy et al., 2003; Maier et al., 1999; Maier, Miller, and West, 1999). Thus, alcohol's teratogenicity interferes with various ontogenetic stages of neural development. Consequently, the pattern of structural and functional abnormalities will vary depending on alcohol exposure during particular critical periods of development, as different aspects of the developing nervous system become more or less vulnerable to alcohol's toxicity.

The genetic background of both the mother and fetus is another important factor that influences the effect of alcohol on the developing fetus. Genes affect the metabolism of alcohol and an organism's functional sensitivity to alcohol. For example, particular alleles for alcohol dehydrogenase (ADH1B*2 and ADH1B*3) allow for a faster alcohol metabolism, thereby reducing the risk of exposure to the fetus (McCarver et al., 1997).

Environmental factors related to prenatal care and nutrition are also important risk modifiers in FASD. Many mothers who drink during pregnancy do not receive proper prenatal care and nutrition. A complex interaction exists between nutrition and alcohol: food affects the rate of alcohol absorption and metabolism (Sedman et al., 1976), but alcohol often alters the requirement for and absorption of nutrients (Morgan and Levine, 1988). Alcohol exposure in combination with low nutrient levels increases the risk for FASD. Other risk factors for FASD include polysubstance abuse, maternal age, ethnicity, and socioeconomic status (Warren and Foudin, 2001).

1.5
Prevalence and Impact of FAS and FASD

Unfortunately, the prevalence of FAS and FASD is not as well understood as might be hoped. Today, epidemiological research into FAS and FASD is constantly challenged by issues related to methodology and questions regarding the diagnostic criteria used (May and Gossage, 2001). In the United States, the overall estimated prevalence of FAS is approximately 0.5 to 2 per 1000 births (May and Gossage, 2001), although within certain groups the prevalence is estimated to be much higher. For example, among the Plain and Plateau culture tribes in the United Sates, the average FAS rate is 9 per 1000 children between the ages of one and four years (May, McCloskey, and Gossage, 2002). Elsewhere, rates among the Southwestern tribes varied from 0.0 to 26.7 per 1000 over the time period of 1969–1982, depending on the specific community studied (May et al., 1983). These high-risk communities are typically of low socioeconomic status, and include a significant proportion of individuals who binge-drink on a frequent basis. The rate of FAS in children aged between five and nine years in the Cape Colored community of the Western Cape Province of South Africa is proposed to be as high as 46.4 per 1000 (May et al., 2000), while Canadian FAS data have estimated a range of prevalence which varies from 0.52 to 14.8 per 1000 (Habbick et al., 1996, Williams, Odaibo, and McGee, 1999). In one county in Washington, USA, the number of first graders with FAS was reported as 3.1 per 1000 (Clarren et al., 2001).

When considering the entire range of prenatal alcohol effects, the incidence of FASD has been estimated to be 9.1 per 1000 births (Sampson et al., 1997), which is approximately one out of every 100 births (May and Gossage, 2001). Health Canada has estimated the incidence of FASD at 9 per 1000 births (Health Canada, 2006).

Perhaps one of the most important rates that must be addressed is the prevalence of FAS in families who already have a child with FAS. Women who have already given birth to a child with FAS are at extremely high-risk of having another affected child if they continue to abuse alcohol, and typically later-born children are more impacted than the older children in the family.

Given these numbers, it is not surprising that FASD is associated with significant social and economic ramifications. FASD can cause longlasting medical and psychological problems, and result in economic costs of billions of dollars. Moreover, individuals with FASD suffer from many physical, mental, behavioral, and educational problems which affect daily functioning and have lifelong implications. As a result, individuals with FASD often experience mental health issues, problems in school and work environments, trouble with the law, substance abuse, inappropriate sexual behavior, and difficulties with independent living, among other challenges (Bertrand et al., 2004; Streissguth et al., 2004).

Lupton, Burd, and Harwood (2004) have approximated the cost of a single individual with FASD to be US$2.0 million throughout the individual's lifetime. More recently, the US National Task Force on FAS estimated the adjusted annual cost of FAS in the US to be approximately US$3.6 billion (Olson et al., 2009). Moreover, as the cost of medical treatment, special education, psychosocial intervention and residential care for individuals with FASD increases, these costs will only continue to rise (Lupton, Burd, and Harwood, 2004). In Canada, the average annual cost per child with FASD in 2009 was estimated at $21 642, with the total annual cost of FASD being $5.3 billion (Stade et al., 2009).

1.6
Prevention

Given the incredible impact of FAS/FASD on both individuals and society as a whole, the prevention of PAE and its effects is crucial. Despite ongoing health warnings, pregnant women continue to use alcohol, particularly in patterns that significantly increase the risk of prenatal injury. In 2004, the Center for Disease Control (CDC) reported that 13% of women in the US continue to use alcohol even after knowledge of their pregnancy, and 3% report binge drinking and/or drinking at levels that are known to produce adverse effects in the developing fetus (Bertrand et al., 2004). Furthermore, approximately 55% of women of childbearing age in the US report drinking alcohol, and 12.4% report binge drinking (Rasmussen et al., 2009). As over 50% of pregnancies in the US are unplanned (Finer and Henshaw, 2006), these women are particularly high-risk. Unaware of their pregnancies, women will likely continue their alcohol use during the early stages of embryonic development. These data suggest that more effort must be made to develop effective, evidence-based prevention strategies to reduce the number of alcohol-exposed pregnancies. The first step of prevention must address the disparity between knowledge and behavior, understanding why some women – despite being aware of FAS – continue to engage in high-risk drinking behaviors.

Different levels of prevention fall along a continuum ranging from universal to selective to indicated intervention; as risk behaviors increase, prevention measures become more targeted and intensive (Barry et al., 2009).

Universal prevention attempts to promote the health of the general public, targeting all members of a population or particular group, regardless of risk. Examples of universal approaches include encouraging the abstinence from alcohol during pregnancy, raising public awareness of FASD, and creating alcohol policy and educational programs that minimize the risks of alcohol consumption during pregnancy. Methods to disseminate information include media campaigns, educational materials, and alcoholic beverage labeling. Research into the impact of the U.S. Federal Beverage Labeling Act in several different populations, including a sample of inner-city African-American pregnant women (Hankin et al., 1996; Hankin et al., 1993; Hankin, Sloan, and Sokol, 1998), has revealed an increased awareness of alcohol beverage warning labels since the law's inception in 1989. However, despite this increasing awareness, drinking rates have not necessarily followed suit. Within this sample, Hankin et al. (1993) observed a slight decrease in alcohol consumption, although the decline was apparent only among lighter drinkers; the warning labels did not have any significant effect on high-risk drinkers (Hankin et al., 1993). Furthermore, whilst there was a significant decrease in drinking behaviors post-label for women who had not previously given birth, no change was evident for those women who had already given birth to a child (Hankin et al., 1996). Ultimately, the observed decrease in drinking rates appears to be only short-lasting, and the effectiveness of the labels may lose their impact as women become habituated to them (Hankin, Sloan, and Sokol, 1998). While universal techniques have been shown to increase awareness of the risks associated with alcohol consumption during pregnancy, insufficient data are available regarding any concrete changes in drinking rates among pregnant or non-pregnant women of childbearing age (Barry et al., 2009).

Selective prevention is directed at individuals who are at greater risk than the rest of the general public of having an alcohol-exposed pregnancy due to risky behaviors, such as women of childbearing age who consume high levels of alcohol. Selective interventions are more specific when compared to universal preventions, and may include outreach to at-risk groups, alcohol screening at doctors' offices, referral, and brief intervention strategies aimed at reducing the mother's drinking and minimizing harm to her potential offspring.

Indicated preventions are aimed specifically at the highest risk individuals, such as binge-drinkers, women who are alcoholics, and women who have already given birth to a child with an FASD. These approaches involve a screening process to identify such individuals and help them minimize or cease their alcohol abuse. Several brief screening instruments have been developed for use in a diversity of populations to identify problematic alcohol use in women (Bertrand et al., 2004). Studies have documented the efficacy of screening and brief interventions in reducing risky drinking behaviors and alcohol-exposed pregnancies (Floyd et al., 2009). These brief interventions include clinical advice and counseling regarding the risks of PAE, encouragement to change behaviors, and strategies and goals for reducing the use of alcohol during pregnancy.

1.7
Interventions

The considerable variability in the type and extent of deficits of FASD has led to the development of efficacious interventions becoming a particular challenge. A diagnosis of FAS or the identification of an FASD does not lend itself to a single effective treatment practice that could target the entire range of neurobehavioral problems which an individual may have (Hannigan and Berman, 2000). The various combinations of physical, mental, behavioral, and learning/educational problems among individuals with PAE highlight the complexity of these disorders, and the necessity for interventions that are problem-specific as well as flexible. Rather than confront the entire scope of the disorder—which may be overwhelming and unfeasible—specific behavioral problems should be identified in order to direct specific treatments (Hannigan and Berman, 2000). For example, recent research into interventions for FASD has demonstrated the efficacy of targeted treatments for social skills impairment associated with prenatal alcohol exposure (Paley and O'Connor, 2009).

In addition to intervening with alcohol-exposed individuals to mitigate the deficits and consequences of PAE, treatment practices should focus on providing education and support to the families and caregivers of these individuals. The consequences of FASD extend far beyond the experiences of affected individuals. Both biological and foster parents often experience high levels of stress associated with dealing with their child's impairments (Paley et al., 2006). Many foster parents have expressed their need for more education about FASD, parenting skills training, social support, and professional services to help raise a child with alcohol-related disabilities (Brown, Sigvaldason, and Bednar, 2004). In fact, it has been shown that educating parents and caregivers to realize that the origins of their child's behavioral problems are rooted in brain changes may help them become more understanding and respond to their child in a more supportive manner (Paley and O'Connor, 2009). Interventions should address the issues of the families of those affected by PAE in order to improve their own adjustment and functioning, as well as that of their child.

Current research supports the efficacy of numerous treatment approaches for individuals with FASD as well as their caregivers (as reviewed in Paley and O'Connor, 2009). These treatments include educational and cognitive interventions, parenting interventions, and adaptive skills training. Children exposed to alcohol during pregnancy present with an array of neuropsychological deficits such as overall lower intelligence performance, impaired learning and memory and executive functioning, attention deficits, and hyperactivity. These impairments often result in educational difficulties, including inferior school performance (Mattson et al., 1998), learning disabilities (Burd et al., 2003), and classroom behavioral problems (Carmichael Olson et al., 1991). Educational interventions are focused on developing teaching strategies that facilitate learning in alcohol-exposed children, such as modifying classroom environments that may interfere with a child's ability to learn, and providing support and resources for teachers to help them adapt their instruction and improve their ability to work with students with FASD (Paley and O'Connor, 2009). Additionally, cognitive and academic interventions aim to help enhance skills that will improve an individual's academic performance, focusing on improving general learning skills and/or specific cognitive or academic domains (Paley and O'Connor, 2009). Some of these interventions include cognitive control therapy (CCT) (Riley et al., 2003), language and literacy training (LLT) (Adnams et al., 2007), rehearsal strategies to improve working memory (Loomes et al., 2008), and socio-cognitive habilitation programs to improve behavioral and math functioning (Kable, Coles, and Taddeo, 2007).

As mentioned above, raising children with FASD can be particularly challenging, since many parenting strategies that may be effective with typically developing children may not be successful with alcohol-exposed children. Parent-focused interventions should develop effective parenting skills, improve the parent–child relationship, decrease parent stress, and increase parental self-efficacy (Paley and O'Connor, 2009). Parent–child interaction therapy (PCIT) (Eyberg and Boggs, 1998), parenting support and management (PSM), and supportive behavioral consultation – particularly families moving forward (FMF) (Bertrand, 2009) – are evidence-based practices that provide parents with the necessary support and skills to manage the difficulties of raising a child with the consequences of PAE, and to improve their relationship with their child. Finally, adaptive skills training helps individuals with FASD to develop important age-appropriate skills that help them become less dependent on others and function more independently in their everyday lives. Adaptive skills training may target a range of functional domains, including communication, social interactions (O'Connor et al., 2006), and safety skills (Coles et al., 2007). Clearly, a multi-faceted approach to the prevention and treatment of FAS/FASD, targeting at-risk and affected individuals as well as their families/caregivers, is important in order to mitigate the detrimental personal and societal effects of PAE.

Acknowledgments

These studies were funded in part by grants R01 AA10417 and U24AA014811 to EPR from the National Institute on Alcohol Abuse and Alcohol, National Institutes of Health, USA.

References

Abel, E.L. (1999) Was the fetal alcohol syndrome recognized by the Greeks and Romans? *Alcohol Alcohol.*, **34** (6), 868–872.

Adnams, C.M., Sorour, P., Kalberg, W.O., Kodituwakku, P., Perold, M.D., Kotze, A., September, S., Castle, B., Gossage, J., and May, P.A. (2007) Language and literacy outcomes from a pilot intervention study for children with fetal alcohol spectrum disorders in South Africa. *Alcohol*, **41** (6), 403–414.

Barry, K.L., Caetano, R., Chang, G., DeJoseph, M.C., Miller, L.A., O'Connor, M.J., Olson, H.C., Floyd, R.L., Weber, M.K., DeStefano, F., Dolina, S., Leeks, K., and National Task Force on Fetal Alcohol Syndrome and Fetal Alcohol Effect (March 2009) *Reducing Alcohol-Exposed Pregnancies: A Report of the National Task Force on Fetal Alcohol Syndrome and Fetal Alcohol Effect*, Centers for Disease Control and Prevention, Atlanta, GA.

Bertrand, J. (2009) Interventions for children with fetal alcohol spectrum disorders (FASDs): overview of findings for five innovative research projects. *Res. Dev. Disabil.*, **30** (5), 986–1006.

Bertrand, J., Floyd, R.L., Weber, M.K., O'Connor, M., Riley, E.P., Johnson, K.A., Cohen, D.E., and National Task Force on Fetal Alcohol Syndrome and Fetal Alcohol Effect (2004) *Fetal Alcohol Syndrome: Guidelines for Referral and Diagnosis*, Centers for Disease Control and Prevention, Atlanta, GA, pp. 1–62.

Bonthius, D.J. and West, J.R. (1988) Blood alcohol concentration and microencephaly: a dose–response study in the neonatal rat. *Teratology*, **37** (3), 223–231.

Bonthius, D.J., Goodlett, C.R., and West, J.R. (1988) Blood alcohol concentration and severity of microencephaly in neonatal rats depend on the pattern of alcohol administration. *Alcohol*, **5** (3), 209–214.

Brown, J.D., Sigvaldason, N., and Bednar, L.M. (2004) Foster parent perceptions of placement needs for children with a fetal alcohol spectrum disorder. *Child. Youth Serv. Rev.*, **27**, 309–327.

Burd, L. and Wilson, H. (2004) Fetal, infant, and child mortality in a context of alcohol use. *Am. J. Med. Genet. C Semin. Med. Genet.*, **127C** (1), 51–58.

Burd, L., Klug, M.G., Martsolf, J.T., and Kerbeshian, J. (2003) Fetal alcohol syndrome: neuropsychiatric phenomics. *Neurotoxicol. Teratol.*, **25** (6), 697–705.

Burton, R. (1621) *The Anatomy of Melancholy*, Vintage, New York, NY.

Calhoun, F. and Warren, K. (2007) Fetal alcohol syndrome: historical perspectives. *Neurosci. Biobehav. Rev.*, **31** (2), 168–171.

Carmichael Olson, H., Sampson, P.D., Barr, H.M., Streissguth, A.P., and Bookstein,

F.L. (1991) Prenatal exposure to alcohol and school problems in late childhood: a longitudinal prospective study. *Dev. Psychopathol.*, **4**, 341–359.

Chudley, A.E., Conry, J., Cook, J.L., Loock, C., Rosales, T., and LeBlanc, N. (2005) Fetal alcohol spectrum disorder: Canadian guidelines for diagnosis. *Can. Med. Assoc. J.*, **172** (Suppl. 5), S1–S21.

Clarren, S.K., Randels, S.P., Sanderson, M., and Fineman, R.M. (2001) Screening for fetal alcohol syndrome in primary schools: a feasibility study. *Teratology*, **63** (1), 3–10.

Coles, C.D., Strickland, D.C., Padgett, L., and Bellmoff, L. (2007) Games that "work": using computer games to teach alcohol-affected children about fire and street safety. *Res. Dev. Disabil.*, **28** (5), 518–530.

Cook, R.T., Keiner, J.A., and Yen, A. (1990) Ethanol causes accelerated G1 arrest in differentiating HL-60 cells. *Alcohol. Clin. Exp. Res.*, **14** (5), 695–703.

Eyberg, S.M. and Boggs, S.R. (1998) Parent-child interaction therapy: a psychosocial intervention for the treatment of young conduct-disordered children, in *Handbook of Parent Training: Parents As Co-Therapists for Children's Behavior Problems*, 2nd edn (eds J.M. Briesmeister and C.E. Schaefer), John Wiley & Sons, Inc., Hoboken, NJ, pp. 61–97.

Finer, L.B. and Henshaw, S.K. (2006) Disparities in rates of unintended pregnancy in the United States, 1994 and 2001. *Perspect. Sex. Reprod. Health*, **38** (2), 90–96.

Floyd, R.L., Weber, M.K., Denny, C., and O'Connor, M.J. (2009) Prevention of Fetal Alcohol Spectrum Disorders. *Ment. Retard. Dev. Disabil. Res. Rev.*, **15** (3), 193–199.

Habbick, B.F., Nanson, J.L., Snyder, R.E., Casey, R.E., and Schulman, A.L. (1996) Foetal alcohol syndrome in Saskatchewan: unchanged incidence in a 20-year period. *Can. J. Public Health*, **87** (3), 204–207.

Hankin, J.R., Sloan, J.J., Firestone, I.J., Ager, J.W., Sokol, R.J., and Martier, S.S. (1993) A time series analysis of the impact of the alcohol warning label on antenatal drinking. *Alcohol. Clin. Exp. Res.*, **17** (2), 284–289.

Hankin, J.R., Firestone, I.J., Sloan, J.J., Ager, J.W., Sokol, R.J., and Martier, S.S. (1996) Heeding the alcoholic beverage warning label during pregnancy: multiparae versus nulliparae. *J. Stud. Alcohol.*, **57** (2), 171–177.

Hankin, J.R., Sloan, J.J., and Sokol, R.J. (1998) The modest impact of the alcohol beverage warning label on drinking during pregnancy among a sample of African-American women. *J. Public Policy Mark.*, **17**, 61–69.

Hannigan, J.H. and Berman, R.F. (2000) Amelioration of fetal alcohol-related neurodevelopmental disorders in rats: exploring pharmacological and environmental treatments. *Neurotoxicol. Teratol.*, **22** (1), 103–111.

Health Canada (2006) It's Your Health: Fetal Alcohol Spectrum Disorder [Internet]. Health Canada, Public Health Agency of Canada [original 2006 Sep], http://www.hc-sc.gc.ca/hl-vs/iyh-vsv/diseases-maladies/fasd-etcaf-eng.php (accessed 13 February 2010).

Hoyme, H.E. (2005) A practical clinical approach to diagnosis of fetal alcohol spectrum disorders: clarification of the 1996 Institute of Medicine Criteria–Reply. *Pediatrics*, **115** (6), 1787–1788.

Hoyme, H.E., May, P.A., Kalberg, W.O., Kodituwakku, P., Gossage, J.P., Trujillo, P.M., Buckley, D.G., Miller, J.H., Aragon, A.S., Khaole, N., Viljoen, D.L., Jones, K.L., and Robinson, L.K. (2005) A practical clinical approach to diagnosis of fetal alcohol spectrum disorders: clarification of the 1996 institute of medicine criteria. *Pediatrics*, **115** (1), 39–47.

Jacobson, J.L., Jacobson, S.W., Sokol, R.J., and Ager, J.W., Jr (1998) Relation of maternal age and pattern of pregnancy drinking to functionally significant cognitive deficit in infancy. *Alcohol. Clin. Exp. Res.*, **22** (2), 345–351.

Jones, K.L. and Smith, D.W. (1973) Recognition of the fetal alcohol syndrome in early infancy. *Lancet*, **302** (7836), 999–1001.

Jones, K.L. and Smith, D.W. (1975) The fetal alcohol syndrome. *Teratology*, **12** (1), 1–10.

Jones, K.L., Smith, D.W., Ulleland, C.N., and Streissguth, P. (1973) Pattern of malformation in offspring of chronic alcoholic mothers. *Lancet*, **1** (7815), 1267–1271.

Kable, J.A., Coles, C.D., and Taddeo, E. (2007) Socio-cognitive habilitation using the math interactive learning experience program for alcohol-affected children. *Alcohol. Clin. Exp. Res.*, **31** (8), 1425–1434.

Lemoine, P., Harousseau, H., Borteyru, J.P., and Menuet, J.C. (1968) Les enfants de parents alcooliques: anomalies observes a propos de 127 cas. *Ouest. Med.*, **21**, 476–482.

Livy, D.J., Miller, E.K., Maier, S.E., and West, J.R. (2003) Fetal alcohol exposure and temporal vulnerability: effects of binge-like alcohol exposure on the developing rat hippocampus. *Neurotoxicol. Teratol.*, **25** (4), 447–458.

Loomes, C., Rasmussen, C., Pei, J., Manji, S., and Andrew, G. (2008) The effect of rehearsal training on working memory span of children with fetal alcohol spectrum disorder. *Res. Dev. Disabil.*, **29** (2), 113–124.

Lupton, C., Burd, L., and Harwood, R. (2004) Cost of fetal alcohol spectrum disorders. *Am. J. Med. Genet. C Semin. Med. Genet.*, **127C** (1), 42–50.

McCarver, D.G., Thomasson, H.R., Martier, S.S., Sokol, R.J., and Li, T. (1997) Alcohol dehydrogenase-2*3 allele protects against alcohol-related birth defects among African Americans. *J. Pharmacol. Exp. Ther.*, **283** (3), 1095–1101.

Maier, S.E., Chen, W., and West, J.R. (1996) The effects of timing and duration of alcohol exposure on development of the fetal brain, in *Fetal Alcohol Syndrome: From Mechanism to Prevention* (ed. E.L. Abel), CRC Press, Boca Raton, FL, pp. 27–50.

Maier, S.E., Miller, J.A., Blackwell, J.M., and West, J.R. (1999) Fetal alcohol exposure and temporal vulnerability: regional differences in cell loss as a function of the timing of binge-like alcohol exposure during brain development. *Alcohol. Clin. Exp. Res.*, **23** (4), 726–734.

Maier, S.E., Miller, J.A., and West, J.R. (1999) Prenatal binge-like alcohol exposure in the rat results in region-specific deficits in brain growth. *Neurotoxicol. Teratol.*, **21** (3), 285–291.

Maier, S.E. and West, J.R. (2001) Drinking patterns and alcohol-related birth defects. *Alcohol Res. Health*, **25** (3), 168–174.

Mattson, S.N., Riley, E.P., Gramling, L., Delis, D.C., and Jones, K.L. (1998) Neuropsychological comparison of alcohol-exposed children with or without physical features of fetal alcohol syndrome. *Neuropsychology*, **12** (1), 146–153.

May, P.A. and Gossage, J.P. (2001) Estimating the prevalence of fetal alcohol syndrome. A summary. *Alcohol Res. Health*, **25** (3), 159–167.

May, P.A., Hymbaugh, K.J., Aase, J.M., and Samet, J.M. (1983) Epidemiology of fetal alcohol syndrome among American Indians of the Southwest. *Soc. Biol.*, **30** (4), 374–387.

May, P.A., Brooke, L., Gossage, J.P., Croxford, J., Adnams, C., Jones, K.L., Robinson, L., and Viljoen, D. (2000) Epidemiology of fetal alcohol syndrome in a South African community in the Western Cape Province. *Am. J. Public Health*, **90** (12), 1905–1912.

May, P.A., McCloskey, J., and Gossage, J.P. (2002) Fetal alcohol syndrome among American Indians: epidemiology, issues, and research, in *Alcohol Use Among American Indians and Alaska Natives: Multiple Perspectives on a Complex Problem* (eds P.D. Mail, S. Heurtin-Roberts, S.E. Martin, and J. Howard), Department of Health and Human Services, Bethesda, MD, pp. 321–369.

Miller, M.W. (1993) Migration of cortical neurons is altered by gestational exposure to ethanol. *Alcohol. Clin. Exp. Res.*, **17** (2), 304–314.

Miller, M.W. (1996) Limited ethanol exposure selectively alters the proliferation of precursor cells in the cerebral cortex. *Alcohol. Clin. Exp. Res.*, **20** (1), 139–143.

Morgan, M.Y. and Levine, J.A. (1988) Alcohol and nutrition. *Proc. Nutr. Soc.*, **47** (2), 85–98.

O'Connor, M.J., Frankel, F., Paley, B., Schonfeld, A.M., Carpenter, E., Laugeson, E.A., and Marquardt, R. (2006) A controlled social skills training for children with fetal alcohol spectrum disorders. *J. Consult. Clin. Psychol.*, **74** (4), 639–648.

Olson, H.C., Ohlemiller, M.M., O'Connor, M.J., Brown, C.W., Morris, C.A., Damus, K., and National Task Force on Fetal Alcohol Syndrome and Fetal Alcohol Effect (2009) A call to action: Advancing Essential Services and Research on Fetal Alcohol Spectrum Disorders–A report of the National Task Force on Fetal Alcohol Syndrome and Fetal Alcohol Effect, March 2009.

Paley, B. and O'Connor, M.J. (2009) Intervention for individuals with fetal alcohol spectrum disorders: treatment approaches and case management. *Ment. Retard. Dev. Disabil. Res. Rev.*, **15** (3), 258–267.

Paley, B., O'Connor, M.J., Frankel, F., and Marquardt, R. (2006) Predictors of stress in parents of children with fetal alcohol spectrum disorders. *J. Dev. Behav. Pediatr.*, **27** (5), 396–404.

Rasmussen, S.A., Erickson, J.D., Reef, S.E., and Ross, D.S. (2009) Teratology: from science to birth defects prevention. *Birth Defects Res. A Clin. Mol. Teratol.*, **85** (1), 82–92.

Riley, E.P. and McGee, C.L. (2005) Fetal alcohol spectrum disorders: an overview with emphasis on changes in brain and behavior. *Exp. Biol. Med.*, **230** (6), 357–365.

Riley, E.P., Mattson, S.N., Li, T.K., Jacobson, S.W., Coles, C.D., Kodituwakku, P.W., Adnams, C.M., and Korkman, M.I. (2003) Neurobehavioral consequences of prenatal alcohol exposure: an international perspective. *Alcohol. Clin. Exp. Res.*, **27** (2), 362–373.

Royal College of Physicians of London (1726) *Annals*, Royal College of Physicians, London, England, p. 253.

Sampson, P.D., Streissguty, A.P., Bookstein, F.L., Little, R.E., Clarren, S.K., Dehaene, P., Hanson, J.W., and Graham, J.M. Jr (1997) Incidence of fetal alcohol syndrome and prevalence of alcohol-related neurodevelopmental disorder. *Teratology*, **56** (5), 317–326.

Sedman, A.J., Wilkinson, P.K., Sakmar, E., Weidler, D.J., and Wagner, J.G. (1976) Food effects on absorption and metabolism of alcohol. *J. Stud. Alcohol*, **37** (9), 1197–1214.

Stade, B., Ali, A., Bennett, D., Campbell, D., Johnston, M., Lens, C., Tran, S., and Koren, G. (2009) The burden of prenatal exposure to alcohol: revised measurement of cost. *Can. J. Clin. Pharmacol.*, **16** (1), e91–e102.

Stockard, C.R. (1910) The influence of alcohol and other anaesthetics on embryonic development. *Am. J. Anat.*, **10** (1), 369–392.

Streissguth, A.P. and Dehaene, P. (1993) Fetal alcohol syndrome in twins of alcoholic mothers: concordance of diagnosis and IQ. *Am. J. Med. Genet.*, **47** (6), 857–861.

Streissguth, A.P., Bookstein, F.L., Sampson, P.D., and Barr, H.M. (1993) *The Enduring Effects of Prenatal Alcohol Exposure on Child Development: Birth Through 7 Years: A Partial Least Squares Solution*, University of Michigan Press, Ann Arbor.

Streissguth, A.P., Bookstein, F.L., Barr, H.M., Sampson, P.D., O'Malley, K., and Young, J.K. (2004) Risk factors for adverse life outcomes in fetal alcohol syndrome and fetal alcohol effects. *J. Dev. Behav. Pediatr.*, **25** (4), 228–238.

Sulik, K.K. (2005) Genesis of alcohol-induced craniofacial dysmorphism. *Exp. Biol. Med. (Maywood)*, **230** (6), 366–375.

Sullivan, W.C. (1899) A note on the influence of maternal inebriety on the offspring. *J. Ment. Sci.*, **45**, 489–503.

Warren, K.R. and Foudin, L.L. (2001) Alcohol-related birth defects–the past, present, and future. *Alcohol Res. Health*, **25** (3), 153–158.

Warren, K.R. and Hewitt, B.G. (2009) Fetal alcohol spectrum disorders: when science, medicine, public policy, and laws collide. *Dev. Disabil. Res. Rev.*, **15** (3), 170–175.

Warren, K.R. and Li, T.K. (2005) Genetic polymorphisms: impact on the risk of fetal alcohol spectrum disorders. *Birth Defects Res. A Clin. Mol. Teratol.*, **73** (4), 195–203.

Williams, R.J., Odaibo, F.S., and McGee, J.M. (1999) Incidence of fetal alcohol syndrome in northeastern Manitoba. *Can. J. Public Health*, **90** (3), 192–194.

Part One
Incidence, Prevalence, and Economic Aspects of FASD

2
Researching the Prevalence and Characteristics of FASD in International Settings

Philip A. May

2.1
Introduction

Currently, the exact prevalence (some refer to this as the "incidence") of Fetal Alcohol Spectrum Disorder (FASD) at birth or at other ages is unknown. Although, the simple view of causation is that, if a woman drinks alcohol during pregnancy, then she will have a child with an FASD. More explicitly the literature indicates that the degree of maternal risk is mainly determined by the so-called "QFT" factors. In other words, a child's traits will be influenced by the *Quantity* of alcohol consumed by the mother during pregnancy, the *Frequency* of alcohol use, and the *Timing* of alcohol intake during gestation of the child (Stratton, Howe, and Battaglia, 1996; May, 1995; Abel, 1998). Unfortunately, however, the situation is not that simple, and this and various other complications make the prevalence of FASD a difficult area to understand.

2.2
Maternal Risk Factors and FASD

The most basic question in maternal risk can be summarized as follows. Is there a threshold for the blood alcohol concentration (BAC) that will cause a particular level of FASD or Fetal Alcohol Syndrome (FAS)? This information cannot be extrapolated to humans from animal studies, and human research is limited because the methods that might be used to determine a threshold scientifically are not ethical; it is simply not acceptable to provide alcohol to a pregnant woman to study this issue.

The above variables are somewhat simplistic for the purposes of determining the occurrence of an FASD birth. Many other maternal traits and cofactors, in addition to alcohol use, will affect not only the degree of FASD of an individual child, but also the incidence and prevalence of FASD within a population. These various cofactors may increase, mediate, or mitigate the risk for FASD in a particular child, in a particular mother, in a particular pregnancy, and in a particular community.

Fetal Alcohol Spectrum Disorder–Management and Policy Perspectives of FASD. Edited by Edward P. Riley,
Sterling Clarren, Joanne Weinberg, and Egon Jonsson
Copyright © 2011 WILEY-VCH Verlag GmbH & Co. KGaA, Weinheim
ISBN: 978-3-527-32839-0

Moreover, they can be as broad as socioeconomic and public health conditions, or as specific as the foods that a woman consumes. But, the point here is that many conditions other than alcohol *do* affect the rate of FASD in any given population.

In this chapter, some known cofactors involved in influencing the prevalence and severity of FASD will be briefly discussed. Whilst this is not an exhaustive list, by any means, it does provide an idea of the difficulty of attempting to determine the prevalence of FASD:

- **Nutrition:** In studies conducted in South Africa, the nutrition of the mother – both lifelong and during pregnancy – has been shown to be a significant factor in the degree to which the features of FASD manifest in the child (May *et al.*, 2004, 2008). Notably, the poorer the nutrition of the mother, the more likely it is that the damage will be severe. The body mass index (BMI) of the mother is also significant (see below).

- **Genetic factors:** Despite not being fully understood even in laboratory animals, genetic factors certainly play a role, particularly with regard to the metabolism of alcohol. Individuals with particular polymorphisms of alcohol dehydrogenase (ADH) and aldehyde dehydrogenase (ALDH), both of which influence the metabolism of alcohol, will tend to drink less and therefore have a lower risk of producing children with FAS (Khaole *et al.*, 2004). But, the inverse is also true, in that other polymorphisms of these isoenzymes do not discourage alcohol consumption.

- **Race and ethnicity:** These are also factors, in that drinking patterns vary among racial and ethnic groups. However, there may also be race-related physiological features (particularly the size of the mother) that influence the rate and severity of FASD.

- **Demographic factors:** These have also been recognized; the older a mother is, the more likely she is to have an affected child. Similarly, the higher the gravidity or parity, or the more pregnancies or births a woman has had, the more likely it is that maternal drinking will result in a child with severe FASD. Birth order is also significant, with later-born children more likely to be damaged by alcohol exposure (May *et al.*, 2000, 2007).

- **Socioeconomic status:** This is very important; in the general population of the US and in other studies from several countries, those in the lower social classes had a much higher rate of FASD (May *et al.*, 2007).

- **Spirituality:** This is an interesting factor; in some populations, women who are more religious and who practice their religion on a regular basis are less likely to be drinkers. Therefore less likely to have children with FAS (May *et al.*, 2000; Viljoen *et al.*, 2002).

- **Familial factors:** These have also been identified; women in families that drink a great deal are themselves more likely to be heavy drinkers, and this often continues throughout pregnancy (May *et al.*, 2000; Viljoen *et al.*, 2005).

At this point, an example will be provided which has been drawn from a large general population study in South Africa. In general, extremely good data are acquired from South African studies, because the women interviewed are very honest when reporting details of their lives and risk factors for FASD. Moreover, the interviewers are highly skilled and utilize innovative questions and methods during the interviews. The primary risk factor was the drinks per drinking day (DDD), or the average number of drinks consumed by the mother on a normal drinking day during the first, second, and third trimesters. In addition, the women's blood alcohol concentration (BAC) was estimated, using the BACCUS method (Markham et al., 1993). The study groups were mothers of children with FAS, mothers of children with partial FAS (PFAS), and a control group, which was selected at random. Among mothers of FAS children, the DDD during the first trimester was 5.7; this meant that on a day that they drank, the women consumed an average of six drinks (see Table 2.1). Mothers of children with PFAS consumed about four DDD, while 24% of the mothers of normal (control) children who drank during pregnancy consumed a similar amount of alcohol as did the mothers of PFAS children. The drinking pattern was identified in all three groups, with only a small decline through all three trimesters (see May et al., 2008).

When the DDD values were translated into an estimated BAC (see Table 2.1), although the mothers of children with PFAS and of control children had lower

Table 2.1 Daily drinking levels, peak blood alcohol concentrations, and body mass indices of South African women: multiple reasons for outcomes.

	Drinking mothers of children		
	With FAS	With PFAS	Normal controls (24%)
1st trimester			
DDD	5.7	3.9	3.8
BAC[b]	0.197 ± 0.17	0.155 ± 0.07	0.122 ± 0.11[a]
2nd trimester			
DDD	5.7	3.2	3.7[a]
BAC[b]	0.200 ± 0.17	0.124 ± 0.09	0.084 ± 0.09[a]
3rd trimester			
DDD	5.5	2.7	3.7
BAC[b]	0.191 ± 0.17	0.102 ± 0.12	0.076 ± 0.09[a]
Body mass index (kg/m^2)	22.5	23.5	27.5

a) $P < 0.05$.
b) Values are mean ± SD.
DDD = average drinks per drinking day.
BAC = blood alcohol concentration (in mg%), estimated by the BACCUS technique (considers person's weight, quantity, and duration of drinking).
Source: May et al., 2008.

BAC-values than did the mothers of FAS children, there was a striking similarity in their BAC-values. So, how did the children of drinking control mothers escape from having PFAS or Alcohol-Related Neurodevelopmental Disorder (ARND)? Clearly, the situation is more complex than simply translated DDD-values into a BAC-value. One reason for such lack of simplicity is that, in South Africa, the size of the mother is a significant factor; typically, mothers who have given birth to children with FASD are shorter and lighter, and although their head circumferences were similar in this particular study, they had a greatly reduced BMI (see Table 2.1). In other words, small mothers were more likely to have an FASD child, whereas the control mothers had an average BMI of 27.4, which was deemed "overweight." This particular cofactor of maternal body size is a very important variable that appears to regulate how – and how much – the alcohol will affect the fetus and, ultimately, the child. Notably, the drinking control mothers drank for slightly fewer days than those who had a child with FAS or PFAS (data not in table). Nonetheless, these data reveal how difficult it is to estimate the prevalence of FASD in a population, based on drinking levels alone.

2.3
Determining the Prevalence of FASD: How the Methods Have Influenced the Rates

Currently, three main methods are available for studying the prevalence of FAS and FASD. First, *surveillance and record collection systems* use existing information and data that are captured, recaptured or derived from birth certificates, registries, or disability clinic records. Second, *clinic-based studies* are usually carried out in prenatal or antenatal clinics, where good maternal histories can be obtained and the children born can be diagnosed either as infants, or during the first six months of their life. Third, *active-case ascertainment systems* use data gathered in specialty, referral clinics in outreach populations or in in-school studies (May et al., 2009).

These different methods produce different rates of estimated FAS prevalence. The approximately 15 surveillance studies that are readily available identified an average of 0.85 children with FAS per 1000, and a median of 0.27 per 1000. The clinic-based studies, of which over 50 have now been reported, show a higher rate of FAS, at 1.8 per 1000 and a median of 1.9. The eight active-case ascertainment studies reviewed for this chapter (which exclude the in-school studies presented below) produced a rate of 15.6 per 1000 and a median of about 9.

This tremendous variation of prevalence rates can also be seen in studies of FASD. Although no surveillance studies of FASD were identified, the clinic-based studies produced an average rate of 6 per 1000, with a median of 5. The active-case ascertainment studies (excluding in-school studies) produced an average of 38 per 1000, with a median of about 19. Although it would appear that unequivocally higher rates are reported from active-case ascertainments, caution is required here as active case ascertainment studies are usually carried out in high-risk populations. Consequently, such prevalence patterns and rates cannot be extrapolated directly to the general population.

The most quoted estimate of the rate of FAS in the extant literature is about 0.5–2 per 1000, or 0.5–3 per 1000, from the Institute of Medicine report (Stratton, Howe, and Battaglia, 1996). For FASD, the rate is about 1.0%, an estimate previously derived by Sampson et al. (1997) from a longitudinal study conducted in Seattle.

2.4
The Prevalence of FASD from In-School Studies

Since 1997, the present author and colleagues have been conducting prevalence studies in schools that provide access for a team of specialists to a general population of children. Those children in first grade (age 6–7 years) are at an optimal age for an accurate diagnosis of FAS, since dysmorphology and growth are well known at that age, and neurobehavioral testing can begin to discriminate a complex array of traits and limitations in such areas as intelligence, executive functioning, motor control, and learning at this age (Aragon et al., 2008). Also key to in-school methods is the fact that the mothers are interviewed and, indeed, several studies have revealed that in retrospective interviews, the mothers generally report a higher alcohol consumption than was claimed at the antenatal clinic (Hannigan et al., 2010). Moreover, an early identification – that is, diagnosis by grade one – makes it possible to give FAS children the life and educational opportunities that they need to prosper later in life. One challenge here is that the access to children in these studies requires the active consent of the parents. Unfortunately, it is not always possible to acquire consent for more than 50% of the children to participate in these studies, and consequently the representativeness of the study might be called into question. However, every in-school study undertaken to date has detected a substantial number of cases of FAS, PFAS, and other affected children who had never been – and may never have been – identified or diagnosed via normal clinical channels or referral systems (Clarren et al., 2001; May et al., 2006).

Data were acquired from South Africa, from Italy, from a western city in the United States, and from a head-start school in a community on the American plains (see May et al., 2009) (Figure 2.1). In the South African studies of FAS, the average prevalence was found to be about 50 per 1000 (or 5%). It would appear that South Africa is a special case, where many forces have converged to create a very high-risk population. Whilst many consider that the South African FASD studies do not apply to North American populations, this may not be the case. In fact, the methods do apply quite directly, and many of the risk factors that exist in South Africa also exist elsewhere, but to a lesser degree; therefore, they also apply to other populations. By using similar methods, and by knowing the relative degree of these risk factors in other populations, the findings from populations such as those in South Africa can be extrapolated to other countries, including North America.

In Italy, prevalence rates of 5.6 and 6.8 per 1000 were found for FAS, while a rate of 2 per 1000 was identified in one wave of pilot screening in the western city

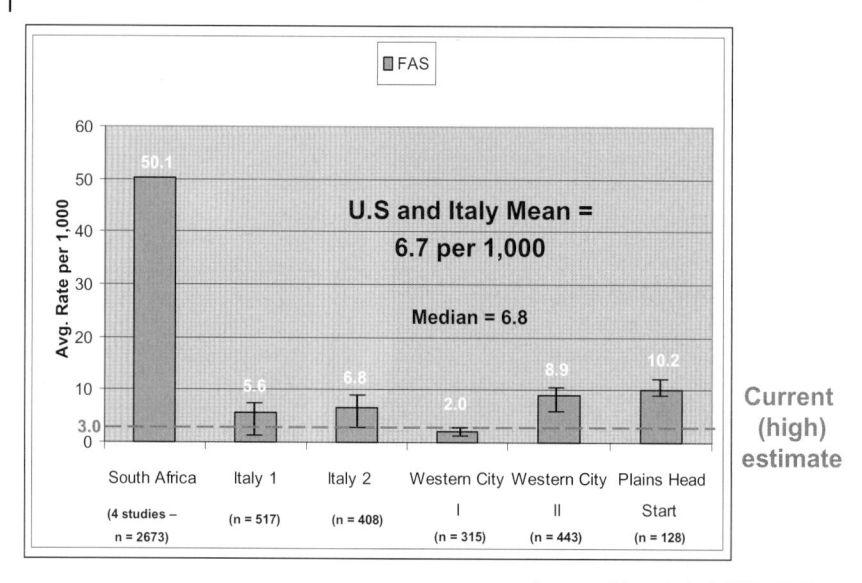

Figure 2.1 Summary of the prevalence of FAS from in-school studies.

Source: May et al., DDRR, 2009.

in the US, and of 8.9 per 1000 in another screening of the same city (Figure 2.1). In the plains head-start school, a rate of 10.2 per 1000 was identified.

Combining data on FAS and PFAS for an estimate of FASD, in South Africa the estimate was about 72 per 1000, or 7%. The rate for FAS and PFAS combined was about 3% in Italy (30 per 1000), and 4% in another wave of the Italian community (both were general-population, middle class communities in Italy). In the western city of the US, the rate was about 1.4%, about 2% in a second wave of the same city, and about 2% in the plains head-start school (Figure 2.2).

2.5
Summary Rates of FASD and Their Meaning

These data suggest that the prevalence of all types of FASD within a population can best be determined through screening in representative, population-based samples, such as in-school studies. In such cases, the signs and symptoms of the spectrum in the children may be examined by dysmorphologists, and the children tested with appropriate and discriminating cognitive and behavioral batteries by psychologists/educational diagnosticians, while their mothers may be interviewed for risk factors during their pregnancies. Thus, a general prevalence can be approximated better than by employing surveillance referral, or clinic-based methods. The reporting of routinely diagnosed cases does not approach the true prevalence

2.5 Summary Rates of FASD and Their Meaning

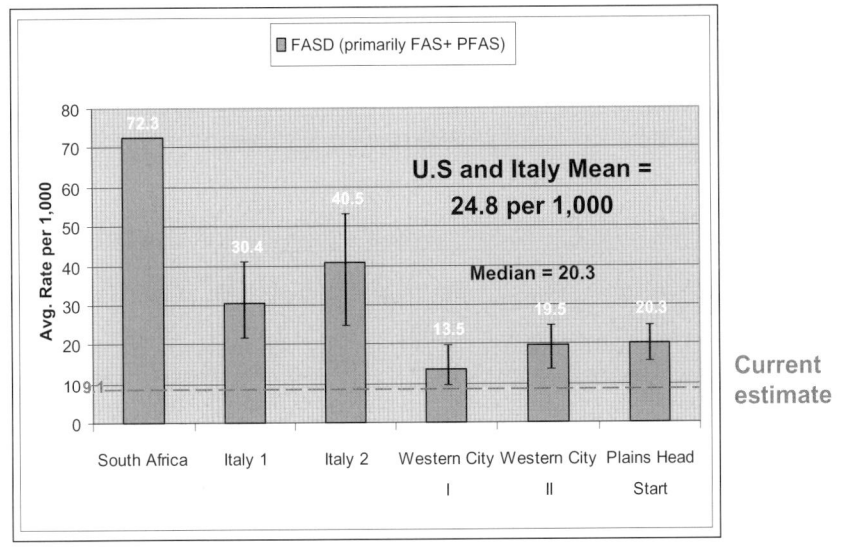

[Bar chart showing FASD prevalence rates per 1,000 across studies: South Africa 72.3, Italy 1 30.4, Italy 2 40.5, Western City I 13.5, Western City II 19.5, Plains Head Start 20.3. U.S and Italy Mean = 24.8 per 1,000. Median = 20.3. Current estimate indicated.]

Source: May et al., DDRR, 2009.

Figure 2.2 Summary of the prevalence of FASD from in-school studies.

of FASD, since many individuals – even those with full-blown FAS or PFAS – may go undetected and undiagnosed for the full extent of their lives.

Whilst previous estimates of the prevalence of FASD and FAS in the general population have been reported as ranging between 0.5 and 3 per 1000, today the prevalence of FAS is believed to be closer to 2 to 7 per 1000. Likewise, for FASD (and specifically for PFAS and FAS, the more dysmorphic forms), the present rates in the general population of developed countries may lie between 2% and 5%, rather than the previous estimate of 1%. At present, the prevalence for the less-dysmorphic forms of FASD, such as Alcohol-Related Neurodevelopment Deficit (ARND), cannot be accurately estimated.

To summarize, does this represent a substantial public health problem? Despite being highly skeptical regarding a high prevalence of FAS, the first 25 years of the present author's career was spent working with high-risk populations rather than with the general population as at present. Consequently, even with previous active-case ascertainment studies with referrals to tertiary screening, many cases of FASD were missed. More recent in-school studies have added a great deal to this area, however, and the prevalence of FASD is now believed to be higher among the general population than previously estimated. It would appear that FASD does indeed represent a very significant public health problem, even within the mainstream of general populations. If FASD cases are not actively sought, then the majority of them will not be identified, especially those with PFAS and ARND who blend into the general populations because their dysmorphology is not classical, and/or their behavior is less dramatically affected. Although, for many years,

FASD has been considered a leading cause of mental deficiency in many modern societies, the data acquired through these in-school studies appear increasingly to prove this statement to be even more accurate than many thought possible.

References

Abel, E.L. (1998) *Fetal Alcohol Abuse Syndrome*, Plenum Press, New York.

Aragon, A.S., Coriale, G., Fiorentino, D., Kalberg, W.O., Buckley, D., Gossage, J.P., Ceccanti, M., Mitchell, E.R., and May, P.A. (2008) Neuropsychological characteristics of Italian children with fetal alcohol spectrum disorders. *Alcohol. Clin. Exp. Res.*, **32**, 1909–1919.

Clarren, S.K., Randels, S.P., Sanderson, M., and Fineman, R.M. (2001) Screening for fetal alcohol syndrome in primary schools: a feasibility study. *Teratology*, **63**, 3–10.

Hannigan, J.H., Chiodo, L.M., Sokol, R.J., Janisse, J., Ager, J., Greenwald, M.K., and Delaney-Black, V. (2010) A 14-year retrospective maternal report of alcohol consumption in pregnancy predicts pregnancy and teen outcomes. *Alcohol*, in press.

Khaole, N.C., Ramchandani, V.A., Viljoen, D.L., and Li, T.K. (2004) A pilot study of alcohol exposure and pharmacokinetics in women with or without children with fetal alcohol syndrome. *Alcohol Alcohol.*, **39**, 503–508.

Markham, M.R., Miller, W.R., and Arciniega, L. (1993) BACCus 2.01: Computer software for quantifying alcohol consumption (A blood alcohol concentration calculating system). *Behav. Res. Meth. Instr. Comp.*, **25**, 420–421.

May, P.A. (1995) A multiple-level, comprehensive approach to the prevention of fetal alcohol syndrome (FAS) and other alcohol-related birth defects (ARBD). *Int. J. Addict.*, **30**, 1549–1602.

May, P.A., Brooke, L.E., Gossage, J.P., Croxford, J., Adnams, C., Jones, K.L., Robinson, L.K., and Viljoen, D.L. (2000) The epidemiology of Fetal Alcohol Syndrome in a South African community in the Western Cape Province. *Am. J. Public Health*, **90**, 1905–1912.

May, P.A., Gossage, J.P., White-Country, M., Goodhart, K., DeCouteau, S., Trujillo, P.M., Kalberg, W.O., Viljoen, D.L., and Hoyme, H.E. (2004) Alcohol consumption and other maternal risk factors for fetal alcohol syndrome among three distinct samples of women before, during, and after pregnancy: the risk is relative. *Semin. Med. Genet.*, **127C**, 10–20.

May, P.A., Fiorentino, D., Gossage, J.P., Kalberg, W.O., Hoyme, H.E., Robinson, L.K., Coriale, G., Jones, K.L., Del Campo, M., Tarani, L., Rome, M., Kodituwakku, P.W., Deiana, L., Buckley, D., and Ceccanti, M. (2006) The epidemiology of FASD in a province in Italy: prevalence and characteristics of children in a random sample of schools. *Alcohol. Clin. Exp. Res.*, **30**, 1562–1575.

May, P.A., Gossage, J.P., Marais, A.S., Adnams, C.M., Hoyme, H.E., Jones, K.L., Robinson, L.K., Khaole, N.C., Snell, C., Kalberg, W.O., Hendricks, L., Brooke, L., Stellavato, C., and Viljoen, D.L. (2007) The epidemiology of fetal alcohol syndrome and partial FAS in a South African community. *Drug Alcohol Depend.*, **88** (2–3), 259–271.

May, P.A., Gossage, J.P., Marais, A.S., Hendricks, L., Snell, C., Tabachnick, B.G., Stellavato, C., Buckley, D.G., Brooke, L., and Viljoen, D.L. (2008) Maternal risk factors for fetal alcohol syndrome and partial fetal alcohol syndrome in South Africa: a third study. *Alcohol. Clin. Exp. Res.*, **32**, 738–753.

May, P.A., Gossage, J.P., Kalberg, W.O., Robinson, L.K., Buckley, D.G., Manning, M., and Hoyme, H.E. (2009) The prevalence and epidemiologic characteristics of FASD from various research methods with an emphasis on in-school studies. *Dev. Disabil. Res. Rev.*, **15**, 176–192.

Sampson, P.D., Streissguth, A.P., Bookstein, F.L., Little, R.E., Clarren, S.K., Dehanne,

P., Hanson, J.W., and Graham, J.M. (1997) Incidence of fetal alcohol syndrome and prevalence of alcohol-related neurodevelopmental disorder. *Teratology*, **56**, 317–326.

Stratton, K.R., Howe, C.J., and Battaglia, F.C. (1996) *Fetal Alcohol Syndrome Diagnosis, Epidemiology, Prevention, and Treatment. Institute of Medicine (Division of Biobehavioral Sciences and Mental Disorders, Committee to Study Fetal Alcohol Syndrome and National Institute on Alcohol Abuse and Alcoholism)*. National Academy Press, Washington, D.C.

Viljoen, D.L., Croxford, J., Gossage, J.P., and May, P.M. (2002) Characteristics of mothers of children with Fetal Alcohol Syndrome in the Western Cape Province of South Africa: a case control study. *J. Stud. Alcohol*, **63**, 6–17.

Viljoen, D.L., Gossage, J.P., Adnams, C.M., Jones, K.L., Robinson, L.K., Hoyme, H.E., Snell, C., Khaole, N., Asante, K.K., Findlay, R., Quinton, B., Brooke, L.E., and May, P.A. (2005) Fetal alcohol syndrome epidemiology in a South African community: a second study of a very high prevalence area. *J. Stud. Alcohol*, **66**, 593–604.

3
Frequency of FASD in Canada, and What This Means for Prevention Efforts

Suzanne C. Tough and Monica Jack

3.1
Introduction

While rates of incidence and prevalence are important when describing the impact of any disease, they may also be helpful when planning prevention efforts. In this respect, Fetal Alcohol Spectrum Disorders (FASD) are no exception, and although much data exist relating to the incidence and prevalence of FASD, there remain some unique challenges that must be considered when interpreting these rates.

3.2
Challenges to Obtaining Accurate Incidence and Prevalence Rates

The term "incidence" describes the number of new cases of FASD that occur during a specific time period within a defined population (May and Gossage, 2001), while "prevalence" includes all existing cases of FASD at any particular point in time (May and Gossage, 2001). Incidence rates (the number of new cases during a defined period of time) are usually reported per 1000 live births, while prevalence is usually reported per 1000 people among a particular population (e.g., per 1000 children in care), although it is easier to understand these rates by expressing them as percentages. In this chapter, incidence and prevalence will be reported in both ways, although in the case of FASD, prevalence rates are more commonly reported than are incidence rates. Prevalence rates can provide a good idea of how many people require resources and support for living with FASD, and also help to determine when – and where – services are most needed.

Although both incidence and prevalence include new cases of FASD, it is difficult to know when to define a case of FASD as a new case (May and Gossage, 2001). A new case of FASD is diagnosed at some time after birth, although the impact of alcohol on the fetus has occurred during the time of pregnancy. However, miscarriages are more common among alcohol-abusing women (May and Gossage, 2001). Thus, the birth rate of FASD may be lower than the prenatal rate of FASD, because fetuses with FASD may be more likely to be spontaneously aborted

---

Fetal Alcohol Spectrum Disorder–Management and Policy Perspectives of FASD. Edited by Edward P. Riley, Sterling Clarren, Joanne Weinberg, and Egon Jonsson
Copyright © 2011 WILEY-VCH Verlag GmbH & Co. KGaA, Weinheim
ISBN: 978-3-527-32839-0

(i.e., miscarried). Thus, the rates of FASD must be considered based on the denominator.

Furthermore, while new cases of FASD are discovered through diagnosis, a diagnosis may take place much later than the onset of FASD. Typically, a "new case" of FASD may go undetected for years, especially when the child has no facial abnormalities. Hence, the issue of a delayed diagnosis can result in an underestimate of the incidence rates.

Although more of a problem in the past, the diagnostic definitions used in incidence and prevalence studies have also varied. For example, a study that incorporates a broader definition of FASD will have higher rates than will a study with a narrow definition of FASD. In Canada, a set of guidelines for the diagnosis of FASD was published in 2005, and this provides standards for diagnosis across Canada (Chudley et al., 2005). These guidelines recommend a comprehensive multidisciplinary diagnostic evaluation for the best diagnosis. However, there is limited capacity and expertise in Canada to universally implement this approach to diagnosis, and hence, current rates of FASD may underestimate the true rate. Considering the issues around diagnosis, reductions in incidence and prevalence rates can mean one of two things: (i) that prevention programs are working; or (ii) that there is an under-identification of the disease.

Aside from diagnosis variations, there are also different ways of studying incidence and prevalence that can affect the rates determined. Currently, there are three main approaches to studying the incidence and prevalence of FASD (May and Gossage, 2001):

- Passive surveillance systems use existing records in a geographic area.
- Clinic-based studies recruit pregnant women, usually through prenatal clinics, and assess the children some time after birth; alternatively, the children are investigated when they attend specialized behavioral or developmental clinics.
- Active case ascertainment methods actively recruit children who may have FASD within a geographic area.

An overview of the advantages and disadvantages of each approach is provided in Table 3.1. Because each method will lead to different estimates of incidence and prevalence, the rates obtained from studies with different designs cannot be directly compared.

Neither can any wide generalization be made from a study in one geographic area to another, nor from a study of one subpopulation to another. The incidence and prevalence of FASD will vary from one geographic area to another, and from one subpopulation to another. As an example, a study of youth in the criminal justice system is not representative of all children in government care, nor of youth in general. Likewise, a study conducted in a small northeastern community of a province is not necessarily representative of all of that province. There may be true variability in FASD and alcohol use during pregnancy between these communities or subpopulations, but there may also variability in the reporting and diagnosis of FASD, or in the ability to track down cases of FASD (Abel, 1995; Williams, Odaibo, and McGee, 1999).

Table 3.1 Approaches to studying FASD incidence and prevalence (May and Gossage, 2001).

Methodology	Advantages	Disadvantages
Passive surveillance systems: Use existing records in a geographic area	• Efficiently use existing resources • Cost less • Easy to conduct	• Difficult to diagnose at birth • Rely on non-specialist physicians to diagnose • Records may be missing critical data
Clinic-based studies: Recruit pregnant women usually through prenatal clinics and assess the children some time after birth	• Allow for postpartum diagnosis through study • Can gather data on history of mother • Large number of pregnancies with variation in alcohol exposure • Greater control and rigor in study design	• Self-selection bias as women at highest risk of delivering an infant with FASD may be least likely to attend prenatal clinics regularly, if at all • Results may not generalize to other clinics if the clinic serves a people or a region with unique demographic characteristics (e.g., low socioeconomic status) • FASD is often diagnosed at birth when many cases will be missed, but children are more accurately diagnosed between 3 and 12 years of age
Active case ascertainment methods: Actively recruit children who may have FASD within a geographic area	• Can diagnose children at appropriate ages • Broad outreach in a community may identify undetected children • Reduce selection bias	• High costs, time- and labor-intensive • May be difficult to gain support and cooperation from the community of interest • May not generalize to other populations, as often high-risk communities are selected

3.3
Incidence of FASD

3.3.1
National Rates in Canada (see Box 3.1)

To date, no official statistics have been released for the incidence of Fetal Alcohol Syndrome (FAS) or FASD in Canada. Government organizations have estimated the incidence of FAS at 1–3 per 1000 live births (0.1–0.3%), and of FASD at 9 per 1000 live births (0.9%) (Alberta Alcohol and Drug Abuse Commission, 2004; Health Canada, 2006). These figures seem to be extrapolated from those of the United States and the Western World (Abel, 1995; Sampson et al., 1997).

> **Box 3.1   How Information on Incidence and Prevalence Rates Were Found**
>
> - Known and available sources papers and reports were included.
> - MEDLINE was searched for articles related to the rates of FAS/FASD prevalence or incidence, using the following search criteria:
>   - Any of: FAS, FASD, fetal alcohol effects
>   - Either of: prevalence, incidence
>   - Any of: Canada, BC, AB, SK, MB, ON, QC, PE, NS, NB, NL, NT, YT, NU
> - When reported rates were in secondary sources (i.e., cited from other sources), references lists were used to obtain primary sources (i.e., articles originally reporting figure) where available.
> - Other sources were also obtained from reference lists.
> - Provincial and territorial government web pages were searched for provincially reported rates.

3.3.2
Provincial Rates in Canada

Some western provinces in Canada have information on incidence from research studies, although some statistics date back as far as the 1970s. Other provinces do not appear to have statistics on the incidence of FASD.

3.3.2.1   British Columbia

An unpublished document[1] reported on the birth defects registry in British Columbia from 1972 to 1980. The incidence of FAS was reported at 0.25 per 1000 live births (0.025%) among non-Aboriginals, and 4.7 per 1000 live births (0.47%) among Aboriginals (Williams, Odaibo, and McGee, 1999).

3.3.2.2   Alberta

Alberta does not have official statistics for FAS or FASD, but reports estimates similar to the national estimates (Alberta's Commission on Learning, 2003).

3.3.2.3   Saskatchewan

In Saskatchewan, individuals known to have FAS were identified by an FAS clinic and clinicians likely to be in contact with individuals with FAS (Habbick et al., 1996). These individuals were then grouped according to the time period in which

---

1) A copy of the original unpublished report [Wong, N. (1983) British Columbia Surveillance Registry, B.C. Ministry of Health, Vancouver, British Columbia] could not be obtained, despite it being cited in the published literature.

they were born, and an incidence rate was calculated using the live births in that given time period. This study estimated the incidence of FAS at 0.589 per 1000 live births (0.0589%) between 1988 and 1992 (Habbick et al., 1996).

3.3.2.4 Manitoba

In a small study in northeastern Manitoba, the incidence of FAS was calculated among all 745 children born at Thompson General Hospital during 1994 (Williams, Odaibo, and McGee, 1999). At the time of the study, the children were aged about 2 years, and their birth records were screened for potential cases of FAS. If such children were identified, their follow-up hospital records or public health records were then screened, and those still identified as a potential case of FAS were examined individually by a pediatrician, for a diagnosis. Subsequently, the estimated incidence of FAS was 7.2 per 1000 live births (0.72%) (Williams, Odaibo, and McGee, 1999). Although the research team involved warned that this study was too small to provide reliable estimates of incidence, the data obtained did indicate that FASD was a problem in this region. Of note, Thompson General Hospital serves a predominantly Aboriginal population in northern Manitoba; consequently, the figures obtained may not be generalizable to other regions of the province.

3.3.2.5 Other Provinces

There do not appear to be any population-based statistics available for FAS or FASD incidence for other Canadian provinces in the research literature, nor publicly available on government web sites.

A summary of the incidence rates reported in Canada and its provinces is provided in Table 3.2.

3.4
Prevalence of FASD

The prevalence of FASD in Canada has been estimated for three main subpopulations: child welfare systems; corrections systems; and Aboriginal communities. A summary of the prevalence rates reported in Canada is provided in Table 3.3.

3.4.1
Child Welfare Systems

Manitoba is the only Canadian province where data were available regarding the prevalence of FAS/FASD among children in care. Data from the existing child welfare information-gathering system suggested that, during 2004, 11% of all children in care in Manitoba had FASD (110 children with FASD per 1000 children in care) (Fuchs et al., 2007). FASD represents one-third of the disabilities among children in care in Manitoba (Fuchs et al., 2007).

Table 3.2 Incidence rates reported in Canada.

Population	Method	Years	Ages	FAS incidence rate	FASD incidence rate
Canada (Alberta Alcohol and Drug Abuse Commission, 2004; Health Canada, 2006)	No official statistics; estimates only	–	–	0.1–0.3%	0.9%
British Columbia (Williams, Odaibo, and McGee, 1999)	Passive surveillance (birth defects registry)	1972–1980	Newborns	0.025% (non-Aboriginals) 0.47% (Aboriginals)	–
Alberta (Alberta's Commission on Learning, 2003)	No official statistics; estimates only	–	–	0.1–0.3%	0.2–0.9%
Saskatchewan (Habbick et al., 1996)	Active case ascertainment (known cases to FAS clinic and clinicians)	1988–1992	All	0.0589%	–
Northeastern Manitoba (Williams, Odaibo, and McGee, 1999)	Clinic-based (all hospital live births)	1994	0–2 years	0.72%	–

Estimates of the incidence of FASD include children with FAS as well as other disorders related to prenatal alcohol exposure.

Although the Canadian Incidence Study of Reported Child Abuse and Neglect did not specifically report on the prevalence of FAS or FASD, it provided some insight into how often case workers become aware of child-functioning problems related to substance abuse in general. In Canada, 3% of substantiated child maltreatment investigations in 2003 involved a child whose case worker had become aware of a diagnosis or indication of a substance-abuse-related birth defect (Trocme et al., 2005). The figure for Alberta for the same time period was 10% (MacLaurin et al., 2005). These figures are thought to be underestimates, however, as case workers did not systematically assess child functioning but rather became aware of child-functioning difficulties during their investigations (Trocme et al., 2005; MacLaurin et al., 2005).

However, in the future, some information on prevalence rates of FASD among children in care in other provinces may become available. One provincial government ministry is known to be developing a new case management information system that may provide the ability to gather information systematically on FASD on children served by their ministry.

Table 3.3 Prevalence rates reported in Canada.

Population	Method	Years	Ages	FAS prevalence rate	FASD prevalence rate
Children in care in Manitoba (Fuchs et al., 2007)	Passive surveillance (child welfare case files)	2004	0–20 years	–	11%
Canadian corrections system (Burd et al., 2003)	Passive surveillance (known cases, no new screening)	2001/02	≥18 years	0.0087%	–
Forensic Inpatient Assessment Unit in British Columbia (Fast, Conry, and Loock, 1999)	Assessed all youth coming into the unit	1995/96	12–18 years	1%	23.3%
A Manitoba First Nation community (Kowlessar, 1997)	Assessment of all children within a birth cohort	1995/96	5–15 years	3.1–6.2%	5.1–10.1%
A British Columbia First Nation community (Robinson, Conry, and Conry, 1987)	Assessed all children in community	1984/85	0–18 years	12.1%	19.0%
Twenty-two communities in Northwest British Columbia (Asante and Nelms-Maztke, 1985)	Active case ascertainment	1984	0–16 years	–	2.5% (Aboriginal) 0.04% (non-Aboriginal)
Fourteen communities in Yukon Territory (Asante and Nelms-Maztke, 1985)	Active case ascertainment	1983	0–16 years	–	4.6% (Aboriginal) 0.04% (non-Aboriginal)

Estimates of the prevalence of FASD include children with FAS as well as other disorders related to prenatal alcohol exposure.

3.4.2
Corrections Systems

The prevalence of FASD in the Canadian corrections system in 2001/2002 was estimated at 0.0087% (0.087 per 1000 people in corrections facilitations or community corrections) (Burd et al., 2003). However, the study relied on the Director of Corrections from each province/territory estimating the number of cases with diagnosed FAS and alcohol-related neurodevelopmental disorders, and no screening for FASD was undertaken among inmates (Burd et al., 2003). The results of this study also showed that the Canadian corrections systems had inadequate staff training for FASD, and a limited ability to screen and identify cases of FASD (Burd et al., 2003).

Among youth in the criminal justice system, a more thorough approach to the identification of FASD was undertaken in an Inpatient Assessment Unit in Burnaby, British Columbia in 1995/1996 (Fast, Conry, and Loock, 1999). The study results showed that 1% of the 287 youths remanded to this psychiatric inpatient assessment unit had FAS, and 23.3% had FASD (Fast, Conry, and Loock, 1999).

3.4.3
Aboriginal Communities

There are no estimates for the prevalence of FASD among all First Nations, Inuit, or Métis children in Canada, although the prevalence of FASD has been examined in a single community or geographic area in some studies (McShane, Smylie, and Adomako, 2009; Buell et al., 2006). In Manitoba in 1995/1996, a study conducted in school-aged children in a First Nation community showed that the prevalence of FAS ranged between 3.1% and 6.2% (31–62 per 1000 children) (Kowlessar, 1997). Moreover, the prevalence of FASD (which included FAS) was between 5.1% and 10.1% (51–101 per 1000 children) (Kowlessar, 1997). This range in prevalence figures resulted from different denominators used in the study. When the entire birth cohort from the Medical Services Branch records was used as the denominator (352 children), this resulted in the lowest prevalence rate of 3.1% for FAS, and of 5.1% for FASD. The use of children from the birth cohort who were on the local school list ($n = 243$) led to an intermediate prevalence rate of 4.5% for FAS, and of 7.4% for FASD. The use of only those children who participated in the study ($n = 178$) led to the highest prevalence rate of 6.2% for FAS, and of 10.1% for FASD. These findings illustrate the challenges associated with determining which denominator allows a better understanding of the population of interest. Regardless of which denominator was used, however, the study results indicated that FAS/FASD was a significant issue among this community.

In British Columbia, a similar study was carried out in 1984/1985 in which all school-aged children in an isolated Aboriginal community were assessed. The study results showed a prevalence rate of 12.1% for children with FAS (121 per 1000 children), and 19.0% for those having FASD (190 per 1000 children) (Robinson, Conry, and Conry, 1987).

Knowledge of the prevalence of FASD in one Aboriginal community helps in planning prevention and treatment efforts for that particular community. However, the findings in one Aboriginal community cannot necessarily be applied to other such communities, as each community is different. The prevalence of FASD among several communities in two geographic areas has been examined in only one study (Asante and Nelms-Maztke, 1985). In this case, in 1983, based on 14 communities in the Yukon, the prevalence of FASD among Aboriginal children was estimated at 4.6% (46 per 1000 children) (Asante and Nelms-Maztke, 1985). Likewise, in 1984, based on 22 communities in northwest British Columbia, the prevalence of FASD among Aboriginal children was estimated at 2.5% (25 per 1000 children) (Asante and Nelms-Maztke, 1985). Notably, both values were substantially higher than those estimated for non-Aboriginal children in these communities (0.04% or 0.4 per 1000 non-Aboriginal children) (Asante and Nelms-Maztke, 1985).

3.5
Rate of Exposure to Risk

It has been indicated by some research groups that simply knowing the incidence of FAS/FASD is insufficient. Thus, Sampson *et al.* (1997) have suggested that if a condition is related to an exposure, then "... information on the rate of this exposure in the population is needed in order to interpret the incidence fully." Although alcohol consumption during pregnancy is the exposure that creates risk for FAS/FASD, there will be a range of alcohol consumption patterns during pregnancy, and it is important to note that variations in the quantity, frequency, and timing of consumption may have different impacts on fetal development. Moreover, these variations in consumption patterns are not reflected in the rates of alcohol consumption during pregnancy that are reported below.

In 2005, across Canada, 11% of mothers reported some consumption of alcohol during pregnancy (Public Health Agency of Canada, 2008). The consumption rates were seen to vary by province, with Quebec having the highest reported rate (18% of mothers reported alcohol consumption during pregnancy), and Newfoundland and Labrador having the lowest rate, at 4% (Figure 3.1) (Public Health Agency of Canada, 2008). While the national reported rate of alcohol consumption in Canada is comparable to that determined by other North American research investigations (Tsai *et al.*, 2007), information obtained from Alberta has shed some light on the limitations of these numbers.

In Alberta in 2002/2003, half of all surveyed first-time mothers reported drinking some alcohol while pregnant, but *before they knew* they were pregnant (Tough *et al.*, 2006a). The patterns of alcohol consumption among these women are shown in Figure 3.2 (Tough *et al.*, 2006a). Most women reported that usually they had no more than two drinks on any occasion, and fewer than nine drinks per week in total (low-risk drinking) (Tough *et al.*, 2006a). However, almost 25% of the women reported binge drinking (consuming five or more drinks during a 24-h period) while pregnant but, again, *before they knew* they were pregnant (Tough *et al.*, 2006a). After recognizing their pregnancy, 18% of first-time mothers reported drinking alcohol, all of whom reported low-risk drinking patterns (Tough *et al.*, 2006a). These data were closer to those reported across North America (Public Health Agency of Canada, 2008; Tsai *et al.*, 2007).

The findings from this Alberta study suggested that women may be less likely to report alcohol consumption during pregnancy when asked *generally* about it, compared to a specific question about consumption *before they knew* they were pregnant. This suggestion was supported by other Alberta research, which found

Canada, 2000–2001, 2003 and 2005

Figure 3.1 Rate of maternal alcohol consumption during pregnancy, by province/region. Reprinted with permission from Public Health Agency of Canada (2008) Canadian Perinatal Health Report, 2008 Edition, Public Health Agency of Canada, Ottawa, ON.

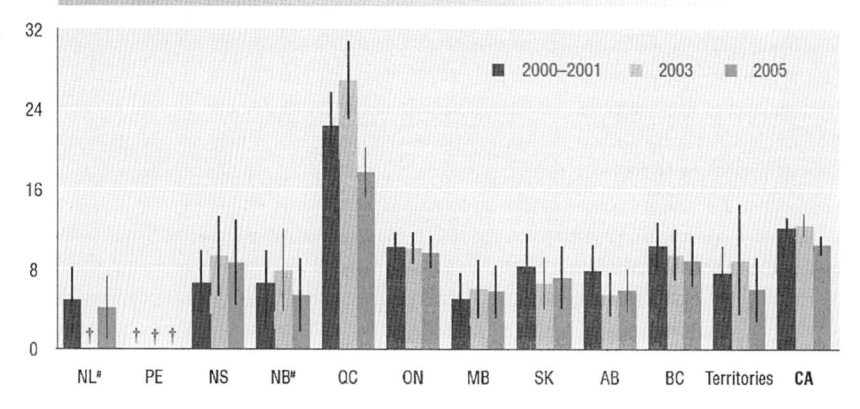

Source: Statistics Canada. Canadian Community Health Survey, 2000–2001, 2003, 2005.
* Women who gave birth in the five years preceding the survey; denominators exclude responses of "do not know" and "not stated," and refusal to answer.
† Estimates not shown because sample size was less than 10.
# High level of sampling variability for 2000–2001 and 2005 data from Newfoundland and Labrador, and 2005 data from New Brunswick.
CI—confidence interval

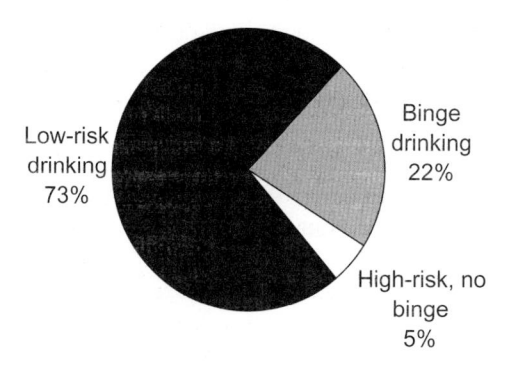

Figure 3.2 Patterns of alcohol consumption among first-time mothers who consumed alcohol before knowing they were pregnant.

- Low-risk drinking 73%
- Binge drinking 22%
- High-risk, no binge 5%

that about half of all women who consumed alcohol before they knew they were pregnant reported that they did not drink during pregnancy when asked a *general* question (Hicks, 2007). For about half of the women, 'drinking during pregnancy' only seemed to mean those who drank alcohol and who knew they were pregnant. Thus, the national estimates of alcohol consumption during pregnancy may miss some women who consume alcohol before knowing they are pregnant, and so may not represent the actual exposure of risk among pregnant women.

Although many women adjust their drinking behavior after discovering they are pregnant, alcohol consumption during an unrecognized pregnancy constitutes a risk for the fetus. Women who reported drinking any alcohol before recognizing their pregnancy were more likely not to be planning the pregnancy (Tough et al., 2006a), and they were also more likely to have a higher income, to smoke, and to be Caucasian (Tough et al., 2006a). Women who reported binge drinking before recognizing they were pregnant were more likely to have unplanned pregnancies, to smoke, and to have a lower self-esteem (Tough et al., 2006b; Naimi et al., 2003).

It is important to remember that alcohol consumption during pregnancy does not necessarily mean that an infant will go on to develop FAS/FASD. While current evidence suggests that heavy drinking creates the greatest risk for FASD, the development of FASD is a complex interaction between maternal alcohol consumption patterns, fetal susceptibility, and maternal characteristics, such as age, nutrition, genetics, biology, and lifestyle (Maier and West, 2001; Hicks, 2007). Recent research has indicated that some women drink alcohol, and drink heavily, before they recognize they are pregnant. This, in turn, suggests a need for widespread health promotion activities to advance knowledge about drinking during an unrecognized pregnancy, and to support optimal behavior and attitudes to prevent FASD.

3.6
Gaps in the Data

Canada and many of its provinces and urban centers do not have any estimates of FASD. Moreover, where estimates of the incidence or prevalence of FAS/FASD are available, the rates have not been tracked over time; indeed, some of this information is between 15 and 30 years old. Furthermore, there are variations in diagnostic criteria, reporting, and study methodologies that have led to discrepancies in reported rates, so that it is difficult to draw comparisons between two or more studies (May and Gossage, 2001). The wide range of incidence and prevalence estimates reported in Canada should alert policy makers and program planners to interpret these rates with caution.

3.7
Policy Considerations

3.7.1
Establish Baseline Rates of FAS/FASD and Track Them Over Time

The ultimate goal is to be reassured that low incidence and prevalence rates of FAS/FASD reflect an impact of prevention efforts, rather than a low identification rate of the disease. Currently, governments may not have good-quality, recent data (or any data at all) upon which to ascertain the impact of prevention efforts. Some government ministries have established processes to identify FASD during the course of service provision, but acknowledge that they have not been documenting cases of FASD well. Better identification and tracking of FAS/FASD cases would enhance the ability of government ministries to understand the impact of prevention efforts. Establishing rates of FAS/FASD and tracking these rates over time is important to determine where the need for resources is, and to ensure an appropriate allocation of resources. Without clear incidence and prevalence data, it is difficult to determine the resource needs at the current time, and especially difficult to make projections for the future (Government of British Columbia and Children's and Women's Health Center of British Columbia, 2003).

3.7.2
Continue with Intervention Efforts

While the establishment of baselines rates of FAS/FASD is important, a lack of baseline rates must not prevent these individuals from receiving help. Importantly, current systems should continue to intervene in individuals with FAS/FASD, and provide services according to their developmental needs.

3.7.3
Assess and Intervene in Areas with Higher FAS/FASD Frequency

While at present there is no comprehensive picture of FAS/FASD incidence and prevalence across the Canadian provinces and communities, the systems of care and subpopulations where the incidence and prevalence of FAS/FASD is of concern are well known. It is also known that individuals with FASD often suffer from secondary disabilities, including mental health problems, incarceration and retention in the justice system, confinement, inappropriate sexual behaviors, alcohol and drug abuse, and school incompletion, all of which may also reduce the likelihood of meaningful employment (Streissguth, 1997; Streissguth et al., 2004). This information points to the following places where it would be possible to assess individuals on a systematic basis for FAS/FASD, and to follow up with intervention when it is detected:

- Children in care
- Youth and adults interacting with the justice system and in the corrections systems
- Specific Aboriginal communities where substance abuse is known to be a problem
- Alcohol and drug treatment facilities
- Systems where mental health issues are prevalent (e.g., homeless shelters)
- Agencies that support the unemployed.

The feasibility and effectiveness of establishing systematic assessment in these systems would need to be explored.

3.7.4
Intervene to Prevent FAS/FASD Where Risk is Higher

Although the frequency of FAS/FASD has not been tracked well in Canada, some additional information is available that might help when targeting prevention efforts. In the same way that systems and communities can be used to identify areas where there is a higher need for FAS/FASD assessment and intervention, systems and communities can help to target prevention efforts. In adulthood, women who are dependent on alcohol are more likely to be unmarried, to have less than a high school education, to experience poverty and unemployment, have a partner that uses substances, have poor social networks, experience domestic violence, and to experience health problems that include poor nutrition, smoking, and drug abuse (Badry *et al.*, 2006; Astley *et al.*, 2000a, 2000b; Alberta's Alcohol and Drug Abuse Commission, 2006).

More than half of a group of women in treatment for substance dependence in Alberta had been arrested on alcohol-related charges, had mental health problems during the past month, and had been involved with child protection in the past three years (Badry *et al.*, 2006). In addition, about one-third had been hospitalized due to alcohol (Badry *et al.*, 2006). Considering these points, women at risk of alcohol use and misuse during pregnancy can be identified in the preconception and prenatal periods, in part by their interaction with the social welfare, justice or health system or women's shelters. Some form of assessment and intervention could be integrated into these systems so as to minimize the risk of an alcohol-exposed pregnancy and to optimize fetal health.

Going back even further, women with alcohol dependence often report childhood experiences of an addicted parent, poverty, sexual abuse, child maltreatment, experience with foster care as a child, poor academic achievement, and family chaos (Badry *et al.*, 2006; Dube *et al.*, 2002; Grella, Hser, and Huang, 2006). This evidence indicates that adverse childhood experiences may be an underlying factor contributing to substance dependence, risky behaviors, and adult disease. Policies and programs that are organized to identify families at risk and intervene early can have a long-term positive impact, and may reduce the probability that intensive intervention strategies will be required at a later time, when costs are higher and the probability of success is reduced (Carneiro and Heckman, 2004). For instance, those children who come into contact with child welfare authorities, or children of parents who have had contact with child welfare as a child, are at higher risk of developing substance abuse issues, and can be identified through databases of government ministries. When children are first identified as needing support from people other than the family, intensive and sustained evidence-informed interventions can provide these children with the supportive environments that may help them become adults who do not abuse substances.

3.7.5
Work Towards Developing a Consistent Message

Although the frequency of FAS/FASD is not clearly known, this does not preclude from the development of consistent messages. Currently, there are conflicting messages about alcohol consumption during pregnancy internationally; some national guidelines say no alcohol is best, while others say some alcohol is acceptable (O'Leary et al., 2007; Alberta Medical Association, 2007). Even within a country, there may be variability in how professionals address and support alcohol abstinence during pregnancy. In Canada, approximately 10% of obstetricians and family physicians do not recommend alcohol abstinence during pregnancy, which differs from national clinical practice guidelines (Tough et al., 2004).

In the meantime, however, what is known can be shared: that *alcohol consumption during pregnancy could lead to FASD for a child*. This message can be communicated in a number of ways to improve public knowledge. Information could be:

- included with marriage licenses, new home warranties, and at mortgage offices to target those women who are at a stage of life where they may be considering pregnancy and parenting;

- integrated into the education curriculum using existing curriculum modules (Sulik and Meeker, 2008); and

- provided through the recommendations of physician and healthcare providers regarding alcohol consumption during pregnancy.

3.7.6
Key Players

There is a clear need for government ministries to allocate resources to establishing surveillance projects or systems that can track cases of FAS/FASD, so that baseline rates can be established. However, government ministries also need to ensure that there are sufficient resources to identify FAS/FASD and support these individuals. Trained personnel are critical to comprehensively assess and treat these individuals. Those government ministries that are responsible for children in care, youth and adults interacting with the justice system and in the corrections systems, specific Aboriginal communities where substance abuse is known to be a problem, alcohol and drug treatment facilities, systems where mental health issues are prevalent (e.g., homeless shelters), and agencies that deal with the unemployed, are key players in this agenda. These ministries and agencies can assist in developing systems to track cases of FAS/FASD, and can also pilot the systematic assessment of individuals for FAS/FASD.

Support from government ministries is also essential to prevention efforts. Social welfare ministries, justice ministries, health ministries, and women's shelters are all places where efforts can be made to identify women at risk of an alcohol-exposed pregnancy, and intervene to minimize this risk. These ministries and organizations could dedicate resources for developing procedures to identify these women in their systems. Pilot projects could be helpful in determining the feasibility and effectiveness of such initiatives. Ministries responsible for public health and education also need to consider what they can do to promote a consistent message that *alcohol consumption during pregnancy could lead to FASD for a child*, regardless of whether the pregnancy is recognized or not.

3.8
Conclusions

With improved identification of FAS/FASD and tracking of FAS/FASD incidence and prevalence, a country is better able to plan for the resources required for services, and to determine if prevention efforts are working. Canada is in a position to improve its tracking of FAS/FASD cases, to systematically assess and intervene with individuals in specific systems where rates of FAS/FASD are higher, to continue intervention efforts, and to use current knowledge to undertake prevention efforts.

Acknowledgments

The authors would like to thank Alberta Innovates–Health Solutions (formerly the Alberta Heritage Foundation for Medical Research) for salary support for Dr Suzanne Tough, and Courtney Crockett for seeking and obtaining much of the literature related to incidence and prevalence rates, and for proofreading this manuscript.

References

Abel, E.L. (1995) An update on incidence of FAS: FAS is not an equal opportunity birth defect. *Neurotoxicol. Teratol.*, **17** (4), 437–443.

Alberta Alcohol and Drug Abuse Commission (2004) *Estimating the Rate of FASD and FAS in Canada*. Women and Substance Use Information Series, Alberta Alcohol and Drug Abuse Commission. Available at: http://www.aadac.com/documents/women_info_estimating_fasd.pdf.

Alberta Alcohol and Drug Abuse Commission (2006) *Women Working Toward Their Goals Through AADAC Enhanced Services for Women (ESW): Technical Report*. Alberta Alcohol and Drug Abuse Commission (AADAC).

Alberta Partnership on Fetal Alcohol Syndrome (2007) *Guideline to the Prevention of Fetal Alcohol Syndrome (FAS)*, Toward Optimized Practice (TOP) Program, Edmonton, AB. Available at: http://www.topalbertadoctors.org/informed_practice/clinical_practice_guidelines/complete%20set/FASD%20Prevention/FASD_prevention_guideline.pdf.

Alberta's Commission on Learning (2003) Every Child Learns, Every Child Succeeds: Report and Recommendations, http://education.alberta.ca/media/413413/commissionreport.pdf (accessed February 16, 2010).

Asante, K. and Nelms-Maztke, J. (1985) *Report on the Survey of Children With*

Chronic Handicaps and Fetal Alcohol Syndrome in the Yukon and Northwest British Columbia, Mills Memorial Hospital, Terrace, BC.

Astley, S.J., Bailey, D., Talbot, C., and Clarren, S.K. (2000a) Fetal alcohol syndrome (FAS) primary prevention through FAS diagnosis: I. Identification of high-risk birth mothers through the diagnosis of their children. *Alcohol Alcohol.*, **35** (5), 499–508.

Astley, S.J., Bailey, D., Talbot, C., and Clarren, S.K. (2000) Fetal alcohol syndrome (FAS) primary prevention through FAS diagnosis: II. A comprehensive profile of 80 birth mothers of children with FAS. *Alcohol Alcohol.*, **35** (5), 509–519.

Badry, D., Trute, B., Benzies, K., Tough, S., and Holmes, E. (2006) The Addiction Severity Index: A Feasibility Study Examining the Scientific Quality and Research Utility of Data Gathered in Three Alberta Sites. Prepared for the Public Health Agency of Canada (Alberta).

Buell, M., Carry, C., Korhonen, M., and Anawak, R. (2006) *Fetal Alcohol Spectrum Disorder: An Environmental Scan of Services and Gaps in Inuit Communities*, National Aboriginal Health Organization, Ottawa, ON.

Burd, L., Selfridge, R.H., Klug, M.G., and Juelson, T. (2003) Fetal alcohol syndrome in the Canadian corrections system. *J. FAS Int.*, **1** (14), 1–10.

Carneiro, P., and Heckman, J.J. (2004) Human capital policy, in *Inequality in America: What Role for Human Capital Policy?* (eds J.J. Heckman and A. Krueger), MIT Press, Cambridge, MA.

Chudley, A.E., Conry, J., Cook, J., Loocke, C., Rosales, T., and LeBlanc, N. (2005) Fetal alcohol spectrum disorder: Canadian guidelines for diagnosis. *Can. Med. Assoc. J.*, **172** (Suppl. 5), S1–S21.

Dube, S.R., Anda, R.F., Felitti, V.J., Edwards, V.J., and Croft, J.B. (2002) Adverse childhood experiences and personal alcohol abuse as an adult. *Addict. Behav.*, **27** (5), 713–725.

Fast, D.K., Conry, J., and Loock, C.A. (1999) Identifying fetal alcohol syndrome among youth in the criminal justice system. *Dev. Behav. Pediatr.*, **20** (5), 370–372.

Fuchs, D., Burnside, L., Marchenski, S., and Mudry, A. (2007) Children with disabilities involved with the child welfare system in Manitoba: current and future challenges, in *Putting A Human Face on Child Welfare: Voices From the Prairies* (eds I. Brown, F. Chaze, D. Fuchs, J. Lafrance, S. McKay, and S.T. Prokop), Prairie Child Welfare Consortium/Centre of Excellence for Child Welfare, pp. 127–145.

Government of British Columbia and Children's and Women's Health Centre of British Columbia (2003) Fetal Alcohol Spectrum Disorder: A Strategic Plan for British Columbia, http://www.mcf.gov.bc.ca/fasd/pdf/fasd_strategic_plan-final.pdf (accessed February 16, 2010).

Grella, C.E., Hser, Y.I., and Huang, Y.C. (2006) Mothers in substance abuse treatment: differences in characteristics based on involvement with child welfare services. *Child Abuse Negl.*, **30** (1), 55–73.

Habbick, B.F., Nanson, J.L., Snyder, R.E., Casey, R.E., and Schulman, A.L. (1996) Foetal alcohol syndrome in Saskatchewan: unchanged incidence in a 20-year period. *Can. J. Public Health*, **87** (3), 204–207.

Health Canada (2006) *It's Your Health: Fetal Alcohol Spectrum Disorder*, Minister of Health.

Hicks, M. (2007) *Meconium Alcohol and Drug Screening*, University of Calgary.

Kowlessar, D. (1997) *An Examination of the Effects of Prenatal Alcohol Exposure on School-Age Children in A Manitoba First Nation Community: A Study of Fetal Alcohol Syndrome Prevalence and Dysmorphology*, Department of Human Genetics, University of Manitoba, Winnipeg, MB.

MacLaurin, B., Trocme, N., Fallon, B., McCormack, M., Pitman, L., Forest, N., Banks, J., Shangreaux, C., and Perrault, E. (2005) *Alberta Incidence Study of Reported Child Abuse and Neglect – 2003 (AIS-2003): Major Findings*, University of Calgary, Calgary, AB.

Maier, S.E. and West, J.R. (2001) Drinking patterns and alcohol-related birth defects. *Alcohol Res. Health*, **25** (3), 168–174.

May, P.A., and Gossage, J. (2001) Estimating the prevalence of fetal alcohol syndrome.

A summary. *Alcohol Res. Health*, **25** (3), 159–167.

McShane, K., Smylie, J., and Adomako, P. (2009) Health of First Nations, Inuit, and Metis Children in Canada, in *Indigenous Children's Health Report: Health Assessment in Action* (eds J. Smylie and P. Adomako), The Centre for Research on Inner City Health. Toronto, ON., pp. 10–65.

Naimi, T.S., Lipscomb, L.E., Brewer, R.D., and Gilbert, B.C. (2003) Binge drinking in the preconception period and the risk of unintended pregnancy: implications for women and their children. *Pediatrics*, **111** (5 Part 2), 1136–1141.

O'Leary, C.M., Heuzenroeder, L., Elliott, E.J., and Bower, C. (2007) A review of policies on alcohol use during pregnancy in Australia and other English-speaking countries, 2006. *Med. J. Aust.*, **186** (9), 466–471.

Public Health Agency of Canada (2008) *Canadian Perinatal Health Report, 2008 Edition*, Public Health Agency of Canada, Ottawa, ON, http://www.phac-aspc.gc.ca/publicat/2008/cphr-rspc/pdf/cphr-rspc08-eng.pdf.

Robinson, G.C., Conry, J.L., and Conry, R.F. (1987) Clinical profile and prevalence of fetal alcohol syndrome in an isolated community in British Columbia. *Can. Med. Assoc. J.*, **137** (3), 203–207.

Sampson, P.D., Streissguth, A.P., Bookstein, F.L., Little, R.E., Clarren, S.K., Dehaene, P., Hanson, J.W., and Graham, J.M., Jr (1997) Incidence of fetal alcohol syndrome and prevalence of alcohol-related neurodevelopmental disorder. *Teratology*, **56** (5), 317–326.

Streissguth, A.P. (1997) *Fetal Alcohol Syndrome: A Guide for Families and Communities*, Paul H. Publishing Co, Baltimore, MD.

Streissguth, A.P., Bookstein, F.L., Barr, H.M., Sampson, P.D., O'Malley, K., and Young, J.K. (2004) Risk factors for adverse life outcomes in fetal alcohol syndrome and fetal alcohol effects. *J. Dev. Behav. Pediatr.*, **25** (4), 228–238.

Sulik, K. and Meeker, M. (2008) Better Safe Than Sorry: Preventing a Tragedy, http://pubs.niaaa.nih.gov/publications/Science/curriculum.html (accessed February 16, 2010).

Tough, S.C., Clarke, M., Hicks, M., and Clarren, S.K. (2004) Clinical practice characteristics and preconception counseling strategies of health care providers who recommend alcohol abstinence during pregnancy. *Alcohol. Clin. Exp. Res.*, **28** (11), 1724–1731.

Tough, S., Tofflemire, K., Clarke, M., and Newburn-Cook, C. (2006a) Do women change their drinking behaviours while trying to conceive? An opportunity for preconception counseling. *Clin. Med. Res.*, **4** (2), 97–105.

Tough, S., Tofflemire, K., Clarke, M., and Newburn-Cook, C. (2006b) Are women changing their drinking behaviours while trying to conceive? An opportunity for pre-conception counselling. *Clin. Med. Res.*, **4** (2), 97–105.

Trocme, N., Fallon, B., MacLaurin, B., Daciuk, J., Felstiner, C., Black, T., Tonmyr, L., Blackstock, C., Barter, K., Turcotte, D., and Cloutier, R. (2005) *Canadian Incidence Study of Reported Child Abuse and Neglect – 2003: Major Findings*, Minister of Public Works and Government Services Canada, http://www.phac-aspc.gc.ca/ncfv-cnivf/pdfs/childabuse_final_e.pdf.

Tsai, J., Floyd, R.L., Green, P.P., and Boyle, C.A. (2007) Patterns and average volume of alcohol use among women of childbearing age. *Matern. Child Health J.*, **11** (5), 437–445.

Williams, R.J., Odaibo, F.S., and McGee, J.M. (1999) Incidence of fetal alcohol syndrome in northeastern Manitoba. *Can. J. Public Health*, **90** (3), 192–194.

4
Costs of FASD

Nguyen Xuan Thanh, Egon Jonsson, Liz Dennett, and Philip Jacobs

4.1
Introduction

Fetal Alcohol Spectrum Disorder (FASD) comes with lifelong disabilities, which vary in terms of their impact on the person's health and quality of life. Like many other disabilities, these may require medical treatment, special education, family support, and other community support services.

The costs of these services for the individual, the family and for society at large, are known only to a certain extent; costs that may occur as a consequence of disability for example at the work place, or in the correctional services.

The costs attributed specifically to FASD are, therefore, indeed difficult to assess. Although several studies have been conducted which consider the cost of FASD, these have used different methods, and have included different components of cost and, therefore, have produced different results. Moreover, studies of the aggregated costs of FASD, at national or regional levels, use different figures for the incidence of FASD, which makes the results essentially incomparable. Such variations in the methodology of different studies make it more challenging to interpret or use the results obtained.

In this chapter, the available studies from the international literature on the cost of FASD are reviewed, analyzed, and synthesized. This is in addition to costing studies available elsewhere, such as in government reports, web sites, and in the gray literature.

The results show that the annual cost of FASD in Canada is about CA$ 6.2 billion (2009 price level), at the most often-quoted incidence rate of nine per 1000 live births (see Chapter 3). This amount is equivalent to about CA$ 180 million per million population; for the province of Alberta, at the same incidence rate, the annual cost is about CA$ 520 million. Of these total costs, healthcare services account for 30%, followed by educational services (24%), social services (19%), correctional services (14%), and others (13%). The annual cost per person with FASD is about CA$ 25 000, and the lifetime cost per person with FASD is about CA$ 1.8 million (2009 price level).

Fetal Alcohol Spectrum Disorder–Management and Policy Perspectives of FASD. Edited by Edward P. Riley, Sterling Clarren, Joanne Weinberg, and Egon Jonsson
Copyright © 2011 WILEY-VCH Verlag GmbH & Co. KGaA, Weinheim
ISBN: 978-3-527-32839-0

To some extent, knowledge concerning the total cost of FASD may be helpful in assessing the magnitude of the disease, while the individual contributions of each cost component may indicate where attention may be focused. The annual and lifetime costs per person with FASD might be useful when assessing the resources required for the prevention of FASD. The discounted incremental lifetime cost per person with FASD, of CA$ 742,000, gives an indication of the monetary value of preventing one case of FASD.

4.2 Methods

4.2.1 Literature Search

Both, subject headings and keywords were used in the search of the following databases up to May 2010: Cochrane Library; MEDLINE (including in process citations); EMBASE; CINAHL; PsycINFO; Web of Science; and CRD Databases (DARE, HTA and NHS EED). Gray literature was located using Google. No date restrictions were applied, but only English language results were retrieved. In order to supplement the electronic searches, *Fetal Alcohol Research* (formerly *Journal of FAS International*) and *Alcohol Research & Health* (formerly *Alcohol Health & Research World*) were hand-searched, as both journals have gaps when they were not included in electronic indices. Reference lists of retrieved articles were also reviewed to identify further studies. Details of the complete search strategy are available in Appendix 4.1.

4.2.2 Inclusion Criteria

In order to interpret and compare the results of the different studies, only well-documented original cost-of-illness studies with clearly described methods, including cost perspectives, cost components, time horizon, and target population, were included:

- *Cost perspectives* answer the question: "who pays for the cost?" A societal perspective means that both direct and indirect costs are included, and/or the costs involve several sectors/stakeholders in society.
- The cost components indicate relative contributions of each sectors/stakeholders to the total cost.
  - *Direct cost* refers to the cost of health care, social services (institutionalization, residence care), special education, and justice system costs.
  - *Indirect cost* refers to productivity losses of patients and caregivers due to the disabilities.
- The *time horizon* is a duration of time in which the costs are incurred that a study covers, and can be a year, or a life time.

- The *target population* comprises the groups of people that a study targets, such as different age groups.

4.2.3
Cost Adjustment

Since the reviewed studies were conducted in different countries and in different years, the original costs were converted to 2009 Canadian dollars (CA$). The Consumer Price Index was used for either Canada or the United States (depending on the study), to adjust for price changes over time, and the exchange rate at mid-2009 to convert US dollars (US$) to CA$.

4.3
Results

4.3.1
Search Results

In total, the search strategy identified 207 references. On reviewing the titles and abstracts, two reviews and 14 original cost-of-illness studies on the costs of FASD were identified. Of these, nine original cost-of-illness studies were well-documented, with clearly described methods, cost components, and so on. These data are summarized in Section 4.3.2, while the methods and results are analyzed and commented in detail in Sections 4.3.3 and 4.3.4, respectively.

4.3.2
Summary of Studies Included in the Review

Among the nine studies, four were Canadian and focused on FASD, while five were US studies and focused on Fetal Alcohol Syndrome (FAS). FASD includes a range of disabilities, of which FAS is the most severe. Notably, both the incidence and prevalence of FASD are greater than those of FAS.

Abel and Sokol (1987) estimated direct costs associated with FAS among children aged 0–21 years in the US at US$ 321 million per year (CA$ 764 million in 2009), using an incidence rate of 1.9 FAS cases per 1000 live births. The costs included hospitalization for growth retardation, surgical repair of organic anomalies (e.g., cleft palate, tetralogy of Fallot), and treatment of sensorineural problems and mental retardation. FAS-related mental retardation alone accounted for 11% of the annual cost for all mentally retarded institutionalized residents in the US. The authors pointed at the fact that the treatment costs for FAS-related problems were about 100-fold higher than the Federal funding for FAS research.

Abel and Sokol (1991a) revised the previous estimate by including "corrections" for background rates of low birth weight, and costs normally incurred for housing and food, regardless of whether an individual required institutionalization, or not. The revised estimate of the annual cost of FAS was substantially lower (US$ 250 million in 1991, equivalent to CA$ 544 in 2009). Of this total cost, mental retardation accounted for almost 60%. The authors concluded that one-quarter of a billion dollars per year was a high economic incremental cost by any reasonable standard, and represented a benchmark against which the costs of potential prevention strategies may be judged.

Abel and Sokol (1991b) conducted a new analysis of the incidence of FAS and its economic impact among children aged 0–21 years in the US. This was a conservative analysis which reflected concern over the possible inclusion of "false positives" in the previous estimates. Using a lower incidence rate of 0.33 FAS cases per 1000 live births (compared to the incidence rate used in the earlier studies of 1.9 per 1000), by adjusting for costs that would be incurred whether cases were FAS or not, and by including estimated costs for anomalies in FAS cases that were not considered in the previous estimates, the authors estimated the incremented annual direct cost associated with this disorder to be US$ 75 million (CA$ 163 million in 2009). About three-fourths of this economic burden was associated with the care of FAS cases with mental retardation.

Harwood and Napolitano (1985) estimated both the total costs of FAS for the US, and the lifetime cost per person with FAS. The components of cost were the direct costs, which included hospitalization for growth retardation, surgical repair of organic anomalies, and treatment of sensorineural problems and mental retardation, and the indirect costs, which included productivity losses of the patients. Using a range of incidence rates that varied from 1 to 1.67 to 5 FAS cases per 1000 live births, the corresponding range of national costs for FAS was estimated at US$ 1.9 to US$ 3.2 and to US$ 9.7 billion in 1980 (CA$ 5.8, CA$ 9.7, and CA$ 29.0 billion in 2009). The lifetime cost per person with FAS was estimated at US$ 596 000 in 1980 (CA$ 1.8 million in 2009) and the 4% discounted lifetime cost per person with FAS was $248 000 in 1980 (CA$ 742 000 in 2009). The authors suggested that the discounted lifetime cost per person with FAS might be considered the maximum economic value that could be spent to prevent a FAS birth.

Klug and Burd (2003) used the North Dakota Claims database to examine the potential cost savings from the prevention of one case of FAS each year in the state of North Dakota, USA. The difference in the annual costs of healthcare for children aged 0–21 years with and without FAS was estimated at US$2342 per person in 1997 (CA$ 3646 in 2009). The authors concluded that the prevention of one case of FAS per year in North Dakota would result in cost savings of US$ 128 810 in 1997 (CA$ 201 000 in 2009) over a period of 10 years.

Fuchs et al. (2009) used the Child and Family Services Administrative database and the population-based data repository at Manitoba Center for Health Policy to estimate the incremental costs of health care (e.g., hospital, physician visits and prescription drugs), education, and social services for children with FASD aged 0–21 years, who resided in a permanent ward in Manitoba, Canada. The study results showed that, compared to the cost of the general population, there was an additional cost of health care of CA$ 1001 (CA$ 1065 in 2009) and for education, CA$ 5166 (CA$ 5498 in 2009), and for subsidized child care costs of CA$ 249 (CA$ 265 in 2009) incurred each year for every child with FASD and in permanent ward.

Stade et al. (2006) surveyed caregivers of children with FASD aged 1–21 years to estimate the annual cost per case, and used a prevalence rate of three FASD cases per 1000 live births to estimate the annual costs of FASD for Canada. The study included direct costs (medical, education, social services, and out-of-pocket) and indirect costs (productivity losses of the caregivers). The results showed that the annual cost per person with FASD in Canada was estimated at CA$ 14 342 (CA$ 16 722 in 2009), while the annual costs of FASD for Canada was estimated at CA$ 344.2 million (CA$ 401 million in 2009). Stade et al. (2009) used a similar method, but a different target population (aged 0–53 years) and a higher prevalence rate (10 FASD cases per 1000), to revise the previous estimates. Subsequently, the revised annual cost per person with FASD in Canada was estimated at CA$ 21 642 (CA$ 22 393 in 2009), and the revised annual costs of FASD for Canada was estimated at CA$ 5.3 billion (CA$ 5.5 billion in 2009).

Thanh and Jonsson (2009) conducted a study which was based on that of Stade et al. (2006) cost per case per year, Alberta birth and death data, and the FASD cost calculator (available at www.online-clinic.com) to estimate the annual long- and short-term costs, and the lifetime cost per person with FASD in Alberta, Canada. By using an incidence rate which ranged from three to nine cases per 1000 live births, the annual long-term cost (incurred by the cohort of children born with FASD each year) was estimated at CA$ 130–400 million (CA$ 150–450 million in 2009), while the annual short-term cost (incurred by people who are presently living with FASD) was estimated at CA$ 48–143 million (CA$ 53–157 million in 2009). The lifetime cost per case was estimated at CA$ 1.1 million (CA$ 1.2 million in 2009).

4.3.3
Summary of Methods Used in the Reviewed Studies

Two main approaches were used in the studies summarized above, namely modeling and observational (Table 4.1). In the modeling approach, the analyst estimates the cost by multiplying the utilization rate of services with the number of people with FAS/FASD, and with the unit cost of services. This type of study relies on data obtained from the literature on the utilization rate, unit cost, and the prevalence or incidence of FAS/FASD, and can be used to compare the burden of illness with those in other countries and diseases. The observational studies use two methods, namely population surveys and databases. The latter approach is best suited for planning purposes, and for estimating the impact of preventive interventions.

Besides these main approaches, four other differences in methodology lead to the costs being variable between the studies. First, the incidence and prevalence rates of FAS and FASD varied substantially between the studies. Abel and Sokol (1991b) used the lowest rate of FAS (0.33 cases per 1000 live births), while Harwood and Napolitano (1985) used the highest rate (5 cases per 1000 live births). With regards to FASD, the ranges of incidence and prevalence rates ranged from three to nine cases per 1000 live births (Thanh and Jonsson, 2009) and three to ten cases

Table 4.1 Methods used in the reviewed studies.

Study	Country or province	Year	Condition	Population	Incidence (I) or Prevalence (P)	Main outcomes	Data sources
Modeling studies							
Abel and Sokol (1987)	USA	1984	FAS	≤21 years	I = 1.9/1000	Nationwide costs	Literature
Abel and Sokol (1991a)	USA	1987	FAS	≤21 years	I = 1.9/1000	Nationwide costs	Literature
Abel and Sokol (1991b)	USA	1987	FAS	≤21 years	I = 0.33/1000	Nationwide costs	Literature
Harwood and Napolitano (1985)	USA	1980	FAS	All ages	I = 1–5/1000	Lifetime cost per case	Literature
Thanh and Jonsson (2009)	Alberta, Canada	2005	FASD	All ages	I = 3–9/1000	Lifetime cost per case	Literature
Observational studies							
Stade et al. (2006)	Canada	2003	FASD	1–21 years	P = 3/1000	Annual cost per case	Survey
Stade et al. (2009)	Canada	2007	FASD	0–53 years	P = 10/1000	Annual cost per case	Survey
Fuchs et al. (2009)	Manitoba, Canada	2006	FASD	≤21 years	N/A	Annual cost per case	Manitoba Center for Health Policy database
Klug and Burd (2003)	North Dakota, USA	1996/7	FAS	≤21 years	N/A	Annual cost per case	North Dakota claims database

per 1000 people (Stade et al., 2006, 2009), respectively. These variations will have a substantial impact on the results of the cost calculations.

Second, the target population is different between studies. Harwood and Napolitano (1985) and Thanh and Jonsson (2009) targeted people at all ages, whereas other authors targeted children aged 0–21 years old, except for Stade et al., whose target populations were children aged 1–21 years old in the earlier (Stade et al., 2006) and people aged 0–53 years old in the later study (Stade et al., 2009).

Third, the studies focus on different conditions within the spectrum of fetal alcohol disorders. Some studies estimate the costs of FAS, while others calculate the costs of FASD. As FAS is medically the most severe condition within FASD, the cost per case with FAS is likely to be higher, whereas the national or provincial cost of FAS is likely to be lower than those of FASD, because the incidence and prevalence rates of FAS are much lower than those of FASD.

Finally, the studies are different in the cost components included for analysis (Table 4.2). The authors included direct cost only, or both direct and indirect costs. Direct costs relate to for services that are paid for directly; the typical components of direct costs are medical and nonmedical services. The latter include social services (e.g., institutionalization, residence care), special education, and justice system costs. None of the studies included all of these cost components; for example, none of the justice system costs was included in any of the studies.

Five of the studies included only direct costs. For example, Klug and Burd (2003) included only medical costs, while Abel and Sokol (1987, 1991a, 1991b) examined the direct costs for medical and institutionalization services. Fuchs *et al.* (2009) included the direct costs for medical, social, and special education services.

Harwood and Napolitano (1985), Stade *et al.* (2006, 2009), and Thanh and Jonsson (2009) included both direct and indirect costs; that is, the value of lost productivity due to illness or care-giving. All of these studies included direct costs for medical, social, and special education services. However, the direct out-of-pocket costs were included only in State *et al.* (2006, 2009), and Thanh and Jonsson (2009), but not by Harwood and Napolitano (1985). However, the components of indirect cost differed between these studies. Productivity losses of caregivers were included by Stade *et al.* (2006, 2009) and by Thanh and Jonsson (2009), but Harwood and Napolitano (1985) included the productivity losses of patients.

4.3.4
Summary of Results of the Reviewed Studies

In general, the studies estimated two types of cost: (i) the annual cost for all persons with FAS/FASD in a population; and/or (ii) the cost per case with FAS/FASD. Cost per case could be either an *annual* cost per case, or a *lifetime* cost per case. Abel and Sokol (1987, 1991a, 1991b) measured annual nationwide costs of FAS for the US, while Stade *et al.* (2006, 2009) and Fuchs *et al.* (2009) examined annual costs per case of FASD for Canada and the province of Manitoba, respectively. Harwood and Napolitano (1985) and Thanh and Jonsson (2009) modeled lifetime costs per case with FAS in the US and with FASD in Canada, respectively.

4.3.4.1 Annual Cost of FAS/FASD for the US, Canada, and the Province of Alberta

A summary of the population-based costs generated in the various studies is provided in Table 4.3. For the US, Abel and Sokol (1991b) estimated annual direct costs for FAS at CA$ 163 million; this was for people aged 0–21 years old, and was based on an incidence rate that was revised significantly downward (to 0.33 FAS cases per 1000 live births) from their previous two studies, in which the rate was 1.9 per 1000 live births and the annual direct costs for FAS were estimated at CA$ 250–321 million.

Table 4.2 Components of cost included in each study.

Cost items	Abel and Sokol (1987)	Abel and Sokol (1991a)	Abel and Sokol (1991b)	Harwood and Napolitano (1985)	Klug and Burd (2003)	Stade et al. (2006)	Stade et al. (2009)	Thanh and Jonsson (2009)	Fuchs et al. (2009)
Direct cost									
Medical services									
Low-birth weight infants	x	x	x	x					
Heart defects	x	x	x	x					
Spina bifida		x	x						
Cleft palate	x	x	x	x					
Serous otitis media with myringotomy		x	x	x					
Sensorineural hearing Loss/audiological deficits	x	x	x	x					
Inguinal hernia		x	x						
Hypospadias		x	x						
Neurotube defects				x					
Kidney defects				x					
ICD-9 code: 760.71					x				
Hospitalizations						x	x	x	x
ER visits						x	x	x	
Specialist visits						x	x	x	x
Nonspecialist visits						x	x	x	x
Other health professional visits						x	x	x	
Prescription drugs									x

4.3 Results

Social services									
Severe mental retardation (24 h/lifetime institutionalization)		x		x	x	x	x		x
Mild-moderate mental retardation					x				
Respite care						x	x	x	
Foster care						x	x	x	
ODSP						x	x	x	
Legal aid						x	x	x	
Special education									
Special education for minimal brain dysfunction					x				x
Home schooling						x	x		
Special schooling						x	x		
Residential program						x	x		
Post-secondary education-Tutor						x	x		
Job education						x	x		
Justice system									
Police									
Court									
Correction services									
Out-of-pocket									
Transportation per visit						x	x		x
Parking						x	x		x
Externalizing behaviors						x	x		x
Indirect costs									
Productivity loss of patients					x				
Productivity loss of caregivers						x			x

4 Costs of FASD

Table 4.3 Annual cost of FAS/FASD for the US, Canada, and the province of Alberta (CA$ million).

Study	Incidence (I) or Prevalence (P)	Direct cost Medical	Direct cost Nonmedical	Total direct cost Original	Total direct cost 2009 CA$	Total indirect cost Original	Total indirect cost 2009 CA$	Total Original	Total 2009 CA$
USA									
Abel and Sokol (1987)	I = 1.9/1000	135	186	321	764	not incl.		321	764
Abel and Sokol (1991a)	I = 1.9/1000	105	145	250	544	not incl.		250	544
Abel and Sokol (1991b)	I = 0.33/1000	17	58	75	163	not incl.		75	163
Harwood and Napolitano (1985)	I = 1/1000	491	934	1427	4267	510	1525	1937	5792
	I = 1.67/1000	699	1684	2382	7122	853	2550	3235	9673
	I = 5/1000	2076	5000	7133	21328	2553	7635	9686	28962
Canada									
Stade et al. (2006)	P = 3/1000	104	212	316	369	28	32	344	401
Stade et al. (2009)	P = 10/1000	1855	1484	4929	5100	371	384	5300	5484
Alberta									
Thanh and Jonsson (2009)	I = 3/1000	42	85	126.04	138	11	12	137	150
	I = 9/1000	124	253	377	414	33	36	410	450

Harwood and Napolitano (1985) had a much higher estimate for the US. This was due mainly to the timelines chosen for the study (lifetime estimates), and also to the higher incidence rates encountered (1–5 FAS cases per 1000 live births). As a result, the direct lifetime costs for their most conservative incidence rate was CA$ 4.3 billion, and the indirect lifetime costs were CA$ 1.5 billion. For the average incidence rate (1.67 FAS cases per 1000 live births), the annual lifetime cost of FAS, including both direct and indirect costs, was estimated at CA$ 9.7 billion.

In one of the Canadian studies, Stade et al. (2009) estimated that the annual total cost for Canada was CA$ 5.5 billion, which included both direct (medical, social and educational services) and indirect costs. The other Canadian study, conducted by Thanh and Jonsson (2009), estimated direct medical costs for FASD in Alberta, at between CA$ 140 million and CA$ 410 million, and indirect costs at between CA$ 12 and CA$ 36 million, depending on whether they used a lower (three FASD cases per 1000 births) or higher (nine FASD cases per 1000 live births) incidence rate. In the latter study, healthcare services accounted for 35% of the direct cost, followed by educational services (28%), social services (22%), and others (15%).

The Canadian estimates of indirect cost are not comparable to the US estimates, as the former included productivity losses of caregivers only (the time that caregivers spent on taking care of people with FASD), while the US estimates included productivity losses only of people with FASD (see Table 4.2). Both estimates are likely underestimations, however, as indirect costs should include productivity losses of people with FASD and of their caregivers.

4.3.4.2 Annual Cost per Case

Four studies estimated the annual cost per case of FAS/FASD (Table 4.4). By surveying the caregivers of people with FASD (aged 1–21 years in the 2006 study, and aged 0–53 years in the 2009 study), Stade et al. (2006, 2009) estimated the annual cost per case at CA$ 17 000 and CA$ 22 000, respectively. The latter study was a revision which used a target population that was more representative than the previous estimate, and therefore may be more reliable. However, this type of cost calculation per case results in the gross cost per case. In addition, people without FASD draw costs from the health services, from educational support, and from family support. The correct cost of FASD – and, for that matter, of any disability or disease – is the *extra* cost or, in economic terms, the incremental cost for a particular disability such as FASD. For comparisons of costs between different disabilities or diseases, however, it is unimportant whether the gross cost or the incremental cost is used. The components of cost in the study conducted by Stade et al. included the cost of medical interventions, education, social services, cost of patient and family, as well as productivity losses and externalizing behaviours (e.g., acts of aggression, such as damage to people/property or stealing). An estimate of the cost per case in different age groups showed that the costs had decreased dramatically in the age group of 18 years or older. This may be an indication of a lack of services for adults with FASD in Canada.

Table 4.4 Annual cost per case with FAS/FASD.

	Canada by Stade et al. (2006, 2009)				Manitoba, Canada by Fuchs et al. (2009)		North Dakota, US by Klug and Burd (2003)	
Population	1–21 years		0–53 years		0–21 years		0–21 years	
Value in	2003 CA$	2009 CA$	2007 CA$	2009 CA$	2006 CA$	2009 CA$	1996/7 US$	2009 CA$
Direct cost	12 053	14 053	18 780	19 432				
Medical	3976	4636	6630	6860	1001	1065	2342	3646
Education	4275	4984	5260	5443	5166	5498		
Social services	2866	3342	4076	4217	249	265		
Out of pocket	936	1091	2814	2912				
Indirect cost	1055	1230	1431	1481				
Total	13 108	15 283	20 211	20 912	6416	6828		
Adjusted total	14 342	16 722	21 642	22 393				

By using administrative databases, Fuchs et al. (2009) and Klug and Burd (2003) estimated the annual medical cost per case of FAS/FASD aged 0–21 years in Manitoba, Canada and in North Dakota, USA at CA$ 1100 and CA$ 3600, respectively (Table 4.4). Fuchs et al. (2009) also estimated the annual costs per case for education and social services (CA$ 5498 and CA$ 265, respectively). These are the incremental costs between children with and without FAS/FASD, and thus best suited for an evaluation of the economic value of investing in effective preventive programs. Klug and Burd (2003) suggested that a prevention program capable of preventing one new case of FAS each year could lead to cost savings, in the health services alone, on the order of CA$ 3600 per year for every year of a lifetime. During a period of 10 years, the cost savings for health services would accumulate to CA$ 201 000.

4.3.4.3 Lifetime Cost per Case

The nondiscounted lifetime cost per person with FAS in the United States was estimated at CA$ 1.8 million (Harwood and Napolitano, 1985) and for FASD at CA$ 1.2 million (Thanh and Jonsson, 2009) (Table 4.5). It is difficult to compare these results because of variations in methodology and the different components of cost used in the two studies. Additionally, Thanh and Jonsson's (2009) lifetime cost per case with FASD was not incremental between persons with and without FASD – mainly because they used non-incremental annual cost per case as estimated by Stade et al. (2006) to estimate the lifetime cost, whereas the estimate of Harwood and Napolitano (1985) reflected incremental costs.

Table 4.5 Lifetime cost per case with FAS/FASD (CAS thousands).

Study	Condition	Nondiscounted	Discounted
		CA$ (2009)	CA$ (2009)
Thanh and Jonsson (2009)	FASD	1208	
Harwood and Napolitano (1985)	FAS	1782	742

The discounted lifetime cost (the present value of the lifetime cost) per person with FAS was estimated at CA$ 742 000, using a discount rate of 4%, by Harwood and Napolitano (1985).

4.4 Discussion

In this chapter, a series of published studies on the costs of FAS/FASD has been identified and analyzed. The authors of these reports presented their estimates in a wide variety of ways, using a broad array of assumptions. These included total annual national or provincial costs, lifetime costs per case, and annual cost per case, while the range of services included medical, social service, and educational (although inclusion varied by study). Indirect costs were included only in some of the studies. Moreover, the studies approached the subject with different methodologies, and covered the issues over a long timespan for a condition the definition of which has been steadily expanding over the years. The incidence and prevalence rates are not known for sure, due partly to the fact that diagnosis may come much later in life than at birth, and that it is difficult to trace the consumption of alcohol during pregnancy. Further, there are few administrative data which can be used; only recently was Manitoba able to integrate social, educational and healthcare databases (Fuchs et al., 2009), to obtain annual estimates of direct costs.

The incidence rates of FAS and FASD have recently been estimated at one to three FAS and nine FASD cases per 1000 live births (Health Canada, 2006). By using such rates, in combination with the findings in the presently reviewed studies, the best estimates of the total cost of FASD were assessed (see Table 4.6). The data in the table include a rough estimate also of the cost incurred in the correctional system of Canada, based on the following points. In Canada, the annual expenditure on adult correctional services is about CA$ 3 billion (Landry and Sinha, 2008), and it has been estimated that 23.3% of youth offenders have FASD (Fast, Conry, and Loock, 1999). Therefore, a conservative estimate of CA$ 700 million (i.e., 23.3% of CA$ 3 billion) can be considered as the annual costs of FASD for the Canadian justice system.

The annual cost of FAS in the US is CA$ 9.7 billion, and the lifetime costs per person with FAS is CA$ 1.8 million, while the discounted incremental lifetime cost per case of FAS is CA$ 742 000.

Table 4.6 Total annual and life-time cost of FASD in Canada (2009 price level).

	Canada	Alberta
Annual cost of FASD (CA$)	5.5 billion	450 million
Estimated annual cost of correctional services (CA$)	700 million	70 million
Total cost at national and provincial level (CA$)	6.2 billion	520 million
Annual cost per person with FASD (CA$)[a]	25 000	
Lifetime cost per person with FASD (CA$)[b]	1.8 million	
Annual incremental direct cost per person aged 0–21 years with FASD (CA$)	6800	

a) Adding correctional costs to the annual cost per person with FASD estimated by Stade et al. (2009).
b) Calculated by using annual cost and methodology from Thanh and Jonsson (2009).

4.5
Conclusion

The policy relevance of knowing the costs of FASD is related to three areas. First, the total costs of FASD may be helpful in assessing the magnitude of the disease, if compared to other diseases. Second, the individual contributions of each cost component can indicate where attention might be focused for prevention and/or services for persons with FASD. Third, the annual and the lifetime costs per person with FASD might be useful in evaluating the need of resources for FASD-prevention programs.

In assessing the benefits of preventive programs, the incremental cost per case may be useful when determining the need of resources for that purpose. To date, there has been no estimate of the annual incremental total cost (direct and indirect) per case. However, the annual incremental healthcare cost per case of FAS was estimated at CA$ 3800 for people aged 0–21 years in North Dakota, USA (Klug and Burd, 2003), while the annual incremental direct cost (including costs for medical, educational, and social services) per case of FASD was estimated at CA$ 6800 for people aged 0–21 years in Manitoba, Canada (Fuchs et al., 2009).

Similarly, the incremental lifetime cost per case of FASD is more useful than the FASD-associated lifetime cost per case for the allocation of resources to prevention. The incremental lifetime cost per case of FASD could, in theory, be considered the maximum amount of money that could be spent to prevent a FASD birth (Harwood and Napolitano, 1985). From an economic perspective, the present value of the lifetime cost of FASD, may provide a better measure to assist in health policy-making on prevention in the field. That measure has been estimated at CA$ 742 000 per case (2009 price level) by Harwood and Napolitano (1985).

4.6
Appendices to Chapter 4

4.6.1
Appendix 1: Search Strategy

The literature search was conducted by the IHE Research Librarian in July 2009, and updated in May 2010.

Database	Edition or date searched	Search terms††
Core Databases		
Cochrane Database of Systematic Review http://www.thecochranelibrary.com	1994–23 July 2009	fetal alcohol in Title, Abstract or Keywords or (alcohol* AND (pregnancy OR fetus OR prenatal)) in Title, Abstract or Keywords AND Cost or economic in Title, Abstract or Keywords (0 results)
Medline (OVID Interface) (in process and other non-indexed citations	1970–23 July 2009	1. fetal alcohol syndrome/ 2. f?etal alcohol.tw. 3. (alcohol* adj3 (birth defects or congenital malformations)).tw. 4. (alcohol* adj3 neurodevelopmental).tw. 5. (alcohol* adj4 prenatal).tw. 6. fasd.tw. 7. fae.tw. 8. arbd.tw. 9. arnd.tw. 10. or/1–9 11. Alcoholism/ 12. exp Alcoholic Beverages/ 13. Alcohol Drinking/ 14. or/11–13 15. fetus/ 16. pregnancy/ 17. 15 or 16 18. 14 and 17 19. 10 or 18 20. exp "Costs and Cost Analysis"/ 21. (cost-benefit or benefit-cost or cost effective* or cost utility).tw.

Database	Edition or date searched	Search terms[††]
		22. cost of illness.tw.
		23. (cost minimization or cost minimization or cost offset*).tw.
		24. ((loss or lost) adj2 productivity).tw.
		25. ((economic or cost*) adj2 (evaluat* or analys* or study or studies or assess* or consequence*)).mp.
		26. or/20–25
		27. 19 and 26
		28. Fetal alcohol syndrome/ec
		29. 27 or 28
		30. limit 29 to english language
		(58 results)
CRD Databases (DARE, HTA & NHS EED) http://www.crd.york.ac.uk/crdweb/	1995–23 July 2009	# 1 fetal AND alcohol OR fetal AND alcohol
		# 2 alcohol* AND (pregnancy OR fetus OR prenatal)
		# 3 #1 OR #2
		(30 NHS EED articles)
		# 4 MeSH Costs and Cost Analysis EXPLODE 1
		# 5 cost*:ti OR economic*:ti OR expenditures:ti OR price:ti OR fiscal:ti OR financial:ti OR burden:ti OR efficiency:ti OR pay:ti OR valuation:ti OR pharmacoeconomic:ti OR spending:ti
		# 6 economic AND evaluat* OR economic AND analys* OR economic AND study OR economic AND studies OR economic AND assess* OR economic AND consequence*
		# 7 cost* AND evaluat* OR cost* AND analys* OR cost* AND assess* OR cost* AND consequence*
		# 8 cost-benefit OR benefit-cost OR cost AND effectiv* OR cost AND utility OR cost AND minimization OR cost AND minimization OR cost AND offset*
		# 9 #4 OR #5 OR #6 OR #7 OR #8
		# 10 #3 AND #9
		(4 Dare and 1 HTA results)

Database	Edition or date searched	Search terms[††]
Embase (Ovid Interface)	1988–23 July 2009	1. Fetal Alcohol Syndrome/ 2. f?etal alcohol.tw. 3. (alcohol* adj3 (birth defects or congenital malformations)).tw. 4. (alcohol* adj3 neurodevelopmental).tw. 5. (alcohol* adj4 prenatal).tw. 6. fasd.tw. 7. fae.tw. 8. arbd.tw. 9. arnd.tw. 10. or/1–9 11. Alcoholism/ 12. alcohol abuse/ 13. exp Alcoholic Beverage/ 14. alcohol consumption/ 15. alcohol/ 16. drinking behavior/ 17. or/11–16 18. Fetus/ 19. pregnancy/ 20. 18 or 19 21. 17 and 20 22. 10 or 21 23. exp economic evaluation/ 24. cost/ 25. ((economic or cost*) adj2 (evaluat* or analys* or study or studies or assess* or consequence*)).tw. 26. cost of illness.tw. 27. pharmacoeconomics/ 28. ((loss or lost) adj2 productivity).tw. 29. health economics/ 30. (cost* or economic* or expenditures or price or fiscal or financial or pay or valuation or pharmacoeconomic or spending).ti. 31. or/23–30 32. 22 and 31 33. limit 32 to english language **(67 results)**

Database	Edition or date searched	Search terms††
PsycINFO (Ovid interface)	1970–23 July 2009 (July Week 3 2009)	1. exp Fetal Alcohol Syndrome/ 2. f?etal alcohol.tw. 3. (alcohol* adj3 (birth defects or congenital malformations)).tw. 4. (alcohol* adj3 neurodevelopmental).tw. 5. fasd.tw. 6. fae.tw. 7. arbd.tw. 8. arnd.tw. 9. or/1–8 10. alcoholism/ or alcohol abuse/ or alcohol drinking patterns/ or alcohol intoxication/ 11. exp alcoholic beverages/ 12. 10 or 11 13. pregnancy/ 14. fetus/ 15. prenatal exposure/ 16. or/13–15 17. 12 and 16 18. 9 or 17 19. "costs and cost analysis"/ 20. ((loss or lost) adj2 productivity).mp. [mp = title, abstract, heading word, table of contents, key concepts] 21. economics/ 22. health care costs/ 23. ((economic or cost*) adj2 (evaluat* or analys* or study or studies or assess* or consequence*)).mp. 24. (cost-benefit or benefit-cost or cost effectiv* or cost utility).mp. [mp = title, abstract, heading word, table of contents, key concepts] 25. (cost minimization or cost minimization or cost offset*).mp. [mp = title, abstract, heading word, table of contents, key concepts] 26. "cost of illness".mp. 27. (cost* or economic* or expenditures or price or fiscal or financial or pay or valuation or pharmacoeconomic or spending).ti. 28. or/19–27 29. 18 and 28 **(18 results)**

Database	Edition or date searched	Search terms[††]
CINAHL (Ebsco interface)	1970–23 July 2009	S1: fetal alcohol
		S2: ((MH "Alcoholic Intoxication") or (MH "Alcohol Abuse (Saba CCC)") or (MH "Alcohol Drinking") or (MH "Alcoholic Beverages+") or (MH "Alcoholism")) and ((MH "Fetus") or (MH "Pregnancy") or Prenatal)
		S3: S1 OR S2
		S4: (MH "Costs and Cost Analysis")
		S5: (MH "Cost Benefit Analysis")
		S6: (MH "Economic Aspects of Illness")
		S7: (MH "Health Care Costs")
		S8: (MH "Economics")
		S9: S4 or S5 or S6 or S7 or S8
		S10: S3 and S9
		(7 results)
Web of Science ISI Interface Licensed Resource	1970–23 July 2009	#1 TS = fetal alcohol
		#2 TI = (cost* or economic* or expenditures or price or fiscal or financial or pay or valuation or pharmacoeconomic or spending)
		#3 TS = (cost effectiveness OR cost benefit OR (productivity AND (loss OR lost)) OR economic analys*)
		#4 #1 AND (#2 OR #3) AND Language = (English)
		(23 results)
Internet searching		
Google http://www.google.com	13 March 2009	"fetal alcohol" cost-effectiveness OR cost-benefit OR economic-analysis OR cost -pubmed (reviewed first 100 links)

Note:
[††] "*", "# ", and "?" are truncation characters that retrieve all possible suffix variations of the root word; for example, surg* retrieves surgery, surgical, surgeon, etc.
Searches separated by semicolons have been entered separately into the search interface.

4.6.2
Appendix 2: Summary of Included Studies

Studies	Place/year/ condition/ population	Incidence (I) or Prevalence (P)	Methods and cost components	Results	Comments/notes
Modeling studies					
Abel and Sokol (1987)	USA/ 1984/ FAS/ 0–21 years	I = 1.9/1000	Direct costs, including hospital, surgical, treatment, and institutionalized costs for: – growth retardation, – organic anomalies – sensorineural problems – mental retardation	– The economic cost associated with FAS was estimated at 321 million per year in the US – FAS-related mental retardation alone may account for 11% of the annual cost for all mentally retarded institutionalized residents	The costs are about 1000-fold higher than current federal funding for FAS research which is necessary to develop cost-effective early identification and prevention strategies
Abel and Sokol (1991a)	USA/ 1987/ FAS/ 0–21 years	I = 1.9/1000	Direct costs, including hospital, surgical, treatment, and institutionalized costs for: – growth retardation, – organic anomalies – sensorineural problems – mental retardation	– The current estimate for annual costs related to FAS in the US was $249.7 million – Mental retardation accounted for almost 60% of the estimated total cost	– This is a revised estimate that included "corrections" for background rates of low birth weight, and costs normally incurred for housing and food, regardless of whether an individual requires institutionalization, or not
Abel and Sokol (1991b)	USA/ 1987/ FAS/ 0–21 years	I = 0.33/1000	Direct costs, including hospital, surgical, treatment, and institutionalized costs for: – growth retardation, – organic anomalies – sensorineural problems – mental retardation	– The annual cost associated with FAS in the US was estimated at $74.6 million. – About three-fourths of this economic burden was associated with care of FAS cases with mental retardation	This more conservative analysis reflects the concern over possible inclusion of "false positives" in the previous estimates

Studies	Place/year/ condition/ population	Incidence (I) or Prevalence (P)	Methods and cost components	Results	Comments/notes
Harwood and Napolitano (1985)	USA/ 1980/ FAS/ all ages	I = 1/1000 I = 1.67/1000 I = 5/1000	Direct costs, including hospital, surgical, treatment, and institutionalized costs for: – growth retardation, – organic anomalies – sensorineural problems – mental retardation Indirect cost: productivity losses of the patients	– The total costs to society of FAS in the US were estimated at US$ 1937, US$ 3236, and US$ 9.687 corresponding to the three incidence rates: 1, 1.67, and 5/1000 – Lifetime cost per case was US$ 596 000 (4% discounted lifetime cost per case was US$ 248 000)	The discounted lifetime cost per case might be considered the maximum economic value that could be spent to prevent an FAS birth
Thanh and Jonsson (2009)	Alberta/ 2005/ FASD/all ages	I = 3/1000 I = 9/1000	Based on Stade's cost per case per year, Alberta birth and death data, and FASD cost calculator at www.online-clinic.com. Direct costs: – medical: – education – social services – out-of-pocket Indirect cost: productivity losses of the caregivers	– The annual long-term cost (incurred by lives of the cohort of children born with FASD each year) in Alberta was estimated at CA$ 130–400 million. – The annual short-term cost (incurred by people who are presently living with FASD) in Alberta was estimated at CA$ 48–143 million. – The lifetime cost per case was estimated at CA$ 1.1million	The results suggest a need for a provincial FASD prevention strategy. The costs of FASD can be used to evaluate the benefits of prevention programs to society

Studies	Place/year/ condition/ population	Incidence (I) or Prevalence (P)	Methods and cost components	Results	Comments/notes
Observational studies					
Stade et al. (2006)	Canada/ 2003/ FASD/ 1–21 years	P = 3/1000	Surveyed caregivers of children with FASD. Direct costs: - medical - education - social services - out-of-pocket Indirect cost: productivity losses of the caregivers	- The annual cost per case in Canada was estimated at CA$ 14 342 - The annual cost of FASD for Canada was estimated at CA$ 344.2 million	- This study did not use a random sample - Individuals residing in institutions such as facilities for disable children were not included. It is possible that individuals with severe disability were not included in this study; thus, the costs may be underestimated
Stade et al. (2009)	Canada/ 2007/ FASD/ 0–53 years	P = 10/1000	Surveyed caregivers of children, youth and adults, with FASD. Direct cost: - medical - education - social services - out-of-pocket Indirect cost: productivity losses of the caregivers	- The annual cost per case in Canada was estimated at CA$ 21 642 - The annual cost of FASD for Canada was estimated at CA$ 5.3 billion.	- This study did not use a random sample - Individuals who were incarcerated at the time of data collection were not included, and thus may have led to a lower cost estimate

Studies	Place/year/ condition/ population	Incidence (I) or Prevalence (P)	Methods and cost components	Results	Comments/notes
Fuchs et al. (2009)	Manitoba/ 2006/ FASD/ 0–21 years	N/A	- Child and Family Services Administrative database - Population-based data repository at Manitoba Center for Health Policy Direct costs: - hospital visits (in- and out-patient) - physician services - prescription drugs - Education - social services	There was an additional CA$ 1001 in healthcare costs, CA$ 5166 in education costs, and CA$ 249 in subsidized child care costs incurred each year for every child with FASD and permanent ward, compared to the general population in Manitoba, Canada	The difference in healthcare costs between this study and the North Dakota study can be explained by the fact that FAS is more severe (costly) conditions than FASD
Klug and Burd (2003)	North Dakota/ 2006–7/ FAS/ 0–21 years	N/A	- North Dakota Health Claims Database - ICD9 code of FAS: 760.71 - Direct medical costs from the Healthcare financing administration dataset: claims made to Medicare, Medicaid, both in- and out-patients	- Annual cost of healthcare for children birth through 21 years of age with FAS in North Dakota, USA was US$ 2842. Per capita, this was US$ 2342 more than the annual average cost of care for children without FAS	Preventing one case of FAS per year would result in a cost saving of US$ 128810 in 10 years and US$ 491820 after 20 years

4.6.3
Appendix 3: Excluded studies that consider the costs of FAS/FASD

Study no.	Excluded studies
Reviews	
1	Bloss, G. (1994) The economic cost of FAS. *Alcohol Health & Research World*, **18** (1), 53–54.
2	Lupton, C., Burd, L., and Harwood, R. (2004) Cost of fetal alcohol spectrum disorders. *Am. J. Med. Genet. C Semin. Med. Genet.*, **127** (1), 42–50.
Studies	
1	DHS Public Health Division, Office of Family Health, Women's and Reproductive Health Section, Oregon FAS Prevention Program (2009) Cost of Fetal Alcohol Spectrum Disorder (FASD) in Oregon. January 2, 2009. Available at http://www.oregon.gov/DHS/ph/wh/docs/fas/cost_of_fasd_in_oregon.pdf (accessed on 22 April 2010).
2	Harwood, H. (2000) Updating Estimates of the Economic Costs of Alcohol Abuse in the United States: Estimates, Update Methods and Data. Report prepared by The Lewin Group. National Institute on Alcohol Abuse and Alcoholism, Bethesda, MD.
3	McDowell Group (2005) Economic costs of alcohol and other drug abuse in Alaska, update. Prepared for the Advisory board on Alcoholism and Drug Abuse Department of Health & Social Services, December 2005.
4	Rice, D.P. (1993) The economic costs of alcohol abuse and alcohol dependence: 1990. *Alcohol. Health Res. World*, **17**, 10–11.
5	Rice, D.P., Kelman, S., and Miller, L.S. (1991) Estimates of economic costs of alcohol and drug abuse and mental illness, 1985 and 1988. *Publ. Health Rep.*, **106** (3), 280–292.

References

Abel, E.L. and Sokol, R.J. (1987) Incidence of fetal alcohol syndrome and economic impact of FAS-related anomalies. *Drug Alcohol Depend.*, **19**, 51–70.

Abel, E.L. and Sokol, R.J. (1991a) A revised estimate of the economic impact of Fetal Alcohol Syndrome. *Recent Dev. Alcohol.*, **9**, 117–125.

Abel, E.L. and Sokol, R.J. (1991b) A revised conservative estimate of the incidence of FAS and its economic impact. *Alcohol. Clin. Exp. Res.*, **15**, 514–524.

Bloss, G. (1994) The economic cost of FAS. *Alcohol. Health Res. World*, **18**, 53–54.

DHS Public Health Division, Office of Family Health, Women's and Reproductive Health Section, Oregon FAS Prevention Program (2009) Cost of Fetal Alcohol Spectrum Disorder (FASD) in Oregon, http://www.oregon.gov/DHS/ph/wh/docs/fas/cost_of_fasd_in_oregon.pdf (accessed 22 April 2010).

Fast, D.K., Conry, R., and Loock, C.A. (1999) Identifying fetal alcohol syndrome among youth in the criminal justice system. *Dev. Behav. Pediatr.*, **20** (5), 370–372.

Fuchs, D., Burnside, L., De Riviere, L., Brownell, M., Marchenski, S., Mudry, A., and Dahl, M. (2009) *Economic Impact of*

Children in Care with FASD and Parental Alcohol Issues Phase 2: Costs and Service Utilization of Health Care, Special Education, and Child Care, Centre of Excellence for Child Welfare, Ottawa. http://www.cecw-cepb.ca/sites/default/files/publications/en/FASD_Economic_Impact_Phase2.pdf.

Harwood, H.J. (2000) *Updating Estimates of the Economic Cost of Alcohol Abuse in the US: Estimates, Update Methods and Data.* Report prepared by the Lewin Group. National Institute on Alcohol Abuse and Alcoholism, Bethesda, MD.

Harwood, H.J. and Napolitano, D.M. (1985) Economic implications of the Fetal Alcohol Syndrome. *Alcohol Health Res. World*, **10**, 38–75.

Health Canada (2006) Fetal Alcohol Spectrum Disorder. It's Your Health, http://www.hc-sc.gc.ca/iyh-vsv/alt_formats/cmcd-dcmc/pdf/fasd-etcaf_e.pdf (accessed 2 April 2007).

Klug, M. and Burd, L. (2003) Fetal Alcohol Syndrome and cumulative cost savings. *Neurotoxicol. Teratol.*, **25**, 763–765.

Landry, L. and Sinha, M. (2008) Adult correctional services in Canada, 2005/2006. *Statistics Canada*, Catalogue no. 85-002-XIE, **28** (6), 1–26. Available at: http://www.statcan.gc.ca/pub/85-002-x/85-002-x2008006-eng.pdf (accessed on 20 May 2010).

Lupton, C., Burd, L., and Harwood, R. (2004) Cost of fetal alcohol spectrum disorders. *Am. J. Med. Genet. C Semin. Med. Genet.*, **127** (1), 42–50.

McDowell Group (2005) Economic Costs of Alcohol and other Drug Abuse in Alaska, December 2005 update. Prepared for the Advisory board on Alcoholism and Drug Abuse Department of Health & Social Services.

Rice, D.P. (1993) The economic costs of alcohol abuse and alcohol dependence: 1990. *Alcohol Health Res. World*, **17**, 10–11.

Stade, B., Ungar, W.J., Stevens, B., Beyene, J., and Koren, G. (2006) The burden of prenatal exposure to alcohol: Measurement of cost. *JFAS Int.*, **4**, e5.

Stade, B., Ali, A., Bennett, D., Campbell, D., Johnston, M., Lens, C., Tran, S., and Koren, G. (2009) The burden of prenatal exposure to alcohol: revised measurement of cost. *Can. J. Clin. Pharmacol.*, **16** (1), e91–e102.

Thanh, N.X. and Jonsson, E. (2009) Costs of fetal alcohol spectrum disorder in Alberta, Canada. *Can. J. Clin. Pharmacol.*, **16**, e80–e90.

Part Two
Causes and Diagnosing of FASD

5
Direct and Indirect Mechanisms of Alcohol Teratogenesis: Implications for Understanding Alterations in Brain and Behavior in FASD

Kristina A. Uban, Tamara Bodnar, Kelly Butts, Joanna H. Sliwowska, Wendy Comeau, and Joanne Weinberg

5.1
Introduction

Experimental animal models have been critical to research in the field of Fetal Alcohol Spectrum Disorders (FASD). Publications by Lemoine *et al.* (1968) and Jones and Smith (1973) which first described Fetal Alcohol Syndrome (FAS), the most severe end of the spectrum of alcohol's adverse effects, provided strong evidence linking maternal drinking to adverse effects in the offspring. Since then, numerous animal models have been developed to explore this link. *In vivo* models, including rodents, chicks, dogs, pigs, sheep and primates, have been complemented over the years by *in vitro* studies. These models provide control over environmental (dose and timing of alcohol exposure, exposure to other drugs, maternal nutrition and health, pre- and postnatal environments) and genetic (sensitivity to alcohol, differences in alcohol absorption, distribution and metabolism) variables that are not possible to control in human studies. The data from this research have, therefore, been critical in demonstrating that alcohol is, in itself, a teratogen. Importantly, the physical, biological, and neurobehavioral effects of prenatal alcohol exposure described in FASD have been paralleled to a large extent in the animal studies. These models thus provide a valuable tool for investigating both direct and indirect mechanisms that underlie alcohol's adverse effects on the brain and behavior. Increasing insight into alcohol's mechanisms of action will ultimately allow for the development of more targeted and effective intervention and treatment strategies.

5.1.1
Mechanisms of Alcohol's Teratogenic Effects

Direct toxic effects of alcohol on the embryo are well known. Alcohol readily crosses the placental and blood–brain barriers, and can thus act directly on developing fetal cells. However, in mammals it is likely that interactions between direct and indirect (maternally mediated) effects of alcohol are responsible for its adverse

---

Fetal Alcohol Spectrum Disorder–Management and Policy Perspectives of FASD. Edited by Edward P. Riley, Sterling Clarren, Joanne Weinberg, and Egon Jonsson
Copyright © 2011 WILEY-VCH Verlag GmbH & Co. KGaA, Weinheim
ISBN: 978-3-527-32839-0

effects (Randall, Ekblad, and Anton, 1990). Moreover, as alcohol is known to act on or modulate many different target molecules, multiple mechanisms, activated at different stages of development or at different dose thresholds of exposure, probably contribute to the diverse phenotypes seen in FASD (Goodlett, Horn, and Zhou, 2005). In this chapter, the major direct and indirect mechanisms underlying alcohol's actions will be described, and some examples provided of these mechanisms. An indirect mechanism, endocrine imbalance, will be discussed at greater length, as this is the focus of research in the authors' laboratory.

5.1.2
Direct Mechanisms of Alcohol's Actions on the Fetus

The results of many studies have provided abundant evidence that alcohol can have direct effects on fetal growth and development. Studies on nonmammalian species, where the organism develops outside the mother and there are no effects of the placenta or the maternal environment, have provided unique insights into the teratogenic effects of alcohol. For example, by using *Caenorhabditis elegans*, a microscopic nematode worm, Davis, Li, and Rankin (2008) showed that chronic exposure to alcohol during larval development temporarily delayed physical growth, slowed development, delayed the onset of reproductive maturity, and decreased both reproductive fertility and longevity. Similarly, acute embryonic exposure of *C. elegans* eggs to high concentrations of alcohol at different stages of development resulted in a lower probability that exposed eggs would hatch into larval worms, at least at some stages. For worms that did hatch, many displayed distinct physical dysmorphologies. Studies in chicks have similarly revealed direct adverse effects of alcohol on the fetus. Pennington *et al.* (1983) reported dose-dependent effects of alcohol on suppression of the rate of cell division in embryonic tissue, resulting in fewer cells/embryo for a given time in gestation. Interestingly, it was suggested that the suppression in cell division may be related to alcohol-induced changes in the metabolism of the *prostaglandins*, which are naturally produced hormone-like fatty acids known to be involved in a wide range of biological processes including the regulation of blood pressure, metabolism, and immune response (see Section 5.1.3.1). Cartwright and Smith (1995) used a chick model to study mechanisms underlying the craniofacial abnormalities observed in FAS. A single alcohol exposure was shown to result in a range of craniofacial abnormalities, depending on the dose and timing of exposure. These studies provided important evidence indicating that the timing of alcohol exposure relative to stage of development could account, in part, for the variations in craniofacial defects observed in FAS.

Importantly, direct effects of alcohol on mammalian embryos have also been observed. For example, Brown, Goulding, and Fabro (1979) showed that rat embryos grown in culture media containing alcohol displayed dose-dependent embryonic retardation of growth and differentiation, with specific reduction in both DNA and protein content. Consistent with these findings, data from *in vivo* studies have shown that both acute and chronic alcohol exposure *in utero* can inhibit protein synthesis and decrease RNA and DNA content in the fetal and neonatal brain (Rawat, 1975). Similarly, Gallo and Weinberg (1986) reported that prenatal and early postnatal alcohol exposure can reduce brain, heart and kidney weights in preweaning offspring due to direct effects on protein and/or DNA content. Alcohol-induced disruptions in the proliferation of stem cell populations are another mechanism of alcohol's direct actions on the fetus, leading to a reduction in the generation of both new neurons and new glial cells (Guerri et al., 1993; Miller, 1992). Neuronal cell damage and/or cell death can also occur through both programmed cell death (apoptosis) (Ikonomidou et al., 2000) and the inhibition or disruption of enzymes that play a role in metabolism in neural tissue (Goodlett, Horn, and Zhou, 2005). Importantly, extensive changes in organ weight, although not detrimental in and of themselves, may be indicative of altered functioning that could potentially increase vulnerability to later diseases/disorders. Of particular interest to both researchers and clinicians, there appear to be fairly large differences in the susceptibility of different brain regions to alcohol, depending on the dose and timing of exposure. The hippocampus, amygdala, and cerebellum show particular sensitivity to the inhibition of protein synthesis by alcohol, resulting in decreased numbers of mature neurons. Furthermore, prenatal exposure to alcohol can result in a disorganized cortical architecture, which ultimately influences the pattern of communication in and across regions involved in higher cognitive function. This suggests one possible mechanism by which prenatal alcohol exposure might produce cognitive deficits in children with FASD (Clarren, 1986). Similarly, alcohol-induced changes in function of the hippocampus and cerebellum are known to be involved in some of the behavioral alterations that occur in FASD (Michaelis, 1990; Guerri, 1998). Whether the latter changes are due to direct or indirect effects of alcohol remains to be determined.

5.1.3
Indirect Mechanisms of Alcohol's Actions on the Fetus

Numerous secondary or indirect mechanisms mediating alcohol's adverse effects on neurobiological and neurobehavioral outcomes have been demonstrated in animal models and *in vitro* studies. These include: nutritional deprivation or deficiencies (e.g., calories, protein, zinc, folate, vitamin A); abnormalities in calcium signaling; altered alcohol–prostaglandin interactions; placental dysmorphology/dysfunction; alcohol-induced circulatory changes in placenta and/or fetus; disrupted cell–cell interactions (cell adhesion); interference with growth factors or other cell signaling mechanisms that mediate cell proliferation, growth, differentiation, migration and maturation; oxidative stress and free radical damage; disruption of neuronal development in specific cell populations (e.g., serotonergic neurons); and disruption of endocrine balance (Randall, Ekblad, and Anton, 1990; Goodlett, Horn, and Zhou, 2005; Michaelis, 1990; Guerri, 1998; Shibley, McIntyre, and Pennington, 1999; West, Chen, and Pantazis, 1994). Here, several of these indirect mechanisms will be discussed, followed by an in-depth discussion of one mechanism, namely the disruption of endocrine balance.

5.1.3.1 Alcohol Effects on Prostaglandins

Seminal studies conducted by Randall and colleagues (Randall *et al.*, 1989; Randall, Anton, and Becker, 1987) have shown that prostaglandins may play a role in mediating the adverse effects of alcohol. Acute alcohol exposure stimulates prostaglandin synthesis, and prostaglandin degradation is impaired following chronic alcohol exposure. If abnormally high levels occur due to such disruption of the balance between synthesis and degradation, prostaglandins themselves may increase fetal mortality during development. Randall's studies in a mouse model have shown that the inhibition of prostaglandin synthesis can attenuate some of the deficits induced by prenatal alcohol exposure. Thus, for example, the treatment of pregnant mice with low-dose aspirin, which inhibits prostaglandin synthesis, attenuated the effects of prenatal alcohol exposure on fetal mortality and reduced the incidence of birth defects by 50%. Clearly, prostaglandins do not mediate all of alcohol's effects on fetal growth and development. However, the data reveal one potential mechanism that could contribute to at least some of alcohol's adverse effects, and suggest one possible direction for the development of therapeutic interventions.

5.1.3.2 Alcohol-Induced Disruption of Cell–Cell Interactions or Cell Adhesion

Cell adhesion molecules (CAMs) are proteins that aid in cell–cell binding and also facilitate interaction between nerve cells and their surrounding environment; both effects are critical in the normal development of the nervous system. One such CAM is the L1 cell adhesion molecule, a member of the immunoglobulin superfamily (Bearer, 2001a; Bearer 2001b; Gubitosi-Klug, Larimer, and Bearer, 2007; Ramanathan *et al.*, 1996; Wilkemeyer *et al.*, 2002). When L1 binds to neurons it activates cell signaling cascades that mediate functions such as cell–cell adhesion, outgrowth of nerve cell processes, and neuron migration. Children with mutations of the gene for L1 may show mental retardation and a variety of brain abnormalities similar to those in children with FAS. Research has shown that alcohol can inhibit the functions of L1, and that antagonists that prevent this alcohol-mediated inhibition may be neuroprotective. *In vivo* studies have tested the effects of two small peptides secreted by glial cells: SALLRSIPA (SAL), an active fragment of activity-dependent neurotrophic factor (ADNF), and NAPVSIPQ (NAP), an active fragment of activity-dependent neuroprotective protein. The data demonstrated that both NAP and SAL are extremely potent and effective antagonists of alcohol inhibition of L1 adhesion. Ongoing studies are being conducted to examine the therapeutic potential of NAP and SAL in attenuating the adverse prenatal alcohol effects on the ability of nerve cells to interact appropriately with each other, or with their environment.

5.1.3.3 Alcohol and Oxidative Stress

Recently reported data have shown that, in addition to its ability to antagonize the alcohol inhibition of L1, NAP may also represent a potential therapeutic treatment for the oxidative stress associated with alcohol exposure (Sari, 2009). Oxidative stress (the interaction of high levels of free radicals or reactive oxygen species with molecules within the cell) is a known mechanism of alcohol-induced cellular injury, and has been shown to play a role in alcohol teratogenesis (LeBel and Bondy, 1991; Kotch, Chen, and Sulik, 1995). Fetal tissues generally have lower activities and levels of oxidative defenses than adult tissues, such that the fetus is highly sensitive to oxidative stress, especially early in gestation. Oxidative stress in the fetus can cause a wide range of problems, from structural malformations to embryonic death, through the effects of free radicals on cell membranes, cytoskeleton, mitochondria, and membrane protein receptors (Henderson et al., 1995). Interestingly, Spong et al. (2001) found that a combined treatment with NAP and SAL prevented fetal death and growth restriction in a mouse model of FAS. Protection from the toxic effects of alcohol appeared to result from a blockade of the induction of free radicals by alcohol. Further support for this possibility derives from the studies of Sari and colleagues (Sari, 2009), who showed that prenatal alcohol-induced cell death is mediated in part through mitochondrial dysfunction. The administration of NAP together with alcohol to pregnant mice on embryonic day 7 reversed alcohol-induced changes in mitochondrial pathways, and thus blocked the increased cell death resulting from alcohol-induced oxidative stress. Moreover, Wilkemeyer et al. (2002) suggested that the loss of L1-mediated cell–cell adhesion might also be linked to oxidative injury through the induction of apoptosis, triggered by a loss of contact of nerve cells with each other, or with their environment. The possibility that NAP could be utilized as a therapeutic agent against the oxidative stress associated with alcohol exposure represents a very exciting translational outcome of this basic research on the mechanisms of alcohol's actions on the embryo.

5.1.3.4 Disruption of Endocrine Balance

Alcohol has long been known to alter endocrine function in the adult organism. Studies conducted in both humans and animal models, involving the acute administration of alcohol or chronic alcohol consumption, have demonstrated effects of alcohol on the secretory activity of the adrenal glands, ovaries, testes, and thyroid, as well as on secretion of numerous pituitary hormones, including growth hormone, prolactin, vasopressin, adrenocorticotropin, and the gonadotropic hormones (for reviews, see Weinberg, 1993a; Weinberg 1993b; Weinberg, Nelson, and Taylor, 1986). Furthermore, it has been shown that alcohol can act directly on the endocrine glands themselves, as well as (or in addition to) having an action on the central aspects of hypothalamic–pituitary function. Interestingly, much of what is known of alcohol-related effects on endocrine function comes from male subjects. It is only relatively recently that studies in clinical settings and with animal models of alcoholism have begun to include female subjects. Even fewer studies have examined the effects of alcohol exposure on endocrine function during pregnancy.

It is known that alterations in both maternal and offspring endocrine function are among the physiological abnormalities produced by maternal alcohol intake. Whether alcohol-induced endocrine imbalances contribute to the etiology of FASD is unknown, but it is certainly a possibility (Anderson, 1981). The effects of alcohol on interactions between the pregnant female and fetus are complex (Weinberg, 1993a, 1993b, 1994; Rudeen and Taylor, 1992), and both direct and indirect effects of alcohol on fetal development are known to occur. Alcohol readily crosses the placenta, and can therefore act directly on developing fetal cells and tissues, including those related to endocrine function. In addition, because the pregnant female and fetus constitute an interrelated functional unit, disruption of the normal hormonal interactions between the pregnant female and the fetus by alcohol can indirectly alter the development of fetal metabolic, physiological, and endocrine functions. Alcohol intake can also affect the female's ability to maintain a successful pregnancy, resulting in miscarriage or, if the fetus is carried to term, possible congenital defects.

By using a rat model, research in the present authors' laboratory has focused on the effects of maternal alcohol consumption on the maternal and offspring hypothalamic–pituitary–adrenal (HPA) axis, an endocrine system involved in multiple metabolic functions and in the ability to respond appropriately to stressors. A particular focus has been on fetal programming. Fetal or early programming refers to the concept that early environmental or nongenetic factors, including pre- or perinatal exposure to stress, drugs or other toxic agents, can permanently organize or imprint physiological and behavioral systems and increase vulnerability to illnesses or disorders later in life (Bakker, Bel, and Heijnen, 2001; Matthews, 2002; Welberg and Seckl, 2001). The HPA axis is highly susceptible to programming during fetal and neonatal development (Matthews, 2002; Welberg and Seckl, 2001), and data have suggested the possibility that ethanol-induced disruption of the reciprocal interconnections between the maternal and neonatal HPA axes may provide a common pathway for fetal programming by a number of different agents (Angelucci *et al.*, 1985). Recently, an examination has been conducted into the hypothesis that the alcohol-induced fetal programming of HPA activity sensitizes neuroadaptive mechanisms that mediate HPA activity and regulation, resulting in hyper-reactivity to subsequent life stressors, and, in turn, an increased vulnerability to illnesses, including mental health disorders such as depression, anxiety, and addiction.

5.1.4
Neurobiology of Stress

Stress can be defined as any stimulus or event that disrupts homeostasis. The typical response to such an event is the activation of both behavioral and physiological responses that facilitate coping, and ultimately act to restore homeostasis (McEwen, 2007). For example, stress-induced activation of the HPA axis results in altered hormone production and release, as well as altered metabolic and immune responses that allow for the reallocation of energy and attention to the immediate threat/event. An adaptive HPA response to stress involves a rapid activation of the stress system, allowing a rapid initiation of appropriate responses, followed by a quick recovery to basal levels and restoration of homeostasis (de Kloet, 2000; Chrousos, 1998). For completeness, it should be noted that there

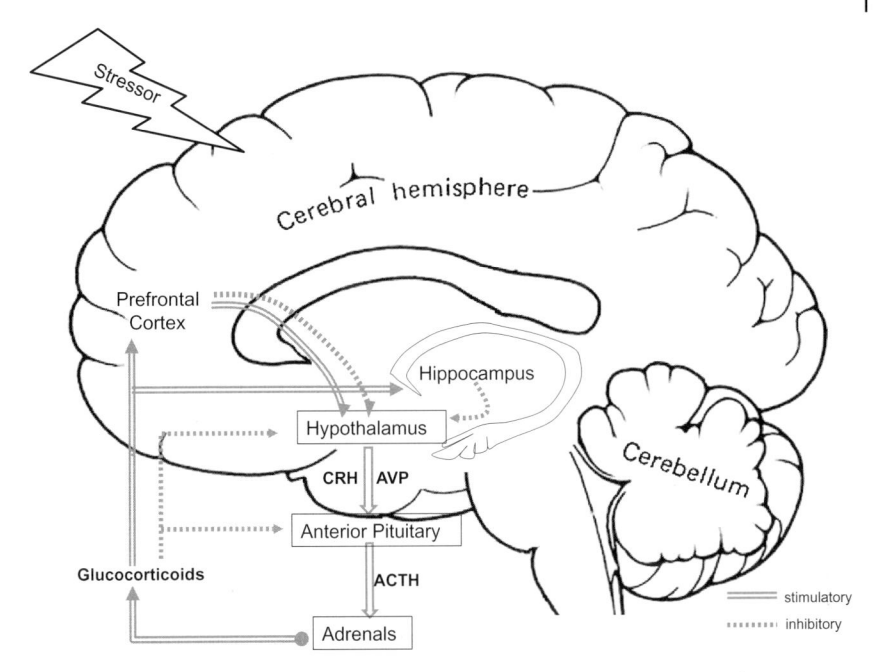

Figure 5.1 Regulation of the hypothalamic–pituitary–adrenal (HPA) axis. Following the presentation of a stressor, corticotropin-releasing hormone (CRH) and arginine vasopressin (AVP) are two hormones released from the hypothalamus, which in turn stimulate the release of adrenal corticotrophin-releasing hormone (ACTH) from the anterior pituitary gland. This cascade of events results in the secretion of stress hormones, known as glucocorticoids, from the adrenal cortex. Glucocorticoids can exert their influence on the central and peripheral nervous systems. In addition, glucocorticoids provide feedback at all levels of the HPA axis, typically inhibiting further glucocorticoid release and returning the system to homeostasis. However, some brain regions, such as certain subregions of the prefrontal cortex, can provide positive feedback in response to high levels of circulating glucocorticoids, and potentiate CRH release from the hypothalamus.

is a second component of the stress system: the autonomic nervous system. The autonomic nervous system has two components: (i) the sympathetic division has an activating role, increasing the heart rate, blood pressure and blood supply to the muscles, for example, and allowing a rapid response to a threat or event; and (ii) the parasympathetic division, which has a calming role. The HPA axis and sympathetic nervous system work together to produce an optimal stress response.

The HPA axis involves a cascade of hormonal responses (Figure 5.1), whereby corticotropin-releasing hormone (CRH) and arginine vasopressin (AVP) are released from the paraventricular nucleus (PVN) of the hypothalamus. CRH is the major releasing hormone, but with increasing or repeated exposure to stressors, CRH neurons also begin to synthesize AVP, which acts to facilitate the actions of CRH. These releasing hormones then stimulate the release of adrenocorticotropic hormone (ACTH) from the anterior pituitary. The ACTH, in turn, stimulates the release of the glucocorticoid hormones [cortisol in humans, corticosterone (CORT) in most rodents] from the adrenal cortex (Figure 5.1). As with all major hormones, the glucocorticoids are released into the bloodstream, where they can exert their influence on a large number of metabolic processes and on both the central and peripheral nervous systems. In addition, glucocorticoids provide negative feedback at all levels of the HPA axis, inhibiting further hormone release and returning the system to homeostasis. Glucocorticoid secretion undergoes a daily rhythm, with peak secretion occurring at the initiation of the waking cycle, that may be enhanced at any point in the day by exposure to a stressor. Moreover, other systems engaged by stressors (e.g., neurotransmitters such as norepinephrine and dopamine; Chrousos, 1998), as well as higher brain regions, including the hippocampus and prefrontal cortex (see further discussion below), play a significant role in the initiation, regulation, and termination of the stress response, and thus influence HPA activity (Ulrich-Lai and Herman, 2009). Importantly, psychological stressors can activate a similar or even greater stress response than physical or physiological stressors, and repeated or chronic activation of the HPA axis has been implicated in a number of mental health disorders (McEwen, 2007; Selye, 1971).

The *hippocampus*, a structure located in the temporal lobe of the brain, has historically been viewed as one of the major players in negative feedback regulation of the HPA stress response, and is an important target for glucocorticoid actions (Sapolsky, Meaney, and McEwen, 1985; Feldman and Weidenfeld, 1999). Binding of glucocorticoids to their receptors in the hippocampus not only provides an inhibitory signal to "turn off" the stress response, but also enhances cell metabolism, thereby increasing available energy to the organism, and modulating the processing of context-specific behaviors (Kloet, 2000).

The prefrontal cortex (PFC) is also known to be a major target for glucocorticoids, and to play an important role in HPA feedback regulation (McEwen, De Kloet, and Rostene 1986). In rats, a frontal midline area of the PFC known as the medial PFC (mPFC) is considered homologous to the dorsolateral PFC of humans, with both areas subserving executive functions. Interestingly, the mPFC has a relatively high density of glucocorticoid receptors, although considerably lower than that found in the hippocampus (Diorio, Viau, and Meaney, 1993). Nonetheless, the mPFC appears to be especially sensitive to high levels of glucocorticoid hormones (Ulrich-Lai and Herman, 2009) and plays an important role in both the initiation and inhibition of the stress response (Diorio, Viau, and Meaney, 1993). Interestingly, research data have suggested that the influence of the mPFC on the HPA axis is region-specific (for reviews, see Sullivan and Gratton, 2002 and Ulrich-Lai and Herman, 2009). In rats, the prelimbic (PL) region of the mPFC facilitates inhibition of HPA axis function, whereas the more ventral infralimbic region appears to be involved in HPA axis activation. As outputs from the mPFC to the hypothalamus are excitatory, modulation of the PVN of the hypothalamus (the locus of control for the HPA axis) must be indirect, relaying through central structures that provide inhibitory inputs to the PVN (for a review, see Ulrich-Lai and Herman, 2009).

Another interesting characteristic of PFC stress regulation is that its role in negative feedback regulation may be stress-specific. For example, Diorio, Viau, and Meaney (1993) reported decreases in plasma ACTH and CORT with restraint (psychological), but not ether (physiological) stress following the administration of exogenous CORT into the dorsal region of the mPFC (PL and anterior cingulate cortex). Conversely, lesions of this same area produced increases of ACTH and CORT, and again were stress-specific. One interpretation of these findings is that PFC regulation of the stress response may be limited to PFC-relevant stimuli and information (Herman and Cullinan, 1997). That is, PFC involvement in negative feedback may be relegated to stressors that require higher levels of processing (using past experience), leaving physiological stress responses to be facilitated by faster-acting pathways in the brainstem. Herman and Cullinan (1997) have proposed a similar idea, suggesting two pathways for mediation of psychological versus physiological stressors.

Relevant to an understanding of the neurobiological mechanisms related to FASD, chronic or uncontrollable stress produces alterations in the stress circuitry that could have implications for ongoing control of the stress response system (Ulrich-Lai and Herman, 2009). Chronic stress increases HPA responsiveness even in the presence of elevated levels of glucocorticoids, implying a tempered control via negative-feedback mechanisms. Altered cortical architecture has been reported in the mPFC following both chronic stress and prenatal alcohol exposure (Radley et al., 2008; Perez-Cruz et al., 2007; Radley et al., 2006; Cook and Wellman, 2004), which could moderate the efficacy of inputs that would typically play a role in terminating the stress response. Thus, prenatal alcohol-related changes in the structure and function of the PFC may underlie and perpetuate an inability to regulate the stress response appropriately, leading to further dysregulation. Adding to the issues of altered mPFC function in HPA regulation, reduced mPFC cell numbers following prenatal exposure to alcohol could also potentially decrease the input from the PFC, thereby further diminishing regulatory mechanisms. Cumulative effects of such a scenario would include diminished mPFC function, further dendritic atrophy, deficits in cognitive processing, and reduced HPA feedback, all of which, in turn, could lead to an increased vulnerability to psychiatric disorders.

5.1.5
FASD and Stress Responsiveness

Not surprisingly, dysregulation in HPA function is associated with several mental health (e.g., depression, anxiety, substance use) and physiological disorders. Results from studies conducted by the present authors and other investigators have shown that prenatal alcohol exposure (PAE) leads not only to HPA dysregulation (reviewed by Weinberg et al., 2008) but also to a reprogramming of the fetal HPA axis, such that HPA tone is increased throughout life (Weinberg et al., 2008). Both basal cortisol levels (Jacobson, Bihun, and Chiodo, 1999; Ramsay, Bendersky, and Lewis, 1996) and cortisol reactivity to stressors (Jacobson, Bihun, and Chiodo, 1999; Haley, Handmaker, and Lowe, 2006) are elevated in children exposed to alcohol *in utero*. Similarly, in rodent models, PAE offspring exhibit increased HPA activation and/or delayed or deficient recovery to basal levels following stress (Weinberg et al., 2008; Lee et al., 1990, 2000; Redei, Clark, and McGivern, 1989; Taylor et al., 1988). Changes in central HPA regulation suggest that PAE results in both increased HPA drive and deficits in feedback regulation, and likely alters HPA function at multiple levels of the axis (Weinberg et al., 2008; Redei, Clark, and McGivern, 1989; Lee and Rivier, 1996; Gabriel et al., 2005; Redei et al., 1993; Glavas et al., 2007).

Across species, sex differences in HPA function are well documented, with females typically showing greater ACTH and glucocorticoid responses, as well as a greater resistance to negative feedback by glucocorticoids compared to males. Consistent with this, sex differences in the effects of PAE on HPA activity in rodent models are also observed; typically, PAE females tend to show a greater HPA responsiveness to acute stressors, whereas males show a greater responsiveness to chronic or prolonged stressors (Weinberg et al., 2008; Lee and Rivier, 1996). However, stressor type and intensity, as well as the hormonal endpoints measured, all influence the response observed (Weinberg et al., 2008; Weinberg, Taylor, and Gianoulakis, 1996). Similarly, sex differences in HPA, autonomic and behavioral reactivity to stressors have been demonstrated in boys and girls following prenatal alcohol exposure (Haley, Handmaker, and Lowe, 2006), suggesting alterations in the normal interactions between the HPA and hypothalamic–pituitary–gonadal (HPG) axes following PAE. The results of studies in animal models have supported this suggestion, providing evidence that PAE alters the development of the HPG axis (Esquifino, Sanchis, and Guerri, 1986; McGivern, Handa, and Raum, 1998), as well as HPG–HPA interactions (Lan et al., 2006, 2009) in both male and female offspring. The bidirectional interaction between the HPG and HPA axes occurs at multiple levels: acute stressors typically activate, whereas repeated or chronic stressors typically inhibit HPG function, and the sex-steroids reciprocally modulate HPA activity (Young, 1995; Viau, 2002).

5.2
Fetal Programming: Programming of the HPA Axis by PAE

As noted, the concept that early environmental factors can permanently organize or imprint physiological and behavioral systems is referred to as "fetal/early programming" (see Bakker, Bel, and Heijnen, 2001; Matthews, 2002; Welberg and Seckl, 2001). The concept of fetal programming was developed originally from studies showing that a low birth weight is associated with an increased biological risk for coronary heart disease in adult life (Barker et al., 1989). Following these studies, both epidemiological data from human populations and experimental studies in animals linked the intrauterine environment with the development of hypertension, diabetes, elevated blood cholesterol and fibrinogen concentrations, polycystic ovarian syndrome, and psychiatric disorders (for a review, see Godfrey and Barker, 2000). These findings led to the hypothesis that common adult diseases may originate during fetal development – that is, the "fetal origins of adult disease" hypothesis. Adult lifestyle factors such as alcohol consumption, smoking, and exercise appear to be additive to early life influences, suggesting that the fetal environment has distinct roles and consequences.

There is today a large body of evidence indicating that stress-related changes in maternal glucocorticoid levels may underlie programming of the fetal HPA axis. Maternal glucocorticoids pass through the placenta to reach the fetus, increasing circulating glucocorticoid levels in the fetus and modifying brain development (Seckl, 2008). However, fetal glucocorticoid levels are much lower than maternal levels due to the presence of a placental enzyme, 11β-hydroxysteroid dehydrogenase type 2, that catalyzes the conversion of active glucocorticoids (cortisol, corticosterone) to inactive inert forms (cortisone, 11-dehydrocorticosterone) (Meaney, Szyf, and Seckl, 2007). This enzyme thus protects the fetus from high levels of maternal glucocorticoids, although 10–20% of maternal glucocorticoids do cross intact to the fetus (Benediktsson et al., 1997; White et al., 1997; Waddell et al., 1998).

As it is difficult to study the action of glucocorticoids in humans, the present understanding of the action of glucocorticoids on the fetus derives primarily from animal studies. It has been shown that exposure of the dam to stress or to glucocorticoids increases maternal glucocorticoid secretion and affects the fetal HPA axis (Cadet et al., 1986). Modulating glucocorticoid levels in stressed dams by adrenalectomy (in essence, removing the main source of glucocorticoid synthesis and release) and the replacement of corticosterone at low basal levels prevents these effects. These findings provide support for the hypothesis that elevations in maternal glucocorticoids mediate prenatal programming of the fetal HPA axis (Barbazanges et al., 1996). It has been suggested (Lan et al., 2009; Viau, 2002; Waddell et al., 1998; Cadet et al., 1986) that the resetting of key hormonal systems by early environmental events may be one mechanism linking early life experiences with long-term health consequences. (Lan et al., 2009; Viau, 2002; Waddell et al., 1998; Cadet et al., 1986). For example, studies by Phillips and coworkers (Phillips et al., 1998, 2000) have demonstrated strong correlations between birth weight, plasma cortisol levels and the development of hypertension and type II diabetes in humans, implicating the intrauterine reprogramming of the HPA axis as a potential mechanism underlying the observed associations between low birth weight and increased risk of disease later in life. The mechanisms underlying fetal programming have important clinical implications, as in North America, approximately 10% of pregnant women are treated with synthetic glucocorticoids in late gestation to promote fetal lung maturation in fetuses at risk of being delivered prematurely (Conference, NCD, 1995).

While maternal glucocorticoids likely play a role in alcohol-induced programming of the fetal HPA axis, and thus in the disturbances in physiology and behavior in PAE, it is unlikely that maternal glucocorticoids alone mediate fetal programming of HPA function. Alcohol consumed by the mother also readily crosses the placenta and can act directly on the fetal HPA axis, which is functional before birth. Thus, programming of the fetal HPA axis likely occurs through a complex interaction between alcohol and the glucocorticoid hormones. Moreover, other factors are also likely to be involved in fetal programming. Of note, the serotonergic (hydroxytryptamine; 5-HT) system, which is critical for normal brain development and HPA function, is known to be affected adversely by alcohol, and there is a growing body of evidence pointing to the involvement of the 5-HT system in programming of the fetal HPA axis (see further discussion of 5-HT below) (Matthews, 2002; Druse, Kuo, and Tajuddin, 1991; Sari and Zhou, 2004; Zhou et al., 2001; Zhou, Sari, and Powrozek, 2005).

5.3
Altered Epigenetic Regulation of Gene Expression: A Possible Mechanism Underlying Fetal Programming of the HPA Axis and Altered Neuroendocrine-Immune Interactions

A potential mechanism underlying HPA programming may involve altered epigenetic regulation of gene expression. Epigenetic mechanisms include stable, yet potentially reversible, modifications in the genome that do not involve alterations in the DNA sequence; the best-known of these include DNA methylation and histone modifications. The majority of such changes are passed on through mitosis, transmitting functional information to daughter cells. DNA methylation, the addition of methyl groups to specific building blocks of DNA, can occur in the promoter or regulatory region of a gene, which may subsequently influence gene expression. For example, if a particular region of DNA becomes more or less accessible to the action of transcription factors or other molecules, this could affect mRNA and protein synthesis (Figure 5.2). Histone modifications can affect

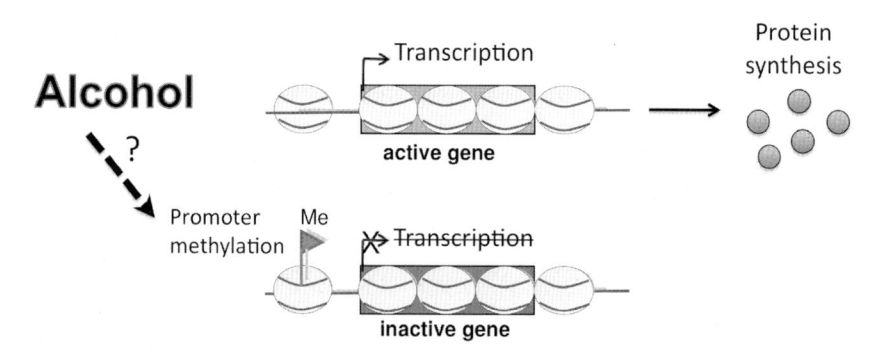

Figure 5.2 Possible mechanism of the epigenetic effects of prenatal alcohol exposure. Alcohol exposure may alter the epigenome, for example, by resulting in increased histone methylation (Me). When methylation occurs in the promoter region of a gene, it may become inaccessible to the molecular machinery involved in gene transcription. When transcription is inhibited, protein synthesis cannot occur, which will subsequently affect the processes and pathways involving the protein in question.

chromatin shape and structure, which may also alter gene accessibility and, subsequently, influence gene expression.

In general, all cells of an organism contain the same genetic code and as a result, epigenetic modifications allow for variations in gene expression profiles, producing functional diversity in different cell and tissue types. Of particular relevance is the plastic nature of epigenetic modifications. That is, the epigenome can be altered by a multitude of environmental factors such as maternal diet or dietary supplements (e.g., protein or fat levels, nutritional deprivation, folic acid, choline and vitamin B12) (Cooney, Dave, and Wolff, 2002), endocrine factors such as naturally occurring hormones or the synthetic estrogen diethylstilbestrol (Newbold, Padilla-Banks, and Jefferson, 2006), postnatal maternal care (Weaver et al., 2004; Weaver, 2007) and, of particular relevance, *in utero* exposure to alcohol (Garro et al., 1991) or other toxic substances. Alcohol is thought to disrupt epigenetic reprogramming by interfering with the methionine cycle, and thus with DNA methylation and histone modification capacity (O'Neil et al., 2007). It is possible that epigenetic mechanisms may underlie at least some of the physiological and behavioral changes that arise following prenatal alcohol exposure (Figure 5.2).

Evidence for epigenetic reprogramming of the HPA axis by early life experience derives from recent studies carried out by Weaver and colleagues (Weaver et al., 2004), which showed that enhanced maternal care is associated with decreased methylation of the gene encoding the glucocorticoid receptor (GR). Decreased methylation leads to an increased expression of GR. As activation of GR is essential for efficient termination of the stress response, increased GR expression will result in enhanced feedback regulation of the HPA axis and ultimately, more effective HPA recovery following exposure to stressful stimuli (Weaver et al., 2004). Although not yet established, it is hypothesized that reprogramming of the HPA axis following *in utero* alcohol exposure is influenced by epigenetic factors resulting from an interplay between genetic influences and environmental factors.

The central nervous system (CNS), the endocrine system (including the HPA axis) and the immune system do not operate independently, but rather are developmentally and functionally intertwined, with shared receptors, hormones and neurotransmitters, as well as interacting regulatory feedback (Bumiller et al., 1999; Haddad, Saade, and Safieh-Garabedian, 2002; Chesnokova and Melmed, 2002). This bidirectional communication underlies the effects of stress on the immune system. In general, acute or short-term exposure to stressors activates the immune system, specifically the innate immune system, stimulates antibody production, and prepares the organism to respond to subsequent immune challenges (for a review, see Dhabhar, 2002). Importantly, immune system activity quickly returns to baseline through the acute actions of glucocorticoids. By contrast, under conditions of chronic stress, sustained high levels of glucocorticoids can result in immunosuppression and a significant decline in immune competence, decreasing the immune system function by over 50% (Dhabhar et al., 1995). White blood cells (leukocytes) express GRs, and long-term increases in plasma glucocorticoids can suppress the immune system by inhibiting leukocyte formation, decreasing leukocyte responsiveness to infection and resulting in leukocyte apoptosis or cell death (for a review, see Sheridan *et al.*, 1994). In addition, leukocytes express receptors for endocrine mediators such as CRH, ACTH, CORT, epinephrine and norepinephrine, and signaling through these receptors affects leukocyte trafficking, proliferation, and survival (Dhabhar *et al.*, 1995; Gonzalo *et al.*, 1993). Owing to the interaction between the immune and endocrine systems, the endocrine effects of PAE may subsequently affect immune competency, and *vice versa*.

Importantly, data have shown that PAE has marked effects on the offspring immune system. Typically, PAE depresses the lactational transfer of immunity and suppresses development of the neonatal immune system (Steven, Stewart, and Seelig, 1992; Seelig, Steven, and Stewart, 1996), producing long-term immune system deficits, many of which persist into adulthood (Zhang, Sliwowska, and Weinberg, 2005). Studies of children with FAS have found decreased immune competence at the level of both innate and adaptive immunity. Specifically, the incidence of common bacterial and upper respiratory tract infections is increased in children exposed prenatally to alcohol, as is the susceptibility to opportunistic infections (Johnson *et al.*, 1981; Church and Gerkin, 1988). Deficits in immune competence have been confirmed in PAE animal models, with the most marked deficits occurring in adaptive immunity; that is, both T- and B-cell responses to immune challenge and the corresponding cytokine (hormone-like products produced by immune cells) and antibody responses are generally depressed (Seelig, Steven, and Stewart, 1996). Although alcohol exposure in these models is usually limited to the gestational and/or lactational period, it is important to note that many of the immune deficits associated with PAE persist well into adulthood. Thus, alterations in neuroendocrine–immune interactions may be at the root of many of the immune deficiencies associated with PAE. Interestingly, similar to what has been shown in relation to HPA responsiveness, many immune deficits associated with PAE are not detected under basal conditions but rather are unmasked following exposure to stress, further implicating PAE-related dysregulation of the HPA axis in FASD-related immune deficiencies.

In the context of the above discussion, it is possible that enduring alcohol-related immune deficiencies may be mediated by fetal/early programming and/or epigenetic reprogramming of the HPA axis. A decreased immune competence following prenatal alcohol exposure may also be mediated by epigenetic modifications at the level of critical immune system genes. The effects of PAE on the epigenome have yet to be elucidated; however, of relevance is the finding that alcohol-induced immune system impairments may increase across generations (Seelig, Steven, and Stewart, 1999). For example, Seelig and colleagues found that first-generation rats, exposed to alcohol 30 days prior to and during gestation and lactation, did not differ in their immune response to *Trichinella spiralis* larvae compared to controls (Seelig, Steven, and Stewart, 1999). Second-generation PAE animals, however, showed a compromised immune response, with decreased levels of T cells and natural killer cells (Seelig, Steven, and Stewart, 1999). This suggests that ethanol is not acting simply as an immunoteratogen, but rather that alterations in immune competency are set up during development, potentially through epigenetic reprogramming and thus are transmissible to offspring, ultimately decreasing the organism's ability to respond to immune challenges in adulthood (Seelig, Steven, and Stewart, 1999). Owing to the integral link between the neuroendocrine and immune systems, it is plausible that epigenetic reprogramming of the HPA axis and immune system are acting in concert, resulting in some of the pathophysiological changes observed in FASD.

5.4
Prenatal Alcohol Exposure: Early Experience, Stress Responsiveness, and Vulnerability to Depression

Alcohol exposure during development is associated with an increased prevalence of depressive and anxiety disorders in patient populations (Famy, Streissguth, and Unis, 1998; Kodituwakku, 2007; O'Connor and Kasari, 2000; O'Connor and Paley, 2006; O'Connor et al., 2002; O'Connor, O'Halloran, and Shanahan, 2000). *Depression* is a highly prevalent, chronic, and recurring illness, and a leading cause of disability. Although effective pharmacological treatments exist for depression, the pathophysiology of the disease is not fully understood, and therefore the mechanisms that might link PAE and depression remain poorly understood. In human populations, depression and anxiety are highly co-morbid, most likely owing to some similarities in etiology and a partial overlap in neurocircuitry. However, they are distinct psychopathologies that can be dissociated from each other, both behaviorally and neurobiologically, and that can be assessed through specific behavioral tests in animal models (Hellemans et al., 2010).

Stress is thought to be a major contributor to the development of psychiatric illnesses, and dysregulation of the HPA axis has been hypothesized to play a central role in the pathogenesis of depression (Holsboer, 2000; Raison and Miller, 2003; Nestler et al., 2002). Initially described over 50 years ago, the origins of HPA dysfunction are evident at multiple structural and functional levels of HPA axis regulation. Dysfunction or dysregulation is manifest as elevated basal cortisol level, a flattening of the diurnal rhythm, and an increased cortisol release following dexamethasone suppression (i.e., dexamethasone, a synthetic glucocorticoid, is less effective in depressed patients than in controls at suppressing HPA activity) (Board, Wadeson, and Persky, 1957; Carroll et al., 1981; Yehuda et al., 1996; Deuschle et al., 1997). Of particular relevance is the finding that prenatal exposure to alcohol results in HPA dysregulation that parallels, in many respects, that seen in patients with depression. However, this is not to suggest that stress *per se* causes depression. Rather, in the context of the Stress Diathesis Hypothesis, it has been suggested that exposure to stressors over the life course may lead to a maladaptive cascade of events and an increased vulnerability to depression, but *only* in the context of an already sensitized HPA axis (Gutman and Nemeroff, 2003; Swaab, Bao, and Lucassen, 2005). Ongoing studies conducted by the present authors include an examination of how exposure to alcohol *in utero* might sensitize the stress system, resulting in hyper-reactivity to subsequent stressful events and an increased vulnerability to depressive-/anxiety-like and substance-use disorders in adulthood.

5.4.1
Interactions Between Central Monoaminergic Neurotransmitters and the HPA Axis

Evidence from patient populations has suggested that dysfunction in monoaminergic [serotonin (5-HT), noradrenaline (NA) and dopamine (DA)] neurotransmitter systems is an underlying pathophysiological feature of depression (Nash *et al.*, 2008; Nutt and Stein, 2006; Bowden *et al.*, 1997; Mann and Malone, 1997; Trivedi *et al.*, 2008; D'Haenen and Bossuyt, 1994). Similarly, abnormalities in the 5-HT, NA, and DA systems have been observed in PAE animal models. For example, decreased brain concentrations of 5-HT, NA, DA or their metabolites have been reported during fetal life and in weanling animals (Druse, Kuo, and Tajuddin, 1991; Druse *et al.*, 1990; Rathbun and Druse, 1985; Blanchard *et al.*, 1993). PAE may also permanently alter 5-HT and DA transmission in discrete brain regions (Blanchard *et al.*, 1993; Zafar *et al.*, 2000). For example, fewer 5-HT neurons and a lower density of 5-HT fibers in the dorsal raphe (Sari and Zhou, 2004; Zhou *et al.*, 2001; Zhou, Sari, and Powrozek, 2005), as well as altered levels of 5-HT$_{1A}$ receptors in the hippocampus have been reported in PAE animals (Sliwowska *et al.*, 2008). Importantly, the prenatal administration of 5-HT1A receptor agonists to pregnant dams has been shown to ameliorate some of the serotonergic deficits resulting from PAE (Tajuddin and Druse, 1993, 2001). On a functional level, the data obtained by the present authors and others have shown that PAE results in physiological and behavioral abnormalities consistent with altered 5-HT and DA receptor function (Hofmann *et al.*, 2002; Hannigan, 1990).

The symptoms of depression do not appear to be mediated by a single, localized brain region, but rather are indicative of dysfunction in a number of different regions that interact to form a neural network (George *et al.*, 1995; Mayberg *et al.*, 1999). A shared characteristic of 5-HT, NA and DA pathways is their modulating action on brain structures within the network, which appear to play a role in the regulation of emotional and cognitive processes (Mayberg *et al.*, 1999; Swerdlow and Koob, 1987). The PFC is a key brain region thought to subserve the emotional and cognitive manifestations of depression; indeed, it is thought that low mood may involve a lack of serotonin and DA signaling, particularly in the medial PFC. The PFC is richly innervated by 5-HT, NA and DA, and increasing 5-HT levels in the frontal cortex through drugs, such as the selective serotonin reuptake inhibitors (SSRIs), can lead to elevations in mood (Bosker, Klompmakers, and Westenberg, 1995; Delgado *et al.*, 1999; Dawson *et al.*, 2000). Similarly, atypical antipsychotics which increase dopaminergic neurotransmission in the prefrontal circuitry are effective mood stabilizers with antidepressant effects (Pira, Mongeau, and Pani, 2004).

Hormones involved in the stress system, such as glucocorticoids, have been found to modulate prefrontal 5-HT and DA neurotransmission, and may provide a common mechanism mediating alterations in neurotransmitter levels (Dinan,

2005). Studies using animal models have demonstrated that chronic stress appears to lower 5-HT and DA levels in the forebrain (Minton et al., 2009; Mizoguchi et al., 2002). Thus, dysfunctional 5-HT and DA neurotransmission could potentially result in prefrontal hypoactivity and subsequent alterations in emotional processing.

5.4.2
FASD, Stress, and Depression

Whereas, links between the HPA axis and depression in children with FASD are not established, animal research exploring this relationship continues to provide data to support a role for stress in the development of depressive-/anxiety-like behaviors following PAE. A series of studies at the authors' laboratory, undertaken in the context of the Stress Diathesis Hypothesis, has tested the hypothesis that HPA programming by prenatal exposure to alcohol sensitizes the HPA system and confers an increased susceptibility to the development of depressive-/anxiety-like disorders if stressors are encountered later in life. Thus, animals from PAE and control conditions were exposed to a modified chronic mild stress (CMS) protocol in adulthood. CMS consisted of exposure for 10 days, twice daily, at random times of day, to a series of mild psychological stressors. Following CMS, the animals were tested in a battery of behavioral tests designed to measure depressive-/anxiety-like behaviors (O'Connor et al., 2002). The study results showed that the modified CMS procedure was effective in increasing depressive-/anxiety-like behaviors in all animals. Importantly, however, the PAE animals showed markedly increased depressive- and anxiety-like behaviors relative to their control counterparts, and did so in a sexually dimorphic manner. PAE males exposed to CMS showed greater anxiety (elevated plus maze), an impaired hedonic responsivity (sucrose contrast test), locomotor hyperactivity (open field), and alterations in affiliative and non-affiliative social behaviors (social interaction) compared to control males. By contrast, while PAE females were similar to males in showing greater anxiety (elevated plus maze) and altered social interactions, they also showed greater levels of "behavioral despair" (Porsolt forced swim test) compared to their control counterparts. Importantly, even on tasks where both PAE males and females showed deficits, they differed in how these were manifested (Hellemans et al., 2008, 2010). These findings support and extend the studies of Redei and colleagues (Slone and Redei, 2002; Wilcoxon et al., 2005), who showed that PAE increased "behavioral despair" in the Porsolt forced swim test. Taken together, these results suggest that fetal programming by PAE permanently alters sensitivity to stressors, and thus may be a predisposing factor for the prevalence of mood disorders in FASD populations.

The PFC may be one brain area that provides a link between stress and some of the behavioral alterations observed in both depression and FASD. In addition to a role in regulating emotion, the PFC is associated with working memory and executive functions (Dalley, Cardinal, and Robbins, 2004). Deficits in PFC-mediated working memory and executive functions are characteristic of patients with both bipolar and unipolar affective disorders (DeBattista, 2005; Porter, Bourke, and Gallagher, 2007), as well as individuals with FASD (Astley et al., 2009a; Vaurio, Riley, and Mattson, 2008). Furthermore, imaging studies have revealed abnormalities in PFC function, both in patients with depression and in those with FASD (Astley et al., 2009b; Blumberg et al., 2003). Dopamine release plays an important role in mediating and modulating cognition in the PFC (Arnsten, 1997; Floresco and Magyar, 2006), and therefore stress-induced alterations in DA may be one factor mediating the neurocognitive deficits frequently observed in both depression and FASD.

5.4.3
Prevention and Treatment of Depression in FASD Populations

In view of the increased rates of mental health problems, with high levels of depression/anxiety and substance use disorders in children and adults with FASD, it is clinically important to identify and to understand the mechanisms by which stressors modify behavioral and physiological phenotypes, leading to an increased vulnerability in those affected individuals. The present authors' data provide support for the Stress Diathesis Hypothesis as a potential explanation for the prevalence of mood disorders in FASD populations. It is suggested that prenatal exposure to alcohol can be viewed as an adverse early life experience that programs neurobiological and neuroadaptive mechanisms, such that vulnerability to subsequent life stressors is increased and, in turn, the vulnerability to depressive and anxiety disorders is also increased. Animal models of FASD will no doubt play a central role in identifying pathophysiological mechanisms involved in both FASD and depressive disorders, and provide a basis for the development of new therapeutic treatments.

Of particular relevance to this discussion, an important line of studies aimed at improving treatment of depression in patient populations is targeting of the HPA axis. There is evidence that normalization of the HPA disturbances typically seen in major depression may be a prerequisite for successful antidepressant treatment (Brouwer et al., 2006; Ising, 2007; Young et al., 2004a). Consistent with this, glucocorticoids interfere with the ability of SSRIs to elevate 5-HT in the frontal cortex, the proposed mechanism of antidepressant action (Gartside, Leitch, and Young, 2003), and indeed it was shown that treatment with a selective glucocorticoid receptor antagonist would reverse these alterations (Johnson et al., 2007). Recent pre-clinical and clinical studies have examined the potential use of anti-glucocorticoid drug augmentation as a method of improving treatment response in severe depression, with promising results (Young et al., 2004b; Flores et al., 2006). For example, in a study examining mifepristone (a GR antagonist) in the treatment of bipolar depression, Young et al. (2004b) found that mifepristone-treated patients experienced significantly greater improvements in cognition (working spatial memory) and mood when compared to placebo-treated patients. In addition, the drug appeared to be well tolerated. Although further research is clearly needed, it is possible that treatment with a selective GR antagonist may provide a future therapeutic strategy in the treatment of mood disorders in general, and in FASD in particular. Studies are currently under way at the authors' laboratory to determine the effect of GR antagonists in an FASD animal model, which will allow for a careful examination of the possible role of GRs in modifying behavioral and physiological phenotypes.

5.5
FASD and Substance Abuse

The prevalence of substance-use disorders is higher in individuals with FASD than in the general population (Baer et al., 1998). It is hypothesized that *in utero* alcohol exposure may cause neurobiological alterations that increase vulnerability to substance-use disorders; however, systematic research to further clarify this relationship in humans is lacking. Importantly, animal models are now providing some insight into the relationship between PAE and substance abuse. For example, the majority of studies using animal models of PAE and alcohol addiction have reported an increased alcohol consumption in adulthood (for review, see Chotro, Arias, and Laviola, 2007). This supports the proposal that neurobiological alterations caused by PAE could contribute, at least in part, to the high prevalence of substance-use disorders in children and adults with FASD. Moreover, although environmental, social, and genetic factors are known to play a role in the development of substance-use disorders (Li et al., 2001; DeRijk and Kloet, 2008), PAE was shown to be a better predictor of problems related to alcohol abuse than was a family history of alcohol abuse or prenatal exposure to other substances (Baer et al., 1998). This evidence further supports the hypothesis that alterations caused by PAE lead to an increased vulnerability to substance-use problems.

5.5.1
Neurobiology of Addiction

There is a large body of research on the neurobiological systems underlying reward and motivation. The *mesocorticolimbic pathway*, one of the major dopaminergic pathways in the brain, is an important part of the neural circuitry involved in drug reward, and is involved in functions, such as motivation, memory, reward, and emotion. All drugs of abuse have been found to increase DA levels or to activate DA neurons along the mesocorticolimbic pathway (Koob and Le Moal, 2008a), making DA a central component to drug reward. Glutamate is another neurotransmitter essential to the rewarding properties of drugs. The glutamate-mediated activation of dopaminergic neurons facilitates DA release, which increases both the motivation to use cocaine (Allen et al., 2007; Kalivas, 2007) and the behavioral response to cocaine (Schenk et al., 1993; Cervo and Samanin, 1995).

Of relevance, studies in animal models have shown that PAE alters the activity of the mesocorticolimbic DA reward system (Blanchard et al., 1993; Carneiro et al., 2005), and changes the activity of glutamate-containing neurons (Dettmer et al., 2003). However, relatively few studies have been conducted to determine how PAE alters these key neurotransmitters, or to identify the link between altered brain circuitry and increased substance-use behavior following PAE.

5.5.2
Stress and Substance Use

There is growing evidence to support the hypothesis that stress interacts with the neurobiological pathways implicated in drug reward. (Koob and Kreek, 2007; Koob and Le Moal, 2008a; Koob, 2008). For example, both, stress and acute drug use can increase extracellular DA levels and the excitability of DA neurons in the mesocorticolimbic pathway (Rouge-Pont et al., 1995; Saal et al., 2003). Furthermore, stress activates the mesocorticolimbic dopaminergic system, and substance abuse causes a stress response in a nondependent user. Thus, dysregulation of the stress system can alter the mesocorticolimbic system, and *vice versa*.

Currently, one of the leading theories of addiction centers on the interaction between the stress system and the mesocorticolimbic dopaminergic system (for a review, see Koob and Le Moal, 2008b). It has been suggested that perturbations in the stress system are observable throughout the transition from casual to uncontrollable and compulsive drug-taking behavior, which is a hallmark of addiction. CRH, the key hormone that drives the stress response, is known to be released from brain areas outside the hypothalamus. Substance-induced alterations in extrahypothalamic CRH release that are specific to various stages of drug use (i.e., anticipation, binge, abstinence) have been described (Koob and Kreek, 2007; Koob and Le Moal, 2008a; Koob, 2009), and appear to be a central component in the development of addiction.

Circulating levels of glucocorticoids have been shown to reflect these central alterations in CRH release in response to substance use, and may provide further insight into vulnerability to addiction. For example, in rats, basal CORT levels correlate with amount of self-administered drug; acute drug use activates, whereas chronic use attenuates HPA activity. Furthermore, CRH is released with an escalation of drug intake during self-administration (Sarnyai, Shaham, and Heinrichs, 2001). Dysregulation of the HPA axis has also been observed with alcohol consumption; HPA activity increases with acute alcohol consumption, but decreases with repeated exposure (Brick and Pohorecky, 1983; Lee et al., 2001; Adinoff et al., 1996). Similarly, increased alcohol consumption, craving, and relapse to drinking have been reported in abstinent alcoholics with high levels of stress and anxiety (Hore, 1971; Kushner, Sher, and Beitman, 1990; Wit, 1996), while periods of heavy drug use can cause a persistent activation of the HPA axis (Lovallo, 2006). Chronic alcohol consumption can also result in a loss of the diurnal rhythm of cortisol secretion, which can be regained at one to four weeks into abstinence (Adinoff et al., 2005a, 2005b, 2005c). Importantly, the magnitude of the decrease in CORT secretion following cessation of use may be predictive of a propensity to relapse, and suggests that individual differences in vulnerability to substance use are indeed related to HPA dysregulation (Lovallo, 2006). Moreover, it has been shown that stress sensitizes healthy individuals to the rewarding effects of drugs (i.e., the response to the drug is increasingly amplified), and can induce relapse after abstinence (Sarnyai, Shaham, and Heinrichs, 2001), further suggesting that HPA alterations provide a pathway for increased vulnerability to addiction and relapse. In the context of this discussion, it can be speculated that dysregulation of the stress system most likely contributes to the increase in vulnerability to substance-use problems among individuals with FASD. However, it is important to note that exposure to stressors does not always have an adverse outcome. The effects of stress on the voluntary consumption of rewarding substances depend on the type of stressor (e.g., predictability, controllability and psychological characteristics) (Pacak et al., 2002). Similarly, mental health disorders, such as depression and addiction, are typically linked to unpredictable, uncontrollable psychological stressors, whereas predictable and controllable stressors can have favorable effects on resiliency (for a review, see Miller, Chen, and Zhou, 2007).

The effects of PAE and stress on vulnerability to drug intake, as well as alterations in the interaction between the stress and reward systems remain understudied areas of research. Clearly, further investigation is needed in individuals with FASD and PAE animal models in order to provide new insights into the prevention, intervention, and treatment of substance-use disorders in this vulnerable population.

5.6
Summary and Policy Considerations

In this chapter, a selective review has been provided of the mechanisms underlying the effects of PAE on behavioral and physiological function. Attention has been focused primarily on animal models, which provide unique tools to investigate both the direct and indirect mechanisms that underlie alcohol's adverse effects on the brain and behavior. In particular, the discussion has been focused on the endocrine mediation of alcohol's effects on the fetus, which is the major area of research in the authors' laboratory.

It is well established that experience can have profound effects on the CNS, particularly early in life when neurobiological systems are still developing and maturing. Indeed, rather than the static process it was once believed to be, brain development is now viewed as an ongoing process that is continuously modified and molded by experiential factors. Moreover, environmental factors that influence brain development at one point will have the potential to shape subsequent developmental processes. In this manner, the final phenotype may be "stamped" or imprinted by early life events, allowing the brain to be adapted to the environment through experience. For this reason, although brain plasticity – the ability of the brain to be modified by experience – certainly extends well into later life, experience appears to have a much more profound and persistent influence on the developing brain.

Of particular relevance, many reports have pointed towards a relationship between depression in adulthood and adverse early life events (Gutman and Nemeroff, 2003; Nemeroff and Vale, 2005). Adverse events can include severe early life trauma, such as physical or sexual abuse or neglect, or less severe events, such as family overcrowding, poor parenting, or marital or family instability. Furthermore, it has been suggested that fetal or early life programming of the HPA axis may provide a final common pathway for early life experiences. It is suggested that prenatal exposure to alcohol can be viewed as an adverse early life experience that programs the neurobiological and neuroadaptive mechanisms involved in regulation of the stress response. Through this programming, vulnerability to subsequent life stressors is increased and, in turn, the vulnerability to illnesses or disorders later in life – including mental health disorders such as depression, anxiety or substance-use disorders – is also increased. Moreover, there is growing support for the possibility that epigenetic mechanisms similar to those described by Weaver and colleagues (Weaver, 2007) could mediate the changes induced by alcohol exposure *in utero*. Further research is needed to elucidate how the interactions between stress and early life history influence vulnerability/resilience to mental health problems in individuals with FASD.

The research reviewed in this chapter suggests that, owing to pre-existing vulnerabilities within the FASD population, policy decisions will differentially impact individuals with FASD compared to individuals with other developmental problems. This unique population exhibits behavioral sequelae that have a physiological and neurobiological basis. Policies based on an understanding of the neurobiological alterations caused by PAE and the specific challenges faced by this unique population will result in the development of more targeted and more effective prevention, intervention, and treatment programs. For example, it is possible that targeting HPA dysregulation through a combination of pharmaceutical and behavioral therapies could improve the quality of life and have a significant impact on behavioral and mental health problems. Importantly, supportive environments, including appropriate support services for individuals with FASD and their caregivers, are critical in enabling FASD individuals to function in society.

Acknowledgments

The research discussed in this chapter was supported by grants from: NIH/NIAA AA007789, NIH/NIMH MH081797 and Coast Capital Savings Depression Research Fund to J.W., Canadian Foundation for Fetal Alcohol Research to J.W. and L.A.M. Galea. K.A.U., J.H.S. and K.B. were supported by IMPART (CIHR Training Grant).

Glossary

Adaptive immune system: includes highly specialized cells and mechanisms involved in the targeted elimination of pathogens. Adaptive immunity takes approximately 3–7 days to become activated; however, unlike innate immunity, it is longlasting and results in an "immunological memory," which allows the body to respond quickly to a second infection by the same pathogen. The adaptive immune system can be divided into two main branches: (i) cell-mediated immunity, which involves the activation of immune cells (e.g., macrophages and T lymphocytes); and (ii) humoral immunity, which is mediated by B lymphocytes and secreted antibodies.

Amygdala: a region located within the medial temporal lobes of the brain. This region is part of the limbic system, and is heavily implicated in the processing and memory of emotional and/or stressful stimuli.

Antibodies: proteins secreted by B cells that bind to and neutralize foreign particles, such as viral, bacterial, and fungal components. Antibodies contain a hypervariable region, which allows for the production of unique antibody variants (each having different hypervariable regions) to selectively bind to a wide array of antigens, targeting them for destruction by the immune system.

B lymphocyte: a type of white blood cell involved in humoral immunity. These cells are produced in the bone marrow and mature in the spleen; new B cells are continually being formed and develop throughout the organism's lifespan. The principle function of the B cell is to become activated in response to specific antigens, and subsequently to secrete antibodies that target the antigen in question. There are multiple subtypes of B cells, including plasma B cells (secrete antibody directly) and memory B cells (long-lived, respond quickly to a second infection by the same pathogen).

Cerebellum: a region located at the bottom of the brain, near the spinal cord. This brain region is implicated in coordination of movement, as well as cognition, attention, language, and emotion. The cerebellum also integrates inputs from many other regions.

Cytokines: signaling molecules secreted by immune system cells as a means of cellular communication. Cytokines include proteins, peptides and glycoproteins and their defining feature is that they have a wide array of immunomodulating effects. For examples, cytokines can enhance cellular or humoral immune responses, trigger the release of additional cytokines, and induce gene transcription. Cytokines are grouped into different categories such as interleukins (e.g., IL-1a, IL-1 β, IL-2, IL-6), which are secreted by T cells and are mainly involved in T- and B-cell development, interferons (e.g., IFN-α, IFN-β, IFN-γ), which are mainly involved in fighting viruses and tumors, and tumor necrosis factors (e.g., TNF-α), which are involved in triggering cell death.

Epigenetics: the study of stable yet potentially reversible modifications in genome information that occur without altering the underlying DNA sequence, and may be triggered in response to the environment, thus connecting external influences to the genome.

Epigenome: the overall epigenetic state of a cell, which includes features such as DNA methylation and histone modifications.

Glial cells (or glia): non-neuronal cells within the brain that provide support, myelination, nutrients, and oxygen to surrounding cells. These cells also help to maintain homeostasis and facilitate neurotransmission.

Hippocampus: a region located in the medial temporal lobe of the brain. This region belongs to the limbic system and is heavily implicated in spatial navigation, learning, and memory. This region also plays a central role in HPA regulation via a dense population of glucocorticoid receptors.

Histones: proteins involved in the tight packaging of DNA in the nucleus of eukaryotic cells. There are four core histones (H2A, H2B, H3 and H4) that assemble as an octamer to form the nucleosome, which is wrapped by a section of DNA, separated by a linker region. The nucleosome is the fundamental repeated unit involved in the packing of DNA.

Innate immune system: this includes the non-specific cells and mechanisms that act as a first line of defense against pathogens. Innate immunity is a critical component of the immune response, as it is activated immediately prior to the onset of adaptive immunity; however, it is neither longlasting nor protective in future encounters with the same pathogen. Important components of the innate immune system include leukocytes such as macrophages, natural killer cells, neutrophils and mast cells, as well as complement proteins.

Leukocytes: white blood cells; critical cells of the immune system involved in defending the body against pathogens.

Mesocorticolimbic pathway: one of the major dopaminergic pathways in the brain that begins in the brainstem, with projections to limbic system structures such as the amygdala and hippocampus, as well as the mPFC. Behaviors associated with reward/motivation and reinforcement are heavily modulated by dopamine via this pathway.

Methylation: the addition of a methyl group (chemical formula CH_3) to a compound. Methyl groups can be added to cytosine residues of CpG dinucleotides (cytosine nucleotides that are adjacent to guanine nucleotides) or to histone proteins, and this can subsequently alter gene transcription.

Mitochondria: often referred to as the "powerhouses" of cells, mitochondria are found inside cells and produce most of the chemical energy.

Neural crest: a transient component of the developing fetus where neurons and glia of the autonomic nervous system are produced, as well as supporting and hormone-producing cells in certain organs.

Oxidative stress: when a cell is unable to effectively deal with reactive oxygen, the production of free radicals and peroxides can occur, that can damage a cell's proteins, DNA, or lipids.

Phosphorylation: the addition of a phosphate group (chemical formula PO_4) to a compound. The phosphorylation of histones affects DNA packing, gene regulation, and DNA repair.

Pre-frontal cortex (PFC): anterior part of the frontal lobes of the brain. The PFC is composed of several subregions, such as the medial PFC (mPFC; comprised of anterior cingulate, prelimbic and infralimbic divisions of the PFC). The PFC is implicated in executive functioning and expression of personality. The PFC is involved in planning of complex cognitive tasks, including decision making and appropriate social interaction.

Prostaglandins: found in many organs and tissues, the prostaglandins are produced by nucleated cells (except lymphocytes). These lipid mediators are synthesized from fatty acids by enzymes, and act upon several types of autocrine and paracrine cells.

T lymphocyte: a type of white blood cell involved in cell-mediated immunity. These cells mature in the thymus and express unique cell-surface receptors (T-cell receptors), allowing them to bind and recognize antigens. There are many different subtypes of T lymphocytes, including T helper cells, cytotoxic T cells, and memory T cells, each serving important roles in the context of cell-mediated immunity.

11β-Hydroxysteroid dehydrogenase type 2 (11β-HSD-2): an enzyme produced by the placenta that catalyzes the conversion of active glucocorticoids into inactive 11-keto products.

Abbreviations

CMS	chronic mild stress
CNS	central nervous system
CORT	corticosterone
CRH	corticotropin-releasing hormone
DA	dopamine
HPA-axis	hypothalamic–pituitary–adrenal axis
HPG-axis	hypothalamic–pituitary–gonadal axis
mPFC	medial prefrontal cortex
PAE	prenatal alcohol exposure
PFC	prefrontal cortex
PL	prelimbic region of mPFC
PVN	paraventricular nuclei
5-HT	serotonin (5-hydroxytryptamine)
11β-HSD	11β-hydroxysteroid dehydrogenase

References

Adinoff, B., Anton, R., Linnoila, M., Guidotti, A., Nemeroff, C.B., and Bissette, G. (1996) Cerebrospinal fluid concentrations of corticotropin-releasing hormone (CRH) and diazepam-binding inhibitor (DBI) during alcohol withdrawal and abstinence. *Neuropsychopharmacology*, **15** (3), 288–295.

Adinoff, B., Junghanns, K., Kiefer, F., and Krishnan-Sarin, S. (2005a) Suppression of the HPA axis stress-response: implications for relapse. *Alcohol. Clin. Exp. Res.*, **29** (7), 1351–1355.

Adinoff, B., Krebaum, S.R., Chandler, P.A., Ye, W., Brown, M.B., and Williams, M.J. (2005b) Dissection of hypothalamic-pituitary-adrenal axis pathology in 1-month-abstinent alcohol-dependent men, Part 2: response to ovine corticotropin-releasing factor and naloxone. *Alcohol. Clin. Exp. Res.*, **29** (4), 528–537.

Adinoff, B., Krebaum, S.R., Chandler, P.A., Ye, W., Brown, M.B., and Williams, M.J. (2005c) Dissection of hypothalamic-pituitary-adrenal axis pathology in 1-month-abstinent alcohol-dependent men, Part 1: adrenocortical and pituitary glucocorticoid responsiveness. *Alcohol. Clin. Exp. Res.*, **29** (4), 517–527.

Allen, R.M., Uban, K.A., Atwood, E.M., Albeck, D.S., and Yamamoto, D.J. (2007) Continuous intracerebroventricular infusion of the competitive NMDA receptor antagonist, LY235959, facilitates escalation of cocaine self-administration and increases break point for cocaine in Sprague-Dawley rats. *Pharmacol. Biochem. Behav.*, **88** (1), 82–88.

Anderson, R.A., Jr (1981) Endocrine balance as a factor in the etiology of the fetal alcohol syndrome. *Neurobehav. Toxicol. Teratol.*, **3** (2), 89–104.

Angelucci, L., Patacchioli, F.R., Scaccianoce, S., Di Sciullo, A., Cardillo, A., and Maccari, S. (1985) A model for later-life effects of perinatal drug exposure: maternal hormone mediation. *Neurobehav. Toxicol. Teratol.*, **7** (5), 511–517.

Arnsten, A.F. (1997) Catecholamine regulation of the prefrontal cortex. *J. Psychopharmacol.*, **11** (2), 151–162.

Astley, S.J., Olson, H.C., Kerns, K., Brooks, A., Aylward, E.H., Coggins, T.E., Davies, J., Dorn, S., Gendler, B., Jirikowic, T., et al. (2009a) Neuropyschological and behavioral outcomes from a comprehensive magnetic resonance study of children with fetal alcohol spectrum disorders. *Can. J. Clin. Pharmacol.*, **16** (1), e178–e201.

Astley, S.J., Richards, T., Aylward, E.H., Olson, H.C., Kerns, K., Brooks, A., Coggins, T.E., Davies, J., Dorn, S., Gendler, B., et al. (2009b) Magnetic resonance spectroscopy outcomes from a comprehensive magnetic resonance study of children with fetal alcohol spectrum disorders. *Magn. Reson. Imaging*, **27** (6), 760–778.

Baer, J., Barr, H., Bookstein, F., Sampson, P., and Streissguth, A. (1998) Prenatal alcohol exposure and family history of alcoholism in the etiology of adolescent alcohol problems. *J. Stud. Alcohol Drugs*, **59** (5), 533–543.

Baer, J.S., Barr, H.M., Bookstein, F.L., Sampson, P.D., and Streissguth, A.P. (1998) Prenatal alcohol exposure and family history of alcoholism in the etiology of adolescent alcohol problems. *J. Stud. Alcohol*, **59**, 533–543.

Bakker, J.M., van Bel, F., and Heijnen, C.J. (2001) Neonatal glucocorticoids and the developing brain: short-term treatment with life-long consequences? *Trends Neurosci.*, **24** (11), 649–653.

Barbazanges, A., Piazza, P.V., Le Moal, M., and Maccari, S. (1996) Maternal glucocorticoid secretion mediates long-term effects of prenatal stress. *J. Neurosci.*, **16** (12), 3943–3949.

Barker, D.J., Osmond, C., Golding, J., Kuh, D., and Wadsworth, M.E. (1989) Growth in utero, blood pressure in childhood and adult life, and mortality from cardiovascular disease. *Br. Med. J.*, **298** (6673), 564–567.

Bearer, C.F. (2001a) L1 cell adhesion molecule signal cascades: targets for ethanol developmental neurotoxicity. *Neurotoxicology*, **22** (5), 625–633.

Bearer, C.F. (2001b) Mechanisms of brain injury: L1 cell adhesion molecule as a target for ethanol-induced prenatal brain injury. *Semin. Pediatr. Neurol.*, **8** (2), 100–107.

Benediktsson, R., Calder, A.A., Edwards, C.R., and Seckl, J.R. (1997) Placental 11 beta-hydroxysteroid dehydrogenase: a key regulator of fetal glucocorticoid exposure. *Clin. Endocrinol. (Oxf.)*, **46** (2), 161–166.

Blanchard, B.A., Steindorf, S., Wang, S., LeFevre, R., Mankes, R.F., and Glick, S.D. (1993) Prenatal ethanol exposure alters ethanol-induced dopamine release in nucleus accumbens and striatum in male and female rats. *Alcohol. Clin. Exp. Res.*, **17** (5), 974–981.

Blumberg, H.P., Leung, H.C., Skudlarski, P., Lacadie, C.M., Fredericks, C.A., Harris, B.C., Charney, D.S., Gore, J.C., Krystal, J.H., and Peterson, B.S. (2003) A functional magnetic resonance imaging study of bipolar disorder: state- and trait-related dysfunction in ventral prefrontal cortices. *Arch. Gen. Psychiatry*, **60** (6), 601–609.

Board, F., Wadeson, R., and Persky, H. (1957) Depressive affect and endocrine functions; blood levels of adrenal cortex and thyroid hormones in patients suffering from depressive reactions. *AMA Arch. Neurol. Psychiatry*, **78** (6), 612–620.

Bosker, F.J., Klompmakers, A.A., and Westenberg, H.G.M. (1995) Effects of single and repeated oral administration of fluvoxamine on extracellular serotonin in the median raphe nucleus and dorsal hippocampus of the rat. *Neuropharmacology*, **34** (5), 501–508.

Bowden, C., Theodorou, A.E., Cheetham, S.C., Lowther, S., Katona, C.L.E., Rufus Crompton, M., and Horton, R.W. (1997) Dopamine D1 and D2 receptor binding sites in brain samples from depressed suicides and controls. *Brain Res.*, **752** (1–2), 227.

Brick, J. and Pohorecky, L.A. (eds) (1983) *The Neuroendocrine Response to Stress and the Effect of Ethanol*, Elsevier, New York.

Brouwer, J.P., Appelhof, B.C., van Rossum, E.F., Koper, J.W., Fliers, E., Huyser, J., Schene, A.H., Tijssen, J.G., Van Dyck, R., Lamberts, S.W., et al. (2006) Prediction of treatment response by HPA-axis and glucocorticoid receptor polymorphisms in major depression. *Psychoneuroendocrinology*, **31** (10), 1154–1163.

Brown, N.A., Goulding, E.H., and Fabro, S. (1979) Ethanol embryotoxicity: direct effects on mammalian embryos *in vitro*. *Science*, **206** (4418), 573–575.

Bumiller, A., Gotz, F., Rohde, W., and Dorner, G. (1999) Effects of repeated injections of interleukin 1beta or lipopolysaccharide on the HPA axis in the newborn rat. *Cytokine*, **11** (3), 225–230.

Cadet, R., Pradier, P., Dalle, M., and Delost, P. (1986) Effects of prenatal maternal stress on the pituitary adrenocortical reactivity in guinea-pig pups. *J. Dev. Physiol.*, **8** (6), 467–475.

Carneiro, L.M.V., Diógenes, J.P.L., Vasconcelos, S.M.M., Aragão, G.F., Noronha, E.C., Gomes, P.B., and Viana, G.S.B. (2005) Behavioral and neurochemical effects on rat offspring after prenatal exposure to ethanol. *Neurotoxicol. Teratol.*, **27** (4), 585–592.

Carroll, B.J., Feinberg, M., Greden, J.F., Tarika, J., Albala, A.A., Haskett, R.F., James, N.M., Kronfol, Z., Lohr, N., Steiner, M., et al. (1981) A specific laboratory test for the diagnosis of melancholia: standardization, validation, and clinical utility. *Arch. Gen. Psychiatry*, **38** (1), 15–22.

Cartwright, M.M. and Smith, S.M. (1995) Increased cell death and reduced neural crest cell numbers in ethanol-exposed embryos: partial basis for the fetal alcohol syndrome phenotype. *Alcohol. Clin. Exp. Res.*, **19** (2), 378–386.

Cervo, L. and Samanin, R. (1995) Effects of dopaminergic and glutamatergic receptor antagonists on the acquisition and expression of cocaine conditioning place preference. *Brain Res.*, **673** (2), 242–250.

Chesnokova, V. and Melmed, S. (2002) Minireview: neuro-immuno-endocrine modulation of the hypothalamic-pituitary-adrenal (HPA) axis by gp130 signaling molecules. *Endocrinology*, **143** (5), 1571–1574.

Chotro, M.G., Arias, C., and Laviola, G. (2007) Increased ethanol intake after prenatal ethanol exposure: studies with animals. *Neurosci. Biobehav. Rev.*, **31** (2), 181–191.

Chrousos, G.P. (1998) Stressors, stress, and neuroendocrine integration of the adaptive response: the 1997 Hans Selye memorial lecture. Stress of Life: From Molecules to Man. *Ann. N. Y. Acad. Sci.*, **851**, 311–335.

Church, M.W. and Gerkin, K.P. (1988) Hearing disorders in children with fetal alcohol syndrome: findings from case reports. *Pediatrics*, **82** (2), 147–154.

Clarren, S.K. (1986) Neuropathology in fetal alcohol syndrome, in *Alcohol and Brain Development* (ed. J.R. West), Oxford University Press, New York, pp. 158–166.

Conference, NCD (1995) Effect of corticosteroids for fetal maturation and perinatal outcomes. *Am. J. Obstet. Gynecol.*, **173**, 253–344.

Cook, S.C. and Wellman, C.L. (2004) Chronic stress alters dendritic morphology in rat medial prefrontal cortex. *J. Neurobiol.*, **60** (2), 236–248.

Cooney, C.A., Dave, A.A., and Wolff, G.L. (2002) Maternal methyl supplements in mice affect epigenetic variation and DNA methylation of offspring. *J. Nutr.*, **132** (Suppl. 8), S2393–S2400.

Dalley, J.W., Cardinal, R.N., and Robbins, T.W. (2004) Prefrontal executive and cognitive functions in rodents: neural and neurochemical substrates. *Neurosci. Biobehav. Rev.*, **28** (7), 771–784.

Davis, J.R., Li, Y., and Rankin, C.H. (2008) Effects of developmental exposure to ethanol on *Caenorhabditis elegans*. *Alcohol. Clin. Exp. Res.*, **32** (5), 853–867.

Dawson, L.A., Nguyen, H.Q., Smith, D.I., and Schechter, L.E. (2000) Effects of chronic fluoxetine treatment in the presence and absence of (±)pindolol: a microdialysis study. *Br. J. Pharmacol.*, **130** (4), 797–804.

DeBattista, C. (2005) Executive dysfunction in major depressive disorder. *Expert Rev. Neurother.*, **5** (1), 79–83.

de Kloet, E.R. (2000) Stress in the brain. *Eur. J. Pharmacol.*, **405** (1–3), 187.

Delgado, P.L., Miller, H.L., Salomon, R.M., Licinio, J., Krystal, J.H., Moreno, F.A., Heninger, G.R., and Charney, D.S. (1999) Tryptophan-depletion challenge in depressed patients treated with desipramine or fluoxetine: implications for the role of serotonin in the mechanism of antidepressant action. *Biol. Psychiatry*, **46** (2), 212–220.

DeRijk, R.H. and de Kloet, E.R. (2008) Corticosteroid receptor polymorphisms: determinants of vulnerability and resilience. *Eur. J. Pharmacol.*, **583** (2–3), 303–311.

Dettmer, T.S., Barnes, A., Iqbal, U., Bailey, C.D.C., Reynolds, J.N., Brien, J.F., and Valenzuela, C.F. (2003) Chronic prenatal ethanol exposure alters ionotropic glutamate receptor subunit protein levels in the adult guinea pig cerebral cortex. *Alcohol. Clin. Exp. Res.*, **27** (4), 677–681.

Deuschle, M., Schweiger, U., Weber, B., Gotthardt, U., Korner, A., Schmider, J., Standhardt, H., Lammers, C.-H., and Heuser, I. (1997) Diurnal activity and pulsatility of the hypothalamus-pituitary-adrenal system in male depressed patients and healthy controls. *J. Clin. Endocrinol. Metab.*, **82** (1), 234–238.

Dhabhar, F.S. (2002) Stress-induced augmentation of immune function–the role of stress hormones, leukocyte trafficking, and cytokines. *Brain Behav. Immun.*, **16** (6), 785–798.

Dhabhar, F.S., Miller, A.H., McEwen, B.S., and Spencer, R.L. (1995) Effects of stress on immune cell distribution. Dynamics and hormonal mechanisms. *J. Immunol.*, **154** (10), 5511–5527.

D'Haenen, H.A. and Bossuyt, A. (1994) Dopamine D2 receptors in depression measured with single photon emission computed tomography. *Biol. Psychiatry*, **35** (2), 128.

Dinan, T.G. (2005) Stress: the shared common component in major mental illnesses. *Eur. Psychiatry*, **20** (Suppl. 3), S326.

Diorio, D., Viau, V., and Meaney, M.J. (1993) The role of the medial prefrontal cortex (cingulate gyrus) in the regulation of hypothalamic-pituitary-adrenal responses to stress. *J. Neurosci.*, **13** (9), 3839–3847.

Druse, M.J., Tajuddin, N., Kuo, A., and Connerty, M. (1990) Effects of in utero ethanol exposure on the developing dopaminergic system in rats. *J. Neurosci. Res.*, **27** (2), 233–240.

Druse, M.J., Kuo, A., and Tajuddin, N. (1991) Effects of *in utero* ethanol exposure on the developing serotonergic system. *Alcohol. Clin. Exp. Res.*, **15** (4), 678–684.

Esquifino, A.I., Sanchis, R., and Guerri, C. (1986) Effect of prenatal alcohol exposure on sexual maturation of female rat offspring. *Neuroendocrinology*, **44** (4), 483–487.

Famy, C., Streissguth, A.P., and Unis, A.S. (1998) Mental illness in adults with fetal alcohol syndrome or fetal alcohol effects. *Am. J. Psychiatry*, **155** (4), 552–554.

Feldman, S. and Weidenfeld, J. (1999) Glucocorticoid receptor antagonists in the hippocampus modify the negative feedback following neural stimuli. *Brain Res.*, **821** (1), 33–37.

Flores, B.H., Kenna, H., Keller, J., Solvason, H.B., and Schatzberg, A.F. (2006) Clinical and biological effects of mifepristone treatment for psychotic depression. *Neuropsychopharmacology*, **31** (3), 628–636.

Floresco, S.B. and Magyar, O. (2006) Mesocortical dopamine modulation of executive functions: beyond working memory. *Psychopharmacology (Berl.)*, **188** (4), 567–585.

Gabriel, K.I., Glavas, M.M., Ellis, L., and Weinberg, J. (2005) Postnatal handling does not normalize hypothalamic corticotropin-releasing factor mRNA levels in animals prenatally exposed to ethanol. *Brain Res. Dev. Brain Res.*, **157** (1), 74–82.

Gallo, P.V. and Weinberg, J. (1986) Organ growth and cellular development in ethanol-exposed rats. *Alcohol*, **3** (4), 261–267.

Garro, A.J., McBeth, D.L., Lima, V., and Lieber, C.S. (1991) Ethanol consumption inhibits fetal DNA methylation in mice: implications for the fetal alcohol syndrome. *Alcohol. Clin. Exp. Res.*, **15** (3), 395–398.

Gartside, S.E., Leitch, M.M., and Young, A.H. (2003) Altered glucocorticoid rhythm attenuates the ability of a chronic SSRI to elevate forebrain 5-HT: implications for the treatment of depression. *Neuropsychopharmacology*, **28** (9), 1572.

George, M.S., Ketter, T.A., Parekh, P.I., Horwitz, B., Herscovitch, P., and Post, R.M. (1995) Brain activity during transient sadness and happiness in healthy women. *Am. J. Psychiatry*, **152** (3), 341–351.

Glavas, M.M., Ellis, L., Yu, W.K., and Weinberg, J. (2007) Effects of prenatal ethanol exposure on basal limbic-hypothalamic-pituitary-adrenal regulation: role of corticosterone. *Alcohol. Clin. Exp. Res.*, **31** (9), 1598–1610.

Godfrey, K.M. and Barker, D.J. (2000) Fetal nutrition and adult disease. *Am. J. Clin. Nutr.*, **71** (Suppl. 5), S1344–S1352.

Gonzalo, J.A., Gonzalez-Garcia, A., Martinez, C., and Kroemer, G. (1993) Glucocorticoid-mediated control of the activation and clonal deletion of peripheral T cells *in vivo*. *J. Exp. Med.*, **177** (5), 1239–1246.

Goodlett, C.R., Horn, K.H., and Zhou, F.C. (2005) Alcohol teratogenesis: mechanisms of damage and strategies for intervention. *Exp. Biol. Med.*, **230** (6), 394–406.

Gubitosi-Klug, R., Larimer, C.G., and Bearer, C.F. (2007) L1 cell adhesion molecule is neuroprotective of alcohol induced cell death. *Neurotoxicology*, **28** (3), 457–462.

Guerri, C. (1998) Neuroanatomical and neurophysiological mechanisms involved in central nervous system dysfunctions induced by prenatal alcohol exposure. *Alcohol. Clin. Exp. Res.*, **22** (2), 304–312.

Guerri, C., Saez, R., Portoles, M., and Renau-Piqueras, J. (1993) Derangement of astrogliogenesis as a possible mechanism involved in alcohol-induced alterations of central nervous system development. *Alcohol Alcohol. Suppl.*, **2**, 203–208.

Gutman, D.A. and Nemeroff, C.B. (2003) Persistent central nervous system effects of an adverse early environment: clinical and preclinical studies. *Physiol. Behav.*, **79** (3), 471–478.

Haddad, J.J., Saade, N.E., and Safieh-Garabedian, B. (2002) Cytokines and neuro-immune-endocrine interactions: a role for the hypothalamic-pituitary-adrenal revolving axis. *J. Neuroimmunol.*, **133** (1–2), 1–19.

Haley, D.W., Handmaker, N.S., and Lowe, J. (2006) Infant stress reactivity and prenatal alcohol exposure. *Alcohol. Clin. Exp. Res.*, **30** (12), 2055–2064.

Hannigan, J.H. (1990) The ontogeny of SCH 23390-induced catalepsy in male and female rats exposed to ethanol *in utero*. *Alcohol*, **7** (1), 11.

Hellemans, K.G., Verma, P., Yoon, E., Yu, W., and Weinberg, J. (2008) Prenatal alcohol exposure increases vulnerability to stress and anxiety-like disorders in adulthood. *Ann. N. Y. Acad. Sci.*, **1144**, 154–175.

Hellemans, K.G.C., Sliwowska, J.H., Verma, P., and Weinberg, J. (2010) Prenatal alcohol exposure: fetal programming and later life vulnerability to stress, depression and anxiety disorders. *Neurosci. Biobehav. Rev.*, **34** (6), 791–807.

Henderson, G.I., Devi, B.G., Perez, A., and Schenker, S. (1995) In utero ethanol exposure elicits oxidative stress in the rat fetus. *Alcohol. Clin. Exp. Res.*, **19** (3), 714–720.

Herman, J.P. and Cullinan, W.E. (1997) Neurocircuitry of stress: central control of the hypothalamo-pituitary-adrenocortical axis. *Trends Neurosci.*, **20** (2), 78.

Hofmann, C.E., Simms, W., Yu, W.K., and Weinberg, J. (2002) Prenatal ethanol exposure in rats alters serotonergic-mediated behavioral and physiological function. *Psychopharmacology (Berl.)*, **161** (4), 379–386.

Holsboer, F. (2000) The corticosteroid receptor hypothesis of depression. *Neuropsychopharmacology*, **23**, 477.

Hore, B.D. (1971) Factors in alcoholic relapse. *Br. J. Addict.*, **66**, 89–96.

Ikonomidou, C., Bittigau, P., Ishimaru, M.J., Wozniak, D.F., Koch, C., Genz, K., Price, M.T., Stefovska, V., Horster, F., Tenkova, T., *et al.* (2000) Ethanol-induced apoptotic neurodegeneration and fetal alcohol syndrome. *Science*, **287** (5455), 1056–1060.

Ising, M. (2007) Bringing basic and clinical research together to an integrated understanding of psychiatric disorders. *J. Psychiatr. Res.*, **41** (1–2), 1.

Jacobson, S.W., Bihun, J.T., and Chiodo, L.M. (1999) Effects of prenatal alcohol and cocaine exposure on infant cortisol levels. *Dev. Psychopathol.*, **11** (2), 195–208.

Johnson, D.A., Grant, E.J., Ingram, C.D., and Gartside, S.E. (2007) Glucocorticoid receptor antagonists hasten and augment neurochemical responses to a selective serotonin reuptake inhibitor antidepressant. *Biol. Psychiatry*, **62** (11), 1228.

Johnson, S., Knight, R., Marmer, D.J., and Steele, R.W. (1981) Immune deficiency in fetal alcohol syndrome. *Pediatr. Res.*, **15** (6), 908–911.

Jones, K. and Smith, D. (1973) Recognition of the fetal alcohol syndrome in early infancy. *Lancet*, **2**, 999–1001.

Kalivas, P. (2007) Cocaine and amphetamine-like psychostimulants: neurocircuitry and glutamate neuroplasticity. *Dialogues Clin. Neurosci.*, **9** (4), 389–397.

Kobor, M. and Weinberg, J. Epigenetics and FASD in: Fetal Alcohol Spectrum Disorders. *Alcohol. Res. Health.*, **32** (4), in press.

Kodituwakku, P.W. (2007) Defining the behavioral phenotype in children with fetal alcohol spectrum disorders: a review. *Neurosci. Biobehav. Rev.*, **31** (2), 192–201.

Koob, G.F. (2008) A role for brain stress systems in addiction. *Neuron*, **59** (1), 11–34.

Koob, G.F. (2009) Neurobiological substrates for the dark side of compulsivity in addiction. *Neuropharmacology*, **56** (Suppl. 1), 18–31.

Koob, G. and Kreek, M.J. (2007) Stress, dysregulation of drug reward pathways, and the transition to drug dependence. *Am. J. Psychiatry*, **164** (8), 1149–1159.

Koob, G.F. and Le Moal, M. (2008a) Addiction and the brain antireward system. *Annu. Rev. Psychol.*, **59** (1), 29–53.

Koob, G.F. and Le Moal, M. (2008b) Review. Neurobiological mechanisms for opponent motivational processes in addiction. *Philos. Trans. R. Soc. Lond. B Biol. Sci.*, **363** (1507), 3113–3123.

Kotch, L.E., Chen, S.Y., and Sulik, K.K. (1995) Ethanol-induced teratogenesis: free radical damage as a possible mechanism. *Teratology*, **52** (3), 128–136.

Kushner, M.G., Sher, K.L., and Beitman, B.D. (1990) The relation between alcohol problems and anxiety disorders. *Am. J. Psychiatry*, 685–695.

Lan, N., Yamashita, F., Halpert, A.G., Ellis, L., Yu, W.K., Viau, V., and Weinberg, J. (2006) Prenatal ethanol exposure alters the effects of gonadectomy on hypothalamic-pituitary-adrenal activity in male rats. *J. Neuroendocrinol.*, **18** (9), 672–684.

Lan, N., Yamashita, F., Halpert, A.G., Sliwowska, J.H., Viau, V., and Weinberg, J. (2009) Effects of prenatal ethanol exposure on hypothalamic-pituitary-adrenal function across the estrous cycle. *Alcohol. Clin. Exp. Res.*, **33** (6), 1075–1088.

LeBel, C.P. and Bondy, S.C. (1991) Oxygen radicals: common mediators of neurotoxicity. *Neurotoxicol. Teratol.*, **13** (3), 341–346.

Lee, S. and Rivier, C. (1996) Gender differences in the effect of prenatal alcohol exposure on the hypothalamic-pituitary-adrenal axis response to immune signals. *Psychoneuroendocrinology*, **21** (2), 145–155.

Lee, S., Imaki, T., Vale, W., and Rivier, C. (1990) Effect of prenatal exposure to ethanol on the activity of the hypothalamic-pituitary-adrenal axis of the offspring: importance of the time of exposure to ethanol and possible modulating mechanisms. *Mol. Cell. Neurosci.*, **1** (2), 168–177.

Lee, S., Schmidt, D., Tilders, F., and Rivier, C. (2000) Increased activity of the hypothalamic-pituitary-adrenal axis of rats exposed to alcohol in utero: role of altered pituitary and hypothalamic function. *Mol. Cell. Neurosci.*, **16** (4), 515–528.

Lee, S., Schmidt, E.D., Tilders, F.J., and Rivier, C. (2001) Effect of repeated exposure to alcohol on the response of the hypothalamic-pituitary-adrenal axis of the rat. *Alcohol. Clin. Exp. Res.*, **25**, 98–105.

Lemoine, P., Harousseau, H., Borteyu, J.P., and Menuet, J.C. (1968) Children of alcoholic parents: abnormalities observed in 127 cases. *Ouest Med.*, **8**, 476–482.

Li, T.-K., Spanagel, R., Colombo, G., McBride, W.J., Porrino, L.J., Suzuki, T., and Rodd-Henricks, Z.A. (2001) Alcohol reinforcement and voluntary ethanol consumption. *Alcohol. Clin. Exp. Res.*, **25** (Suppl. 1), 117S–126S.

Lovallo, W.R. (2006) Cortisol secretion patterns in addiction and addiction risk. *Int. J. Psychophysiol.*, **59** (3), 195–202.

McEwen, B.S. (2007) Physiology and neurobiology of stress and adaptation: central role of the brain. *Physiol. Rev.*, **87** (3), 873–904.

McEwen, B.S., De Kloet, E.R., and Rostene, W. (1986) Adrenal steroid receptors and actions in the nervous system. *Physiol. Rev.*, **66** (4), 1121–1188.

McGivern, R.F., Handa, R.J., and Raum, W.J. (1998) Ethanol exposure during the last week of gestation in the rat: inhibition of the prenatal testosterone surge in males without long-term alterations in sex behavior. *Neurotoxicol. Teratol.*, **20** (4), 483–490.

Mann, J.J. and Malone, K.M. (1997) Cerebrospinal fluid amines and higher-lethality suicide attempts in depressed inpatients. *Biol. Psychiatry*, **41** (2), 162.

Matthews, S.G. (2002) Early programming of the hypothalamo-pituitary-adrenal axis. *Trends Endocrinol. Metab.*, **13** (9), 373–380.

Mayberg, H.S., Liotti, M., Brannan, S.K., McGinnis, S., Mahurin, R.K., Jerabek, P.A., Silva, J.A., Tekell, J.L., Martin, C.C., Lancaster, J.L., et al. (1999) Reciprocal limbic-cortical function and negative mood: converging PET findings in depression and normal sadness. *Am. J. Psychiatry*, **156** (5), 675–682.

Meaney, M.J., Szyf, M., and Seckl, J.R. (2007) Epigenetic mechanisms of perinatal programming of hypothalamic-pituitary-adrenal function and health. *Trends Mol. Med.*, **13** (7), 269–277.

Michaelis, E.K. (1990) Fetal alcohol exposure: cellular toxicity and molecular events involved in toxicity. *Alcohol. Clin. Exp. Res.*, **14** (6), 819–826.

Miller, G.E., Chen, E., and Zhou, E.S. (2007) If it goes up, must it come down? Chronic stress and the hypothalamic-pituitary-adrenocortical axis in humans. *Psychol. Bull.*, **133** (1), 25–45.

Miller, M.W. (1992) The effects of prenatal exposure to ethanol on cell proliferation and neuronal migration, in *Development of the Central Nervous System: Effects of Alcohol and Opiates* (ed. M.W. Miller), Alan R. Liss, New York, pp. 47–69.

Minton, G.O., Young, A.H., McQuade, R., Fairchild, G., Ingram, C.D., and Gartside, S.E. (2009) Profound changes in dopaminergic neurotransmission in the prefrontal cortex in response to flattening of the diurnal glucocorticoid rhythm: implications for bipolar disorder. *Neuropsychopharmacology*, **34** (10), 2265.

Mizoguchi, K., Yuzurihara, M., Ishige, A., Sasaki, H., and Tabira, T. (2002) Chronic stress impairs rotarod performance in rats: implications for depressive state. *Pharmacol. Biochem. Behav.*, **71** (1–2), 79.

Nash, J.R., Sargent, P.A., Rabiner, E.A., Hood, S.D., Argyropoulos, S.V., Potokar, J.P., Grasby, P.M., and Nutt, D.J. (2008) Serotonin 5-HT1A receptor binding in people with panic disorder: positron emission tomography study. *Br. J. Psychiatry*, **193** (3), 229–234.

Nemeroff, C.B. and Vale, W.W. (2005) The neurobiology of depression: inroads to treatment and new drug discovery. *J. Clin. Psychiatry*, **66** (Suppl. 7), 5–13.

Nestler, E.J., Barrot, M., DiLeone, R.J., Eisch, A.J., Gold, S.J., and Monteggia, L.M. (2002) Neurobiology of depression. *Neuron*, **34** (1), 13.

Newbold, R.R., Padilla-Banks, E., and Jefferson, W.N. (2006) Adverse effects of the model environmental estrogen diethylstilbestrol are transmitted to subsequent generations. *Endocrinology*, **147** (Suppl. 6), S11–S17.

Nutt, D.J. and Stein, D.J. (2006) Understanding the neurobiology of comorbidity in anxiety disorders. *CNS. Spectrums*, **11** (10, Suppl. 12), 13–20.

O'Connor, M.J. and Kasari, C. (2000) Prenatal alcohol exposure and depressive features in children. *Alcohol. Clin. Exp. Res.*, **24** (7), 1084–1092.

O'Connor, M.J. and Paley, B. (2006) The relationship of prenatal alcohol exposure and the postnatal environment to child depressive symptoms. *J. Pediatr. Psychol.*, **31** (1), 50–64.

O'Connor, M.J., Shah, B., Whaley, S., Cronin, P., Gunderson, B., and Graham, J. (2002) Psychiatric illness in a clinical sample of children with prenatal alcohol exposure. *Am. J. Drug Alcohol Abuse*, **28** (4), 743–754.

O'Connor, T.M., O'Halloran, D.J., and Shanahan, F. (2000) The stress response and the hypothalamic-pituitary-adrenal axis: from molecule to melancholia. *Q. J. Med.*, **93** (6), 323–333.

O'Neil, R., Lan, N., Innis, S., Devlin, A., Ellis, L., Chan, B., and Weinberg, J. (2007) Metabolic effects of prenatal ethanol exposure and epigenetic reprogramming of the HPA axis. *Alcohol. Clin. Exp. Res.*, **31**, 100A.

Pacak, K., Tjurmina, O., Palkovits, M., Goldstein, D.S., Coch, C.A., Hoff, T., Goldsmith, P., and Chrousos, G.P. (eds) (2002) *Chronic Hypercortisolemia Inhibits Dopaminergic Activity in the Nucleus Accumbens*, Taylor & Francis Inc., New York.

Pennington, S.N., Boyd, J.W., Kalmus, G.W., and Wilson, R.W. (1983) The molecular mechanism of fetal alcohol syndrome (FAS). I. Ethanol-induced growth suppression. *Neurobehav. Toxicol. Teratol.*, **5** (2), 259–262.

Perez-Cruz, C., Muller-Keuker, J.I., Heilbronner, U., Fuchs, E., and Flugge, G. (2007) Morphology of pyramidal neurons in the rat prefrontal cortex: lateralized dendritic remodeling by chronic stress. *Neural. Plast.*, **2007**, 46276.

Phillips, D.I., Barker, D.J., Fall, C.H., Seckl, J.R., Whorwood, C.B., Wood, P.J., and Walker, B.R. (1998) Elevated plasma cortisol concentrations: a link between low birth weight and the insulin resistance syndrome? *J. Clin. Endocrinol. Metab.*, **83** (3), 757–760.

Phillips, D.I., Walker, B.R., Reynolds, R.M., Flanagan, D.E., Wood, P.J., Osmond, C., Barker, D.J., and Whorwood, C.B. (2000) Low birth weight predicts elevated plasma cortisol concentrations in adults from 3 populations. *Hypertension*, **35** (6), 1301–1306.

Pira, L., Mongeau, R., and Pani, L. (2004) The atypical antipsychotic quetiapine increases both noradrenaline and dopamine release in the rat prefrontal cortex. *Eur. J. Pharmacol.*, **504** (1–2), 61.

Porter, R.J., Bourke, C., and Gallagher, P. (2007) Neuropsychological impairment in major depression: its nature, origin and clinical significance. *Aust. N. Z. J. Psychiatry*, **41** (2), 115–128.

Radley, J.J., Rocher, A.B., Miller, M., Janssen, W.G.M., Liston, C., Hof, P.R., McEwen, B.S., and Morrison, J.H. (2006) Repeated stress induces dendritic spine loss in the rat medial prefrontal cortex. *Cereb. Cortex*, **16** (3), 313–320.

Radley, J.J., Rocher, A.B., Rodriguez, A., Ehlenberger, D.B., Dammann, M., McEwen, B.S., Morrison, J.H., Wearne, S.L., and Hof, P.R. (2008) Repeated stress alters dendritic spine morphology in the rat medial prefrontal cortex. *J. Comp. Neurol.*, **507** (1), 1141–1150.

Raison, C.L. and Miller, A.H. (2003) When not enough is too much: the role of insufficient glucocorticoid signaling in the pathophysiology of stress-related disorders. *Am. J. Psychiatry*, **160** (9), 1554–1565.

Ramanathan, R., Wilkemeyer, M.F., Mittal, B., Perides, G., and Charness, M.E. (1996) Alcohol inhibits cell-cell adhesion mediated by human L1 [published erratum appears in *J. Cell Biol.*, 1996; 133 (5): 1139–1140. *J. Cell Biol.*, **133** (2), 381–390.

Ramsay, D.S., Bendersky, M.I., and Lewis, M. (1996) Effect of prenatal alcohol and cigarette exposure on two- and six-month-old infants' adrenocortical reactivity to stress. *J. Pediatr. Psychol.*, **21** (6), 833–840.

Randall, C.L., Anton, R.F., and Becker, H.C. (1987) Alcohol, pregnancy, and prostaglandins. *Alcohol. Clin. Exp. Res.*, **11** (1), 32–36.

Randall, C.L., Anton, R.F., Becker, H.C., and White, N.M. (1989) Role of prostaglandins in alcohol teratogenesis. *Ann. N. Y. Acad. Sci.*, **562**, 178–182.

Randall, C.L., Ekblad, U., and Anton, R.F. (1990) Perspectives on the pathophysiology of fetal alcohol syndrome. *Alcohol. Clin. Exp. Res.*, **14** (6), 807–812.

Rathbun, W. and Druse, M.J. (1985) Dopamine, serotonin and acid metabolites in brain regions from the developing offspring of ethanol-treated rats. *J. Neurochem.*, **44**, 57–62.

Rawat, A.K. (1975) Ribosomal proteins synthesis in the fetal and neonatal rat brain as influenced by maternal ethanol consumption. *Res. Commun. Chem. Pathol. Pharmacol.*, **12** (4), 723–732.

Redei, E., Clark, W.R., and McGivern, R.F. (1989) Alcohol exposure in utero results in diminished T-cell function and alterations in brain corticotropin-releasing factor and ACTH content. *Alcohol. Clin. Exp. Res.*, **13** (3), 439–443.

Redei, E., Halasz, I., Li, L.F., Prystowsky, M.B., and Aird, F. (1993) Maternal adrenalectomy alters the immune and endocrine functions of fetal alcohol-exposed male offspring. *Endocrinology*, **133** (2), 452–460.

Rouge-Pont, F., Marinelli, M., Le Moal, M., Simon, H., and Piazza, P.V. (1995) Stress-induced sensitization and glucocorticoids. II. Sensitization of the increase in extracellular dopamine induced by cocaine depends on stress-induced corticosterone secretion. *J. Neurosci.*, **15** (11), 7189–7195.

Rudeen, P. and Taylor, J. (1992) Fetal alcohol neuroendocrinopathies, in *Alcohol and Neurobiology: Brain Development and Hormone Regulation* (ed. R.R. Watson), CRC Press, Boca Raton, pp. 109–138.

Saal, D., Dong, Y., Bonci, A., and Malenka, R.C. (2003) Drugs of abuse and stress trigger a common synaptic adaptation in dopamine release. *Neuron*, **38** (2), 577–582.

Sapolsky, R.M., Meaney, M.J., and McEwen, B.S. (1985) The development of the glucocorticoid receptor system in the rat limbic brain. III. Negative-feedback regulation. *Brain Res.*, **350** (1–2), 169–173.

Sari, Y. (2009) Activity-dependent neuroprotective protein-derived peptide, NAP, preventing alcohol-induced apoptosis in fetal brain of C57BL/6 mouse. *Neuroscience*, **158** (4), 1426–1435.

Sari, Y. and Zhou, F.C. (2004) Prenatal alcohol exposure causes long-term serotonin neuron deficit in mice. *Alcohol. Clin. Exp. Res.*, **28** (6), 941–948.

Sarnyai, Z., Shaham, Y., and Heinrichs, S.C. (2001) The role of corticotropin-releasing factor in drug addiction. *Pharmacol. Rev.*, **53** (2), 209–243.

Schenk, S., Valadez, A., McNamara, C., House, D.T., Higley, D., Bankson, M.G., Gibbs, S., and Horger, B.A. (1993) Development and expression of sensitization to cocaine's reinforcing properties: role of NMDA receptors. *Psychopharmacology (Berl.)*, **111** (3), 332–338.

Seckl, J.R. (2008) Glucocorticoids, developmental 'programming' and the risk of affective dysfunction. *Prog. Brain Res.*, **167**, 17–34.

Seelig, L.L., Jr Steven, W.M., and Stewart, G.L. (1996) Effects of maternal ethanol consumption on the subsequent development of immunity to *Trichinella spiralis* in rat neonates. *Alcohol. Clin. Exp. Res.*, **20** (3), 514–522.

Seelig, L.L., Jr Steven, W.M., and Stewart, G.L. (1999) Second generation effects of maternal ethanol consumption on immunity to *Trichinella spiralis* in female rats. *Alcohol Alcohol.*, **34** (4), 520–528.

Selye, H. (1971) Hormones and resistance. *J. Pharm. Sci.*, **60** (1), 1–28.

Sheridan, J.F., Dobbs, C., Brown, D., and Zwilling, B. (1994) Psychoneuroimmunology: stress effects on pathogenesis and immunity during infection. *Clin. Microbiol. Rev.*, **7** (2), 200–212.

Shibley, I., McIntyre, T., and Pennington, S. (1999) Review. Experimental models used to measure direct and indirect ethanol teratogenicity. *Alcohol Alcohol.*, **34** (2), 125–140.

Sliwowska, J.H., Lan, N., Yamashita, F., Halpert, A.G., Viau, V., and Weinberg, J. (2008) Effects of prenatal ethanol exposure on regulation of basal hypothalamic-pituitary-adrenal activity and hippocampal 5-HT1A receptor mRNA levels in female rats across the estrous cycle. *Psychoneuroendocrinology*, **33** (8), 1111–1123.

Slone, J.L. and Redei, E.E. (2002) Maternal alcohol and adrenalectomy: asynchrony of stress response and forced swim behavior. *Neurotoxicol. Teratol.*, **24** (2), 173.

Spong, C.Y., Abebe, D.T., Gozes, I., Brenneman, D.E., and Hill, J.M. (2001) Prevention of fetal demise and growth restriction in a mouse model of fetal alcohol syndrome. *J. Pharmacol. Exp. Ther.*, **297** (2), 774–779.

Steven, W.M., Stewart, G.L., and Seelig, L.L. (1992) The effects of maternal ethanol consumption on lactational transfer of immunity to *Trichinella spiralis* in rats. *Alcohol. Clin. Exp. Res.*, **16** (5), 884–890.

Sullivan, R.M., and Gratton, A. (2002) Prefrontal cortical regulation of hypothalamic-pituitary-adrenal function in the rat and implications for psychopathology: side matters. *Psychoneuroendocrinology*, **27** (1–2), 99–114.

Swaab, D.F., Bao, A.-M., and Lucassen, P.J. (2005) The stress system in the human brain in depression and neurodegeneration. *Ageing Res. Rev.*, **4** (2), 141.

Swerdlow, N. and Koob, G. (1987) Dopamine, schizophrenia, mania, and depression: toward a unified hypothesis of cortico-striato-pallido-thalamic function. *Behav. Brain Sci.*, **10**, 197–245.

Tajuddin, N.F. and Druse, M.J. (1993) Treatment of pregnant alcohol-consuming rats with buspirone: effects on serotonin and 5-hydroxyindoleacetic acid content in offspring. *Alcohol. Clin. Exp. Res.*, **17** (1), 110–114.

Tajuddin, N.F. and Druse, M.J. (2001) A persistent deficit of serotonin neurons in the offspring of ethanol-fed dams: protective effects of maternal ipsapirone treatment. *Brain Res. Dev. Brain Res.*, **129** (2), 181–188.

Taylor, A.N., Branch, B.J., Van Zuylen, J.E., and Redei, E. (1988) Maternal alcohol consumption and stress responsiveness in offspring. *Adv. Exp. Med. Biol.*, **245**, 311–317.

Trivedi, M.H., Hollander, E., Nutt, D., and Blier, P. (2008) Clinical evidence and potential neurobiological underpinnings of unresolved symptoms of depression. *J. Clin. Psychiatry*, **69** (2), 246–258.

Ulrich-Lai, Y.M. and Herman, J.P. (2009) Neural regulation of endocrine and autonomic stress responses. *Nat. Rev. Neurosci.*, **10** (6), 397–409.

Vaurio, L., Riley, E.P., and Mattson, S.N. (2008) Differences in executive functioning in children with heavy prenatal alcohol exposure or attention-deficit/hyperactivity disorder. *J. Int. Neuropsychol. Soc.*, **14** (1), 119–129.

Viau, V. (2002) Functional cross-talk between the hypothalamic-pituitary-gonadal and -adrenal axes. *J. Neuroendocrinol.*, **14** (6), 506–513.

Waddell, B.J., Benediktsson, R., Brown, R.W., and Seckl, J.R. (1998) Tissue-specific messenger ribonucleic acid expression of 11beta-hydroxysteroid dehydrogenase types 1 and 2 and the glucocorticoid receptor within rat placenta suggests exquisite local control of glucocorticoid action. *Endocrinology*, **139** (4), 1517–1523.

Weaver, I.C. (2007) Epigenetic programming by maternal behavior and pharmacological intervention. Nature versus nurture: let's call the whole thing off. *Epigenetics*, **2** (1), 22–28.

Weaver, I.C., Cervoni, N., Champagne, F.A., D'Alessio, A.C., Sharma, S., Seckl, J.R., Dymov, S., Szyf, M., and Meaney, M.J. (2004) Epigenetic programming by maternal behavior. *Nat. Neurosci.*, **7** (8), 847–854.

Weinberg, J. (1993a) Neuroendocrine effects of prenatal alcohol exposure. *Ann. N. Y. Acad. Sci.*, **697**, 86–96.

Weinberg, J. (1993b) Prenatal alcohol exposure: endocrine function of offspring, in *Alcohol and the Endocrine System* (ed. S. Zakhari), NIH Press, Bethesda, pp. 363–382.

Weinberg, J. (1994) Recent studies on the effects of fetal alcohol exposure on the endocrine and immune systems. *Alcohol Alcohol. Suppl.*, **2**, 401–409.

Weinberg, J., Nelson, L.R., and Taylor, A.N. (1986) Hormonal effects of fetal alcohol exposure, in *Alcohol and Brain Development* (ed. J. West), Oxford University Press, New York, pp. 310–342.

Weinberg, J., Taylor, A.N., and Gianoulakis, C. (1996) Fetal ethanol exposure: hypothalamic-pituitary-adrenal and beta-endorphin responses to repeated stress. *Alcohol. Clin. Exp. Res.*, **20** (1), 122–131.

Weinberg, J., Sliwowska, J.H., Lan, N., and Hellemans, K.G. (2008) Prenatal alcohol exposure: foetal programming, the hypothalamic-pituitary-adrenal axis and sex differences in outcome. *J. Neuroendocrinol.*, **20** (4), 470–488.

Welberg, L.A. and Seckl, J.R. (2001) Prenatal stress, glucocorticoids and the programming of the brain. *J. Neuroendocrinol.*, **13** (2), 113–128.

West, J.R., Chen, W.J., and Pantazis, N.J. (1994) Fetal alcohol syndrome: the vulnerability of the developing brain and possible mechanisms of damage. *Metab. Brain Dis.*, **9** (4), 291–322.

White, P.C., Mune, T., Rogerson, F.M., Kayes, K.M., and Agarwal, A.K. (1997) 11 beta-Hydroxysteroid dehydrogenase and its role in the syndrome of apparent mineralocorticoid excess. *Pediatr. Res.*, **41** (1), 25–29.

Wilcoxon, J.S., Kuo, A.G., Disterhoft, J.F., and Redei, E.E. (2005) Behavioral deficits associated with fetal alcohol exposure are reversed by prenatal thyroid hormone treatment: a role for maternal thyroid hormone deficiency in FAE. *Mol. Psychiatry*, **10** (10), 961.

Wilkemeyer, M.F., Menkari, C.E., Spong, C.Y., and Charness, M.E. (2002) Peptide antagonists of ethanol inhibition of L1-mediated cell-cell adhesion. *J. Pharmacol. Exp. Ther.*, **303** (1), 110–116.

de Wit, H. (1996) Priming effects with drugs and other reinforcers. *Exp. Clin. Psychopharmacol.*, **4**, 5–10.

Yehuda, R., Teicher, M.H., Trestman, R.L., Levengood, R.A., and Siever, L.J. (1996) Cortisol regulation in posttraumatic stress disorder and major depression: a chronobiological analysis. *Biol. Psychiatry*, **40** (2), 79.

Young, E.A. (1995) The role of gonadal steroids in hypothalamic-pituitary-adrenal axis regulation. *Crit. Rev. Neurobiol.*, **9** (4), 371–381.

Young, E.A., Altemus, M., Lopez, J.F., Kocsis, J.H., Schatzberg, A.F., deBattista, C., and Zubieta, J.-K. (2004a) HPA axis activation in major depression and response to fluoxetine: a pilot study. *Psychoneuroendocrinology*, **29** (9), 1198.

Young, A.H., Gallagher, P., Watson, S., Del-Estal, D., Owen, B.M., and Nicol Ferrier, I. (2004b) Improvements in neurocognitive function and mood following adjunctive treatment with mifepristone (RU-486) in bipolar disorder. *Neuropsychopharmacology*, **29** (8), 1538.

Zafar, H., Shelat, S.G., Redei, E., and Tejani-Butt, S. (2000) Fetal alcohol exposure alters serotonin transporter sites in rat brain. *Brain Res.*, **856** (1–2), 184.

Zhang, X., Sliwowska, J.H., and Weinberg, J. (2005) Prenatal alcohol exposure and fetal programming: effects on neuroendocrine and immune function. *Exp. Biol. Med. (Maywood)*, **230** (6), 376–388.

Zhou, F.C., Sari, Y., Zhang, J.K., Goodlett, C.R., and Li, T. (2001) Prenatal alcohol exposure retards the migration and development of serotonin neurons in fetal C57BL mice. *Brain Res. Dev. Brain Res.*, **126** (2), 147–155.

Zhou, F.C., Sari, Y., and Powrozek, T.A. (2005) Fetal alcohol exposure reduces serotonin innervation and compromises development of the forebrain along the serotonergic pathway. *Alcohol. Clin. Exp. Res.*, **29** (1), 141–149.

6
Genetic Factors Contributing to FASD
Albert E. Chudley

6.1
Introduction

A *teratogen* is defined as any environmental factor that can produce a permanent abnormality in structure or function, restriction of growth, or death of the embryo or fetus (Frias and Gilbert-Barness, 2008). Teratogens comprise medications, drugs, chemicals, and maternal conditions or diseases, including congenital infections. Among humans, Fetal Alcohol Spectrum Disorder (FASD) and its most visible subset, Fetal Alcohol Syndrome (FAS), is perhaps the most common teratogenic disorder (Abel and Sokol, 1986).

FASD occurs as a result of the embryo and fetus being exposed to alcohol. A simplistic view is that a teratogen acts as a toxic agent that directly disrupts normal prenatal development, and results in morphologic or functional impairments in the exposed individual. This, of course, is not the reality as the mechanism of teratogenesis is highly complex.

The basic principles of teratogenesis include critical periods of development, in which the greatest risk is usually during the period of the origin and development of organs (organogenesis). These may be very narrow and specific, as with thalidomide embryopathy[1] (day 20 to day 32 post conception), or broad as with FASD (two weeks post conception to term). Exposure during the early blastocyst stage after conception to the start of gastrulation (the 14–16-day period when the three

---

1) "Thalidomide is a sedative, hypnotic, and anti-inflammatory medication. It was sold from 1957 to 1961 in almost fifty countries, under at least forty names, to pregnant women, as an antiemetic to combat morning sickness and as an aid to help them sleep. Unfortunately, inadequate tests were performed to assess the drug's safety, with catastrophic results for the children of women who had taken thalidomide during their pregnancies. From 1956 to 1962, approximately 10 000 children were born with severe malformations, including phocomelia, because their mothers had taken thalidomide during pregnancy. In 1962, in reaction to the tragedy, the United States Congress enacted laws requiring tests for safety during pregnancy before a drug can receive approval for sale in the US. Other countries enacted similar legislation, and thalidomide was not prescribed or sold for decades." Source: http://www.medic8.com/medicines/Thalomid.html

Fetal Alcohol Spectrum Disorder–Management and Policy Perspectives of FASD. Edited by Edward P. Riley, Sterling Clarren, Joanne Weinberg, and Egon Jonsson
Copyright © 2011 WILEY-VCH Verlag GmbH & Co. KGaA, Weinheim
ISBN: 978-3-527-32839-0

embryonic layers appear and organogenesis begins) was believed to result in an "all-or-nothing" phenomenon; that is, either the exposure was lethal or the organism survived intact (Prolifka and Friedman, 1999). This belief has recently been challenged, however, as there is animal evidence that teratogenic exposure from conception to the end of gastrulation (four weeks) may result in multiple congenital anomalies (Opitz, 2007). The risk of alcohol consumption is also influenced by the timing, pattern, and frequency of exposure, the dose of the exposure, the presence of co-occurring drug exposures, maternal weight and age, stress, and nutritional status (May et al., 2009).

Alcohol-induced adverse effects result from a broad range of complex interactions between environmental, behavioral, social, and genetic factors. There is also a high ethnic and inter-individual variability in the occurrence and gravity of alcohol related pathologies, which often is not correlated to the amount of alcohol intake.

Although FASD has been described in individuals from all ethnic backgrounds, the prevalence of the condition varies among populations, with the highest in Aboriginal populations in North America and mixed-race individuals in South African (see Chapter 2, and references therein). Genetic factors, and the differences in genetic backgrounds between ethnic groups, most likely influence the extent and degree of embryonic and fetal damage resulting from prenatal alcohol exposure, though much remains to be investigated in this complex multifactorial disorder.

The main aim of this chapter will be to review the research investigations into the role of genetic factors that either influence or contribute to the development of FASD.

6.2
The Evidence

The evidence that genetic factors contribute to FASD is derived from several observations and investigations that involve both animal and human studies. Not all children born to women who drink heavily during pregnancy will develop FASD. Indeed, in the author's clinical experience (at the Children's Hospital, Winnipeg, Canada), it is estimated that approximately half of all alcohol-exposed fetuses show no substantial neurodevelopmental impairments or birth defects.

Studies conducted in animals have demonstrated genetic strain differences in the frequency and types of birth defect after prenatal alcohol exposure (Chernoff, 1977; Chernoff 1980; Boehm et al., 1997; Warren and Li, 2005). For example, when Chernoff (Chernoff, 1977) performed genetic (diallelic) crosses among three different mouse strains that were referred to as the CBA, C3H and C57BL, the extent of alcohol-related teratogenesis was more closely correlated with the genetics of the mothers rather than the fetus. Malformations were greater with CBA than with C3H and C57BL mothers, respectively, supporting the view that genetic differences between strains within the same species are responsible for modifying the phenotypic expression. In a more recent report, two different inbred strains of mice (C57BL/6 and DBA/2) were exposed to alcohol while *in utero*, in identical amounts and timing (Ogawa et al., 2005). Whilst the exposure to alcohol caused the normal heart and brain/spinal development to be specifically compromised in C57BL/6 mice, in the DBA/2 mice there was a specific decrease in the number of body segment parts and the development of branchial bars (regions that develop certain facial structures). Despite both strains being exposed to the same dose and timing of alcohol, each had a unique vulnerability in specific organs, with alcohol-related teratogenicity being greater in DBA/2 mice than in C57BL/6 mice. These differences in organ response were considered to result from genetic influences, rather than from maternal influences or the conditions of alcohol exposure.

Subsequent studies in human twins showed an expected higher concordance (that a given trait occurs in both twins) of phenotype in identical versus nonidentical twins. The first report of discordance (that a trait occurs in only one twin) in a nonidentical twin set was made in 1975 (Christoffel and Salafsky, 1975), when one twin was shown to have full FAS while the other twin had a less severe presentation. In 1993, Streissguth and Dehaene (Streissguth and Dehaene 1993) studied 16 pairs of twins who had been heavily exposed to alcohol during pregnancy, and who showed a concordance for diagnosis to be total (5/5) for monozygotic (MZ) twins, but only partial (7/11) for dizygotic (DZ) twins. In two of the DZ pairs, one twin had FAS, while the other had Fetal Alcohol Effect (FAE), whereas in two other DZ pairs one twin had no diagnosis while one had FAE. The IQ scores were most similar within pairs of MZ twins, and least similar within pairs of DZ twins discordant for an alcohol-related diagnosis. Streissguth and Dehaene concluded that the difference in phenotypic expression of the teratogenic effects of alcohol was due to modulating effects of the genes. Yet, differences in the intrauterine environment (including variable placentation) might also have contributed to the discordance, further enhancing the complexity and role of alcohol as a teratogen during pregnancy.

Genetic differences in biotransformation enzymes and in target proteins can affect the individual susceptibility to drugs and environmental chemicals. In recent years, the field of *ecogenetics* has emerged from the older area of pharmacogenetics, and indicates how genetic polymorphisms may represent risk factors for a number of diseases associated with exposure to toxic chemicals (Costa, 2000). Genetic differences in alcohol metabolism, in the context of prenatal exposure to alcohol, are examined in the following section. Current findings support the view that these differences are most likely important when defining the variable phenotypes and risks seen in different populations.

6.3
Genetic Factors in Alcohol Metabolism

The elimination of alcohol is dependent on the activity of enzymes in many tissues, with the majority of the metabolism occurring in the liver. Enzyme activity can vary between individuals, depending on differences in the genes that encode the enzymes which either increase the efficiency of the process, or slow it down (Warren and Li, 2005; Gemma, Vichi, and Testai, 2006). Typically, a person will have two copies of each gene, one inherited from each parent. Individuals may show variations in how they metabolize drugs, since one member of the gene pair or allele may have a different level of activity or efficiency than other member of the gene or allele. Initially, alcohol is oxidized by the enzyme alcohol dehydrogenase (ADH) to produce acetaldehyde, which is oxidized in turn to acetate, under the action of the enzyme acetaldehyde dehydrogenase (ALDH) (Bosron and Li, 1987; Hurley, Edenberg, and Li, 2002) (Figure 6.1).

In addition, about 10% of the alcohol is metabolized by a microsomal enzyme P4502E1 or cytochrome P450 (*CYP2E1*) (Agarwal, 2001). All of these enzymes occur as several "isoenzymes" (enzymes that differ in amino acid sequence, but catalyze the same chemical reaction), encoded by multigene families (Table 6.1).

Specific alleles (or polymorphisms or variations) at the loci *ADH1B* (previously *ADH2*), *ADH1C* (previously *ADH3*), and *ALDH2* can increase the level of acetaldehyde (Agarwal, 2001). This causes an adverse response to alcohol consumption, seen as the "flushing response" that is characterized by an elevated blood flow,

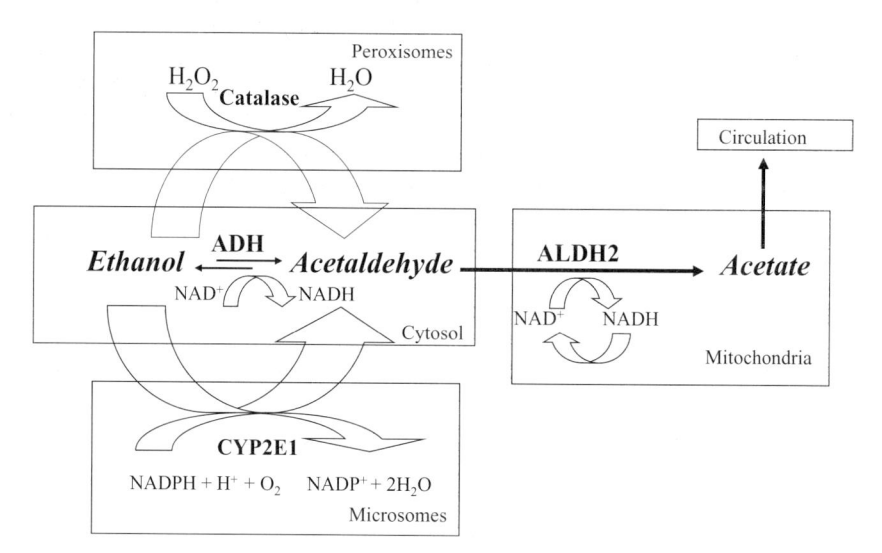

Figure 6.1 Oxidative pathways of alcohol metabolism. The enzymes alcohol dehydrogenase (ADH), cytochrome P450 2E1 (CYP2E1) and catalase all contribute to the oxidative metabolism of alcohol. ADH, present in the fluid of the cell (i.e., cytosol), converts alcohol (i.e., ethanol) to acetaldehyde. This reaction involves an intermediate carrier of electrons, nicotinamide adenine dinucleotide (NAD$^+$), which is reduced by two electrons to form NADH. Catalase, located in cell bodies called peroxisomes, requires hydrogen peroxide (H$_2$O$_2$) to oxidize alcohol. CYP2E1, present predominantly in the cell's microsomes, assumes an important role in metabolizing ethanol to acetaldehyde at elevated ethanol concentrations. Acetaldehyde is metabolized mainly by aldehyde dehydrogenase 2 (ALDH2) in the mitochondria to form acetate and NADH.

Table 6.1 Human alcohol dehydrogenase (ADH) isoenzymes.

Gene nomenclature		Protein	K_m (mM)	V_{max} (min^{-1})	Tissue
New	Former				
ADH1A	ADH1	α	4.0	30	Liver
ADH1B*1	ADH2*1	β$_1$	0.05	4	Liver, lung
ADH1B*2	ADH2*2	β$_2$	0.9	350	
ADH1B*3	ADH2*3	β$_3$	40.0	300	
ADH1C*1	ADH3*1	γ$_1$	1.0	90	Liver, stomach
ADH1C*2	ADH3*2	γ$_2$	0.6	40	
ADH4	ADH4	π	30.0	20	Live, cornea
ADH5	ADH5	χ	>1000	100	Most tissues
ADH7	ADH7	σ(μ)	30.0	1800	Stomach
ADH6	ADH6		?	?	Liver, stomach

dizziness, an accelerated heart rate, sweating, and nausea. Individuals who suffer flushing tend to avoid heavy drinking; this will tend to reduce the likelihood of a woman who carries this allele having a child affected by FASD.

The *ADH1B* variations occur at different frequencies, depending on the ethnic group studied. The *ADH1B*1* form is predominant in Caucasian and Black populations, while the *ADH1B*2* frequency is higher in Chinese and Japanese populations, and occurs in 25% of people with a Jewish ancestry. African Americans (Thomasson, Beard, and Li, 1995) and Native Americans (Wall et al., 1996) with the *ADH1B*3* allele metabolize alcohol at a faster rate than those with *ADH1B*1*.

The *ALDH2* gene has one major polymorphism, resulting in allelic variants *ALDH2*1* and *ALDH2*2*; the latter has a very low activity. *ALDH2*2* is present in about 50% of the Taiwanese, Han Chinese and Japanese populations (Shen et al., 1997), and shows virtually no acetaldehyde-metabolizing activity *in vitro*. People with one copy, and especially those with two copies of the *ALDH2*2* allele, will have increased acetaldehyde levels after alcohol consumption, and thus experience the flushing response.

Since polymorphisms of ADH and ALDH2 determine peak blood acetaldehyde levels, they will influence vulnerability to alcohol dependence. A fast ADH or a slow ALDH will elevate acetaldehyde levels and reduce alcohol drinking. Women with these genotypes are at a very low risk of alcoholism, and therefore are at a low risk of having FASD children.

Although several *CYP2E1* polymorphisms have been identified, very few studies have been undertaken to determine the effect on alcohol metabolism. The presence of the rare c2 allele was associated with a higher alcohol metabolism in Japanese alcoholics (Ueno et al., 1996). Several studies have reviewed the possible relationship of *ADH1B* polymorphisms and alcohol teratogenic risk. For example, Viljoen et al. (2001) studied a mixed-ancestry population in South Africa in which there was a high rate of alcoholism and FAS. The *ADH1B*2* allele frequency for both the FAS children and their mothers was lower, in a statistically significant manner, than in the controls.

Such findings support a possible protective role of the *ADH1B*2* allele in either the mother or the fetus, and are consistent with the hypothesis that the more efficient variant of ADH would provide protection, perhaps through an elevation of acetaldehyde levels resulting in a reduced maternal alcohol consumption. Alternatively, it was suggested that there might be an actual "gene of influence" in linkage disequilibrium with *ADH1B*, though this is yet to be discovered.

In African-Americans, the presence of at least one copy of maternal *ADH1B*3* allele was found to be protective for two specific outcomes: (i) reductions in the infants' birth weight and birth length; and (ii) lower scores on the Bayley Mental Development Index (MDI) at 12 months of age (McCarver et al., 1997). The presence of at least one copy of the *ADH1B*3* allele in the child was also found to be protective from lower scores on the MDI. Das et al. (2004) examined facial morphology from photographs of infants taken at one year of age, and found the use of alcohol in pregnancy prior to the first prenatal visit to be associated with a statistically significant reduction in palpebral fissure length (the distance between the inner and outer parts of the eye), the inner canthal distance (distance between the inner aspect of both eyes), and the nasal bridge to mouth length when neither the mother nor infant possessed an *ADH1B*3* allele.

In another US study on African-American children (Croxford et al., 2003), the maternal genotype *ADH1B*3* allele, but not the child's genotype, was correlated with protection from alcohol teratogenesis, using several developmental growth and neurodevelopmental measures. These findings were, therefore, consistent with those of others (McCarver et al., 1997) for the maternal genotype. The statistically significant association found elsewhere (McCarver et al., 1997) for the absence of a fetal *ADH1B*3* allele and the Bayley MDI was not found by Croxford et al. (2003). In contrast to these results, Stoler et al. (2002) found a higher proportion of affected children with the *ADH1B*3* allele compared to unaffected children. Clearly, further studies are required to validate the role of these polymorphisms and the risk for alcohol teratogenesis.

In a study of North America Indians, Mulligan et al. (2003) showed – by both linkage and association analysis – the presence of several *ADH1C* alleles and a neighboring microsatellite marker that affected the risk of alcohol dependence, and were also related to binge drinking. These data strengthened the support for *ADH* as a candidate locus for alcohol dependence.

The metabolism of alcohol leads to the generation of reactive oxygen species (ROS) that can cause tissue damage. However, enzymes exist which have a strong antioxidant function and protect cells against the natural byproducts of lipid peroxidation and oxidative stress. Genetic polymorphisms that reduce enzyme activity can exacerbate the effects of alcohol on embryonic cells and fetal tissues. One such group of enzymes is the glutathione *S*-transferases (GSTs), which are Phase II xenobiotic-metabolizing enzymes that act as a highly efficient detoxification system.

In humans, genetic differences have been described in the *GSTM1*, *GSTT1*, and *GSTP1* genes. Among the described differences or polymorphisms at the *GSTM1* locus on chromosome 1p13.3, the most extensively studied encodes for a gene deletion (*GSTM1*-null genotype), and results in a complete absence of GSTM1 enzyme activity. The frequency of the *GSTM1*-null genotype ranges from 23 to 62% in different populations, and is approximately 50% in Caucasians (Hayes and Strange, 2000). For the *GSTT1* locus, located on chromosome 12q11.2, one polymorphism has been described. The *GSTT1*-null genotype represents a gene deletion, and is associated with an absence of functional enzyme activity. The frequency of these null genotypes ranges from 16 to 64% in different populations, and is approximately 20% in Caucasians (Verlaan et al., 2003).

6.4
Some Genetic Factors

Alcohol can have adverse effects on normal developmental pathways. For example, it may affect multiple cellular events that include synthesis and degradation, gene regulation, transcription, translation and expression, protein–protein interactions, protein modifications (such as phosphorylation), molecular transport and secretion, chromosome assembly, biogenesis, and various enzymatic activities (Uddin, Treadwell, and Singh, 2005). Based on animal models, dozens – if not scores – of genes are affected by acute exposure to alcohol (Murphy et al., 2002; Treadwell and Singh, 2004).

To date, no human genome-wide association or linkage studies have been conducted for FAS. Previously, Lombard et al. (2007) employed a computational candidate gene selection method that identified genes which might play a role in alcohol teratogenesis. By using a modification of the methodology (termed "convergent functional genomics"), which combines data from human and animal studies, Lombard and coworkers identified a short list of high-probability candidate genes. Also included were additional lines of evidence, in the presence of limited expression studies in an animal model, and an absence of FAS linkage studies. From a list of 87 genes, the group prioritized key biological pathways that were significantly over-represented among the top-ranked candidate genes. These pathways include the transforming growth factor β (TGF-β) signaling pathway, the mitogen-activated protein kinase (MAPK) signaling pathway, and the "Hedgehog" signaling pathway.

The genes in this pathway play pivotal roles during embryogenesis and development, and have a potential role in the distinct characteristics associated with FAS – that is, CNS dysfunction, craniofacial abnormalities, and growth retardation. Of these abnormalities, CNS dysfunction is the most severe and permanent consequence of *in utero* alcohol exposure, and is the only feature present in all diagnostic categories of FASD. These observations also indicate that the TGF-β signaling pathway is an important consideration, as it is essential for both fetal and CNS development. Alcohol inhibits such TGF-β-regulated processes such as cortical cell proliferation and neuronal migration, disrupts axonal (the major extension of a nerve cell) growth, and upregulates cell adhesion molecule expression (Miller and Luo, 2002). The TGF-β signaling pathway interacts with alcohol (and/or its metabolic breakdown products), such that alcohol may have a detrimental effect on the efficiency of this developmentally essential pathway.

The MAPK pathway transmits many signals, leading to growth, differentiation, inflammation, and apoptosis responses (Krens, Spaink, and Snaar-Jagalska, 2006). This pathway is very complex and includes many protein components. Typically, the MAPK pathway components are involved in the regulation of meiosis, mitosis and post-mitotic functions, and also in cell differentiation. The MAPK signaling pathway can be activated by a variety of stimuli, as well as external stress factors, such as alcohol (Aroor and Shukla, 2004). By using a mouse model of FAS, the experimental manipulation of second-messenger pathways (that also impact on the MAPK pathway) caused a complete reversal of the action of ethanol on neuronal migration, both *in vitro* and *in vivo* (Kumada, Lakshmana, and Komuro, 2006).

The Hedgehog signaling pathway was also identified as containing several genes within the candidate list. This signaling pathway is a highly conserved and key regulator of embryonic development, with knock-out mouse models lacking components of this pathway having been observed to develop malformations in the CNS, musculoskeletal system, gastrointestinal tract, and lungs (Ingham and McMahon, 2001). FAS animal models have a similar craniofacial phenotype to mouse models treated with antibodies that block Hedgehog signaling components, specifically the sonic hedgehog (Shh) molecule (Chen *et al.*, 2000; Ahlgren and Bronner-Fraser, 1999). Alcohol resulted in a significant decrease in Shh levels in the developing embryo, as well as a decrease in the level of other transcripts involved in Shh signaling. The addition of Shh following alcohol exposure led to fewer apoptotic (dead or dying) cranial neural crest cells, and a decrease in craniofacial anomalies (Ahlgren, Thakur, and Bronner-Fraser, 2002). Hence, genes in the Hedgehog signaling pathway may also be contributors to the FAS phenotype.

6.5
Epigenetics, the Environment and Nutrition

Epigenetics refers to changes in phenotype (appearance) or gene expression caused by mechanisms other than changes in the underlying DNA sequence. In fact, many of the effects of alcohol on gene expression detailed above are most likely due to epigenetic effects, rather than to direct changes in DNA sequences. *Genomic imprinting* represents one of the most intriguing subtleties of modern genetics, and is one component of epigenetics. Here, the term "imprinting" refers to parent-of-origin-dependent gene expression as a result of epigenetic modification, leading to the control of gene expression as dictated by parental inheritance (Surani, 1998).

The presence of imprinted genes can cause cells with a full parental complement of functional autosomal genes to specifically express one allele, but not the other, and this will result in a monoallelic expression of the imprinted loci. Genomic imprinting plays a critical role in fetal growth and behavioral development, and is regulated by DNA methylation and chromatin structure. Environmental factors are capable of causing epigenetic changes in DNA that can potentially alter imprint gene expression, and that can also result in genetic diseases including cancer and behavioral disorders. An understanding of the contribution of imprinting to the regulation of gene expression would represent an important step in evaluating environmental influences on human health and disease (Jirtle et al., 2000).

Studies conducted in animals have shown that *in utero* alcohol exposure inhibits fetal DNA methylation. One of the well-known features of imprinted genes is differential allele-specific DNA methylation, and this is usually found in regions known as "differentially methylated regions." These include imprinting control regions, and it has been suggested that all clusters of imprinted genes have imprinting control regions, which are differentially methylated (Delaval and Feil, 2004). The expression of many genes is regulated through the methylation of DNA (Lim and van Oudenaarden, 2007; Valles et al., 1997; Garro et al., 1991). Since DNA methylation and imprinting play an important role in the regulation of gene expression during development, alcohol-associated alterations in fetal DNA methylation may contribute to the developmental abnormalities seen in FAS.

In the context of other environmental factors, poor nutrition may also lead to altered gene expression. Women who drink alcohol and have diets lacking in certain nutrients and vitamins (especially folic acid), may experience an exacerbation of the biological and teratogenic effects of alcohol. This is not a new concept, however, with investigators first suggesting over 30 years ago that poor nutrition might contribute to FAS (Lin, 1981).

An additional control mechanism for gene expression is that of microRNAs (miRNAs). These are small noncoding RNAs, each of about 22 nucleotides, that regulate gene expression by targeting mRNAs in a sequence-specific manner, leading to the induction of translational repression and/or mRNA degradation (Ambros and Chen, 2007). Up to 30% of all genes within the human genome might be regulated by miRNAs. Several studies performed in rodents, using DNA microarray analysis, have identified a small number of genes which are significantly altered by prenatal alcohol exposure (Hard et al., 2005; Shankar et al., 2006). These genes are mainly associated with stress and external stimulus responses, transcriptional regulation, cellular homeostasis, and protein metabolism. Certain miRNAs were not only downregulated by ethanol in mouse cortical neurons *in vitro*, but also were shown to exhibit both synergistic and antagonistic interactions.

Folic acid (FA) is essential for nucleic acid synthesis, amino acid metabolism, and protein synthesis. It is well-known that FA supplementation in young women can prevent intrauterine growth restriction, neural tube defects, and other congenital anomalies (Eskes, 1997; De Wals et al., 2007), while studies in experimental animals have shown that FA can also ameliorate toxicity induced by alcohol

(Gutierrez et al., 2007; Xu, Tang, and Li, 2008; Yanaguita et al., 2008). The mechanism of this protective effect remains unclear, but proteomic analysis has indicated that FA can modulate the alcohol-revised proteins involved in energy production, signal pathways, and protein translation (Xu, Tang, and Li, 2008).

When Wang et al. (2009) exposed mouse embryos in culture to alcohol, many anomalies were identified. For example, coincubation with FA blocked alcohol-induced teratogenesis, associated with upregulation of the *Hoxa1* gene, and downregulation of a particular miRNA, miR-10a. The mechanism of the fetal protective effect of FA is unclear, but it may be related to the important role of FA in DNA and RNA synthesis, and in the transfer of methyl groups in the amino acid methylation cycle. FA may also have beneficial effects on fetal development via multiple biochemical pathways and molecular targets, and further studies are required to confirm these possibilities. It has been suggested that miRNAs and their target genes might also be important in the pathogenesis of FAS.

Recent investigations have focused on the effects of choline and its role in preventing the teratogenic effects of alcohol. Choline, which plays a number of important roles during development, contains three methyl groups; consequently, it serves as a methyl donor, influencing the methionine–homocysteine cycle, whereby choline methylates homocysteine to form methionine. A reduction in choline levels would lead to increased concentrations of homocysteine, which are associated with an increased risk for birth defects (Hobbs et al., 2005). Choline also serves as an epigenetic factor by influencing DNA methylation and subsequent gene expression (Niculescu, Craciunescu, and Zeisel, 2006). It also acts as a precursor to phosphatidylcholine and sphingomyelin, the major constituents of cell membranes; thus, choline is capable of influencing the structural integrity and signaling functions of cell membranes (Zeisel, 2006; Zeisel and Niculescu, 2006). Finally, choline also serves as a precursor to acetylcholine, which has a role not only as a neurotransmitter but also as a neurotrophic factor.

Thomas, Abou, and Dominguez (2009) studied pregnant rats exposed to ethanol, and determined whether choline supplementation during ethanol exposure would effectively reduce fetal alcohol effects. The rather impressive results showed that choline supplementation significantly attenuated the effect of ethanol on both birth and brain weight, on incisor emergence, and most behavioral measures. In fact, the behavioral performance of ethanol-exposed rats treated with choline was very similar to that of controls. The possibility that early dietary supplements could reduce the severity of some fetal alcohol effects might have important implications for the children of women who drink alcohol during pregnancy.

6.6
Conclusions, and Some Policy Recommendations

- Although not conclusive, it is probable that allelic variants and other genetic differences (in the face of prenatal alcohol consumption) influence the risk for, and incidence of, FASD, and these variations likely explain the differences between ethnic population groups. This knowledge may be important in identifying high-risk groups, and focusing prevention in a more targeted manner.

- Some genetic differences may be protective, resulting in less alcohol exposure (flushing effect). This group would be at low risk for having affected children.
- Certain genotypes may increase drinking (increasing the likelihood for binge drinking and alcoholism), resulting in a higher incidence of FASD.
- Further research in genetic, ecogenetic and epigenetic factors for FASD may lead to a better recognition of at-risk individuals, and the development of more effective preventive strategies.
- Standardized approaches to estimate accurately the maternal alcohol intake, and an accurate categorization of outcomes, is essential in any future research.
- The investigation of a single or a small number of candidate genes involved in alcohol response may not identify the actual mechanisms of alcohol action in the brain, nor clarify all of the risk factors. Future research must be broad in nature, comprehensive, and interdisciplinary in scope in order to identify at-risk populations and develop effective prevention, education, and treatment strategies. Notwithstanding the enormous biological interest in FASD, and the importance in understanding the mechanisms of alcohol effects, studies conducted to identify the genetic risk factors for FASD are unlikely to have any immediate impact on the prevalence, treatment, or prevention of FASD. This is not because the scientific information cannot be informative or predictive of risk, but that other human idiosyncrasies and social factors are causative. Consequently, until such factors are addressed and ameliorated, FASD will remain a scourge in many communities worldwide.

6.7
Glossary

- **Acetaldehyde:** an organic chemical compound (chemical formula CH_3CHO) produced when ethanol is oxidized by the enzyme alcohol dehydrogenase (ADH). When high levels of acetaldehyde occur in the blood, symptoms of facial flushing, light headedness, palpitations, nausea, and general "hangover" symptoms result.
- **Acetaldehyde dehydrogenase (ALDH):** a group of enzymes that catalyze the oxidation of aldehydes. By oxidizing acetaldehyde into non-toxic acetic acid, ALDH2 plays a crucial role in maintaining low blood levels of acetaldehyde during alcohol metabolism.
- **Alcohol dehydrogenase (ADH):** a group of enzymes that catalyze the oxidation of alcohols and convert them to aldehydes. In the liver, ADH converts ethanol to acetaldehyde, which in turn is oxidized by acetaldehyde dehydrogenase (ALDH).

- **Allele:** one of two or more alternative forms of a gene or DNA sequence at the corresponding sites (loci) on homologous chromosomes.
- **Apoptosis:** programmed cell death, a regulated process of cell death caused by intracellular programming. Apoptosis typically benefits an organism, unlike cell death caused by injury.
- **Autosomal gene:** a gene on a chromosome that is not a sex-determining (X or Y) chromosome. Humans have 22 pairs of autosomes and one pair of sex chromosomes.
- **Blastocyst:** a spherical structure produced by cell division in early embryogenesis, 5 to 7 days following fertilization and before implantation. The blastocyst comprises an outer layer of cells, which form the placenta; a cluster of cells on the interior, which form the embryo; and a fluid-filled cavity.
- **Branchial bar:** a region of an embryo that develops during the fourth week and gives rise to specialized structures in the head and neck, including some facial features (also called pharyngeal arches).
- **Cell adhesion molecule:** a protein on the surface of a cell that is involved in binding the cell with other cells or with the extracellular matrix.
- **Choline:** a water-soluble nutrient usually grouped within the B-complex vitamins. Choline supplementation during an alcohol-exposed pregnancy has been shown to attenuate the effects of alcohol on the developing fetus, particularly reducing the hyperactivity and learning deficits associated with prenatal alcohol exposure.
- **Chromatin:** the structural building block of a chromosome, consisting of a complex of DNA and protein in eukaryotic cells.
- **Cytochrome:** iron-containing protein that transfers electrons during cellular respiration and photosynthesis.
- **Diallelic:** having two alleles, or polymorphic variants, at a particular gene locus.
- **Dizygotic:** arising from two separate fertilized eggs (zygotes), as is the case with nonidentical, "fraternal" twins.
- **DNA methylation:** a biochemical and reversible modification of DNA in which methyl groups are enzymatically added to or removed from the nucleotide cytosine; associated with silencing of DNA sequences.
- **Ecogenetic:** genetic factors that influence susceptibility to toxins.
- **Embryogenesis:** the formation of the embryo during early prenatal development; a process involving multiplication of cells and their subsequent growth, movement and differentiation into tissues and organs.
- **Embryopathy:** a developmental abnormality of an embryo or fetus.

- **Epigenetics:** the study of heritable changes in gene function that occur without a change in the underlying DNA sequence. Both, DNA methylation and histone modifications specifically modify the way that genes are expressed.
- **Gastrulation:** a phase early in the development of an embryo during which the cells of the embryo are reorganized to form three germ layers (ectoderm, mesoderm, and endoderm), which give rise to the different organs. The timing of gastrulation is different in different organisms.
- **Genetic strain:** a genetic variant or subtype of an organism; e.g., a mouse strain is a group of mice that are genetically uniform.
- **Genomic imprinting:** Parent-of-origin-dependent gene expression, in which only one of the two inherited copies (alleles) of a gene is expressed and the phenotype that results from this mono-allelic expression depends on whether the allele that is expressed came from the mother or from the father.
- **Genotype:** the entire genetic makeup of an organism; also the pair of alleles at a locus for an individual.
- **Inner canthal distance:** the distance between the inner part of the left eye and the inner part of the right eye.
- **Knockout mouse:** a laboratory mouse in which a gene has been inactivated, or "knocked out," by replacing it or disrupting it with an artificial piece of DNA. Knocking out a gene enables researchers to study what the gene normally does. Knockout mice are frequently named after the gene that has been deactivated.
- **Methylation:** the addition of a methyl group (chemical formula CH_3) to a compound. The addition of methyl groups to histone proteins or to cytosine nucleotides in DNA can alter gene transcription.
- **Monozygotic:** arising from the same fertilized ovum (zygote), as is the case with identical twins; monozygotic twins are genetically identical.
- **Lipid peroxidation:** the oxidative degradation of lipids, a process whereby free radicals (reactive atoms that have one or more unpaired electrons) "steal" electrons from the lipids in cell membranes, resulting in cell damage. Lipid peroxidation is stimulated in the body by certain toxins and infections.
- **Locus (pl. loci):** the specific location of a gene or DNA sequence on a chromosome.
- **Meiosis:** a type of cell division that a diploid germ cell undergoes to produce gametes (sperm or eggs) that carry half the normal number of chromosomes.
- **MicroRNA genes:** small noncoding genes that regulate protein production by binding to sites on messenger ribonucleic acid (mRNA) transcripts, resulting in repression or degradation of the mRNA at that site.

- **Microsatellite marker:** a locus at which a short DNA sequence (1–6 base pairs) is repeated one or more times. Microsatellites are usually located in noncoding sequences, but some are in functional regions and involved in the regulation of gene activity and metabolic processes.

- **Microsomal enzyme:** one of a group of enzymes that play a role in the metabolism of many drugs.

- **Mitosis:** the process whereby a eukaryotic cell (cell with a nucleus) divides to produce two genetically identical daughter cells.

- **Morphology:** the form and structure of an organism or one of its parts.

- **Neural crest:** a transient component of the developing fetus where neurons and glia of the autonomic nervous system are produced as well as supporting and hormone-producing cells in certain organs.

- **Null genotype:** genotype in which one of the alleles is a variant in which the entire gene is absent.

- **Oxidative stress:** a condition in which a cell is unable to deal effectively with reactive oxygen, giving rise to free radicals and peroxides, which can damage proteins, DNA, and lipids.

- **Organogenesis:** the process by which individual organs and structures are formed from the germ layers of an embryo; organogenesis involves both cell movements and cell differentiation.

- **Oxidation:** loss of electrons from an atom or molecule when hydrogen is removed or oxygen is added; the opposite of reduction.

- **Palpebral fissure length:** distance between the inner part of the eye to the outer part of the eye.

- **Phenotype:** the observable characteristics of an organism, which result from an interaction of genetic inheritance and environment. By contrast, the genotype is the genetic constitution (genome) of an individual.

- **Phosphorylation:** the addition of a phosphate group (chemical formula PO_4) to a compound. Phosphorylation of histones affects DNA packing, gene regulation, and DNA repair.

- **Placentation:** the formation of the placenta inside the uterus; in the case of dizygotic (fraternal) twins, two placentas form, whereas monozygotic (identical) twins develop with one shared placenta.

- **Polymorphism:** a variation in DNA sequence of an allele that is too common to be due to new mutation. A polymorphism must have a frequency of at least 1% in the population.

- **Proteomic:** (protein + genomic) the study of the structure and function of the entire set of proteins expressed by a cell, tissue, or organism.

- **Reactive alcohol species:** also free radical, a chemical species that contains one or more unpaired electrons (electrons that occupy an atomic or molecular orbital by themselves) and can damage cells, proteins and DNA by altering their chemical structure; produced through natural biological processes and introduced from external pollutants.

- **Signaling pathway:** a series of sequential, coupled intracellular events that are initiated by the binding of a signaling molecule, such as a hormone or growth factor, to a receptor on the cell membrane and result in a specific cellular response. By means of these signal-transduction pathways, cells respond to their external environment and to other cells.

- **Teratogen:** any agent or environmental factor that can interfere with the development of an embryo or fetus and result in a permanent abnormality in structure or function, a restriction of growth, or the death of the embryo or fetus. The classes of teratogens are radiation, maternal infections, chemicals, and drugs.

- **Teratogenesis:** the development of malformations or functional defects in the embryo or fetus.

- **Xenobiotic metabolism:** a detoxifying process by which compounds that are foreign to the body, such as poisons or drugs, are altered by enzymes.

References

Abel, E.L. and Sokol, R.J. (1986) Fetal alcohol syndrome is now a leading cause of mental retardation. *Lancet*, **2**, 1222.

Agarwal, D.P. (2001) Genetic polymorphisms of alcohol metabolizing enzymes. *Pathol. Biol.*, **49**, 703–709.

Ahlgren, S.C., and Bronner-Fraser, M. (1999) Inhibition of sonic hedgehog signaling *in vivo* results in craniofacial neural crest cell death. *Curr. Biol.*, **9**, 1304–1314.

Ahlgren, S.C., Thakur, V., and Bronner-Fraser, M. (2002) Sonic hedgehog rescues cranial neural crest from cell death induced by ethanol exposure. *Proc. Natl Acad. Sci. USA*, **99**, 10476–10481.

Ambros, V. and Chen, X. (2007) The regulation of genes and genomes by small RNAs. *Development*, **134**, 1635–1641.

Aroor, A.R. and Shukla, S.D. (2004) MAP kinase signaling in diverse effects of ethanol. *Life Sci.*, **74**, 2339–2364.

Boehm, S.L. II, Lundahl, K.R., Caldwell, J., and Gilliam, D.M. (1997) Ethanol teratogenesis in the C57BL/6J, DBA/2J, and A/J inbred mouse strains. *Alcohol*, **14**, 389–395.

Bosron, W.F. and Li, T.K. (1987) Catalytic properties of human liver alcohol dehydrogenase isoenzymes. *Enzyme*, **37**, 19–28.

Chen, S.Y., Periasamy, A., Yang, B., Herman, B., Jacobson, K., and Sulik, K.K. (2000) Differential sensitivity of mouse neural crest cells to ethanol induced toxicity. *Alcohol*, **20**, 75–81.

Chernoff, G.F. (1977) Fetal alcohol syndrome in mice: an animal model. *Teratology*, **15**, 223–230.

Chernoff, G.F. (1980) The fetal alcohol syndrome in mice: maternal variables. *Teratology*, **22**, 71–75.

Christoffel, K.K. and Salafsky, I. (1975) Fetal alcohol syndrome in dizygotic twins. *J. Pediatr.*, **87**, 963–967.

Costa, L.G. (2000) The emerging field of ecogenetics. *Neurotoxicology*, **21**, 85–89.

Croxford, J., Jacobson, S.W., Carr, L., Sokol, R.J., Li, T.K., and Jacobson, J.L. (2003) Protective effects of the *ADH2*3* allele in African American children exposed to alcohol during pregnancy. *Alcohol. Clin. Exp. Res.*, **27** (Suppl.), A39.

Das, U.G., Cronk, C.E., Martier, S.S., Simpson, P.M., and McCarver, D.G. (2004) Alcohol dehydrogenase 2*3 affects alterations in offspring facial morphology associated with maternal ethanol intake in pregnancy. *Alcohol. Clin. Exp. Res.*, **28**, 1598–1606.

De Wals, P., Tairou, F., Van Allen, M.I., Uh, S.H., Lowry, R.B., Sibbald, B., Evans, J.A., Van den Hof, M.C., Zimmer, P., Crowley, M., Fernandez, B., Lee, N.S., and Niyonsenga, T. (2007) Reduction in neural-tube defects after folic acid fortification in Canada. *N. Engl. J. Med.*, **357** (2), 135–142.

Delaval, K. and Feil, R. (2004) Epigenetic regulation of mammalian genomic imprinting. *Curr. Opin. Genet. Dev.*, **14**, 188–195.

Eskes, T.K. (1997) Folates and the fetus. *Eur. J. Obstet. Gynecol. Reprod. Biol.*, **71**, 105–111.

Frias, J.L. and Gilbert-Barness, E. (2008) Human teratogens: current controversies. *Adv. Pediatr.*, **55**, 171–211.

Garro, A.J., McBeth, D.L., Lima, V., and Lieber, C.S. (1991) Ethanol consumption inhibits fetal DNA methylation in mice: implications for the fetal alcohol syndrome. *Alcohol. Clin. Exp. Res.*, **15**, 395–398.

Gemma, S., Vichi, S., and Testai, E. (2006) Individual susceptibility and alcohol effects: biochemical and genetic aspects. *Ann. Ist. Super. Sanita.*, **42**, 8–16.

Gutierrez, C.M., Ribeiro, C.N., de Lima, G.A., Yanaguita, M.Y., and Peres, L.C. (2007) An experimental study on the effects of ethanol and folic acid deficiency, alone or in combination, on pregnant Swiss mice. *Pathology*, **39**, 495–503.

Hard, M.L., Abdolell, M., Robinson, B.H., and Koren, G. (2005) Gene-expression analysis after alcohol exposure in the developing mouse. *J. Lab. Clin. Med.*, **145**, 47–54.

Hayes, J.D. and Strange, R.C. (2000) Glutathione S-transferase polymorphisms and their biological consequences. *Pharmacology*, **61**, 154–166.

Hobbs, C.A., Cleves, M.A., Melnyk, S., Zhao, W.S., and James J. (2005) Congenital heart defects and abnormal maternal biomarkers of methionine and homocysteine metabolism. *Am. J. Clin. Nutr.*, **81**, 147–153.

Hurley, T.D., Edenberg, H.J., and Li, T.K. (2002) Pharmacogenomics of alcoholism, in *Pharmacogenomics: The Search for Individualized Therapies* (eds J. Licinio and M.A. Wong), Wiley-VCH Verlag GmbH, Weinheim, Germany, pp. 417–441.

Ingham, P.W. and McMahon, A.P. (2001) Hedgehog signaling in animal development: paradigms and principles. *Genes Dev.*, **15**, 3059–3087.

Jirtle, R.L., Sander, M., Barrett, J.C., and Carl, J. (2000) Genomic imprinting and environmental disease susceptibility. *Environ. Health Perspect.*, **108**, 271–278.

Krens, S.F., Spaink, H.P., and Snaar-Jagalska, B.E. (2006) Functions of the MAPK family in vertebrate-development. *FEBS Lett.*, **580**, 4984–4990.

Kumada, T., Lakshmana, M.K., and Komuro, H. (2006) Reversal of neuronal migration in a mouse model of fetal alcohol syndrome by controlling second-messenger signalings. *J. Neurosci.*, **26**, 742–756.

Lim, H.N. and van Oudenaarden, A. (2007) A multi-step epigenetic switch enables the stable inheritance of DNA methylation states. *Nat. Genet.*, **39**, 269–275.

Lin, G.W. (1981) Fetal malnutrition: a possible cause of the fetal alcohol syndrome. *Prog. Biochem. Pharmacol.*, **18**, 115–121.

Lombard, Z., Tiffin, N., Hofmann, O., Bajic, V.B., Hide, W., and Ramsay, M. (2007) Computational selection and prioritization of candidate genes for Fetal Alcohol Syndrome. *BMC Genomics*, **8**, 389, http://www.biomedcentral.com/1471-2164/8/389.

May, P.A., Gossage, J.P., Kalberg, W.O., Robinson, L.K., Buckley, D., Manning, M., and Hoyme, H.E. (2009) Prevalence and epidemiologic characteristics of FASD from various research methods with an emphasis on recent in-school studies. *Dev. Disabil. Res. Rev.*, **15** (3), 176–192.

McCarver, D.G., Thomasson, H.R., Martier, S.S., Sokol, R.J., and Li, T. (1997) Alcohol dehydrogenase-2*3 allele protects against alcohol-related birth defects among African Americans. *J. Pharmacol. Exp. Ther.*, **283**, 1095–1101.

Miller, M.W. and Luo, J. (2002) Effects of ethanol and transforming growth factor beta (TGF beta) on neuronal proliferation and nCAM expression. *Alcohol. Clin. Exp. Res.*, **26**, 1281–1285.

Mulligan, C.J., Robin, R.W., Osier, M.V., Sambuughin, N., Goldfarb, L.G., Kittles, R.A., Hesselbrock, D., Goldman, D., and Long, J.C. (2003) Allelic variation at alcohol metabolism genes (*ADH1B, ADH1C, ALDH2*) and alcohol dependence in an American Indian population. *Hum. Genet.*, **113**, 325–336.

Murphy, B.C., Chiu, T., Harrison, M., Uddin, R.K., and Singh, S.M. (2002) Examination of ethanol responsive liver and brain specific gene expression, in the mouse strains with variable ethanol preferences, using cDNA expression arrays. *Biochem. Genet.*, **40**, 395–410.

Niculescu, M.D., Craciunescu, C.N., and Zeisel, S.H. (2006) Dietary choline deficiency alters global and gene-specific DNA methylation in the developing hippocampus of mouse fetal brains. *FASEB J.*, **20**, 43–49.

Ogawa, T., Kuwagata, M., Ruiz, J., and Zhou, F.C. (2005) Differential teratogenic effect of alcohol on embryonic development between C57BL/6 and DBA/2 mice: a new view. *Alcohol. Clin. Exp. Res.*, **29** (5), 855–863.

Opitz, J.M. (2007) Development: clinical and evolutionary considerations. *Am. J. Med. Genet.*, **143**, 2853–2861.

Prolifka, J.E. and Friedman, J.M. (1999) Clinical teratology: identifying teratogenic risks in humans. *Clin. Genet.*, **56**, 409–420.

Shankar, K., Hidestrand, M., Liu, X., Xiao, R., Skinner, C.M., Simmen, F.A., Badger, T.M., and Ronis, M.J. (2006) Physiologic and genomic analyses of nutrition-ethanol interactions during gestation: implications for fetal ethanol toxicity. *Exp. Biol. Med. (Maywood)*, **231**, 1379–1397.

Shen, Y.C., Fan, J.H., Edenberg, H.J., Li, T.K., Cui, Y.H., Wang, Y.F., Tian, C.H., Zouh, C.F., Zhou, R.L., Wang, J., Zhao, Z.L., and Xia, G.L. (1997) Polymorphism of ADH and ALDH genes among four ethnic groups in China and effects upon the risk for alcoholism. *Alcohol. Clin. Exp. Res.*, **21**, 1272–1277.

Stoler, J.M., Ryan, L.M., and Holmes, L.B. (2002) Alcohol dehydrogenase 2 genotypes, maternal alcohol use, and infant outcome. *J. Pediatr.*, **141**, 780–785.

Streissguth, A.P. and Dehaene, P. (1993) Fetal alcohol syndrome in twins of alcoholic mothers: concordance of diagnosis and IQ. *Am. J. Med. Genet.*, **47**, 857–861.

Surani, M.A. (1998) Imprinting and the initiation of gene silencing in the germ line. *Cell*, **93**, 309–312.

Thomas, J.D., Abou, E.J., and Dominguez, H.D. (2009) Prenatal choline supplementation mitigates the adverse effects of prenatal alcohol exposure on development in rats. *Neurotoxicol. Teratol.*, **31**, 303–311.

Thomasson, H.R., Beard, J.D., and Li, T.K. (1995) ADH2 gene polymorphisms are determinants of alcohol pharmacokinetics. *Alcohol. Clin. Exp. Res.*, **19**, 1494–1499.

Treadwell, J.A. and Singh, S.M. (2004) Microarray analysis of mouse brain gene expression following acute ethanol treatment. *Neurochem. Res.*, **29**, 357–369.

Uddin, R.K., Treadwell, J.A., and Singh, S.N. (2005) Towards unraveling ethanol-specific neuro-metabolomics based on ethanol responsive genes *in vivo*. *Neurochem. Res.*, **30**, 1179–1190.

Ueno, Y., Adachi, J., Imamichi, H., Nishimura, A., and Tatsuno, Y. (1996) Effect of the cytochrome P-450IIE1 genotype on ethanol elimination rate in alcoholics and control subjects. *Alcohol. Clin. Exp. Res.*, **20** (Suppl.), A17–A21.

Valles, S., Pitarch, J., Renau-Piqueras, J., and Guerri, C. (1997) Ethanol exposure affects glial fibrillary acidic protein gene expression and transcription during rat brain development. *J. Neurochem.*, **69**, 2484–2493.

Verlaan, M., Morsche, R.H., Roelofs, H.M., Laheij, R.J., Jansen, J.B., Peters, W.H., and Drenth, J.P. (2003) Glutathione S-transferase Mu null genotype affords protection against alcohol induced chronic pancreatitis. *Am. J. Med. Genet. A*, **120**, 34–39.

Viljoen, D.L., Carr, L.G., Foroud, T.M., Brooke, L., Ramsay, M., and Li, T.K. (2001) Alcohol dehydrogenase-2*2 allele is associated with decreased prevalence of fetal alcohol syndrome in the mixed-ancestry population of the Western Cape Province, South Africa. *Alcohol. Clin. Exp. Res.*, **25** (12), 1719–1722.

Wall, T.L., Garcia-Andrade, C., Thomasson, H.R., Cole, M., and Ehlers, C.L. (1996) Alcohol elimination in Native American Mission Indians: an investigation of interindividual variation. *Alcohol. Clin. Exp. Res.*, **20**, 1159–1164.

Wang, L.-L., Zhang, Z., Li, Q., Yang, R., Pei, X., Xu, Y., Wang, J. Zhou, S.-F., and Li, Y. (2009) Ethanol exposure induces differential microRNA and target gene expression and teratogenic effects which can be suppressed by folic acid supplementation. *Hum. Reprod.*, **24**, 562–579.

Warren, K.R. and Li, T.-K. (2005) Genetic polymorphisms: impact on the risk of fetal alcohol spectrum disorders. *Birth Defects Res. A*, **73**, 195–203.

Xu, Y., Tang, Y., and Li, Y. (2008) Effect of folic acid on prenatal alcohol-induced modification of brain proteome in mice. *Br. J. Nutr.*, **99**, 455–461.

Yanaguita, M.Y., Gutierrez, C.M., Ribeiro, C.N., Lima, G.A., Machado, H.R., and Peres, L.C. (2008) Pregnancy outcome in ethanol-treated mice with folic acid supplementation in saccharose. *Childs Nerv. Syst.*, **24**, 99–104.

Zeisel, S.H. (2006) The fetal origins of memory: the role of dietary choline in optimal brain development. *J. Pediatr.*, **149**, S131–S136.

Zeisel, S.H. and Niculescu, M.D. (2006) Perinatal choline influences brain structure and function. *Nutr. Rev.*, **64**, 197–203.

7
Diagnosis of FASD: An Overview
Gail Andrew

Fetal Alcohol Spectrum Disorder (FASD) is the umbrella term currently used to denote the complex patterns of difficulty in function and abnormalities that can result from Prenatal Alcohol Exposure (PAE) (Chudley *et al.*, 2005; Stratton, Howe, and Battaglia, 1996; Bertrand *et al.*, 2004; Benz, Rasmussen, and Andrew, 2009). There may be physical defects associated with PAE in many organ systems, depending on the timing of the alcohol exposure in the pregnancy and the vulnerability of the developing tissues (Stratton, Howe, and Battaglia, 1996; Clarren and Smith, 1978; Aase, Jones, and Clarren, 1995; Abel, 1998a). For example, the sentinel findings of the facial dysmorphology seen in full Fetal Alcohol Syndrome (FAS) develop with alcohol exposure days 19 to 21 very early in the gestation, and often before the pregnancy is recognized by the birth mother (Sulik, Johnston, and Webb, 1981). However, in most clinical studies of children exposed prenatally to alcohol, only about 10% have significant facial features; in most individuals with PAE, FASD is an "invisible disability," with no physical markers. The brain dysfunction that results from the organic brain damage from the teratogen alcohol is the common presentation (Rasmussen *et al.*, 2008; Riley and McGee, 2005). The maternal use of alcohol in the pregnancy is not diagnostic of FASD, but identifies a risk factor for brain damage. The resulting deficits in function are a result of many complex interactions involving maternal and fetal genetics, the maternal–fetal environment (such as maternal nutrition and stress), and postnatal factors that may be supportive or may compound the brain damage (Chudley *et al.*, 2005; Stratton, Howe, and Battaglia, 1996; Bertrand *et al.*, 2004).

The prevalence of FASD has been estimated at nine per 1000 births, with higher numbers reported in certain populations (Alberta Alcohol and Drug Abuse Commission, 2004; May *et al.*, 2009; Druschel and Fox, 2007). Active surveillance is difficult, as there is not adequate access to trained diagnostic services for most individuals with PAE (Clarren and Lutke, 2008; Peadon *et al.*, 2008). Moreover, many affected individuals are diagnosed with Attention Deficit Hyperactivity Disorder, Cognitive Impairment, Learning Disability, or Mental Health Disorders. The connection of that individual's difficulties back to the PAE may not be made, which would result in a missed opportunity to understand that person from the perspective of organic brain damage, and to provide support based on the model

Fetal Alcohol Spectrum Disorder–Management and Policy Perspectives of FASD. Edited by Edward P. Riley, Sterling Clarren, Joanne Weinberg, and Egon Jonsson
Copyright © 2011 WILEY-VCH Verlag GmbH & Co. KGaA, Weinheim
ISBN: 978-3-527-32839-0

of brain damage that focuses on changing the environment to support the individual rather than changing them. There are also complex issues around obtaining the history of alcohol use in the pregnancy. For example, if a judgmental approach is used with the birth mother, it may result in blaming and shaming, without considering the complex psychosocial and mental health factors associated with the substance abuse. Mothers may not report substance use as they fear the apprehension of their child into the child protection system (Poole, 2007, 2008; Tough, Clarke, and Cook, 2007). The impact of the disability presents across multiple systems of health, child welfare, social services, education, justice, addictions, employability, and homelessness.

The cost of FASD to society is considerable, with a recent Canadian study estimating the adjusted annual cost to be CA$21 642 for each person affected (Stade et al., 2009). The Institute of Health Economics in Alberta (Thanh and Jonsson, 2009) reported the cost of FASD in Alberta to be CA$ 000000. There is also the invisible cost of the burden on the family and caregivers, as well as the loss of the potential of an individual with FASD (Streissguth et al., 2004; Streissguth et al., 1999; Connor and Streissguth, 1996).

7.1
History of Diagnosing FASD

Although details of the harm resulting from prenatal alcohol exposure can be found in the biblical literature and in classic art, the first medical reports were provided by Lemoine in France in 1968, and Jones and Smith in the United States in 1973 (Jones et al., 1973). These initial studies were based on children who had common physical features of the face of FAS pre- and postnatal growth deficiencies (Astley, 2010; Figure 7.1), and problems in cognition and behavior who had been born to alcoholic mothers. Subsequently, children without the physical features, but with a history of PAE and central nervous system dysfunction have been identified; subsequently, the term Fetal Alcohol Effects (FAE) was used to describe this group. The term partial Fetal Alcohol Syndrome (pFAS) is applied if some of the facial features are present along with neurodevelopmental difficulties, and prenatal exposure to alcohol has been confirmed (Chudley et al., 2005; Stratton, Howe, and Battaglia, 1996; Bertrand et al., 2004; Sokol and Clarren, 1989; Clarren and Smith, 1978). Although the term FAE is no longer used, it is often found in past literature (Aase, Jones, and Clarren, 1995). In the longitudinal follow-up studies conducted by Streissguth (Streissguth et al., 1999), those individuals who were diagnosed with FAE rather than FAS demonstrated a poorer life outcome as they were diagnosed later and did not qualify for support services.

In 1996, the Institute of Medicine developed the term Alcohol-Related Neurodevelopmental Disorder (ARND) to describe the group that did not have the physical features, but showed a "... complex pattern of behavioral or cognitive abnormalities

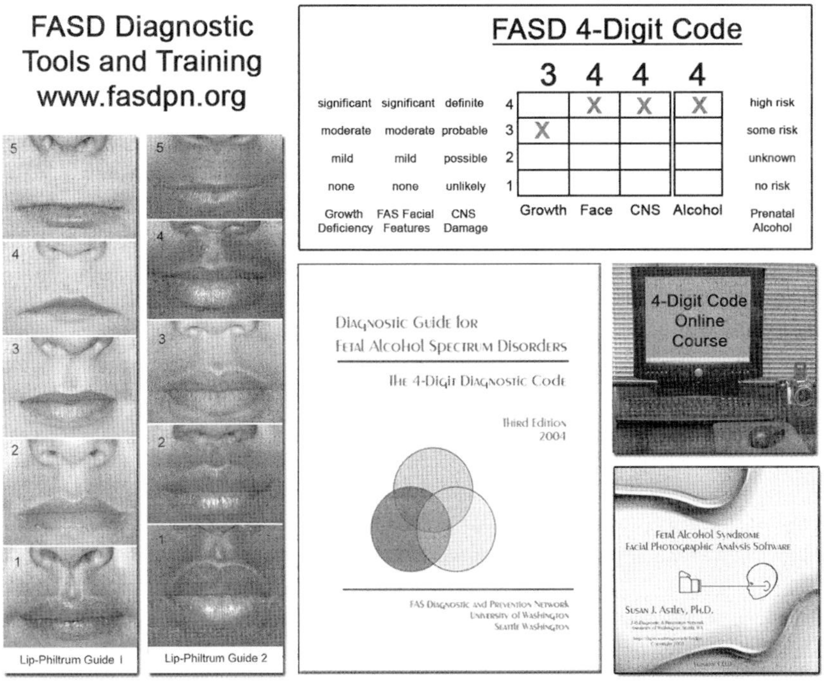

Figure 7.1 Physical features of the face of FAS pre- and postnatal growth deficiencies.
© 2010, Susan Astley PhD, University of Washington.

inconsistent with developmental level and not explained by genetic predisposition, family background or environment alone." (Stratton, Howe, and Battaglia, 1996). Likewise, the term Alcohol-Related Birth Defect (ARBD) was developed to describe the malformations in other organ systems that had occurred as a result of the teratogen alcohol causing disruption during fetal development. This differentiation is important to recognize when providing healthcare to individuals with FASD, and can be missed when the drivers for assessment are disruptive behaviors, learning difficulties, and maladaptation (Stratton, Howe, and Battaglia, 1996; Abel, 1998a; Burd et al., 2007; Hofer and Burd, 2009). For example, the incidence of congenital heart disease is a significant health risk (Burd et al., 2007). Currently, FASD serves as an "umbrella" term for the FAS, pFAS, and ARND categories (Chudley et al., 2005).

To date, the major question has been how best to measure and categorize the brain dysfunction that has resulted from the PAE. Typically, the brain damage may present as cognitive deficits, learning disabilities, attentional difficulties, hyperactivity and impulsivity, motor deficits, functional communication impairments, memory deficits, difficulties in regulation and adaptation in response to the environment, and also as deficits in executive functions such as planning, flexible thinking, shifting and inhibiting, and judgment. This type of assessment requires a trained multidisciplinary team (see Section 7.4). Today, further research is required in the area of brain dysfunction from PAE involving multidisciplinary research teams, including neurobiologists, neuropyschologists and neuroimaging specialists, to provide evidence for the brain basis for these dysfunctions (Astley, 2004; Astley and Clarren, 2000; Rasmussen et al., 2008; Streissguth et al., 2004; Rasmussen, Horne, and Witol, 2006; Coles et al., 2002; Riley and McGee, 2005; Mattson and Riley, 1998; Mattson and Roebuck, 2002; Mattson, Calarco, and Lang, 2006; Rasmussen and Bisanz, 2009; Rasmussen, 2005; Burden et al., 2005; Kodituwakku, 2007; Streissguth, 2007; Crocker et al., 2009; Burden et al., 2009; Santhanam et al., 2009; Whaley, O'Connor, and Gunderson, 2001; Riley et al., 2003; McGee et al., 2008; Aragon et al., 2008; Fryer et al., 2007; Pei et al., 2008; Greenbaum et al., 2009; Jirikowic, Olson, and Kartin, 2008). The diagnostic process will be informed by this research.

7.2
How Does Prenatal Alcohol Exposure Cause Damage?

The role of alcohol ingested during pregnancy by the mother, in causing direct fetal cell damage or disruption in the normal sequential process of cell differentiation, migration and maturation, has been demonstrated using animal models (Abel, 1996; Sulik, Johnston, and Webb, 1981; Weinberg, 1998; Weinberg et al., 2008; Sakar et al., 2007; Lee et al., 2008; Abel, 1998b; Camarillo and Miranda, 2008). In one study, the characteristic lip and philtrum (the area from below the nose to the upper lip) development seen in the facial features of FAS was found to result from exposure to alcohol during days 19 to 21 in human gestation, as extrapolated from an animal model. The midfacial underdevelopment results from a disruption of the fetal frontal forebrain development (Sulik, Johnston, and Webb, 1981). There is research evidence of the effect of alcohol on neuron damage at the cellular level (Weinberg, 1998); disruption in neurotransmitter release and binding have also been suggested (Lee et al., 2008). Disruption in the fetal hypothalamic–pituitary–adrenal (HPA) axis (which regulates the response of the body to acute and chronic stress) as a result of *in utero* alcohol exposure has also been reported (Weinberg, 1998; Weinberg et al., 2008; Sakar et al., 2007). The resultant high levels of cortisol may have significant pre- and postnatal implications for both brain and endocrine function.

Alcohol use by the pregnant mother also has an indirect effect on the development of the fetus, and into the early postnatal period. Alcohol used by the pregnant mother can alter the placenta blood vessels via changes in prostaglandins, and this in turn can affect oxygen and nutrient flow to the fetus, impacting on growth and cell maturation (Abel, 1998b). Alcohol has a significant impact on maternal nutrition and health, and may be associated with the use of other potential teratogenic drugs. For example, the infants in subsequent pregnancies are more severely affected if the mother continues to use alcohol (May et al., 2008).

Whilst alcohol use by the male partner does not directly cause damage to fetal cells, it may cause problems *indirectly* by contributing to maternal stress, with subsequent damage being mediated through the HPA stimulation and increased cortisol production in both mother and fetus. Whereas, in animal studies, the amount of exposure to alcohol, timing of the exposure in gestation, nutrition and level of stress in the environment can be controlled, in the "human" situation multiple complex and compounding variables within the context of life are introduced. Such factors include maternal, paternal and fetal genetics, and also epigenetics. (Note: Epigenetic changes result from interactions between the environment and the gene in which gene functions are turned off or on. Such changes in function may be passed to the offspring, without any alteration in the gene DNA sequence; Haycock, 2009.) Quite often, the mother may also have used other teratogens, such as cocaine, methamphetamines, and prescription drugs. The impact of the postnatal environment has been investigated by Bruce Perry and other research groups working in the area of child maltreatment (Anda *et al.*, 2006; Oppenheim and Goldsmith, 2007).

It has been shown previously, that early deprivation, abuse and a lack of appropriate stimulation can result in brain damage. If the brain has already been exposed to alcohol *in utero*, it is even more vulnerable to this postnatal insult. Adverse postnatal environments also represent missed opportunities to provide interventions in critical periods of brain development. Thus, the diagnosis of FASD requires careful consideration of all these interrelated complex factors.

7.3
Screening for FASD

Screening helps to identify those individuals who require further assessment for a diagnosis of FASD. Notably, the screening tools must be valid, specific, and sensitive for the condition, and also applicable to the population of interest. In addition, access must be available to diagnostic services in a timely manner for those who screen positive. There also needs to be access to supports and interventions following a diagnosis. Currently, a range of different screening tools for FASD is being explored, but further research is required.

In recent years, neuroimaging studies in individuals with PAE have contributed greatly to the evidence that alcohol causes brain damage. Although most clinical structural magnetic resonance imaging (MRI) investigations are reported as normal, the use of more advanced research techniques, such as structural MRI and brain volumetric studies, have demonstrated a reduction in the corpus callosum (which connects the left and right cerebral hemispheres), the cerebellum (which plays an important role in motor control and language) and the caudate (an important area in emotional regulation) (Archibald *et al.*, 2001; Spadoni *et al.*, 2007; Wozniak *et al.*, 2002). Astley has demonstrated a decrease in frontal lobe volume with increasing severity of dysmorphic facial features (Astley *et al.*, 2009a, 2009b), while diffusion tensor imaging MRI studies have identified reductions in white matter pathways connecting from the frontal lobes (Sowell et al., 2008a; Lebel et al., 2008; Sowell et al., 2008b; Wozniak et al., 2006; Fryer et al., 2009; Wozniak et al., 2002; Lebel et al., 2010; Guerri et al., 2009). Functional MRI studies have demonstrated less activation in the prefrontal area with increasing complexity of task. Some of these neuroimaging studies have been correlated with neuropsychological testing that can link deficits in structure with function (Lebel et al., 2010; Astley et al., 2009a, 2009b). At present, there is no distinctive "MRI picture" for FASD; however, neuroimaging remains an informative method of research tool, despite being of minimal value for screening or diagnosis.

Although attempts have also been made to identify biological markers for PAE, no definitive tests have emerged to date. Whilst a facial dysmorphology with a flat philtrum, a thin upper lip and short palprebral fissures (the length of eye slits) is characteristic of FAS, it is important to develop a differential diagnosis for other genetic syndromes that cause similar facial features, as well as confirming any prenatal exposure to alcohol. Clinical measurements or digital photography and computer-based analyses of the face may easily be carried out by experienced physicians who have been trained in this area (Clarren et al., 1987; Astley and Clarren, 2001; Jones, Robinson, and Bakhireva, 2006; Clarren et al., 2010; Astley, 2010). There are new norms of palprebral fissure size for the Canadian population that have recently been published (Clarren et al., 2010). The limitation here is that in most clinic populations, less than 10% of patients have significant facial features. In more high-risk populations (such as children in the fostercare system) the proportion with positive facial features is much higher (Astley et al., 2002). Screening within the school system for dysmorphic features of FAS (May et al., 2009) must be correlated with any history of PAE. There remains a need to define the level of brain dysfunction for each individual with the facial features. This, in turn, requires access to a multidisciplinary team that has been trained in FASD diagnosis.

A laboratory analysis of the newborn *meconium* (the first bowel movement by the newborn) for evidence of fatty acid ethyl esters (FAEE); this serves as a marker for both acute and chronic alcohol intake and has been shown to correlate with PAE (Peterson et al., 2008; Gareri et al., 2008; Pichini et al., 2008; Burd and Hofer, 2008). Unfortunately, however, the meconium is not produced until week 13 of gestation; consequently, if the alcohol exposure has occurred within the first 13 weeks, this will result in a negative FAEE test and give a false reassurance. A positive meconium FAEE test, however, will indicate prenatal exposure to alcohol and identify if an infant is at risk. This is not diagnostic of FASD and, as a result, the developmental follow-up of the infant will be required for an extended period, most likely into middle childhood or adolescence, in order to identify any dysfunction resulting from brain damage and to confirm a diagnosis of FASD. Whilst other screening tests have been proposed, including checklists of behavior, there is at present no evidence of their specificity or sensitivity for FASD (Canadian Association of Pediatric Health Centers, 2008).

One biological marker currently of interest is the measurement of saccadic eye movements (fast eye movements), that depend on oculomotor neural circuitry brain pathways which involve the frontal cortex, parietal cortex, and basal ganglia. This eye-tracking difficulty may serve as an objective marker of brain damage in FASD (Green et al., 2009). The test requires a program on a laptop computer that is easily transportable to screen individuals. Any individuals who screen positive will require further functional assessments to define their own patterns of brain weakness and strengths. At this point in clinical practice, however, the most important screening point is to confirm a history of PAE in the pregnancy, based on reliable sources, and also to seek evidence that the individual is indeed struggling in terms of function (Jacobson et al., 2002). FASD is a lifelong disability, and will present in different ways at different life stages and societal expectations. Often, the major challenge for adults is to link their difficulties back to the PAE.

7.4
The Diagnostic Process

The use of guidelines for the diagnosis of FASD has been proposed to provide consistency and reproducibility of diagnosis across clinical centers. An accurate diagnosis is essential in order to obtain data relating to the prevalence and patterns of brain dysfunction. Changes in the incidences of FASD over time is essential to measure the impact of programs to prevent alcohol use in pregnancy on reducing FASD. Without a rigorous diagnosis, it would be difficult to determine which intervention or strategy might be of benefit and should be replicated at other sites. This will have major implications for the policy makers with regards to planning resources. In 1996, the Institute of Medicine brought together experts in the field to consider "Diagnosis, Epidemiology, Prevention and Treatment of Fetal Alcohol Syndrome" (Stratton, Howe, and Battaglia, 1996). This led to a description of the four key features of alcohol exposure, growth deficiency, facial features and brain dysfunction, as well as a critical recognition to examine the multifactorial ameliorating and exacerbating factors. Subsequently, in 1997, Astley and Clarren, from the Diagnosis and Prevention of FAS Network at the University of Seattle in Washington, developed the 4-Digit Code (Astley, 2004; Astley, 2010; Astley and Clarren, 2000), which provided an objective and quantitative approach to the four key features (growth, face, brain function, and alcohol exposure), as well as rating both the pre- and postnatal factors. In this case, a 4-point Likert scale was used to rate the features as either normal or absent (level 1) up to a severe presentation of the feature or definite evidence of brain damage (level 4) (Astley, 2004; Astley, 2010) (Figure 7.2). Since then, as further information on brain function in FASD has been acquired, the 4-Digit Code has been continually revised, most recently in 2004. The Canadian Guidelines for Diagnosis were published in 2005, following a consultative process with experts in the field of FASD (Chudley et al., 2005), and the details of a similar process in the United States were published in 2004, by the Centers of Disease Control (Bertrand et al., 2004). Subsequently, other clinical groups have investigated diagnostic approaches involving more extensive dysmorphology analysis (Hoyme, May, and Kalberg, 2005; Astley, 2006). Currently, in

FASD 4-Digit Code

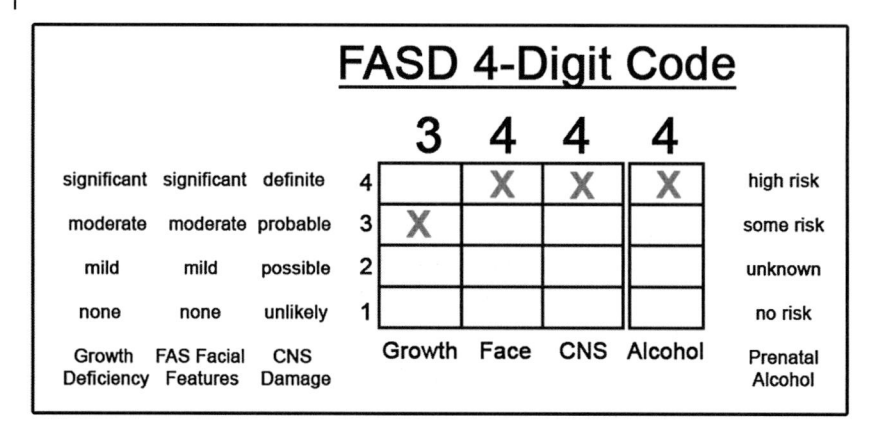

				3	4	4	4	
significant	significant	definite	4		X	X	X	high risk
moderate	moderate	probable	3	X				some risk
mild	mild	possible	2					unknown
none	none	unlikely	1					no risk
Growth Deficiency	FAS Facial Features	CNS Damage		Growth	Face	CNS Alcohol		Prenatal Alcohol

Figure 7.2 The 4-Digit Code, as developed by the FAS Network at the University of Seattle in Washington. © 2010, Susan Astley PhD, University of Washington.

western Canada the 4-Digit Code is used in most clinics that provide a diagnosis of FASD (Canada Northwest FASD Research Network).

The current best practice in the diagnosis of FASD requires a multidisciplinary team input to the diagnostic process, in order to define the aspects of brain function (Chudley et al., 2005; Canada Northwest FASD Research Network). By consensus with expert multidisciplinary teams, eight domains or areas of brain function to be assessed have been identified, including: (i) intellectual; (ii) academic achievement; (iii) attention; (iv) motor, sensory, visual and spatial; (v) communication at both basic and higher levels of functional and social communication; (vi) memory, including encoding, retrieval and working memory; (vii) executive functions such as judgment, inhibition and initiation, mental flexibility, switching, problem solving, planning and sequencing; and (viii) adaptive functions, such as the ability to function that impacts independent living, employability, and not being victimized. These areas of dysfunction are often out of keeping with the tested intellectual ability; hence the need for more complex testing by members of the multidisciplinary team.

Definite evidence of brain damage includes microcephaly (small head circumference; below 2 SD), severe mental impairment, cerebral palsy, definite abnormalities on neuroimaging, and seizures, although other causes for these neurological disorders must be ruled out. The core assessment team is comprised of a Psychologist/Neuropsychologist, a Speech and Language Pathologist, an Occupational Therapist, a Social Worker, and a Physician:

- The *Social Worker* has a major role in assisting the individual and family through the entire diagnostic process. This involves review of past psychosocial history, assessment of current stability of placement, and the needs of both child and caregiver. After diagnosis, the Social Worker works with the family to ensure that the diagnosis leads to positive changes and supports.

- The *Psychologist* and *Speech Language Pathologist* assess from basic to higher level brain functions in their respective areas.

- The *Occupational Therapist* consultation is important especially in the younger age group to assess motor skills, motor planning and sensory difficulties.

- The *Physician* on the team has a core role, including dysmorphology, physical growth, neurological findings, and basic health determinants that can impact function. The *Physician* also provides an assessment of behavior and mental health and contributes to the psychosocial assessment, along with the *Psychologist* and *Social Worker*.

The mental illnesses of individuals with FASD have been considered not to be part of the primary organic brain damage from the alcohol, but rather to be a secondary disability as a result of not being diagnosed at an early age, not being provided with good supports, having multiple placements, and being victims of neglect and abuse (Clark et al., 2004; Streissguth and O'Malley, 2000; Barr et al., 2006; Connor and Streissguth, 1996; Lynch et al., 2003). There may be some mental disorders that are part of the brain damage, and further research is required in this area. Mental illness and involvement with the law may emerge as the presentation of adolescents and adults with FASD.

Formulating the diagnosis requires a synthesis of all information gathered from school, home and community and objective testing and observation on the assessment day, a consideration of all the other life events, and genetics that could be impacting the presentation and ruling out other causes in a differential diagnosis approach. There is also need to consider other co-morbid conditions that are compounding the situation. The assessment requires moving from basic testing to a more complex assessment of executive functions. The Canada Northwest FASD Research Network, established by the combined interests and commitment to FASD from the Canadian Provinces and Territories of Manitoba, Sasaktchewan, Alberta, British Columbia, Yukon, North West Territories and Nunavet, has addressed many aspects of FASD from Diagnosis, Prevention and Intervention. One task has been to examine the tools in each domain that form part of the battery of tests for an assessment for FASD. A consensus process was carried out with the Canada Northwest FASD Diagnostic Clinic Teams to determine the best tools to define each of the core domains (Canada Northwest FASD Research Network). This is likely to evolve as research provides further understanding of the deficits in the brain damage of FASD, and how best such deficits can be measured (Rasmussen et al., 2008; Streissguth et al., 2004; Rasmussen, Horne, and Witol, 2006; Coles et al., 2002; Riley and McGee, 2005; Mattson and Riley, 1998; Mattson and Roebuck, 2002; Mattson, Calarco, and Lang, 2006; Rasmussen and Bisanz, 2009; Rasmussen, 2005; Burden et al., 2005; Kodituwakku, 2007; Streissguth, 2007; Crocker et al., 2009; Burden et al., 2009; Santhanam et al., 2009; Whaley, O'Connor, and Gunderson, 2001; Riley et al., 2003; McGee et al., 2008; Aragon et al., 2008; Fryer et al., 2007; Pei et al., 2008; Greenbaum et al., 2009; Jirikowic, Olson, and Kartin, 2008). In order to increase the diagnostic capacity, newly emerging diagnostic teams must be trained in the diagnostic process. Training will need to be conducted by established training teams that will provide ongoing mentoring as the new teams acquire experience in the field.

The diagnostic assessment represents the first step in understanding the individual with FASD, and must lead to positive supports and interventions. There is, however, a great deal of work to be done both before and after the diagnostic assessment. If there is not a community readiness before the assessment to support the child, youth or adult, then the diagnosis will lead to blaming of the birth mother and her cultural or socioeconomic group. There is also the risk of simply labeling the individual with FASD, and giving up on them. The community may not be prepared to put in place the supports for women to prevent alcohol use in pregnancy, while addressing all of the complex psychosocial and mental health factors that can influence "why women drink" (Poole, 2007, 2008). The diagnosis of an individual with FASD must lead to positive interventions to support that person across areas of education, medical, mental health, employability and living needs. Borrowing from the model of services recommended for individuals with traumatic brain damage (e.g., after automobile accidents), it is well known that the environment must be modified to support the individual, while not expecting the individual with brain injury to change. This basic principle must be applied, likewise, to the brain injury of FASD. In addition, supports for caregivers must be put in place so as to prevent caregiver "burn out" (Brown and Bednar, 2003); this is a major problem when so many children and youth with FASD are in the foster-care system (Brown, Sigvaldason, and Bednar, 2005; Burnside and Fuchs, 2009) or living with extended family members. The caregiver needs help to understand the disability of FASD, to put in place any appropriate expectations, and to prepare for the natural life transitions that result in a need for even more support in FASD. Many adults with FASD do not have a significant advocate to support them, and part of the diagnostic process will be to connect them to a system of support (see also Chapter 16).

7.5
FASD Across the Lifespan

The brain damage from the prenatal alcohol exposure is static and happens in utero. However, progression from infancy to adulthood is a dynamic process with developmental expectations and transitions at multiple levels. Forming neural connections in response to input from the environment is part of normal development. This may not progress as expected in the brain already damaged in utero. In the very early years of life and if the child is living in a supportive environment, the child may present as quite functional. As the need to become more independent, make decisions, problem solve, use judgment, do more complex mental flexibility naturally occurs with growing up, individuals with FASD become more dysfunctional. They are often described as being like much younger than their age in spite of physical maturation. FASD is a lifelong disability and in a model of best practice an individual with PAE would have access to longitudinal surveillance of their needs for support for learning, behavioral and emotional regulation and daily living functions. Another variable is that their caregivers may change, especially for children who are part of the foster care system. In children in care of their birth parents or extended family members, multiple moves may occur around the physical and mental health challenges of their biological family members. If the birth mother also has FASD, the parenting is further compounded by her own disability (Lynch et al., 2003). Predicting and supporting the needs of the caregiver at key life points is necessary and there are promising models of mentoring and coaching. Research is needed to identify what are the best supports at different ages and to evaluate the effectiveness by improved outcomes. A brief review will be provided of how FASD can present at key life stages.

7.5.1
Diagnosis in the Neonatal Period and Early Infancy

The neonatal period and early infancy are critical times to identify the risk factors of PAE, for both child and birth mother (Jacobson et al., 2002). The newborn may present with a small-for-gestational age growth pattern, and possible ARBDs that require medical monitoring. The characteristic facial features may be less evident in the newborn (Coles, Platzman, and Smith, 1992; Stoler and Holmes, 2004), although in severe cases there will be evidence of microcephaly (head circumference below 2SD). Some infants may have difficulties with neurological regulation, showing jitteriness, a poor coordination of suck and swallow, and disorganized movement patterns. This situation is much more common in infants exposed *in utero* to both drugs and alcohol, and is referred to as Neonatal Abstinence Syndrome (NAS) (American Academy of Pediatrics, 1998). Infants with physiological variations in digestion and respiration will require a medical and sensory environmental approach in a special care nursery. The symptoms may persist for several weeks, and not be immediately recognized if there is no suspicion of substance exposure *in utero*. Consequently, a careful discharge planning must be carried out for early infant care, in order to optimize development and prevent failure to thrive. The newborn period represents the best time to review the maternal records of substance use and other compounding factors that might lead to the woman remaining at risk for continued substance abuse; these include domestic violence, poverty, and a lack of positive support systems. The mother must be connected with supports so as to prevent alcohol exposure in subsequent pregnancies (Poole, 2007, 2008), and there is indeed some evidence that programs based on a mentoring model can help the mother to address this complex issue (Grant et al., 2005). Unfortunately, however, mothers are often very reluctant to admit to the use of alcohol or drugs at the time of delivering their baby, as this will most likely result in child welfare involvement and loss of the child to the foster-care system. There is a need for support programs for these high-risk mothers, the aim being to keep the family unit together (Poole, 2007, 2008), yet unfortunately many of these vulnerable infants and mothers are discharged from the delivery hospital without supports in place. This sets the infant up for ongoing feeding difficulties, a failure to thrive, a less than optimal stimulation and, in the worst case, neglect and abuse. Those children who do enter the foster-care system may also be vulnerable if their caregiver does not have access to information relating to the PEA (Brown, Sigvaldason, and Bednar, 2005). The caregiver must know how to set up the environment in order to avoid overstimulation and to provide facilitated developmental opportunities. The child with PAE should be provided with anticipatory monitoring for delays in development by a collaborative and integrated system of care involving health, infant mental health, and early intervention.

7.5.2
Diagnosing in Toddlerhood

In toddlerhood and during the preschool years, children may present with delays in acquiring early milestones in motor, communication, and play skills, although this is neither specific nor diagnostic of FASD. The children may have a disorganized approach to motor skills and play. They may present as hyperactive with a short attention span, and with easy distractability and impulsivity that is suggestive of an emerging pattern of attention deficit hyperactivity disorder (ADHD) (Rasmussen *et al.*, 2008; Crocker *et al.*, 2009). They may also have a disordered language pattern, where their *understanding* of language is worse than their *use* of language (O'Leary *et al.*, 2009). They may have difficulties with sensory regulation and social boundaries, although in a child with multiple placements and attachment issues it is difficult to determine whether this is FASD or an attachment disorder (Jirikowic, Olson, and Kartin, 2008; Oppenheim and Goldsmith, 2007). The paradox during the early years is that the evidence for brain damage is not specific, and is often compounded by a less-than supportive environment, just at the time when maximal brain stimulation can make the most impact. This dilemma can be solved by ensuring that all children at risk with PAE receive supports that are guided by their functional needs, and not by a medical diagnosis (Premji *et al.*, 2007). The challenge to the intervention systems is to identify the best practices for this age group (O'Leary *et al.*, 2009; Peadon *et al.*, 2009). Caregivers must provide appropriate stimulation, yet not assume that the child has FASD. The child at risk will require continued developmental monitoring, and must be ensured stability with their caregiver. Recent investigations have shown that this can help the child to reach their full potential, and reduce any secondary disabilities resulting from multiple moves, as well as being subject to neglect and not understood (Streissguth *et al.*, 2004).

7.5.3
Diagnosing in School Age

At the age of school entry into kindergarten (5–6 years), many children with PAE begin to experience difficulties in learning. They might also have difficulties with attending, and be diagnosed as having ADHD, which although being a common comorbidity of FASD is also a common childhood developmental disorder (Crocker et al., 2009). ADHD is, of itself, not diagnostic of FASD unless accompanied by definite evidence of brain damage. Further difficulties with understanding language and the children expressing themselves may be evident; they are often described as "very talkative" but do not always connect their comments or provide relevant details for their conversation to be followed by an unfamiliar person. They may even present as more competent than they really are, by their good basic language, but when they are older a formal assessment of their functional use and higher-order language will identify these core deficits. Difficulties in learning and self-regulation can be identified at this young age, but are not specific for evidence of brain damage and a diagnosis of FASD. A child in a chaotic environment or multiple moves with different caregivers may have many of these characteristics from environmental factors alone (Stoler and Holmes, 2004). Testing instruments to identify brain damage are limited in this age group.

Recent investigations have shown that, at the ages of 8–10 years, in grades 3 to 4, those children with PAE begin to show more significant difficulties (Rasmussen et al., 2008). This is the natural time for a child to be able to do think more independently, to problem-solve and organize and plan for themselves, and to handle more abstract concepts for learning. Hence, at this time they need to engage in more complex communication and social functions, and to self-regulate. These are the executive functions that are deficient in PAE-mediated brain damage (Rasmussen et al., 2008; Streissguth et al., 2004; Rasmussen, Horne, and Witol, 2006; Coles et al., 2002; Riley and McGee, 2005; Mattson and Riley, 1998; Mattson and Roebuck, 2002; Mattson, Calarco, and Lang, 2006; Rasmussen and Bisanz, 2009; Rasmussen, 2005; Burden et al., 2005; Kodituwakku, 2007; Streissguth, 2007; Crocker et al., 2009; Burden et al., 2009; Santhanam et al., 2009; Whaley, O'Connor, and Gunderson, 2001; Riley et al., 2003; McGee et al., 2008; Aragon et al., 2008; Fryer et al., 2007; Pei et al., 2008; Greenbaum et al., 2009; Jirikowic, Olson, and Kartin, 2008). The child is unable to keep up academically, especially in the language demands of learning and in math skills. They are often socially immature, have difficulties understanding social cues, and have poor perspective-taking, all of which increase the risk for victimization. They have more difficulties during unstructured times and when they are out of routine. Frequent comments from caregivers include: poor judgment, not learning from consequences, needing constant reminders even for daily living activities, lying or not connecting thoughts, mood swings and difficulties with self-regulation, problems with planning, and sequencing needed in many tasks. Objective testing using standardized assessment tools across multiple domains by a multidisciplinary team, including neuropsychological tests with the child, can identify the presence of brain damage. This direct assessment process with the child is collaborated with standardized rating scales and descriptions of the child at home, school and other environments, so as to identify any significant impairment in function. This supports the FASD diagnosis (Chudley et al., 2005; Stratton, Howe, and Battaglia, 1996; Bertrand et al., 2004; Benz, Rasmussen, and Andrew, 2009; Rasmussen et al., 2008; Streissguth et al., 1999).

There are often emerging mental health difficulties that may be primary to the organic brain damage from the PAE, or have been a result of the early adverse life experiences that compound the FASD diagnosis. All of these difficulties must be equally addressed in the interventions and supports that emerge from the diagnostic process. Supports for education, social functioning and personal safety must be provided, based on the individual's pattern of strengths and deficits. The shift to the model of supports for brain damage is important if the caregivers are to understand the need to set up the environment for support, and must result in a shift from the child "will not" to "cannot." Many children with FASD are often given a diagnosis of Oppositional Defiant Disorder, before their disability is understood from a brain damage perspective. Further research is required to provide the evidence for best practices in school programming and community living for these children.

7.5.4
Diagnosing in Adolescence and Adulthood

Throughout adolescence and into adulthood, there is increasing evidence that the individual with FASD is unable to function in life independently, and that this is out of keeping with intellectual ability. Mental health issues of depression, anxiety and externalizing difficulties of aggression and poor mood regulation are often the presenting symptoms (Howell *et al.*, 2006; Clark *et al.*, 2004; Streissguth and O'Malley, 2000; Barr *et al.*, 2006; Connor and Streissguth, 1996; Lynch *et al.*, 2003). There may also be issues of victimization, of being taken advantage by others of in terms of involvement in criminal activity, and of sexual exploitation. Difficulties with money, time management and planning lead to repeated failures in day-to-day life. Individuals with FASD are at high risk for substance abuse themselves, given their genetic factors and the tendency to self-medicate with alcohol and drugs, all of which adds to the complexity of FASD diagnosis in late adolescents and adults. The adult may also be disconnected from their birth family, making access to early history very difficult, especially the history of PAE. In adolescents and adults, the physical characteristics of growth deficiency and facial features are less evident, and the diagnosis will depend on the identification of organic brain damage and history of PAE. The assessment needs to include mental health issues, the ability to function independently, and the potential for employability within a supportive environment. With age, there is evidence of increasing gaps in function and the ongoing need for a high level of supervision. At the present time, there is limited capacity for this type of assessment and supports after diagnosis for the older adolescent and adult.

7.6
Implications of a Diagnosis of FASD

The diagnosis of FASD at any age must lead to supports across key areas and systems that include education, medical, mental health, social, judicial and caregiver aspects. Educational supports include early intervention programs, school supports that meet the individual's strengths and challenges, transitions to more functional life skills and potential employability, and not focusing on academic aspects (Peadon et al., 2009; Premji et al., 2007; Olson et al., 2007; Coles, Kable, and Taddeo, 2009; Bertrand, 2009; Kable, Coles, and Taddeo, 2007; Loomes et al., 2007; Keil et al., 2010). Medical systems need to address basic health maintenance and prevention of mental health issues and addictions, and the appropriate medical management of any comorbid mental diagnosis such as ADHD and depression must be provided (Frankel et al., 2006). Social supports should be in place to assist the person with FASD to be a participant in their community in a safe and meaningful way, such as housing and recreation. Individuals with FASD may become parents themselves, and experience difficulties in using appropriate judgment in meeting the day-to-day needs of their children, without a high level of support (Rutman and van Bibber, 2010).

For those children who need to be in the foster-care system, stability of placement and minimal moves with caregivers who understand FASD are critical. Caregivers must be educated about FASD and anticipatory guidance around transitions. Likewise, respite for caregivers may be needed to prevent "burnout," given the level of support that will be required to put in place (Brown and Bednar, 2004). The best practice would be to provide such resources in the local community, with respect to cultural and spiritual values, and this may best be achieved with a network of services in communities, mentored by expert teams. The need to focus on preventing alcohol use in future pregnancies by a system of primary, secondary, and tertiary preventions should be part of the community imitative, with breaking the multigenerational impact of PEA as the ultimate goal (Poole, 2008).

7.7
Conclusion and Future Directions

Although brain damage caused by PEA is static and neither degenerative nor progressive, the difficulties in function become more evident across the person's lifespan, as there are the natural expectations in society to function independently. Thus, a longitudinal follow-up, using different assessment tools for different age groups, is needed to provide information on functions at key transition points within the system of care. This will not only provide information on the types of interventions and supports required, but also provide caregivers and individuals with FASD with a realistic picture of their strengths and weaknesses. The diagnosis of an individual with FASD is a "Diagnosis for Two," as it potentially identifies a birth mother who was drinking during that pregnancy. Indeed, she may still be dealing with complex issues in her life that may place future pregnancies at risk for alcohol exposure, and also impact on her own health. Reaching out to her with a positive support system can be preventive at many levels (Clarren and Salmon, 2010). Clearly, diagnosis, intervention and prevention in FASD are interlinked and require a cross-sectorial collaborative approach.

7.8
Policy Considerations

In the diagnosis of FASD, there are several areas of for research and policy consideration. As there appears to be a commitment to make a difference in this preventable disability at all levels, from government, researchers, service providers, community and caregivers, the following recommendations in particular should be paid attention:

1) A need to develop accurate and cost-effective neurobiological markers and/or functional assessment tools for FASD across the lifespan.

2) A need to increase diagnostic capabilities by training multidisciplinary teams for FASD diagnosis, assessment and development of recommendations for interventions and support.

3) A need for sustainable funding and mentoring for these emerging teams to insure accuracy and accessibility of diagnosis.

4) A need to develop screening for developmental disorders during the early years, that would include FASD.

5) A need for research into promising, better or best practices in interventions and supports for FASD across the lifespan.

6) A need to implement interventions and resources in local communities to insure connectedness to family and culture (the right place, for the right person, at the right time).

7) Funding for longitudinal follow-up across the systems of care, including transition planning, so that individuals do not "fall between the cracks."

8) A need to develop a continuous support plan for children who enter into the foster-care system, including training of their care providers, minimizing moves, and supporting their developmental and emotional needs and forming healthy attachments.

9) Prevention strategies to include access to best practice models of care for high-risk birth mothers, that address not only the drinking but also all of the factors of *why* she drinks, including health, mental health, and psychosocial determinants.

References

Aase, J.M., Jones, K.L., and Clarren, S.K. (1995) Do we need the term "FAE"? *Pediatrics*, **95**, 428–430.

Abel, E.L. (ed.) (1996) *Fetal Alcohol Syndrome: from Mechanism to Prevention*, CRC Press, New York.

Abel, E.L. (ed.) (1998a) *Fetal Alcohol Abuse Syndrome*, Plenum Press, New York, Chapters 5 to 9 on ARBDs, pp. 49–110.

Abel, E.L. (1998b) Mechanisms, in *Fetal Alcohol Abuse Syndrome*, Plenum Press, New York, pp. 183–210.

Alberta Alcohol and Drug Abuse Commission (2004) Estimating the rate of FASD and FAS in Canada, http://www.aadac.com/547_1224.asp (version current at January 2009, accessed).

American Academy of Pediatrics (1998) Statement on neonatal drug withdrawal. *Pediatrics*, **101** (6), 1078–1088.

Anda, R.F., Felitti, V.J., Bremmer, J.D., Walker, J.D., Whitfield, C., Perry, B.D., Dube, S.R., and Giles, W.H. (2006) The enduring effects of abuse and related adverse experiences in childhood. *Eur. Arch. Psychiatry Clin. Neurosci.*, **265**, 174–186.

Aragon, A.S., Coriale, G., Florentino, D., Kalberg, W.O., Buckley, D., Gossage, P., Ceccanti, M., Mitchel, E.R., and May, P.A. (2008) Neuropsychological characteristics of Italian children with fetal alcohol spectrum disorders. *Alcohol. Clin. Exp. Res.*, **32** (111), 1909–1919.

Archibald, S.L., Fennema-Notestine, C., Gamst, A., Riley, E.P., Matson, S., and Jernigan, T.L. (2001) Brain dysmorphology in individuals with severe prenatal alcohol exposure. *Dev. Med. Child Neurol.*, **43** (3), 148–154.

Astley, S.J. (2004) *Diagnostic Guide for Fetal Alcohol Spectrum Disorder: The 4-Digit Diagnostic Code*, 3rd edn, University of Washington, Seattle, http://depts.washington.edu/fasdpn.

Astley, S.J. (2006) Comparison of the 4-digit code with the Hoyme diagnostic guidelines for fetal alcohol spectrum disorders. *Pediatrics*, **118**, 1532–1545.

Astley, S.J. (2010) Profile of the first 1,400 patients receiving diagnostic evaluations for fetal alcohol spectrum disorder at the Washington State fetal alcohol spectrum disorder diagnostic and prevention network. *Can. J. Clin. Pharmacol.*, **17** (1), e132–e164.

Astley, S.J. and Clarren, S.K. (2000) Diagnosing the full spectrum of fetal alcohol exposed individuals: introducing the 4-digit diagnostic code. *Alcohol*, **35**, 400–410.

Astley, S.J. and Clarren, S.K. (2001) Measuring the facial phenotype of individuals with prenatal alcohol exposure: correlation with brain dysfunction. *Alcohol. Alcohol*, **36**, 147–159.

Astley, S., Stachowiak, J., Clarren, S.K., and Clausen, C. (2002) Application of the fetal alcohol syndrome facial photographic screening tool in a foster care population. *J. Pediatr.*, **141**, 712–717.

Astley, S., Olson, H.C., Kerns, K., Brooks, A., Aylward, E.H., Coggins, T., et al. (2009a) Neuropsychological and behavioral outcomes from a comprehensive magnetic resonance study of children with fetal alcohol spectrum disorders. *Can. J. Clin. Pharmacol.*, **16** (1), e178–e201.

Astley, S., Richards, T., Alyward, E.H., Olson, H.C., Kerns, K., Brooks, A., Coggins, T., et al. (2009b) Magnetic resonance spectroscopy outcomes from a comprehensive magnetic resonance study of children with fetal alcohol spectrum disorders. *Magn. Reson. Imaging*, **27** (6), 760–778.

Barr, H.M., Bookstein, F.L., O'Malley, K.D., Connor, P.D., Huggins, J.E., and Streissguth, A.P. (2006) Binge drinking during pregnancy as a predictor of psychiatric disorders on the structured clinical interview for DSM-IV in young adult offspring. *Am. J. Psychiatry*, **163**, 1060–1065.

Benz, J., Rasmussen, C., and Andrew, G. (2009) Diagnosing fetal alcohol spectrum disorder: history, challenges and future directions. *Pediatr. Child Health*, **14** (4), 231–237.

Bertrand, J. (2009) Interventions for children with fetal alcohol spectrum disorders (FASDs): overview of findings for five innovative research projects. *Res. Dev. Disabil.*, 1–21. doi: 10.1016/j.ridd.2009.02.003.

Bertrand, J., Floyd, L., Webber, M.K., et al. (2004) *National Task Force on FAS/FAE: Guidelines for Referral and Diagnosis*, Centers for Disease Control and Prevention, Atlanta GA. (See also 2005: *MMWR Recomm Rep.*, **54**, 1–4.)

Brown, J.D. and Bednar, L.M. (2003) Parenting children with fetal alcohol spectrum disorder: a concept map of needs. *Dev. Disabil. Bull.*, **31**, 130–154.

Brown, J.D. and Bednar, L.M. (2004) Challenges of parenting children with a fetal alcohol spectrum disorder:

a concept map. *J. Fam. Soc. Work*, **8** (3), 1–18.

Brown, J.D., Sigvaldason, N., and Bednar, L.M. (2005) Foster parent perceptions of placement needs for children with fetal alcohol spectrum disorder. *Child Youth Serv. Rev.*, **27**, 309–327.

Burd, L. and Hofer, R. (2008) Biomarkers for detection of prenatal alcohol exposure: a critical review of fatty acid ethyl esters in meconium. *Birth Defects Res.*, **82** (7), 487–493.

Burd, L., Deal, E., Rios, R., Adickes, E., Wynne, J., and Klug, M.G. (2007) Congenital heart defects and fetal alcohol spectrum disorders. *Congenit. Heart Dis.*, **2** (4), 250–255.

Burden, M.J., Jacobson, S.W., Sokol, R.J., and Jacobson, J.L. (2005) Effects of prenatal alcohol exposure on attention and working memory at 7.5 years of age. *Alcohol. Clin. Exp. Res.*, **29**, 443–453.

Burden, M.J., Andrew, C., Saint-Amour, D., et al. (2009) The effects of fetal alcohol syndrome on response execution and inhibition: an event-related potential study. *Alcohol. Clin. Exp. Res.*, **33** (11), 1994–2004.

Burnside, L. and Fuchs, D. Impact of FASD in the Child Welfare System, http://www.cecw-cepb.ca/catalogue (accessed 10 May 2010).

Camarillo, C. and Miranda, R.C. (2008) Ethanol exposure during neurogenesis induces persistent effects on neural maturation: evidence from an ex vivo model of fetal cerebral cortical neuroepithelial progenitor maturation. *Gene Expr.*, **14** (3), 159–171.

Canadian Association of Pediatric Health Centers (2008) Developing a National Screening Tool Kit for FASD, http://www.caphc.org/programs_fasd.html (accessed 10 May 2010).

Chudley, A.E., Conroy, J., Cook, J.L., Loock, C., Rosales, T., and LeBlanc, N. (2005) Fetal alcohol spectrum disorder. Canadian guidelines for diagnosis. *Can. Med. Assoc. J.*, **172**, S1–S21.

Clark, E., Lutke, J., Minnes, P., and Ouellette-Kuntz, K. (2004) Secondary disabilities among adults with fetal alcohol spectrum disorders in British Columbia. *J. FAS Int.*, **2** (e13), 1–12.

Canada Northwest FASD Research Network. www.canfasd.ca (accessed 10 May 2010).

Clarren, S.K. and Lutke, J. (2008) Building clinical capacity for fetal alcohol spectrum disorder in western and northern Canada. *Can. J. Clin. Pharmacol.*, **15**, 223–237.

Clarren, S.K. and Salmon, A. (2010) Prevention of fetal alcohol spectrum disorder: proposal for a comprehensive approach. *Expert Rev. Obest. Gynecol.*, **5** (1).

Clarren, S.K. and Smith, D.W. (1978) The fetal alcohol syndrome. *N. Engl. J. Med.*, **298**, 1063–1067.

Clarren, S.K., Sampson, P.D., Larsen, J., et al. (1987) Facial effects of fetal alcohol exposure: assessments by photographs and morphometric analysis. *Am. J. Med. Genet.*, **26**, 651–666.

Clarren, S.K., Chudley, A.E., Wong, L., Friesen, J., and Brant, R. (2010) Normal distribution of palpebral fissure lengths in Canadian school age children. *Can. J. Clin. Pharmacol.*, **17** (1), e67–e77.

Coles, C.D., Kable, J.A., and Taddeo, E. (2009) Math performance and behavior problems in children affected by prenatal alcohol exposure: intervention and follow up. *J. Dev. Behav. Pediatr.*, **30** (1), 7–15.

Coles, C.D., Platzman, K., Lynch, M., and Frides, D. (2002) Auditory and visual sustained attention in adolescents prenatally exposed to alcohol. *Alcohol. Clin. Exp. Res.*, **26** (2), 263–271.

Coles, C.D., Platzman, K.A., and Smith, I. (1992) Effects of cocaine and alcohol use in pregnancy on neonatal growth pattern and neurobehavioral status. *Neurotoxicol. Teratol.*, **14**, 23–33.

Connor, P.D. and Streissguth, A.P. (1996) Effects of prenatal exposure to alcohol across the lifespan. *Alcohol Health Res. World*, **20**, 170–174.

Crocker, N., Vaurio, L., Riley, E.P., and Mattson, S.N. (2009) Comparison of adaptive behavior in children with heavy prenatal alcohol exposure or attention deficit/hyperactivity disorder. *Alcohol. Clin. Exp. Res.*, **33** (11), 2015–2023.

Druschel, C.M. and Fox, D.J. (2007) Issues in estimating the prevalence of fetal alcohol syndrome: examination of two counties in New York State. *Pediatrics*, **119**, e384–e390.

Frankel, F., Paley, B., Marquardt, R., and O'Connor, M. (2006) Stimulants, neuroleptics and children's friendship training for children with fetal alcohol spectrum disorders. *J. Child Adolesc. Psychopharmacol.*, **16** (6), 777–789.

Fryer, S.L., McGee, C.L., Matt, G.E., Riley, E.P., and Mattson, S.N. (2007) Evaluation of psychopathological conditions in children with heavy prenatal alcohol exposure. *Pediatrics*, **119** (3), e733–e741.

Fryer, S.L., Schweinsburg, B.C., Bjorquist, O.A., Frank, L.R., Mattson, S.N., Spadoni, A.D., and Riley, E.P. (2009) Characteristics of white matter microstructure in fetal alcohol spectrum disorders. *Alcohol. Clin. Exp. Res.*, **33** (3), 514–521.

Gareri, J., Lynn, H., Handley, M., Rao, C., and Koren, G. (2008) Prevalence of fetal ethanol exposure in a regional population-based sample by meconium analysis of fatty acid ethyl esters. *Ther. Drug Monit.*, **30**, 239–245.

Grant, T.M., *et al.* (2005) Preventing alcohol and drug exposed births in Washington State: intervention findings from three Parent-Child Assistance Program sites. *Am. J. Drug Alcohol Abuse*, **31** (3), 471–490.

Green, C.R., Mihic, A.M., Brien, D.C., Armstrong, I.T., Nikkel, S.M., Stade, B.C., Rasmussen, C., Munoz, D.P., and Reynolds, J.N. (2009) Oculomotor control in children with fetal alcohol spectrum disorders assessed using a mobile eye-tracking laboratory. *Eur. J. Neurosci.*, **29**, 1302–1309.

Greenbaum, R.L., Stevens, S.A., Nash, K., Koren, G., and Rovet, J. (2009) Social cognitive and emotion processing abilities of children with fetal alcohol spectrum disorders: a comparison with attention deficit hyperactivity disorder. *Alcohol. Clin. Exp. Res.*, **33** (10), 1656–1670.

Guerri, C., Bazinet, A., and Riley, E.P. (2009) Foetal alcohol spectrum disorders and alterations in brain and behavior. *Alcohol Alcohol.*, **44** (2), 108–114.

Haycock, P. (2009) Fetal alcohol spectrum disorders: the epigenetic perspective. (Review). *Biol. Reprod.*, **81** (4), 607–617.

Hofer, R. and Burd, L. (2009) Review of published studies of kidney, liver and gastrointestinal birth defects in fetal alcohol spectrum disorders. *Birth Defects Res.*, **85** (3), 179–183.

Howell, K.K., Lynch, M.E., Platzman, K.A., Smith, G.H., and Coles, C.D. (2006) Prenatal alcohol exposure and ability, academic achievement, and school functioning in adolescence: a longitudinal follow up. *J. Pediatr. Psychol.*, **31** (1), 116–126.

Hoyme, H.E., May, P.A., and Kalberg, W.O. (2005) A practical clinical approach to the diagnosis of fetal alcohol spectrum disorders: Clarification of the 1996 Institute of Medicine Criteria. *Pediatrics*, **115**, 39–47.

Jacobson, S.W., Chiodo, L.M., Sokol, R.J., and Jacobson, J.L. (2002) Validity of maternal report of prenatal alcohol, cocaine and smoking in relation to neurobehavioral outcome. *Pediatrics*, **109** (5), 815–825.

Jirikowic, T., Olson, H.C., and Kartin, D. (2008) Sensory processing, school performance and adaptive behavior of young school-age children with fetal alcohol spectrum disorders. *Phys. Occup. Ther. Pediatr.*, **28** (2), 117–135.

Jones, K.L., Robinson, L.K., and Bakhireva, L.N. (2006) Accuracy of the diagnosis of physical features of fetal alcohol syndrome by pediatricians after specialized training. *Pediatrics*, **118**, 1734–1738.

Jones, K.L., Smith, D.W., Ulleland, C.N., and Streissguth, A.P. (1973) Pattern of malformation in offspring of chronic alcoholic mothers. *Lancet*, **1**, 1267–1271.

Kable, J.A., Coles, C.D., and Taddeo, E. (2007) Socio-cognitive habilitation using math interactive learning experience program for alcohol–affected children. *Alcohol. Clin. Exp. Res.*, **31** (8), 1425–1434.

Keil, V., Paley, B., Frankel, F., and O'Connor, M.J. (2010) Impact of a social skills intervention on the hostile attributions of children with prenatal exposure to alcohol. *Alcohol. Clin. Exp. Res.*, **34** (2), 231–241.

Kodituwakku, P.W. (2007) Defining behavioral phenotype in children with fetal alcohol spectrum disorders: a review. *Neurosci. Biobehav. Rev.*, **31**, 192–201.

Lebel, C., Rasmussen, C., Wyper, K., Andrew, G., Yager, J., and Beaulieu, C. (2008) Brain diffusion abnormalities in children with fetal alcohol spectrum disorder. *Alcohol. Clin. Exp. Res.*, **32** (10), 1732–1740.

Lebel, C., Rasmussen, C., Wyper, K., Andrew, G., and Beaulieu, C. (2010) Brain microstructure is related to mathematical ability in children with fetal alcohol spectrum disorder. *Alcohol. Clin. Exp. Res.*, **34** (2), 354–363.

Lee, S., Choi, I., Kang, S., and Rivier, R. (2008) Role of various neurotransmitters in mediating the long–term endocrine consequences of prenatal alcohol exposure. *Ann. N. Y. Acad. Sci.*, **1144**, 176–188.

Loomes, C., Rasmussen, C., Pei, J., Manji, S., and Andrew, G. (2007) The effect of rehearsal training on working memory span of children with fetal alcohol spectrum disorder. *Res. Dev. Disabil.*, **29**, 113–124.

Lynch, M.E., Coles, C.D., Corley, T., and Falek, A. (2003) Examining delinquency in adolescents differentially exposed to alcohol: the role of proximal and distal risk factors. *J. Stud. Alcohol*, **64**, 678–686.

McGee, C.L., Schonfeld, A.M., Roebuck-Spencer, T.M., Riley, E.P., and Mattson, S.N. (2008) Children with heavy prenatal alcohol exposure demonstrate deficits on multiple measures of concept formation. *Alcohol. Clin. Exp. Res.*, **32** (8), 1388–1397.

Mattson, S. and Riley, E.P. (1998) A review of neurobehavioral deficits in children with fetal alcohol syndrome or prenatal exposure to alcohol. *Alcohol. Clin. Exp. Res.*, **22** (2), 279–294.

Mattson, S.N. and Roebuck, T.M. (2002) Acquisition and retention of verbal and nonverbal information in children with heavy prenatal alcohol exposure. *Alcohol. Clin. Exp. Res.*, **26**, 875–882.

Mattson, S., Calarco, K.E., and Lang, A.R. (2006) Focused and shifting attention in children with heavy prenatal alcohol exposure. *Neuropsychology*, **20** (3), 361–369.

May, P.A., Gossage, J.P., Marais, A.S., Hendricks, L.S., Snell, C.L., Tabachnick, B.G., *et al.* (2008) Maternal risk factors for fetal alcohol syndrome and partial fetal alcohol syndrome in South Africa, a third study. *Alcohol. Clin. Exp. Res.*, **32** (5), 738–753.

May, P.A., Gossage, J.P., Kalberg, W.O., *et al.* (2009) Prevalence and epidemiological characteristics of FASD from various research methods with emphasis on recent in school studies. *Dev. Disabil. Res. Rev.*, **15**, 176–192.

O'Leary, C., Zubric, S.R., Taylor, C.L., Dixon, G., and Bower, C. (2009) Prenatal alcohol exposure and language delay in 2 year old children: the importance of dose and timing on risk. *Pediatrics*, **123**, 547–554.

Olson, H.C., Jirikowic, T., Kartin, D., and Astley, S.J. (2007) Responding to the challenge of early intervention for fetal alcohol spectrum disorders. *Infants Young Child.*, **20**, 172–189.

Oppenheim, D. and Goldsmith, D.F. (eds) (2007) *Attachment Theory in Clinical Work with Children*, Guildford Press.

Peadon, E., Fremantle, E., Bower, C., and Elliott, E.F. (2008) International survey of diagnostic services for children with fetal alcohol spectrum disorders. *BMC Pediatr.*, **8**, 12–18.

Peadon, E., Rhys-Jones, B., Bower, C., and Elliott, E. (2009) Systematic Review of Interventions for Children with Fetal Alcohol Spectrum Disorders, http://www.biomedcentral.com/1471-2431/9/35 (accessed 10 May 2010).

Pei, J.R., Rinaldi, C.M., Rasmussen, C., Massey, V., and Massey, D. (2008) Memory patterns of acquisition and retention of verbal and nonverbal information in children with fetal alcohol spectrum disorders. *Can. J. Clin. Pharmacol.*, **15** (1), e44–e56.

Peterson, J., Kirchner, H.L., Xue, W., Minnes, S., and Singer, L.T. (2008) Fatty acid ethyl esters in meconium are associated with poorer neurodevelopmental outcomes to two years of age. *J. Pediatr.*, **152**, 788–792.

Pichini, S., Pellegrini, M., Gareri, J., Koren, G., Garcia-Algar, O., Vall, O., *et al.* (2008)

Liquid chromatography-tandem mass spectrometry for fatty acid ethyl esters in meconium: assessment of prenatal exposure to alcohol in two European cohorts. *J. Pharm. Biomed. Anal.*, **48** (3), 927–933.

Poole, N. (ed.) (2007) *Double Exposure: A Better Practices Review on Alcohol Intervention during Pregnancy*, Act Now BC-Healthy Choices in Pregnancy, www.hcip-bc.org (accessed 10 May 2010).

Poole, N. (2008) Fetal Alcohol Spectrum Disorder (FASD) Prevention: Canadian Perspectives, http://www.publichealth.gc.ca/fasd (accessed 10 May 2010).

Premji, S., Benzies, K., Serrett, K., and Hayden, K.A. (2007) Researched interventions for children and youth with a fetal alcohol syndrome. *Child Care Health Dev.*, **33**, 389–397.

Rasmussen, C. (2005) Executive functioning and working memory in fetal alcohol spectrum disorder. *Alcohol. Clin. Exp. Res.*, **29**, 1359–1367.

Rasmussen, C. and Bisanz, J. (2009) Executive functioning in children with fetal alcohol spectrum disorders. Profiles and age related differences. *Child Neuropsychol.*, **15** (3), 201–215.

Rasmussen, C., Horne, K., and Witol, A. (2006) Neurobehavioral functioning in children with fetal alcohol spectrum disorder. *Child Neuropsychol.*, **12**, 1–16.

Rasmussen, C., Andrew, G., Zwaigenbaum, L., and Tough, S. (2008) Neurobehavioral outcomes of children with fetal alcohol spectrum disorders: a Canadian perspective. *Pediatr. Child Health*, **13** (3), 185–191.

Riley, E.P., Mattson, S.N., Ting-Kai, L., Jacobson, S.W., Coles, C.D., Kodituwakku, P.W., Adnams, C.M., and Korkman, M.I. (2003) Neurobehavioral consequences of prenatal alcohol exposure: an international perspective. *Alcohol. Clin. Exp. Res.*, **27** (2), 362–373.

Riley, E.P. and McGee, C.L. (2005) Fetal alcohol spectrum disorder: an overview with emphasis of changes in brain and behavior. *Exp. Biol. Med.*, **230**, 357–365.

Rutman, D. and Van Bibber, M. (2010) Parenting with fetal alcohol spectrum disorder. *Int. J. Ment. Health. Addict.*. doi: 10.1007/s11469-009-9264-7.

Sakar, D.K., Kuhn, J., Marano, J., et al. (2007) Alcohol exposure during the developmental period induces beta-endorphin neuronal death and causes alteration in the opioid control of stress axis function. *Endocrinology*, **148**, 2828–2834.

Santhanam, P., Li, Z., Hu, X., Lynch, M.E., and Coles, C.D. (2009) Effects of prenatal alcohol exposure on brain activation during an arithmetic task: an fMRI study. *Alcohol. Clin. Exp. Res.*, **33** (11), 1901–1908.

Sokol, R.J. and Clarren, S.K. (1989) Guidelines for terminology describing the impact of prenatal alcohol on the offspring. *Alcohol. Clin. Exp. Res.*, **13**, 597–598.

Sowell, E., Matson, E., Kan, E., Thompson, P., Riley, E.P., and Toga, A. (2008a) Abnormal cortical thickness and brain-behavior correlation in individuals with heavy prenatal alcohol exposure. *Cereb. Cortex*, **18**, 136–144.

Sowell, E., Johnson, A., Kan, E., Van Horn, J.D., Toga, A.W., O'Connor, M.J., and Bookheimer, S.Y. (2008b) Mapping white matter integrity and neurobehavioral correlates in children with fetal alcohol spectrum disorders. *J. Neurosci.*, **28** (6), 1313–1319.

Spadoni, D., McGee, C.L., Fryer, S.L., and Riley, E.P. (2007) Neuroimaging and fetal alcohol spectrum disorders. *Neurosci. Biobehav. Rev.*, **31** (2), 239–245.

Stade, B., Ali, A., Bennett, D., Campbell, D., Johnston, M., Lens, C., Tran, S., and Koren, G. (2009) The burden of prenatal exposure to alcohol: revised measurement of cost. *Can. J. Clin. Pharmacol.*, **16** (1), e91–e102.

Stoler, J.M. and Holmes, L.B. (2004) Recognition of facial features of fetal alcohol syndrome in the newborn. *Am. J. Genet. C Semin. Med. Genet.*, **127**, 21–27.

Stratton, K.R., Howe, C.J., and Battaglia, F.C. (1996) *Fetal Alcohol Syndrome: Diagnosis, Epidemiology, Prevention and Treatment*, National Academy Press, Washington DC, Institute of Medicine.

Streissguth, A.P. (2007) Offspring effects of prenatal alcohol exposure from birth to 25 years: the Seattle prospective longitudinal study. *J. Clin. Psychol. Med. Settings,* **14**, 81–101.

Streissguth, A.P. and O'Malley, K. (2000) Neuropsychiatric implications and long term consequences of fetal alcohol spectrum disorders. *Semin. Neuropsychiatry,* **5**, 177–190.

Streissguth, A.P., Barr, H.M., Bookstein, F.L., Sampson, P.D., and Olson, H.C. (1999) The long term neurocognitive consequences of prenatal alcohol exposure: a 14 year study. *Psychol. Sci.,* **10**, 186–190.

Streissguth, A.P., Bookstein, F.L., Barr, H.M., Sampson, P.D., O'Malley, K., and Young, J. (2004) Risk factors for adverse life outcomes in fetal alcohol syndrome and fetal alcohol effects. *J. Dev. Behav. Pediatr.,* **25** (4), 228–238.

Sulik, K., Johnston, M., and Webb, M. (1981) Fetal alcohol syndrome: embryogenesis in a mouse model. *Science,* **214**, 936–938.

Thanh, N.X. and Jonsson, E. (2009) Costs of fetal alcohol spectrum disorder in Alberta, Canada. *Can. J. Clin. Pharmacol.,* **16** (1), e80–e90.

Tough, S., Clarke, M., and Cook, J. (2007) Fetal alcohol spectrum disorder: prevention approaches among Canadian physicians by proportion of native/aboriginal patients: practices during the preconception and prenatal periods. *Matern. Child Health,* **11**, 385–393.

Weinberg, J. (1998) Hyper responsiveness to stress: differential effects of prenatal ethanol on males and females. *Alcohol. Clin. Exp. Res.,* **12**, 647–652.

Weinberg, J., Sliwowska, J.H., Lan, N., and Hellemans, K.G. (2008) Prenatal alcohol exposure: fetal programming, the hypothalamic-pituitary-adrenal axis and sex differences in outcome (review). *J. Endocrinol.,* **20** (4), 470–488.

Whaley, S., O'Connor, M., and Gunderson, B. (2001) Comparison of adaptive functioning of children prenatally exposed to alcohol to a nonexposed clinical sample. *Alcohol. Clin. Exp. Res.,* **25**, 1018–1024.

Wozniak, J.R., Muetzel, R.L., Mueller, B.A., McGee, C.L., Freerks, M.A., Ward, E.E., Nelson, M.L., Chang, P.N., and Lim, K.O. (2002) Microstructural corpus callosum anomalies in children with prenatal exposure to alcohol. *Alcohol. Clin. Exp. Res.,* **33** (10), 1825–1835.

Wozniak, J.R., Mueller, B.A., Chang, P.N., Muetzel, R.L., Caros, L., and Lim, K.O. (2006) Diffusion tensor imaging in children with fetal alcohol spectrum disorders. *Alcohol. Clin. Exp. Res.,* **30** (10), 1799–1806.

Part Three
Prevention Policies and Programs

8
FASD: A Preconception Prevention Initiative

Lola Baydala, Stephanie Worrell, and Fay Fletcher

8.1
Introduction

The basic architecture of the human brain is constructed through an ongoing process that begins before birth and continues into adulthood. Like the construction of a home, the building process begins with laying the foundation, framing the rooms and wiring the electrical system, in a predictable sequence. Early experiences shape how the brain is built; a strong foundation in the early years increases the probability of positive outcomes, whereas a fragile foundation increases the likelihood of later difficulties. Early prenatal exposure to the toxic effects of alcohol significantly affects the architecture of the developing brain, and is the cause of Fetal Alcohol Spectrum Disorder (FASD) (Rasmussen *et al.*, 2008). Preventing substance use during pregnancy will aid in the development of a solid foundation, by ensuring that the developing brain is free from the toxic effects of alcohol. It will also increase the possibility that all children will begin life with the full potential of becoming productive and responsible citizens.

8.2
Prevention Strategies

In 2008, the Public Health Agency of Canada (PHAC) published a summary of Canadian perspectives on FASD that included a four-part model for FASD prevention and the promotion of women and children's health (Public Health Agency of Canada, 2008). The first level of prevention includes universal strategies that raise public awareness about the risks of prenatal alcohol use through policy and health promotion activities that engage people at the community level:

- Level one prevention strategies include information sheets, media campaigns, booklets, information lines, and websites.
- Level two prevention strategies involve discussions of alcohol use and its related risks with all women of child-bearing years, and their support networks.

Fetal Alcohol Spectrum Disorder–Management and Policy Perspectives of FASD. Edited by Edward P. Riley, Sterling Clarren, Joanne Weinberg, and Egon Jonsson
Copyright © 2011 WILEY-VCH Verlag GmbH & Co. KGaA, Weinheim
ISBN: 978-3-527-32839-0

- Level three prevention includes specialized prenatal programs that provide care and treatment for girls and women who are using alcohol during pregnancy.

- Level four prevention provides support to new mothers to help them maintain healthy lifestyle changes during their pregnancy.

School-based substance-use prevention programs can be classified as preconception FASD initiatives that have the potential to prevent children and youth from using substances that affect their health and contribute to high-risk behaviors. School-based programs include discussions of alcohol use and related risks, and encompass level one and level two prevention strategies. The most effective school-based substance-use programs incorporate three stages of knowledge: (i) resistance skills training that help student's say "no" to drug and alcohol use; (ii) factual information about the risks of drug and alcohol use; and (iii) social and personal self-management skills that support a child's self-esteem (Botvin and Griffin, 2004). A number of substance-use prevention programs are available for implementation as a part of the school curriculum; however, the amount of evidence supporting the effectiveness of the prevention program is not always considered when deciding which program to implement (West and O'Neal, 2004).

8.2.1
The National Registry of Evidence-Based Programs and Practice

The National Registry of Evidence-Based Programs and Practices (NREPP) is a searchable database of evidence-based interventions, including school-based substance-use prevention programs, for the prevention and treatment of mental health and substance-use disorders (NREPP, 2010). The NREPP provides regularly updated reports of evidence-based programs that have been identified. These reports include information about the intervention and its targeted outcomes, the quality of the research to support the program, and whether the program is ready for broad-based dissemination. The NREPP provides research quality ratings as indicators of the strength of the evidence to support the outcomes of a particular intervention, with higher scores indicating stronger evidence. The quality of the research to support an intervention is evaluated by independent reviewers using six criteria and a rating scale of zero to four, with four being the highest possible rating. These criteria include: (i) reliability of measures; (ii) validity of measures; (iii) intervention fidelity; (iv) missing data and attrition; (v) potential confounding variables; and (vi) appropriateness of analysis.

8.2.2
LifeSkills Training

Based on the evidence provided by the NREPP and a review of the literature by the research team, the LifeSkills Training (LST) program, developed by Gilbert J. Botvin, was found to be the most extensively researched and effective school-based substance-use prevention program available. The LST program has been evaluated in numerous randomized controlled trials involving hundreds of schools/sites and thousands of students in suburban, urban, and rural communities (see http://www.lifeskillstraining.com/evaluation.php). Ratings on the quality of the program have consistently ranged from 3.9 to 4.0. A broad dissemination of the LST program began in 1995, since which time over three million students have participated in the program in 32 countries (NREPP, 2008).

The LST program is a generic program that has been proven to be highly effective with a number of different program providers and with students from different geographic regions, and socioeconomic and racial-ethnic backgrounds (Botvin et al., 1995, 2001, 1989a, 1989b). However, despite the overwhelming success of the program it has never been implemented with Canadian Aboriginal children and youth. The aim of this chapter is to demonstrate critical factors for success in implementing the LST program within the context of a First Nations community. These factors include:

- The laddering of grants from seed grants, to community-driven grants, to major funding as the relationship and project goals evolve.
- The importance and priority given to the process of program adaptation.
- A community-based participatory research approach that values capacity building for all members of the research team.
- Adherence to the guidelines set out by the Canadian Institute of Health Research (CIHR) for research involving Aboriginal people (CIHR, 2007).

8.2.3
The Alexis Working Committee

In 2005, the Alexis Nakota Sioux Nation invited researchers from the University of Alberta to work with them to implement a culturally appropriate school-based program that would empower their children and youth to resist substance use (Baydala et al., 2009). In order to secure funding for the project, it was also important that the program chosen was evidence-based. The possibility of implementing the LST program was presented to the community and, after discussions with community members and leaders, a decision was made to make adaptations to the original LST program. The specific goals for the community were to: (i) review and culturally adapt the original LST program; (ii) implement and evaluate the effectiveness of the modified program to ensure that fidelity was maintained after the adaptations were completed; and (iii) restore and preserve the cultural beliefs, values and language of the Alexis Nakota Sioux Nation. Early in the project, the Alexis Working Committee was established to ensure that these goals were met. The Alexis Working Committee, which included members of the Alexis Nakota Sioux Nation and the University of Alberta, completed terms of reference that defined its composition and clarified members' roles and responsibilities. The primary role of the Alexis Working Committee was to oversee general operations of the project, including the acquisition of funding, the management of expenditures and timelines, the interpretation of findings and recommendations, and the presentation of research results to stakeholders in the community and to academic audiences through conference presentations and manuscript preparation.

Over the next five years, the Alexis Working Committee secured a series of grants, making possible the adaptation, delivery, and evaluation of the community's LifeSkills training program "Nimi Icinohabi." In 2005, the Alexis Working Committee applied for a seed grant that was "... intended to support the planning and development of a potential research project" (Alberta Centre for Child, Family and Community Research, 2010a). The grant allowed the committee to continue to build relationships for research, to consider the readiness of the community to engage in a large research project, complete a literature review, address methodological issues, and design a pilot study for baseline data collection. Major accomplishments supported by the seed grant included community presentations, the submission of a larger community-driven research grant to sustain the research project, and a Band Council Resolution[1] indicating the Chief and Council's support of the program adaptations, delivery, and evaluation. Letters of support from community members and support for the project from community Elders were collected and were considered equally important. The Alexis Working Committee's application for a community-driven research grant, defined by the integral involvement of the community in the design and implementation of the research project (Alberta Centre for Child, Family and Community Research, 2010b), was successful and allowed for the gathering of baseline data, which included delivery and evaluation of a pilot program, as well as ongoing program adaptations.

Over the next 12 months, the Alexis Working Committee had two goals:

1) To prepare for the delivery of the adapted program as an integrated part of the school curriculum.
2) To deliver and evaluate the adapted program.

8.2.4
The Adaptations Committee

In order to meet the first goal, an Adaptations Committee was assembled and met weekly. Aboriginal ways of knowing, including cultural ceremonies, prayer, storytelling, and the appreciation of people's own life stories became the underpinnings of the program adaptations. Where appropriate, some of the key words in the program were translated into the Isga[2] language. To ensure that the concepts remained similar to those in the original evidence-based curriculum, the Isga words were back-translated from Isga to English. For example, in the original program the first lesson is entitled "Self-Esteem" (Botvin, 1999). A direct translation of this phrase does not exist in the Isga language, and the meaning of the phrase itself is not encompassed by Nakota Sioux ways of knowing. Community

---

1) A BCR is the authority mechanism by which the elected representatives of a Band Council authorize an action.
2) Isga, also known as Stoney, is the ancestral language of the people of the Alexis Nakota Sioux Nation.

Elders agreed upon the phrase "Norauzi Ahogipa," which translates to "inner spirit" in English, to represent the concepts that were a part of this lesson. A cultural activity or ceremony, chosen by the community Elders working with the Adaptations Committee, was added to each program module. The inclusion of a cultural activity and/or discussion of a cultural ceremony as a part of each lesson connected the teachings of the original program to the cultural teachings of the Nakota Sioux people. A community artist created culturally appropriate images to replace those in the original program. Student art, completed by the children who attended the community school, was also included throughout the program.

8.2.5
Community Member Participation

Community members, including Elders, parents, school staff, and representatives from each community agency, were invited to a three-day workshop prior to the implementation and delivery of the culturally adapted LST program. During the first two days of the workshop, a qualified LST trainer, organized through National Health Promotion Associates (see http://www.lifeskillstraining.com), provided training that was aimed at informing the participants of the content of the original LST program, and at preparing the program providers within the community to deliver the curriculum with content and process fidelity. The workshop included an explanation of the rationale for the prevention approach, a description of the original curriculum materials, and a session-by-session overview of the curriculum. Following the two-day training workshop, an additional one-day workshop provided an opportunity for the broader community to give their input as cultural adaptations were reviewed and additional revisions were made.

The adapted program was delivered by a community program provider once a week for eight weeks, as part of the regular school curriculum. Each module was delivered over 2 h on two separate days, instead of over 1 h as recommended in the original program. The extra hour allowed for children to either participate in or to discuss a cultural ceremony. Outcomes were measured and evaluated to ensure that the fidelity of the program was maintained. The success of the project throughout the duration of the seed and community-driven research grants provided evidence for using a community-based participatory research approach (Israel et al., 1998).

8.3
Research Relationships

The research relationship was based on co-learning and shared leadership throughout all phases of the project. During the adaptation, implementation and evaluation process, academic and community members of the Alexis Working and Adaptations Committees assumed and relinquished leadership roles according to their knowledge, skills, and areas of expertise, and what they believed they were best able to contribute to the project. For some members this was assisting in gathering consensus on difficult decisions that needed to be made; for others, it was understanding the research process, and for some it was understanding and sharing community protocols with the rest of the committee members. This movement in and out of leadership roles was encouraged and supported by all committee members, and led to the bidirectional building of capacity of all committee members (Fletcher, McKennitt, and Baydala, 2007).

8.3.1
Capacity Building

Capacity building was an important part of the project, and contributed to the success of the program adaptations as well as to the community's willingness to deliver the program as a part of the regular school curriculum (Smith *et al.*, 2003). For university team members, capacity included the knowledge and skills that were necessary for the development of meaningful and effective community partnerships. Community team members built academic capacity by identifying community members who were critical to the success of the project. They provided direction to university team members, ensuring that community protocol was followed, and they also directed the creation, ownership and control of specific local cultural knowledge and resources. In addition, the community represented the interests and concerns of its members at committee meetings, they identified community resources and experts, and provided guidance on the best way to take findings back to the community. University team members built the capacity of community members by making a long-term commitment to build the research relationship, by securing funds to support the project through multiple phases, and by sharing with the community the importance of adhering to funding and other academic requirements. University members provided opportunities for community members to access training in a number of areas including research methodology, and they offered invitations for the community to present their work at local and national research meetings. The Community Capacity Building Tool (PHAC, 2007) has previously been used to document longitudinal growth in community capacity; however, a similar tool for the documentation of growth in academic capacity is not yet available (Fletcher, McKennitt, and Baydala, 2007).

8.4
The CIHR Guidelines for Research Involving Aboriginal People

The CIHR Guidelines for Research Involving Aboriginal People provided a frame of reference for decisions regarding the research goals, objectives, measures and knowledge mobilization, as well as the relationship that developed between community and university team members (CIHR, 2007). These guidelines were prepared by the Ethics Office of the CIHR, in conjunction with its Institute of

Aboriginal Peoples' Health, to assist researchers and institutions in carrying out ethical and culturally competent research involving Aboriginal people. They include 15 articles designed to ensure that the research partnerships between community and academic institutions are equitable and collaborative, and that they honor and respect ceremony and cultural knowledge within the community. The guidelines also support the principles of Ownership, Control, Access and Possession (OCAP), as sanctioned by the First Nations Information Governance Committee of the National Aboriginal Health Organization (First Nations Centre, 2007). The application of these guidelines and principles were critical in ensuring that accurate and meaningful cultural adaptations were made to the LST program. Observing the principles and guidelines also ensured that the final adapted LST program was acceptable to community members and leaders.

Research has shown that culturally adapting evidence-based prevention programs can significantly improve engagement and acceptability of the program by the community where the program will be delivered (Burhansstipanov et al., 2000; Capp, Deane, and Lambert, 2001; Crozier Kegler and Halinka Malcoe, 2004; Majumdar, Chambers, and Roberts, 2004; Paradis et al., 2005; Rowley et al., 2000). Furthermore, there is an ethical imperative to ensure that interventions developed for one culture do not negatively impact the cultural values, competence or language of another culture where the program will be delivered. In addition, a community involved in the adaptation and implementation of a prevention program is more likely to feel a sense of ownership and empowerment, which are important components of community capacity (Gosin et al., 2003; Hawkins, Cummins, and Marlatt, 2004; Mohatt et al., 2004). Cultural adaptations to the LST program were also important for children who received the program, as they provided an opportunity to learn more about their Nakota Sioux culture and, through this knowledge, to strengthen their sense of identity. Previous research has shown that culturally adapted prevention programs aid in the development of a stronger identity and cultural pride, which in turn function as protective factors against substance use (Okamoto et al., 2006; Whitbeck, 2006; Wright and Zimmerman, 2006). A strong sense of self and knowledge of one's community and culture are important components of childhood resiliency, especially for minority and marginalized children who may be struggling with how they fit into mainstream society, particularly when they are coming from backgrounds rooted in colonialism (Duran and Duran, 1995).

8.5
Summary

FASD can be prevented at multiple levels. Preconception initiatives, such as the one described in this chapter, provide school-age children with culturally appropriate knowledge and resources that help them resist substance use and its associated high-risk behaviors. Highly effective evidence-based substance-use prevention programs for school-age children and youth exist, and should be made available to all school-age children. An ability to capitalize on available grants in ways that fostered the development of research relationships with individuals and communities affected by FASD, a recognition of the importance of incorporating the cultural beliefs, values, language and visual images of the community where the program will be delivered, and a community-based participatory research approach have all contributed to the successful implementation and evaluation of a preconception FASD prevention strategy with First Nations children and youth.

Acknowledgments

The authors wish to acknowledge the work of the community members and Elders from the Alexis Nakota Sioux Nation, who made significant contributions to adapting and implementing the prevention program.

References

Alberta Centre for Child, Family and Community Research (2010a) Seed Grants, 24 March 2010; http://www.research4children.com/admin/contentx/default.cfm?PageId=7686 (accessed 26 March 2010).

Alberta Centre for Child, Family and Community Research (2010b) Community-Driven Research Grants Description, 11 February 2010; http://www.research4children.com/admin/contentx/default.cfm?PageId=7683 (accessed 26 March 2010).

Baydala, L.T., Sewlal, B., Rasmussen, C., Alexis, K., Fletcher, F., Letendre, L., and Kootenay, B. (2009) A culturally adapted drug and alcohol abuse prevention program for Aboriginal children and youth. *Prog. Community Health Partnership Res., Educ. Action*, **3**, 37–46.

Botvin, G.J. (1999) *LifeSkills Training Level One: Grades 3/4 Student Guide*, Princeton Health Press, White Plains, NY.

Botvin, G.J., Batson, H., Witts-Vitale, S., Bess, V., Baker, E., and Dusenbury, L. (1989a) A psychosocial approach to smoking prevention for urban black youth. *Public Health Rep.*, **104**, 573–582.

Botvin, G.J., Dusenbury, L., Baker, E., James-Ortiz, S., and Kerner, J. (1989b) A skills training approach to smoking prevention among Hispanic youth. *J. Behav. Med.*, **12**, 279–296.

Botvin, G.J., and Griffin, K.W. (2004) Life skills training: empirical findings and future directions. *J. Prim. Prev.*, **25**, 211–231.

Botvin, G.J., Griffin, K.W., Diaz, T., and Ifill-Williams, M. (2001) Drug abuse prevention among minority adolescents: posttest and one-year follow-up of a school-based preventive intervention. *Prev. Sci.*, **2**, 1–13.

Botvin, G.J., Schinke, S.P., Epstein, J.A., Diaz, T., and Botvin, E.M. (1995) Effectiveness of culturally focused and generic skills training approaches to alcohol and drug abuse prevention among minority adolescents: two-year follow-up results. *Psychol. Addict. Behav.*, **9**, 183–194.

Burhansstipanov, L., Dignan, M.B., Wound, D..B., Tenney, M., and Vigil, G. (2000) Native American recruitment into breast cancer screening: the NAWWA project. *J. Cancer Educ.*, **15**, 28–32. doi: 10.1080/08858190009528649.

Canadian Institutes of Health Research (CIHR) (2007) CIHR Guidelines for Health Research Involving Aboriginal People, http://www.irsc.gc.ca/e/documents/ethics_aboriginal_guidelines_e.pdf (accessed 4 April 2010).

Capp, K., Deane, F.P., and Lambert, G. (2001) Suicide prevention in Aboriginal communities: application of community gatekeeper training. *Aust. N. Z. J. Public Health*, **25**, 315–321.

Crozier Kegler, M., and Halinka Malcoe, L. (2004) Results from a lay health advisor intervention to prevent lead poisoning among rural Native American children. *Am. J. Public Health*, **94**, 1730–1735.

Duran, E. and Duran, B. (1995) *Native American Postcolonial Psychology*, State University of the New York Press, Albany, NY.

First Nations Centre (2007) *OCAP: Ownership, Control, Access and Possession*, Sanctioned by the First Nations Information Governance Committee, Assembly of First Nations, National Aboriginal Health Organization, Ottawa. http://www.naho.ca/firstnations/english/documents/FNC-OCAP_001.pdf (accessed 10 April 2010).

Fletcher, F., McKennitt, D., and Baydala, L. (2007) Community capacity building: an Aboriginal exploratory case study. *Pimatisiwin*, **5**, 9–32.

Gosin, M.N., Dusman, P.A., Drapeau, A.E., and Harthun, M.L. (2003) Participatory action research: creating an effective prevention curriculum for adolescents in the Southwestern US. *Health Educ. Res.*, **18**, 363–379. doi: 10.1093/her/cyf026.

Hawkins, E.H., Cummins, L.H., and Marlatt, G.A. (2004) Preventing substance abuse in American Indian and Alaska Native youth: promising strategies for healthier communities. *Psychol. Bull.*, **130**, 304–323. doi: 10.1037/0033-2909.130.2.304.

Israel, B., Schulz, A., Parker, E., and Becker, A. (1998) Review of community-based research: assessing partnership approaches to improve public health. *Annu. Rev. Public Health*, **19**, 173–202.

Mohatt, G.V., Hazel, K.L., Allen, J., Stachelrodt, M., Hensel, C., and Fath, R. (2004) Unheard Alaska: culturally anchored participatory action research on sobriety with Alaska Natives. *Am. J. Community Psychol.*, **33**, 263–273.

Majumdar, B.B., Chambers, T.L., and Roberts, J. (2004) Community-based culturally sensitive HIV/AIDS education for Aboriginal adolescents: implications for nursing practice. *J. Transcult. Nurs.*, **15**, 69–73. doi: 10.1177/1043659603260015.

National Registry of Evidence-Based Programs and Practices (2008) LifeSkills Training, September; http://www.nrepp.samhsa.gov/viewintervention.aspx?id=109 (accessed 4 April 2010).

National Registry of Evidence-Based Programs and Practices (2010, March 18), http://www.nrepp.samhsa.gov/viewall.aspx (accessed 31 March 2010).

Okamoto, S.K., LeCroy, C.W., Tann, S.S., Dixon Rayle, A., Kulis, S., Dustman, P., and Berceli, D. (2006) The implications of ecologically based assessment for primary prevention with Indigenous youth populations. *J. Prim. Prev.*, **27**, 155–170. doi: 10.1007/s10935-005-0016-6

Paradis, G., Lévesque, L., Macaulay, A.C., Cargo, M., McComber, A., Kirby, R., and Potvin, L. (2005) Impact of a diabetes prevention program on body size, physical activity, and diet among Kanien'kehá:ka (Mohawk) children 6 to 11 years old: 8-year results from the Kahnawake schools diabetes prevention project. *Pediatrics*, **115**, 333–339. doi: 10.1542/peds.2004-0745

Public Health Agency of Canada (2007) Community Capacity Building Tool, http://www.phac-aspc.gc.ca/canada/regions/ab-nwt-tno/documents/CCBT_English_web_000.pdf (accessed 4 April 2010).

Public Health Agency of Canada (2008) Fetal Alcohol Spectrum Disorder (FASD) Prevention: Canadian Perspectives. (Cat No.: HP5-73/2008), http://www.phac-aspc.gc.ca/fasd-etcaf/cp-pc-eng.php (accessed 19March 2010).

Rasmussen, C., Andrew, G., Zwaigenbaum, L., and Tough, S. (2008) Neurobehavioral outcomes of children with fetal alcohol spectrum disorders: a Canadian perspective. *Paediatr. Child Health*, **13**, 185–191.

Rowley, K.G., Daniel, M., Skinner, K., Skinner, M., White, G.A., and O'Dea, K. (2000) Effectiveness of a community-directed "healthy lifestyle" program in a remote Australian Aboriginal community. *Aust. N. Z. J. Public Health*, **24**, 136–144.

Smith, N., Baugh Littlejohns, L., and Roy, D. (2003) Measuring Community Capacity: State of the Field Review and Recommendations for Future Research, http://www7.albertahealthservices.ca/resources/documents/reports/MeasuringCommunityCapacity-StateoftheFieldReviewandRecommendationsforFutureResearch.pdf (accessed 3 April 2010).

West, S.L. and O'Neal, K.K. (2004) Project D.A.R.E. outcome effectiveness revisited. *Am. J. Public Health*, **94**, 1027–1029.

Whitbeck, L. (2006) Some guiding assumptions and a theoretical model for developing culturally specific preventions with Native American people. *J. Community Psychol.*, **34**, 183–192. doi: 10.1002/jcop.20094.

Wright, J.C. and Zimmerman, M.A. (2006) Culturally sensitive interventions to prevent youth violence, in *Prevention Youth Violence in a Multicultural Society* (eds N.G. Guerra and E. Phillips), American Psychological Association, Washington, DC, pp. 221–247.

9
Bringing a Women's Health Perspective to FASD Prevention
Nancy Poole

9.1
Introduction

Over the past decade, Canada has become a world leader in the development of a broad public health response to women's use of alcohol in pregnancy. Canadian prevention specialists have articulated and put into practice a framework for Fetal Alcohol Spectrum Disorder (FASD) prevention, collaborated on the development of provincial and national strategic plans, and formed multi-sectoral networks on FASD prevention. These achievements can be attributed, in part, to the influence of experts in women's health promotion, maternity care and women's substance-use research on the Canadian efforts to prevent FASD. Their involvement has generated a fundamental shift in attitude towards a women's health perspective on FASD prevention, in which the well-being of mothers *and* children is seen as inseparable. This chapter describes the contribution to the FASD prevention field of groundbreaking research that established linkages between the determinants of women's substance use and barriers to treatment and support for women at risk of having a child with FASD.

Initially, FASD was framed as a child health issue in Canada, as is still the common practice in most countries. Pediatricians and other child health professionals, along with foster parents and adoptive parents advocated for funding and support from public health and social service systems to address the challenges faced by those with FASD and their families. While this focus was both appropriate and important for the provision of treatment and other supports for children with FASD and their families, a critical element of public health policy was missing. What could be done to reduce the prevalence of FASD? How could pregnant women with alcohol problems and related health and social concerns be supported? The *prevention* of FASD involves health promotion, harm reduction and treatment services for at-risk women, yet this territory is often not the expertise of child welfare experts, child and family advocates, and child health specialists.

Women's health advocates in Canada were concerned about disturbing trends in public health campaigns and legal interventions. Researchers began to take note of how often judgmental and unsympathetic attitudes and practices had been employed in prevention efforts (Greaves, 1996; Greaves et al., 2002b; Rutman et al., 2000). Prevention materials produced during the 1980s were explicitly blaming of pregnant women who did not stop smoking and drinking out of concern for their children. In the 1990s, overtly judgmental messaging was replaced with more subtle, yet equally unsympathetic, prevention campaigns with unequivocal demands, such as "Pregnant? No alcohol!" (Health Canada, 2001). These prevention messages tended to oversimplify the challenges facing pregnant women, both in their exclusive focus on the use of alcohol and the implication that it is a simple matter of will power for any woman to "just say no" to drinking during pregnancy.

In fact, women whose alcohol use has progressed to dependency, and those dealing with emotional and health problems associated with intimate partner violence, inadequate housing or nutrition, mental health issues, and lack of support from partners and families, may be unable to stop on their own, even when they want to. Yet, prevention campaigns placed the responsibility solely on the woman, and did not signal how the healthcare system was prepared to help. Signaling where help is available can be critical, given the research evidence on the overwhelming barriers experienced by pregnant women and mothers with alcohol and related health problems (UNODC, 2004). Fears of prejudicial or punitive treatment from healthcare providers, including the risk of losing custody of their children, are well grounded (Poole and Isaac, 2001). Women's health advocates also took note of the imbalance in the focus of legal, media, and public concern for the health of the fetus, and rarely on the health of the woman herself (Armstrong, 2003; Boyd and Marcellus, 2007; Poole, 2003). It was certain that prevention, harm reduction, and treatment programs and policies could be developed which reflected a concern about mothers *and* children, and promoted better health and well-being for all.

9.2
Applying Gender-Based Analysis to FASD Prevention

In 2002, with funding from the Canadian Institutes for Health Research, the British Columbia Centre of Excellence for Women's Health (BCCEWH) sponsored a three-day workshop in Vancouver to examine the issues related to prevention of FASD from a women's health perspective. Twenty-eight women's health researchers and service providers came together with the goal of developing a research agenda for a women-centered approach to FASD prevention (Greaves, Poole, and Cormier, 2002). To that end, a sex, gender and diversity based analysis (SGDBA) (Clow et al., 2009) was applied to the issue of FASD prevention. Questions were asked—and subsequently answered—with the intention to build knowledge about the nature and scope of women's substance use in pregnancy and to improve policy and treatment responses. These included:

1) How do factors such as poverty, tobacco use and mental health problems interface with a woman's substance use during pregnancy, and with the biological processes affecting her overall health, including an increased risk of giving birth to a child with FASD?

2) What specific ethical considerations need to be attended to regarding women's alcohol use in pregnancy, so that surveillance and screening practices, for example, do not retraumatize women and/or prevent women from accessing pre- and postnatal care?

3) How might messaging and support be tailored to address the needs of diverse women, beginning with diversity in level of risk, through various combinations of community awareness campaigns, individual messaging, and supportive services, policies and community health promotion activities? What combination will be effective in preventing or reducing women's substance use problems, and specifically, substance use in pregnancy?

4) How should women-centered care be defined and measured?

5) How does child welfare and related public policy affect women's choices, and how might policy be conceptualized as both women- and child-centered, rather than a dichotomy?

The researchers and prevention specialists involved in applying this SGDBA have contributed to Canada's leadership on FASD prevention in three key areas, namely:

- articulating a framework for women-centered prevention;
- undertaking research that clarified and extended the key components underlying the framework; and
- advocating for improved policy and service responses.

This work, which has created a new environment of understanding, care and compassion for women at risk of having a child affected by FASD, is described below.

9.3
Developing a Framework for Women-Centered Prevention Practice

Central to the question of what influences a woman's alcohol use in pregnancy is the research on factors beyond drinking, that have been associated with increased risk of FASD. This includes identification of the determinants of women's substance use overall, along with gender-specific barriers to prenatal care and substance use treatment. By turning the wide lens of SGDBA on current views of what FASD prevention could look like, a conceptualization emerged (see Figure 9.1) that moved the focus away from women's alcohol use alone to include evidence on the determinants of, and influences on, women's drinking and access

Prevention of FASD - It's not only about alcohol

- Poverty
- Exposure to violence
- Racial discrimination
- Mother's nutrition
- Mother's use of other drugs
- Resilience
- Policy on mothering
- Mother's Alcohol Use
- Mother's access to prenatal care
- Mother's overall health
- Genetics
- Context/Isolation
- Mother's stress level
- Age
- Experience of loss

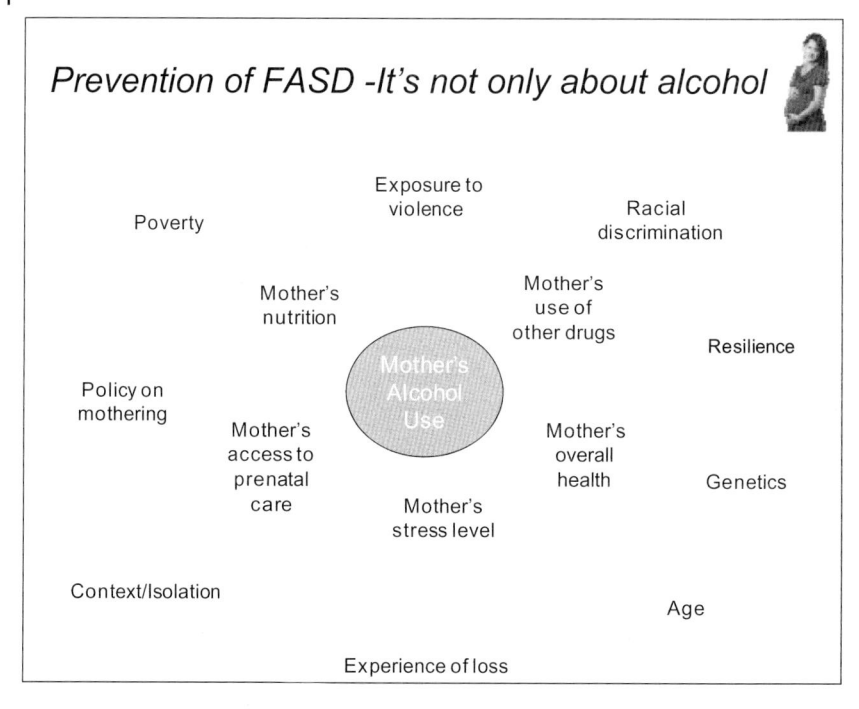

Figure 9.1 Moving the focus from alcohol use alone to the determinants of, and influences on, women's drinking and access to assistance.

to assistance. A study conducted by Astley and a research team from Washington state (Astley *et al.*, 2000), which documented the very difficult lives of birth mothers of FASD children, was foundational to this conceptualization. Evidence of the birth mothers' isolation, devastating experiences of violence, pressure to drink by unsupportive partners, and mental health problems, such as agoraphobia[1], demonstrated the needs of women at this highest level of risk. In particular, the study underlined how prevention efforts, in order to be effective, must be informed by an understanding of women's experience of trauma – in essence, a commitment to be less confrontational and more compassionate in both messaging and service responses. Further, the study illustrated how important it is to remove the gross misconception of women's drinking as an act to deliberately harm children, and how necessary it is to develop effective outreach strategies that welcome women at highest risk.

To assist service providers in British Columbia in shifting to a new paradigm of response to women at risk of having a child with FASD, a values-based framework with three fundamental principles was developed. The prevention support

---

1) Agoraphobia relates to anxiety about being in places or situations outside the home alone; such as being in a crowd, or standing in a line, and traveling in a bus, train, or automobile (abbreviated from DSM-IV).

that is extended should arise from a stance that is: women-centered; oriented towards harm reduction; and collaborative in its commitment to listen to and work with vulnerable women to improve their health and the health of their children.

9.3.1
Women-Centered Care

This principle brings the focus of public health support to the mother–child unit. Emphasis is placed on the woman's health from preconception through pregnancy and post-partum, supporting her interest in making positive changes, and her confidence in her ability to do so (Urquhart and Poole, 2008; Urquhart, Poole, and Horner, 2008). As a first step, the negative and demeaning societal condemnations of substance use during pregnancy need to be acknowledged, so that service providers can assist women in dealing with the stigma they are likely to have experienced (Greaves et al., 2005). Adopting a women-centered approach demands a paradigm shift – away from seeing mothers who use substances as unfit, uncaring and unworthy of support, and towards a positive perspective on women's capacity for change. FASD prevention demands that we see – and are hopeful about – women's strengths, and that we support substance users by removing barriers to treatment and involving women in determining what they need to better care for themselves and their children.

Service providers must recognize diversity in risk levels and determinants of substance use and make a number of service options available. It is of critical importance that services be nonjudgmental, in order to encourage women to access the best possible care and treatment. The Canada Prenatal Nutrition Program, for example, which reaches over 50 000 women annually, enhances intersectional collaboration to support the needs of at-risk pregnant women through a community development approach that "... aims to reduce the incidence of unhealthy birth weights, *improve the health of both infant and mother* [italics added] and encourage breastfeeding" (Public Health Agency of Canada, 2007). In Vancouver, the Sheway program – which works with pregnant women and new mothers in the Downtown Eastside neighborhood, using a one-stop model – are based on "... the recognition that the health of women and their children is linked to the conditions of their lives and their ability to influence these conditions" (Poole, 2000). Utilizing a women-centered and harm-reduction approach, Sheway offers services that are welcoming, nonjudgmental, respectful of First Nations culture and traditions, and supportive of women's self-determination, empowerment and choices.

9.3.2
Harm-Reduction Orientation

In the context of drinking during pregnancy, a harm-reduction approach means that service providers are willing to discuss goals other than abstinence, despite the known risks of substance use. This requires considering all aspects of harm that contribute to, or result from, women's alcohol use. Practitioners oriented toward harm reduction seek to discuss, promote, and facilitate actions that reduce harm in many areas of a woman's life, not only her substance use. This requires recognizing the interrelatedness of issues and being flexible enough to work from where women "are at," without judgment. Service providers must develop the skills and awareness to be able to tolerate women's needs to go at a slower pace than may be optimal. In effect, harm reduction obliges those who offer assistance to manage their own urgency, so that achievable change can happen and be maintained. Women who use substances are central to defining the level, type, and pace of change they can make. Systemic changes needed to minimize harm and support at-risk women are often revealed in this process.

9.3.3
Collaborative Care

This principle involves transformation, shifting FASD prevention from top-down expertise about women's health, to a partnership based on concern about the well-being of mothers and children. Public health practitioners do not confront women who drink with a proscriptive agenda, but do encourage women to express their ideas for reducing harm in their own lives and safeguarding their children. A collaborative approach requires service providers to be strategic in how they ask questions and to leave room for (often imperfect) suggestions for change to come forward from women who seek assistance. In this way, ideas are more likely to be shared and considered meaningfully for their potential to be put into action. Motivational interviewing, with its attention to empathic support and elicitation of ideas for change, has made a significant contribution to the practice of collaborative care (Miller and Rollnick, 2002).

This three-part framework was shared and revised through discussion with over 3000 Canadian professionals in conference, rounds, and community-based training settings in 2006–2009. A wide range of professionals – including physicians, public health nurses, acute care maternity nurses, midwives, doulas, mental health counselors, addictions service providers, anti-violence workers, child protection workers, other social workers, Aboriginal service providers, infant development workers, prenatal program providers and others – participated in this process. Their input helped to demonstrate just how important and how challenging it was for them to bring this paradigm into their practice (Poole and Greaves, 2009).

9.4
Evidencing the Framework

There remains a substantive gap in the literature regarding the effectiveness of FASD prevention approaches. Moreover, the prevention programming which has been studied has reported mainly on narrow outcomes, such as increased awareness of FASD or women's reduction in alcohol use alone. Having identified the centrality of a women's health determinants approach to FASD prevention, women's health advocates, service providers, policy makers and researchers continue their collective efforts to articulate the evidence through research.

9.4.1
Research on Women-Centered, Trauma-Informed Care

Canadian researchers interested in women's substance use have contributed to the understanding of women-centered care in FASD prevention by studying the outcomes of integrated services. For example, a multi-methods study undertaken in British Columbia investigated the connections between stressors, substance use and the experience of violence among women who accessed help from domestic violence shelters in that province (Poole et al., 2008). The study found that women generally decreased their use of alcohol and stimulants in the period following their shelter stay. Supports aimed at promoting improvements in women's health, income, housing and related issues were found to have a pivotal impact in helping women restructure their lives and reduce their use of substances. The study identified a range of social and structural factors that influence violence and the use of alcohol/other substances and deepened understanding of women's experience. Therefore, opportunities for positive change and support are missed when individual stress is seen as the key determinant for use of alcohol and other substances, and this is the sole focus of intervention. Given the complex relationship of life stressors, substance use and violence demonstrated by this study, the alcohol- and drug-specific work undertaken by shelter staff, as well as the broader work on health determinants, appeared to foster positive change. The findings underlined the importance for domestic violence shelters to acknowledge this positive impact of their current work, as well as to consider how to build on it, so that all women accessing shelter services are able to freely discuss their substance use, make connections with other stressors in their lives, and find support to heal and make pragmatic life changes.

Researchers interested in Aboriginal women's substance use examined the skills and traits that substance use treatment providers have found to be important in assisting Aboriginal women on their healing journeys from addiction, trauma and experience of stigma (Dell and Clark, 2009). Staff in treatment centers and treatment participants employed a holistic conceptualization of support for women in reclaiming their identity. Key elements in this approach included showing empathy for the struggles that women face, being accepting and nonjudgmental about women's pasts, communicating in nonhierarchical ways, supporting links to spirituality and culture, and recognizing the impact of trauma on Aboriginal women's health. This study reinforces the importance of pulling the focus from individual responsibility for change, and underlines how critical it is for those supporting change in women's substance use and addiction to address women's lives in context. A similar message emerged from a research study employing virtual focus groups with Aboriginal service providers and health system planners (Poole, Gelb, and Trainor, 2009), which emphasized the need for a holistic approach to care (mental, physical, emotional and spiritual) within Aboriginal communities that also reflects the uniqueness of each woman's experience.

9.4.2
Research on Harm-Reduction Practice

The development of a prevention framework from a women's health perspective highlighted how outcomes beyond a narrow focus on substance use—such as access to prenatal care, nutritional support, and adequate housing—contribute to women's health and thereby reduce the risk of having a child with FASD. This aspect of harm-reduction practice has been the focus of evaluations conducted of Canadian prenatal programming, including Sheway in Vancouver (Poole, 2000), the Maxxine Wright program in Surrey (Robinson *et al.*, November 2003), Women and Children's Healing and Recovery in Yellowknife (Four Worlds Center for Development Learning, July 2003), Breaking the Cycle in Toronto (Motz *et al.*, 2006), and another 18 Early Childhood Development Addiction Initiative programs across Ontario (McGuire *et al.*, May 31, 2006). These studies have all contributed to the present understanding of the importance of combining pragmatic support (such as providing hot meals, vitamins and bus tickets, and advocacy on housing and other basic needs) with outreach services and programming to help women to heal physically, emotionally and spiritually.

An example of the effectiveness of structural change in the provision of prenatal and postnatal care for women and children can be found in a recent study of housing supports extended to women accessing the Sheway program in Vancouver. The impact of a small, short-term financial supplement to enhance access to housing provided by the government department responsible for income assistance was assessed. The study found that pregnant women and new mothers who received even minimal additional assistance were able to access "family-friendly" housing and, as a result, were more likely to retain custody of their children (Salmon, 2010). Funding for a longitudinal cohort study is being sought to further identify the impact of this financial intervention on healthy child development, custody status, ability to retain housing after the top-up is discontinued, maternal health and well-being, and transitions to employment or schooling.

In a systematic review of evidence obtained from peer-reviewed literature on the topic of interventions aimed at supporting women to reduce their use of alcohol in the childbearing years, researchers at the British Columbia Centre of Excellence for Women's Health added two additional steps to the traditional systemic review process (Parkes *et al.*, 2008). The researchers contexualized the findings in the larger literature on women's alcohol use, and invited feedback on the key findings from wide audiences of practitioners, health system planners and researchers. Recommendations arising from the *Double Exposure* review (Parkes *et al.*, 2008) include recommendations for practice, research, knowledge translation, policy and structural change, that take the responsibility for change beyond the individual women to invite engagement by service providers, health system planners, decision makers and researchers in reducing systemic level harms.

9.4.3
Research on Collaborative Care

A study of Canadian newspaper coverage of pregnant women or mothers dealing with substance use, mental illness, or violence in abusive relationships (Greaves et al., 2002b) confirmed the pressing need for respectful and collaborative care as a response to "mothers under duress". The newspaper portrayal[2] of these women was sensational. Pregnant women or mothers with substance use issues, mental illness or in abusive relationships were the subject of news stories, but the news stories typically portrayed children as being at risk *and* their mothers as the cause of that risk. Mothers with substance-use problems were held most responsible for their own and their children's situation, and were portrayed most unsympathetically, as willful and abusive. Distinctions were also seen in the degree to which the social, medical or legal systems were held responsible when things went tragically wrong. In the reporting on women with mental illnesses, there was some recognition that the system was failing to provide adequate support and was, therefore, somewhat responsible when a woman's unrecognized or untreated mental illness was associated with her harming her children. However, the system was unlikely to be blamed in connection with women's substance use and addictions. Indeed, discussion of the treatment and support accessed or available was mentioned in only one instance, illustrating how the media contribute to an understanding of substance-using mothers that emphasizes individual responsibility and overlooks how service responses and policies both shape and constrain mothering. In such a stigmatizing environment, it is critical to provide welcoming treatment and support that empowers women to improve their own health and the health of their children.

Both, prenatal and postnatal service providers have taken steps designed to encourage women's participation through collaborative, blame-free programming. The Breaking the Cycle program in Toronto, for example, has implemented a mother–child, relationship-based service model that draws on the literature on attachment theory regarding the benefits of therapeutic approaches which introduce safety, acceptance, reliability, consistency, structure and caring (Leslie, DeMarchi, and Motz, 2007). Staff frame their work as a partnership with women seeking assistance, participating together in growth-promoting, therapeutic relationships based on respect, mutuality and empathy.

In the larger addictions field, service providers have employed Motivational Interviewing practice, which is grounded in collaborative, rather than provider-driven, agendas for support and change (Rollnick, Miller, and Butler, 2008). American researchers have evidenced this approach in the context of prevention of FASD, initially with college-aged women and subsequently with women in a

---

2) One year of Canadian newspaper coverage of the three situations of mothering under duress was studied . Specifically, 60 articles were reviewed that appeared from 1st May 1999 to 31st April 2000 in *The Globe and Mail*, *The National Post* and *The Vancouver Sun*, as well as the general newspaper coverage of mothers, mothering and related issues, which numbered over 500 articles.

number of other contexts (Floyd *et al.*, 2007; Ingersoll *et al.*, 2005). This research involved the provision of a counseling session that included collaborative discussion of a woman's confidence and readiness to change, as well as the opportunity for her to choose personal goals with respect to changing binge drinking and/or using contraception. Researchers found that this brief, but overtly collaborative, invention could lead to a significant reduction in the risk of alcohol-exposed pregnancies. Canadian researchers have advocated for the introduction of this promising research model in discussions with service providers and health system planners, and work is currently under way in Manitoba to study its implementation.

The need for collaborative practice was also identified in a participatory action research project involving local women and healthcare workers in four locations in Canada, which had the goal of identifying culturally appropriate methods of helping Aboriginal women to reduce their alcohol consumption during pregnancy and postpartum (George *et al.*, 2007). The process, which was led by the Intertribal Health Authority on Vancouver Island, as part of this Healthy Communities, Mothers and Children study, produced the OAR(S) tools (Intertribal Health Authority, May 2006). This approach:

> "... invites women to Own, Act and Reflect on their wellness. It is designed to be proactive and draws on the strengths, abilities and self-knowledge of women, recognizing that they are experts in their journey toward health and wellness. The OAR(S) model provides women with supportive navigation for their journey in a similar way to the oar of a canoe when it is in the hand of its journeyer." (p. 1)

As such, the approaches to pre- and postnatal support with Aboriginal women who drink alcohol developed during the course of the Healthy Communities Mothers and Children research project are examples of culturally relevant, collaborative care.

9.5
Conclusions

The integration of a values-based framework for FASD prevention by women's health advocates, researchers on women's substance use, and community-based practitioners, and other participants in local, provincial and national strategic planning processes, continues to strengthen the link between the health of children and the health of their mothers in Canada. Research is ongoing to further evidence improved FASD prevention responses, from a perspective that is women-centered, oriented to harm reduction, and collaborative. Women's health researchers are currently involved in multi-jurisdictional research networks which are supported by policy makers and designed to seek evidence for an improved FASD prevention response (see www.canfasd.ca/) and to promote its translation into policy and practice. Involvement in such research-to-action processes, brings attention to, and action on, women-centered prevention initiatives, balanced with diagnostic and intervention strategies for people with FASD and their families and caregivers.

There is a very long way to go to effect the paradigm shift towards a compassionate, health-oriented response to pregnant women and mothers who have alcohol problems, in policy, healthcare, social service and media realms in Canada. Similarly, the work to design, implement and study best practice in holistic, women-centered responses is substantial. Based on their contributions to date, the involvement of experts in women's health promotion, maternity care and women's substance use will bring an important women's health perspective to this essential work on prevention of FASD.

References

Armstrong, E.M. (2003) *Conceiving Risk, Bearing Responsibility: Fetal Alcohol Syndrome & the Diagnosis of Moral Disorder*, Johns Hopkins University Press, Baltimore, MD.

Astley, S.J., Bailey, D., Talbot, C., and Clarren, S.K. (2000) Fetal Alcohol Syndrome (FAS) primary prevention through FASD diagnosis II: a comprehensive profile of 80 birth mothers of children with FAS. *Alcohol Alcohol.*, **35** (5), 509–519.

Boyd, S.C. and Marcellus, L. (2007) *With Child: Substance Use during Pregnancy, A Woman-Centred Approach*, Fernwood Publishing, Halifax, NS.

Clow, B., Pederson, A., Haworth-Brockman, M., and Bernier, J. (2009) *Rising to the Challenge: Sex and Gender-Based Analysis for Health Planning, Policy and Research in Canada*, Atlantic Centre of Excellence for Women's Health, Halifax, NS.

Dell, C.A. and Clark, S. (2009) The Role of the Treatment Provider in Aboriginal Women's Healing from Illicit Drug Abuse, http://www.coalescing-vc.org/virtualLearning/community5/documents/Cmty5_InfoSheet2.pdf (accessed 30 May 2009).

Floyd, R.L., Sobell, M., Velasquez, M.M., Ingersoll, K., Nettleman, M., Sobell, L., et al. (2007) Preventing alcohol exposed pregnancies: a randomized control trial. *Am. J. Prev. Med.*, **32** (1), 1–10.

Four Worlds Centre for Development Learning (2003) *Making the Path by Walking It: A Comprehensive Evaluation of the Women and Children's Healing and Recovery Program (WCHRP) Pilot*, WCHRP, Yellowknife, NT.

George, M., Masotti, P., MacLeod, S., Van Bibber, M., Loock, C., Fleming, M., et al. (2007) Bridging the research gap: aboriginal and academic collaboration in FASD prevention. The Healthy Communities, Mothers and Children Project. *Alaska Med.*, **49** (Suppl. 2), 139–141.

Greaves, L. (1996) *Smoke Screen: Women's Smoking and Social Control*, Fenwood Publishing, Halifax, Canada.

Greaves, L., Poole, N., and Cormier, R. (2002a) *Fetal Alcohol Syndrome and Women's Health: Setting A Women-Centred Research Agenda*, BC Centre of Excellence for Women's Health, Vancouver, BC.

Greaves, L., Varcoe, C., Poole, N., Morrow, M., Johnson, J., Pederson, A., et al. (2002b) *A Motherhood Issue: Discourses on Mothering under Duress*, Status of Women Canada, Ottawa, ON.

Greaves, L., Cormier, R., Devries, K., Botorff, J., Johnson, J., Kirkland, S., and Aboussafy, D. (2005) *Expecting to Quit: Best Practices in Smoking Cessation During Pregnancy*. British Columbia Centre of

Excellence for Women's Health, Vancouver, BC.

Health Canada (2001) Fetal Alcohol Syndrome (FAS) "Pregnant? No alcohol!" Campaign, http://www.hc-sc.gc.ca/ahc-asc/activit/marketsoc/camp/fas-saf-eng.php (accessed 15 February 2010).

Ingersoll, K.S., Ceperich, S.D., Nettleman, M.D., Karanda, K., Brocksen, S., and Johnson, B.A. (2005) Reducing alcohol exposed pregnancy risk in college women: initial outcomes of a clinical trial of a motivational intervention. *J. Subst. Abuse Treat.*, **29** (3), 173–180.

Intertribal Health Authority (2006) *Own, Act and Reflect: A Guide for My Healing Journey*, May 2006, ITHA.

Leslie, M., DeMarchi, G., and Motz, M. (2007) Breaking the cycle: an essay in three voices, in *With Child; Substance Use during Pregnancy: A Woman-Centred Approach* (eds S.C. Boyd and L. Marcellus), Fernwood Publishing, Halifax, NS, pp. 91–104.

McGuire, M., Zorzi, R., McGuire, M., and Engman, A. (2006) *Early Childhood Development Addiction Initiative: Final Evaluation Report v.2*, Ministry of Health and Long-Term Care: Mental Health and Addiction Branch, May 31, 2006, Toronto, ON.

Miller, W.R. and Rollnick, S. (2002) *Motivational Interviewing: Preparing People for Change*, 2nd edn, The Guilford Press, New York.

Motz, M., Leslie, M., Pepler, D.J., Moore, T.E., and Freeman, P.A. (2006) Breaking the cycle: measures of progress 1995–2005. *J. FAS Int., Spec. Suppl.*, **4**, e22.

Parkes, T., Poole, N., Salmon, A., Greaves, L., and Urquhart, C. (September 2008) *Double Exposure: A Better Practices Review on Alcohol Interventions During Pregnancy*, British Columbia Centre of Excellence for Women's Health, Vancouver, BC.

Poole, N. (2000) *Evaluation Report of the Sheway Project for High Risk Pregnant and Parenting Women*, British Columbia Centre of Excellence for Women's Health, Vancouver BC.

Poole, N. (2003) *Mother and Child Reunion: Preventing Fetal Alcohol Spectrum Disorder by Promoting Women's Health*, British Columbia Centre of Excellence for Women's Health, Vancouver, BC.

Poole, N. and Isaac, B. (2001) *Apprehensions: Barriers to Treatment for Substance-Using Mothers*. British Columbia Centre of Excellence for Women's Health, Vancouver, BC.

Poole, N., Gelb, K., and Trainor, J. (2009) Preventing FASD through Providing Addictions Treatment and related Supports for First Nations and Inuit Women in Canada – Summary, http://coalescing-vc.org/virtualLearning/community5/documents/Cmty5_InfoSheet1.pdf (accessed 25 January 2010).

Poole, N. and Greaves, L. (2009) Mother and Child Reunion: achieving balance in policies affecting substance-using mothers and their children. *Womens Health Urban Life*, **8** (1), 54–66.

Poole, N., Greaves, L., Jategaonkar, N., McCullough, L., and Chabot, C. (2008) Substance use by women using domestic violence shelters. *Subst. Use Misuse*, **43** (9), 1129–1150.

Public Health Agency of Canada (2007) The Canada Prenatal Nutrition Program: A Decade of Promotion the Health of Mothers, Babies and Communities, http://www.phac-aspc.gc.ca/dca-dea/programs-mes/cpnp-epit_nuji-eng.php (accessed 14 February 2010).

Robinson, E., Berg, P., McGowan, B., and Tombs, B. (2003) *Maxxine Wright Place Project for High Risk Pregnant and Early Parenting Women*, Atira Women's Resource Society, White Rock, BC.

Rollnick, S., Miller, W.R., and Butler, C.C. (2008) *Motivational Interviewing in Health Care*, Guilford Press, New York.

Rutman, D., Callahan, M., Lundquist, A., Jackson, S., and Field, B. (2000) *Substance Use and Pregnancy: Conceiving Women in the Policy Process*, Status of Women Canada, Ottawa, ON.

Salmon, A. (2010) The impact of MEIA-funded support on Sheway clients. Paper presented at the Network Action Team on

FASD Prevention Biannual Meeting, Victoria, BC.

UNODC (2004) *Substance Abuse Treatment and Care for Women: Case Studies and Lessons Learned*, United Nations Office on Drugs and Crime, Vienna, Austria.

Urquhart, C. and Poole, N. (2008) *Supporting Change: Preventing Fetal Alcohol Spectrum Disorder [DVD]. In 2C Visual Communications (Producer)*, British Columbia Centre of Excellence for Women's Health, Canada.

Urquhart, C., Poole, N., and Horner, E. (2008) Using an integrated framework to support women in childbearing years on reducing harms associated with alcohol and tobacco use. Paper presented at the First International Conference on Motivational Interviewing, June 9.

10
Next Steps in FASD Primary Prevention

Robin Thurmeier, Sameer Deshpande, Anne Lavack, Noreen Agrey, and Magdalena Cismaru

10.1
Introduction

Primary prevention strategies for Fetal Alcohol Spectrum Disorder (FASD) have been implemented since the condition was first recognized and defined in 1973. Primary prevention can be defined as strategies used to intercede before the causes of diseases can impact an individual (Cohen, Chavaz, and Chehimi, 2007). Such strategies are considered proactive, and should be focused on populations rather than on individuals. In the past, Health Canada has defined primary prevention as prevention efforts that focus on the individual, societal systems, and behavior change (Roberts and Nanson, 2001). Poole (2008) outlines the three main goals of primary prevention within the area of FASD as:

- raising awareness about the risks of consuming alcohol during pregnancy;
- providing information about where those who need support will find help; and
- promoting community involvement in building FASD awareness.

The benefits of FASD primary prevention include providing women with evidence-based health information, and connecting community members to create service systems to address the community wide impacts of substance use (Poole, 2008). However, one of the main challenges of primary prevention is to find a voice among the plethora of commercial media vying for audience attention (Glick et al., 2008). Although, unfortunately, primary prevention communication has often been "... relegated to a message in a brochure or in a few moments during a medical visit" (Cohen, Chavaz, and Chehimi, 2007), its importance can be found in its very definition.

Primary prevention strategies include activities such as population health promotion, alcohol control measures, multicomponent awareness approaches, and other education efforts. FASD primary prevention strategies can be divided into four categories: educational; legal; community development (Roberts and Nanson, 2001); and social marketing.

Fetal Alcohol Spectrum Disorder–Management and Policy Perspectives of FASD. Edited by Edward P. Riley, Sterling Clarren, Joanne Weinberg, and Egon Jonsson
Copyright © 2011 WILEY-VCH Verlag GmbH & Co. KGaA, Weinheim
ISBN: 978-3-527-32839-0

- **Educational strategies**, such as awareness campaigns, have enjoyed some success in preventing FASD (Basford *et al.*, 2004). The populations targeted by these programs include those who are willing to change behaviors once they are made aware of the positive impact of the desired behavior, and the negative aspects of the current behavior. The strategies employed within educational programs include print and multimedia materials such as brochures, posters and public service announcements (Poole, 2008). The underlying assumption is that these individuals have the motivation, opportunity, and ability to achieve the desired behavior (Basford *et al.*, 2004).

- **Legal strategies** are system-wide policy changes, such as legislating warning labels on bottles of alcohol. These strategies reach a wide audience, but are considered passive as they tend to rely on the audience's ability to read, understand, and follow the warning (Hankin, 1994).

- **Community development strategies** can lead the public to become informed and more willing to buy into and participate in more comprehensive and coordinated prevention strategies (Oliver, White, and Edwards, 1998). These strategies include developing a community understanding and the response to women's substance use. Community development strategies are often related to the social determinants of health (such as housing, partner violence, socioeconomic status). Examples include intensive prenatal clinics or training programs for healthcare professionals (Hankin, 1994). These efforts tend to impact on smaller audiences, but are more intense and active in reaching this audience (Hankin, 1994). The goal is to create system-level changes (Poole, 2008).

- **Social marketing strategies** go beyond creating media campaigns to include the creation and promotion of a favorable environmental opportunity that makes enacting the new behavior desirable (Basford *et al.*, 2004).

An overview of FASD campaign outcomes in north-western Canada will be presented in this chapter, with discussions focusing on behavior change theory, evaluation, and social marketing strategies in FASD primary prevention. It is hoped that the conclusion will provide the reader with some important considerations for future FASD prevention campaign development.

10.2
Current State of FASD Primary Prevention in North-Western Canada

In 2007, a review of current FASD primary prevention resources in north-western Canada was undertaken to create a scan of what has been achieved in the area over the past 30 years. Since the development of the terminology for Fetal Alcohol Syndrome (FAS) in 1973, both government- and community-based organizations have been developing messages which suggest that women should not consume alcohol during pregnancy. The collection of FASD campaigns was reviewed in the context of a 1996 Health Canada developed a document entitled, "Joint Statement: Prevention of Fetal Alcohol Syndrome (FAS) and Fetal Alcohol Effects (FAE) in

Canada." This influential Health Canada document provided recommendations for FASD prevention efforts as follows:

> "Prevention efforts should target women before and during their childbearing years, as well as those who influence such women, including their partners, families, and the community. All efforts should be: family-centred and culturally sensitive, to address the pregnant woman as well as her partner and family in the context of their community; and comprehensive, to draw on all services appropriate to the often complex social, economic and emotional needs of these women." (Health Canada, 1996, p. 5)

Subsequently, FASD campaigns were reviewed to determine their adherence to these Health Canada recommendations, with campaign resource materials being separated by type and placed into one of four categories of: print; multimedia; reports; and miscellaneous:

- *Print materials* included resources such as posters, brochures, fact sheets and information cards, booklets, and facilitator/curriculum guides.
- *Multimedia* (audio/video) resources included television and radio public service announcements or advertisements, as well as longer videos.
- *Reports* included research-related papers developed by government and research networks.
- The *miscellaneous* category included unique prevention items such as bracelets, pens, postcards, pins, and mascots.

The distribution of these resource types, by province and territory, is shown graphically in Figure 10.1. The data showed that, among all the provinces, Alberta

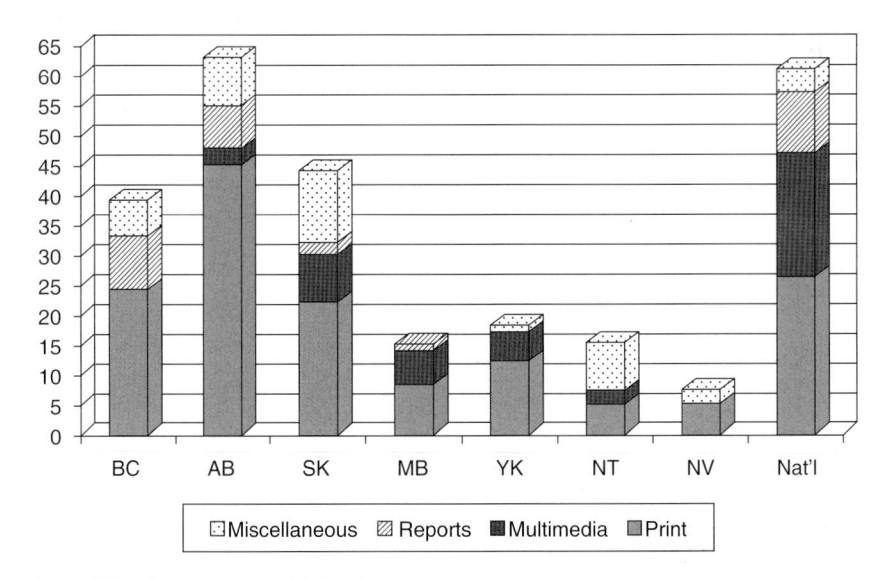

Figure 10.1 Resource type collected by region.

had the largest absolute number of different resources, followed by Saskatchewan and British Columbia. The number of different materials intended for national distribution was approximately equal to the number produced by the province of Alberta. Print materials comprised the largest category of items, followed by multimedia (audio/video) items.

As shown in Table 10.1, the message strategies and themes of these resources have ranged from being informative to being directed at specific target audiences and reflecting different types of advice. Subsequently, in an attempt to determine the effectiveness of these resources, evaluations of past work were also collected from three campaigns conducted across north-western Canada:

- The Born Free Campaign (Alberta Children Services)
- The Mother Kangaroo Campaign (Saskatchewan Prevention Institute)
- The With Child/Without Alcohol Campaign (Manitoba Liquor Control Commission)

Details of these three campaigns are briefly summarized in the following sections.

10.2.1
The Born Free Campaign

In 1999–2000, the Alberta Children Services developed and initiated the "Born Free" campaign, profiling FASD. The intention of the campaign was to increase awareness of the importance of preventing FAS/FAE by abstaining from alcohol use during pregnancy. The partners included the Alberta Restaurant and Foodservices Association, the Alberta Liquor Store Association, the Alberta Gaming and Liquor Commission, 17 regional FAS committees, and Boston Pizza. This initiative included signage in liquor stores with the message, "When you are pregnant, NO alcohol is best." The participating restaurants also provided free nonalcoholic beverages for pregnant customers (Gavin, 2001). The images associated with the message included a pregnant woman holding a drink, while the message was promoted through television advertisements, posters, radio announcements, and printed advertisements.

The evaluation component included a pre- and post-campaign survey of 800 residents. The results showed that initial levels of awareness and support were high, and remained high, with the public recall of information initially 61% but increasing to 73% post-campaign (Roberts and Nanson, 2001). The specific message recall post-campaign was 68% (Burgoyne, 2006). However, when all participating Boston Pizza locations reviewed their experiences with the campaign eight months later, the responses were mixed. This was because restaurants in smaller cities had participated more actively in high-profile community events, whereas those in larger locations had agreed only to a minimal support (e.g., decals in the washroom), mainly because they were concerned about sending mixed messages – that is, supporting the campaign while serving alcoholic beverages (Gavin, 2001).

Table 10.1 Message strategies and themes.

Strategies and themes	Messaging
Informative *These messages provide general information about FASD, not focused on any group in particular*	• No amount of alcohol is safe during pregnancy (AB) • FAS: It's 100% preventable (AB) • No alcohol is best when pregnant (AB) • 101 ways to have a healthy baby (AB) • The word is out on drinking and pregnancy (BC) • FASD is preventable (SK) • There is no safe time, kind, or amount of alcohol (SK) • FAS: A preventable birth defect (MB)
Women-focused *These messages are specifically directed towards women who are pregnant or trying to become pregnant, and reflect the direct link between mother and child*	• You wouldn't give alcohol to your newborn, so why your unborn? (AB) • Really, your baby doesn't need a cold one (AB) • Pregnancy and alcohol? Don't take the chance (AB) • I'm having what she's having (AB) • Alcohol and pregnancy don't mix (AB, BC) • She is not drinking alone (BC) • How can I be a mom? (BC) • How about a drink…kid? (SK) • Let the traditional spirits guide you (SK) • Listen to the inner voice (SK) • Drinking alcohol during pregnancy can harm the baby (SK) • With Child/Without Alcohol (MB) • Healthy moms and babies (NV) • Promise to my unborn child (NT) • For baby's sake … (YK) • Don't drink if you're pregnant (Nat'l) • Pregnant? No alcohol. (Nat'l)
Partner/Friend/Community-focused *These messages indicate that partners, friends, and the community have a responsibility in preventing FASD*	• Friends helping friends (AB) • Help a pregnant friend avoid alcohol (AB) • Who is supporting her alcohol-free pregnancy? (AB) • Working together: We can prevent FASD (BC) • Making a difference one family at a time (BC) • Dads make a difference too (BC) • Protect our children (SK) • Take action (SK) • Healing choices (MB) • Keep families strong (NT) • Together our community can prevent FAS (NT) • This is OUR baby (YK) • Caring together (Nat'l) • Stop FAS/FAE now! (Nat'l) • Together we can prevent FAS (Nat'l)
Effects of FASD *These messages highlight the effects of FASD*	• FASD is a life sentence (AB) • The many faces of FAS (YK)

BC, British Columbia; AB, Alberta; SK, Saskatchewan; MB, Manitoba; YK, Yukon; NT, Northwest Territories; NV, Nunavut; Nat'l, National.

With regards to the target audience, the evaluation acknowledged one weakness of the campaign – that it failed to reach high-risk populations because the message had been targeted more towards the general population. The campaign did not provide any specialized messaging that highlighted coping strategies for those in high-risk populations who may have required more assistance.

10.2.2
The Mother Kangaroo Campaign

The Saskatchewan Prevention Institute developed the "Mother Kangaroo" campaign, which ran from 2006 to 2008. The campaign included television commercials, radio spots, posters, and information cards which featured a mother kangaroo with a partner, a group of friends, and a baby Joey. Each of these scenarios provided messages that encouraged having an alcohol-free pregnancy, and illustrated forms of support by friends, family, and the community. The posters were hung in provincial liquor stores, as well as restaurants and public health buildings.

A pre- and post-campaign evaluation was completed. The geographically representative pre-campaign survey (with a sample size of 401) was conducted in May 2005 to determine public awareness and knowledge of FASD, and to measure the recall of a previous advertising campaign. The results of a previous campaign, which had focused on community support for alcohol-free pregnancies, showed the aided campaign recall to be 69.1%, and unaided recall to be 19.2% (Norsask Consumer Interviewing Services Ltd., 2005). Awareness of the effects of alcohol was very high, with 97% of respondents aware that alcohol use during pregnancy could cause lifelong disabilities in the child (Norsask Consumer Interviewing Services Ltd., 2005). The large majority of participants (70.3%) believed that there was no safe *amount* of alcohol that could be consumed during pregnancy, nor any safe *time* to consume it (88.8%). Whilst awareness of the effects of alcohol during pregnancy was high, the indication to participate in preventive activities was low. Only 28.2% of male respondents and 47.5% of women respondents indicated they had completed, or were planning to complete, an activity aimed at preventing alcohol consumption during pregnancy, such as abstaining from alcohol use or providing non-alcoholic beverages for someone who might be pregnant (Norsask Consumer Interviewing Services Ltd., 2005).

The post-campaign survey was completed in June 2006. A geographically representative sample of 400 Saskatchewan residents completed the survey to test recall of the "Mother Kangaroo" campaign, and to track any changes in attitudes or perceptions about the effects of alcohol use during pregnancy. In total, 71% of the respondents recalled seeing advertisements about the use of alcohol during pregnancy, but in terms of awareness and attitudes about the effects of alcohol used during pregnancy the levels remained unchanged, with 97% of respondents agreeing that alcohol had an effect on the fetus (Fast Consulting, 2006). The number of respondents participating in preventative behaviors was increased from the previous survey, with 44% indicating that they had supported a pregnant woman's choice not to drink, while 42% had told others about the harmful effects of alcohol.

Almost half (49%) indicated they would offer nonalcoholic beverages, and show support by not drinking alcohol themselves (47%).

10.2.3
The With Child/Without Alcohol Campaign

In 2006, the Manitoba Liquor Control Commission developed and implemented a province-wide "With Child/Without Alcohol" campaign that included Public Service Announcements (PSAs) through television and radio, as well as posters, brochures, a website, and an information kit. The aim of the campaign was to increase awareness of FASD within the general public. In April 2006, 400 post-campaign surveys were conducted in a random sample of Manitoba residents aged between 18 and 45 years, 75% of whom were female. The survey was undertaken to measure campaign recall and FASD awareness following instigation of the campaign.

Unaided recall of the advertisement was relatively high; 80% of respondents with a post-secondary education, and 64% of those without, indicated that they had seen or heard an advertisement about alcohol consumption during pregnancy (ChangeMakers, 2006). A total of 73% recalled seeing a PSA on television, compared to only 7% of those who had seen a poster in a restaurant or bar. When determining the awareness of FAS, 93% of respondents maintained that they had heard of FAS (ChangeMakers, 2006), and 95% of them were aware that drinking alcohol while pregnant could lead to FAS (ChangeMakers, 2006). No indicators of behavior change were measured, however.

10.2.4
Summary of Results

Each of these three campaign evaluations highlighted an increased knowledge and awareness of FASD among the general public. Although the methods and resources used to impart this knowledge were successful. it was unclear whether the knowledge and awareness had been increased uniformly across the various demographic groups. Indeed, an evaluation of the Alberta campaign acknowledged that the messages had not reached all of the target groups with equal effect. Social marketing strategies (as discussed below) might help campaign developers to create messages and opportunities that will reach specific target audiences to a greater degree.

While increases in knowledge and awareness of FASD were seen as positive outcomes, the evaluations revealed little about behavioral change. Yet, this was an important consideration, as the intended outcome of these campaigns was to reduce or eliminate alcohol consumption among pregnant women. Although an evaluation of the "Mother Kangaroo" campaign did contain a behavioral component, the results were not especially promising, with less than 50% of the participants in each evaluation indicating that they had taken part in or had supported FASD-prevention activities (e.g., choosing to abstain from alcohol during pregnancy, offering alcohol-free beverages at gatherings). Clearly, further information is required concerning the behavioral outcomes of these campaigns. Nonetheless, the incorporation of a theoretical model into the design, evaluation, and implementation of a campaign may be advantageous in creating a positive behavioral change.

It should also be noted that the above three campaign evaluations were all that could be found during a scan of resources available across north-western Canada. Clearly, there appears to be a surprising lack of evaluation among previous FASD primary prevention campaigns!

10.3
Campaign Evaluation: What Is It and Why It Is Important

The evaluation of a campaign essentially involves determining the efficiency and success of the campaign strategies employed. A variety of prevention strategies, including FASD prevention, should be evaluated for effectiveness (Burgoyne, 2006). The evaluation of prevention strategies can provide health promotion research groups, government officials and campaign developers with a means to learn from their past experiences (The World Bank, 2004). For example, LaChausse (2006) determined that a FASD education campaign targeted towards adolescents was less effective than it might have been, mainly because it lacked principles that had been shown as effective in changing other health behaviors, such as identifying behavioral and/or skill changes intended to change behavioral intentions regarding abstention from alcohol during pregnancy. A research-based approach that incorporates a sound theoretical framework can provide a relevant insight, and help to define a specific health problem, to describe the intended audience, and to identify effective behavioral solutions (O'Sullivan *et al.*, 2003; Scammon, Mayer, and Smith, 1991).

The most effective messages are often determined through formative and pretest research surveys (Walker *et al.*, 2005). Research at the formative research stage is crucial to determining the target audience, and several demographic, psychographic and behavioral factors must be considered when selecting a target population. Demographic factors include age range, socioeconomic status, and education level, while psychographic factors include values and lifestyle. Finally, behavioral factors include user status, user rate, and readiness stage (Kotler and Lee, 2008). Once the target audience has been selected, they must be consulted in order to gain insights with regards to the effectiveness of campaign strategies, such as the focus of a campaign strategy, the types of messages, types of images, mode of communication, and optimal timing in order to ensure campaign effectiveness.

FASD prevention campaigns must also seek a fine balance, in order to avoid creating undue anxiety in women who have inadvertently exposed their unborn babies to alcohol. At the same time, campaigns need to create an awareness of the risks associated with particular patterns of behavior, especially among particular subpopulations (Roberts and Nanson, 2001). It is important to note that preventive campaigns should be comprehensive, collaborative, community-based, culturally appropriate, and nonjudgmental (Oliver, White, and Edwards, 1998). This is particularly true of FASD primary prevention, as there are often misconceptions regarding who is more at risk of consuming alcohol during pregnancy. Whilst most programs continue to target younger women with lower levels of education, income, and status, it is the women who are older, well-educated and have a higher income who often exhibit the same level of risks of drinking alcohol while pregnant (Dell and Roberts, 2006). It is important that prevention campaigns are developed to better target this latter group, and to remove any misconceptions that these women might hold about who is at risk of having a child impacted by alcohol use during pregnancy. If these older, well-educated, higher-income women believe that FASD cannot happen to their children, but only to the children of women who are typically considered "high-risk," then they are less likely to adhere to prevention strategies.

Findings from past campaigns also confirm that awareness-only campaigns are not always effective in influencing audience behavior. As a result, in spite of previous FASD-prevention efforts, alcohol is still consumed in about 11% of all pregnancies. Hence, the failure of previous campaigns might also be attributed to the failure of campaign strategies to address the reasons *why* women drink during pregnancy. Such reasons include (Environics Research Group Limited, 2006; Floyd, Decoufle, and Hungerford, 1999; Network Action Team #3, not dated):

- Some women are unaware of pregnancy early on and consume alcohol during this time.

- Some women are unaware of the extent of fetal damage that alcohol can cause.

- Some women who drank alcohol during pregnancy have children who appear healthy, leading themselves and others to underestimate the possible harm of alcohol consumption during pregnancy.

- Alcohol use may be the norm within a social group or work culture, so abstaining from alcohol use may be difficult in certain social settings.

- Alcohol may be used to cope with difficult life situations, such as poverty or addiction.

The above list illustrates that awareness-only campaigns can address the first three reasons, but not the next two. In order to address the final two reasons, women must be presented with an attractive opportunity (in the form of a tangible product or an intangible service) in the environment (Rothschild, 1999). In other words, managers must consider using social marketing strategies in their campaigns. An example of an attractive tangible product might be an alcohol-free drink that tastes like gin or beer, whereas an example of an attractive intangible service is the provision of alcohol-free socializing opportunities. Social marketers do not always use mass media to promote opportunities; sometimes, they restrict their promotion efforts to community-based vehicles to achieve their objectives (Deshpande and Basil, 2006). For example, Rothschild (1999) has argued that when the audience does not carry out the intended behavior due to a lack of awareness, then message-only strategies are appropriate. However, when the behavior is not carried out due to barriers or a lack of opportunity in the environment, then a more comprehensive social marketing effort might represent a more appropriate strategy.

When these prevention strategies are correctly evaluated and strategically designed to incorporate sound health behavior theory, the goal of improving health through behavioral change can be achieved (O'Sullivan et al., 2003). Theory-based evaluation is useful in the area of health promotion, as it provides a more in-depth understanding of the workings of a program. Frameworks such as Protection Motivation Theory (PMT) can be employed in such contexts. These theoretical frameworks facilitate mapping out the causal factors that are judged important in the success of a program. In this way, critical success factors can be assessed and early feedback collected to determine the likelihood of success of any particular program or project (The World Bank, 2004).

The applications of social marketing strategies and PMT to improve the effectiveness of FASD prevention among pregnant women are elaborated on in the following sections.

10.4
Incorporating Social Marketing Strategies

Social marketing is defined as a tool which is used to influence individual behavior by reducing barriers and offering benefits that are superior to those offered by current behaviors (Kotler and Lee, 2008). In return for adopting desirable behaviors, social marketing offers benefits that are proximal, guaranteed, and explicit in nature. Evidence suggests that social marketing initiatives can have a positive impact on reducing alcohol consumption among pregnant women, as well as changing the attitudes of pregnant women towards alcohol (Deshpande et al., 2005; Yanovitzky and Stryker, 2001). Social marketing is a strategy that is commonly used in primary prevention, because it allows the target audience the "opportunity" to perform desired health behaviors (Deshpande et al., 2005). The theory behind social marketing supports the concept that a well-designed primary prevention program can impact behavior in a positive manner by influencing a target audience to accept a healthy behavior, or reject or abandon an unhealthy behavior, for the benefit of the individual and the community (Cismaru, 2006; Stockburger, 2003).

Primary prevention measures such as social marketing start the campaign process by identifying a specific target population. In the FASD prevention context, this could include women who drink during pregnancy, women who do not know they are pregnant, adolescents, healthcare professionals, and male partners (Basford et al., 2004). Social marketing requires intensive research to develop an extensive understanding of the target audience. One of the key principles of social marketing is to include the target audience in the campaign development process (Mengel et al., 2005). Audience orientation enables the creation of an exchange offer that is consistent with audience needs, and tailored towards that specific group. Thus, marketers can use the input of the target audiences to determine attractive tangible and/or intangible opportunities in the environment, and to determine strategies that reduce barriers and enhance benefits in relation to these opportunities. In order to promote these opportunities, the target audience inputs would also allow managers to create a culturally sensitive message, to determine an appropriate language, color and design, to identify and reduce anything that might detract from the message, and to determine how to distribute the message (Mengel et al., 2005).

10.5
Creating Behavioral Change: Protection Motivation Theory

Health behavior models, which are developed to account for sociodemographic variations in health behavior, can be used when implementing social marketing strategies to better determine how to achieve this behavioral change. In this chapter, attention is focused only on the promotional aspects of social marketing strategy. Thus, questions need to be asked, such as what messages are the most effective in changing behavior, and what factors make campaigns more persuasive and effective (Cismaru, 2006). As there has been little or no attempt to include a theoretical perspective in social marketing strategies associated with FASD in the past, the variables that underlie the decision to drink alcohol while pregnant need to be identified. Similarly, appropriate measures to evaluate behavioral change need to be identified or developed.

In order to better understand the factors that influence the persuasiveness of public health communication, researchers commonly use Protection Motivation Theory (PMT) as a theoretical framework (Maddux and Rogers, 1983; Prentice-Dunn and Rogers, 1986; Rogers, 1983). PMT provides a comprehensive method of explaining the influencing factors of health behaviors that is well accepted and widely used (Boer and Seydel, 1996). Social marketing campaigns which are aimed at preventing alcohol consumption during pregnancy seek to influence many of the cognitions that are the focus of PMT (Leas and McCabe, 2007; Simons-Morton et al., 2006).

Very often, PMT is used as a framework to explore the reasons why people engage in harmful behavior(s), and to guide initiatives aimed at changing health-related behaviors (Maddux and Rogers, 1983; Prentice-Dunn and Rogers, 1986; Rogers 1983; Floyd, Prentice-Dunn, and Rogers, 2000; Milne, Sheeran, and Orbell, 2000), including under-age drinking, drinking and driving, smoking cessation, and obesity across several different audiences (Floyd, Prentice-Dunn, and Rogers, 2000; Milne, Sheeran, and Orbell, 2000). For example, when examining the impact of anti-smoking advertisements on adolescents, Pechmann et al. (2003) found that messages which reflected social disapproval and severe health risks while also increasing confidence to cigarettes, created conditions where youth followed through with positive behavior changes.

PMT was developed to examine the effectiveness of fear-based messaging compared to messaging that also increases an individual's coping skills and confidence level (Prentice-Dunn and Rogers, 1986). PMT theory suggests that the motivation to select positive health behaviors is maximized when:

- the health threat is viewed as severe (perceived severity);
- the person feels vulnerable to, or is affected by, the health threat (perceived vulnerability);
- the suggested change is considered to lead to the desired outcome (response efficacy);
- the costs of making the change in terms of money, effort, inconvenience, or embarrassment are small; and
- the person feels confident in their own ability to complete the suggested change (self-efficacy) (Rogers, 1975, 1983; Milne, Sheeran, and Orbell, 2000).

These factors are divided into two processes: the threat-appraisal process; and the coping-appraisal process (Wu et al., 2005). The theory focuses on an individual's ability to judge the probability of a harmful event occurring to him or her, the severity of the event, the efficacy of a preventative response, his or her perceived ability to successfully complete coping behaviors, the costs of adopting these behaviors, and the intrinsic and extrinsic rewards associated with the unhealthy behaviors (Armitage and Conner, 2000; Murgraff, White, and Phillips, 1999).

- **Threat appraisal process:** This evaluates the perceived intrinsic and extrinsic rewards of the behavior, the severity of the danger, and vulnerability to the danger (Wu et al., 2005). *Perceived vulnerability* refers to the subjective perception of the risk of a negative event occurring; for example, if the perceived vulnerability to a particular health problem is high, then the intention of changing behavior increases (Cismaru, 2006; Armitage and Conner, 2000). Thus, if a woman believes that alcohol use during pregnancy can harm her child, she may be more likely to stop drinking alcohol. Perceived vulnerability includes having an understanding of the risks, or a likelihood of negative outcomes, that are associated with a particular maladaptive behavior (Branco and Kaskkutas, 2001). *Perceived severity* is the feeling or belief concerning the seriousness of the maladaptive behavior (Cismaru, 2006); this could include perceptions about whether alcohol use during pregnancy will have serious, lifelong consequences for the child. As perceived severity increases, a woman is more likely to change behavior and follow recommended health strategies.

- **Coping appraisal process:** This comprises the perceived response efficacy and perceived self-efficacy factors. Individuals evaluate their ability to cope with or avert dangerous behaviors, and also balance the costs and benefits of the protective behavior with the costs and benefits of the maladaptive behavior (Wu et al., 2005). *Perceived response efficacy* refers to an individual's belief that the recommended health behaviors will be effective in reducing maladaptive behaviors and outcomes. It is expected that the more likely an individual is to believe they can achieve the goals of the health recommendation, the more likely they are to adopt the recommended behaviors (Cismaru, 2006). For example, if a male partner believes he can influence positive change in his partner's drinking patterns during pregnancy through changing his own behavior, he will be more likely to attempt healthy behavior changes. *Perceived self-efficacy* refers to a belief in an ability to perform the recommended behavioral change. Perceived cost is the sum of all the barriers to performing the behavior change (Cismaru, 2006; Armitage and Conner, 2000); this cost can include actual monetary costs as well as individual, social, and legal consequences.

The two appraisal strategies—threat appraisal and coping appraisal—combine to form protection motivation. The adoption of health behaviors is then a temporal process, from motivation to decision to action (Wu et al., 2005). The above process becomes a balancing act as threat appraisals and coping appraisals are assessed and enacted. If the balance tips too far one way, then the health goals may be seen as unachievable. According to PMT, campaigns that focus primarily on negative consequences are less likely to have an impact than campaigns that focus on positive behaviors and changes in attitudes, such as modeling appropriate skills, demonstrating benefits of positive behaviors, and promoting support and participation in behavioral change (DeJong, 2002).

Experts argue that fear-based messages could backfire, making the negative behavior more resistant to change (DeJong, 2002), especially when the level of fear is perceived to be too high. In a qualitative study of focus groups of high-risk women in the US, Branco and Kaskutas (2001) found that fear-based messages were considered unbelievable and exaggerated. The images were instantly discounted because they were seen as too simplistic, judgmental, and unrealistic, and reinforced doubts about public health warnings and the relevance to real-life situations. The women had many misconceptions about alcohol use and safety during pregnancy that the messages did not address. The women also suggested using messages that showed situations where the social support of healthy pregnancies would have a more positive impact. Others (Mengel et al., 2005) found that pictures showing harm to children were seen as a "scare tactic," and inappropriate for the group of women that they were attempting to reach.

Recent research has shown that PMT is an appropriate model to use within the context of FASD primary prevention (Cismaru et al., 2010). To illustrate the use of PMT principles in the context of FASD, consider the following example of a poster aimed at reducing alcohol consumption during pregnancy. "Take Action," which was developed by the Saskatchewan Prevention Institute (see Figure 10.2), states that, "Drinking alcohol during pregnancy can cause Fetal Alcohol Spectrum Disorder. There is no safe time, kind or amount of alcohol." This information about the possible negative consequences of drinking while pregnant is meant to increase perceived vulnerability and severity for the reader. The same poster states that, "Plan not to drink alcohol during pregnancy or when breast feeding." This

Figure 10.2 The "Take Action" poster.

is a specific recommendation that may involve costs (inconvenience) and response efficacy. The statement, "Talk to others about the harmful effects of alcohol during pregnancy," is meant to increase the perceived response efficacy of the recommendations. Finally, the poster contains a statement saying, "Support a pregnant woman's choice not to drink alcohol"; this statement attempts to reduce the perceived costs of abstaining from alcohol, as many consumers might otherwise believe that drinking alcohol is the social norm. It suggests that others can also take action to decrease the likelihood of alcohol use during pregnancy.

10.6
Future Considerations for Health Promoters and Policy Makers

Much time and effort has been expended into FASD primary prevention over the past 30 years. Although very few evaluations have been undertaken, the outcomes do show that current efforts can increase the awareness and knowledge of FASD. While this can be viewed as being successful with regards to awareness development, relatively little is known about creating behavioral change. Women continue to consume alcohol during pregnancy for a number of reasons. An intended behavioral change is expected to occur when awareness increases, but this is not always the case.

As a next step, health promoters, policy developers, and research groups must focus more on how to promote alcohol abstinence among pregnant women. When developing a FASD prevention campaign, the following items might be considered:

- **Community needs:** Who is the target audience? Who most needs information/community support to abstain from alcohol? Which partners can be brought on board to create community support?

- **Campaign implementation:** What are the underlying objectives of the campaign – to address awareness, behavior change, or both? How will these goals be reached? Can social marketing strategies be used to create the intended outcomes? What opportunities need to be provided in the environment that will attract the audience to practice the desired behavior? What are the critical message(s) this audience needs to hear? What will be the definition of success for the campaign? For how long will the campaign run? How much will the campaign cost? How will the funding be obtained?

- **Behavioral change:** If a goal of the campaign is behavioral change – what behavioral change is the campaign trying to achieve, and how will this change be made? What theoretical constructs will be used to develop and explain the health promotion process? PMT has been shown to be an appropriate theory to use within the context of FASD.

- **Evaluation:** How will the goals of the campaign be evaluated? What evaluation strategies are in place pre-campaign, during campaign, and post-campaign? And how will the outcomes be disseminated?

Clearly, consideration of these elements, including the theories and tools of Social Marketing and PMT, will help contribute to greater success in developing social marketing campaigns for FASD prevention.

References

Armitage, C.J. and Conner, M. (2000) Social cognition models and health behaviour: a structured review. *Psychol. Health*, **15**, 173–189.

Basford, D.L., Thorpe, K., William, R., and Cardwell, K. (2004) *State of Evidence: Fetal Alcohol Spectrum Disorder (FASD) Prevention*, Alberta Centre for Child, Family & Community Research, Lethbridge, AB.

Boer, H. and Seydel, E.R. (1996) Protection motivation theory, in *Predicting Health Behavior: Research and Practice with Social Cognition Models* (eds M. Conner and P. Norman), Open University Press, Buckingham, pp. 95–120.

Branco, E.I. and Kaskkutas, P.H. (2001) "If it burns going down ...": how focus groups can shape fetal alcohol syndrome prevention. *Subst. Use Misuse*, **36**, 333–345.

Burgoyne, W. (2006) *What We Have Learned: Key Canadian FASD Awareness Campaigns*, Public Health Agency of Canada, Health Canada, Ottawa, ON.

ChangeMakers (2006) *Summary of Advertising Awareness Study*, Manitoba Liquor Control Commission, Winnipeg, MB.

Cismaru, M. (2006) *Using Protection Motivation Theory to Increase the Persuasiveness of Public Service Communications*, The Saskatchewan Institute of Public Policy, University of Regina, Regina, SK.

Cismaru, M., Deshpande, S., Thurmeier, R., Lavack, A., and Agrey, N. (2010) Preventing fetal alcohol spectrum disorder: the role of protection motivation theory. *Health Mark Q*, **27**, 66–85.

Cohen, L., Chavaz, V., and Chehimi, S. (eds) (2007) *Prevention Is Primary: Strategies for*

Community Well-Being, Jossey-Bass, San Francisco.

DeJong, W. (2002) The role of mass media campaigns in reducing high-risk drinking among college students. *J. Stud. Alcohol*, (Suppl. 14), 182–192.

Dell, C. and Roberts, G. (2006) *Alcohol Use and Pregnancy: An Important Canadian Public Health and Social Issue*, Public Health Agency of Canada, Health Canada, Ottawa, ON.

Deshpande, S. and Basil, M. (2006) Lessons from research on social marketing for mobilizing adults for positive youth development, in *Mobilizing Adults for Positive Youth Development: Lessons from the Behavioral Sciences on Promoting Socially Valued Activities* (eds E. Gil Clary and J.E. Rhodes), Kluwer/Plenum, New York, pp. 211–231.

Deshpande, S., Basil, M., Basford, L., Thorpe, K., Piquette-Tomei, N., Droessler, J., et al. (2005) Promoting alcohol abstinence among pregnant women: potential social change strategies. *Health Mark Q*, **23**, 45–67.

Environics Research Group Limited (2006) *Alcohol Use during Pregnancy and Awareness of Fetal Alcohol Syndrome and Fetal Alcohol Spectrum Disorder: Results of a National Survey. Final Report*, Public Health Agency of Canada, Ottawa, ON.

Fast Consulting (2006) *Saskatchewan Prevention Institute: FASD Prevention – Post-Campaign Survey*, Saskatchewan Prevention Institute, Saskatoon, SK.

Floyd, L.R., Decoufle, P., and Hungerford, D.W. (1999) Alcohol use prior to pregnancy recognition. *Am. J. Prev. Med.*, **17**, 101–107.

Floyd, D., Prentice-Dunn, S., and Rogers, R. (2000) A meta-analysis of research on Protection Motivation Theory. *J. Appl. Soc. Psychol.*, **30**, 407–429.

Gavin, K. (2001) *Born Free: Options for Expansion*, The Alberta Alcohol and Drug Abuse Commission, Edmonton, AB.

Glick, D., Prelip, M., Myerson, A., and Eilers, K. (2008) Fetal alcohol syndrome prevention using community-based narrowcasting campaigns. *Health Promot. Pract.*, **9**, 93–103.

Hankin, J.R. (1994) FAS prevention strategies: passive and active measures. *Alcohol Health Res. World*, **18**, 62–66.

Health Canada (1996) *Joint Statement: Prevention of Fetal Alcohol Syndrome (FAS) and Fetal Alcohol Effects (FAE) in Canada*, Health Canada, Ottawa, ON.

Kotler, P. and Lee, N. (eds) (2008) *Social Marketing: Influencing Behaviors for Good*, Sage, Thousand Oaks, CA.

LaChausse, R.G. (2006) The effectiveness of a multimedia program to prevention fetal alcohol syndrome. *Health Promot. Pract.*, **7**, 1–5.

Leas, L. and McCabe, M. (2007) Health behaviors among individuals with schizophrenia and depression. *J. Health Psychol.*, **12**, 563–579.

Maddux, J.E. and Rogers, R.W. (1983) Protection motivation and self-efficacy: a revised theory of fear appeals and attitude change. *J. Exp. Soc. Psychol.*, **19**, 469–479.

Mengel, M.B., Ullione, M., Wedding, D., Jones, E.T., and Shurn, D. (2005) Increasing FASD knowledge by a targeted media campaign: outcome determined by message frequency. *J. FAS Int.*, **3**, 1–14.

Milne, S., Sheeran, P., and Orbell, S. (2000) Prediction and intervention in health-related behavior: a meta-analytic review of Protection Motivation Theory. *J. Appl. Soc. Psychol.*, **30**, 106–143.

Murgraff, V., White, D., and Phillips, K. (1999) An application of protection motivation theory to riskier single-occasion drinking. *Psychol. Health*, **14**, 339–350.

Network Action Team #3. Conversations with women about alcohol use during pregnancy. In development. Canada Northwest FASD Research Network.

Norsask Consumer Interviewing Services Ltd (2005) *Saskatchewan Prevention Institute: FASD Prevention Pre-Campaign Survey*, Saskatchewan Prevention Institute, Saskatoon, SK.

O'Sullivan, G.A., Yonkler, J.A., Morgan, W., and Merritt, A.P. (2003) *A Field Guide to Designing A Health Communication Strategy*, John Hopkins Bloomberg School of Public Health/Centre for Communication Program, Baltimore, MD.

Oliver, C., White, H., and Edwards, M. (1998) *Fetal Alcohol Syndrome: A Hopeful Challenge for Children, Families, and Communities*, Health Canada, Ottawa, ON.

Pechmann, C., Guangzhi, Z., Goldberg, M., and Reibling, E. (2003) What to convey in antismoking advertisements for adolescents: the use of Protection Motivation Theory to identify effective message themes. *J. Marketing*, **67**, 1–18.

Poole, N. (2008) *Fetal Alcohol Spectrum Disorder (FASD) Prevention: Canadian Perspectives*, Public Health Agency of Canada, Ottawa, ON.

Prentice-Dunn, S. and Rogers, R. (1986) Protection motivation theory and preventive health: beyond the Health Belief Model. *Health Educ. Res.*, **1**, 153–161.

Roberts, G. and Nanson, J. (2001) *Best Practices: Fetal Alcohol Syndrome/Fetal Alcohol Effects and the Effects of Other Substance Use during Pregnancy*, Health Canada, Ottawa, ON.

Rogers, R.W. (1975) A Protection Motivation Theory of fear appeals and attitude change. *J. Consum. Psychol.*, **91**, 93–114.

Rogers, R.W. (1983) Cognitive and physiological processes in fear appeals and attitude change: a revised theory of Protection Motivation, in *Social Psychophysiology* (eds J.T. Cacioppo and R.E. Petty), Guilford Press, New York, pp. 153–176.

Rothschild, M.L. (1999) Carrots, sticks, and promises: a conceptual framework for the management of public health and social issue behaviours. *J. Mark.*, **63**, 24–37.

Scammon, D.L., Mayer, R.N., and Smith, K.R. (1991) Alcohol warnings: how do you know when you have had one too many? *J. Public Policy Market.*, **10**, 214–228.

Simons-Morton, B.G., Hartons, J.L., Leaf, W.A., and Preusser, D.F. (2006) Increasing parent limits on novice young drivers: cognitive mediation of the effect of persuasive messages. *J. Adolesc. Res.*, **21**, 83–105.

Stockburger, J. (2003) *Substance Abuse Related Special Needs in Canada: Best Practices for Prevention*, Centre of Excellence for Children and Adolescents with Special Needs, University of Northern British Columbia, Prince George, BC.

The World Bank (2004) *Monitoring & Evaluation: Some Tools, Methods, & Approaches*, The International Bank for Reconstruction & Development, Washington, DC.

Walker, D.S., Darling, C.S., Sherman, A., Wybrecht, B., and Kyndely, K. (2005) Fetal alcohol spectrum disorders prevention: an exploratory study of women's use of, attitudes toward, and knowledge about alcohol. *J. Am. Acad. Nurs. Pract.*, **17**, 187–193.

Wu, Y., Stanton, B.F., Li, X., Galbraith, J., and Cole, M.L. (2005) Protection motivation theory and adolescent drug trafficking: relationship between health motivation and longitudinal risk involvement. *J. Pediatr. Psychol.*, **30**, 127–137.

Yanovitzky, I. and Stryker, J. (2001) Mass media, social norms, and health promotion efforts: a longitudinal study of media effects on youth binge drinking. *Commun. Res.*, **28**, 208–239.

11
Preventing FASD: The Parent–Child Assistance Program (PCAP) Intervention with High-Risk Mothers

Therese M. Grant

11.1
Introduction

With alcohol being legally and widely available it is not surprising that, among pregnant women, the drinking of alcohol is more prevalent than illicit drug use (Centers for Disease Control and Prevention, 2004; Office of Applied Studies, 2002). The irony here is that alcohol is a known teratogen, the neurobehavioral effects of which are more harmful than those of cocaine and other illicit substances of abuse (Jacobson, Jacobson, and Sokol, 1994a, 1994b; Coles *et al.*, 1992; Stratton, Howe, and Battaglia, 1996). Heavy prenatal alcohol exposure puts children at risk for a range of lifelong neurodevelopmental disorders termed Fetal Alcohol Spectrum Disorders (FASD), including the permanent birth defect Fetal Alcohol Syndrome (FAS). Moreover, any problems associated with FASD are exacerbated when a child is raised in a compromised home environment by a mother who has an untreated alcohol abuse disorder.

11.2
FASD Prevention

As a preventable birth defect, FASD is a compelling public health issue. A widely accepted public health approach to FASD prevention was first published in 1996 by the Institute of Medicine (Stratton, Howe, and Battaglia, 1996), and this has recently been updated by the US Department of Health and Human Services (Barry *et al.*, 2009):

- *Universal prevention activities* promote general knowledge about pregnancy alcohol use and healthy practices; examples include warning signs, public health messages, bottle labels, and newspaper articles.
- *Selective prevention activities* involve screening women for alcohol use, training healthcare professionals, working with family members of pregnant women who abuse alcohol, developing biomarkers, brief interventions, and referrals.

Fetal Alcohol Spectrum Disorder–Management and Policy Perspectives of FASD. Edited by Edward P. Riley, Sterling Clarren, Joanne Weinberg, and Egon Jonsson
Copyright © 2011 WILEY-VCH Verlag GmbH & Co. KGaA, Weinheim
ISBN: 978-3-527-32839-0

- *Indicated prevention activities* focus directly on individuals at the highest risk for adverse outcomes, notably pregnant women who are heavy drinkers, and women who have already given birth to a child with FASD and continue to drink.

The Parent–Child Assistance Program (PCAP, which originally was developed at the University of Washington, Seattle and is now widely used in North America) is an example of an "Indicated Prevention" model that works with the group at highest risk for delivering a child with FASD, namely those women who abuse alcohol during pregnancy, or who have already delivered a child with FASD and are continuing to drink.

In October 2009, the Institute of Health Economics (IHE) in Alberta, Canada issued a consensus statement on "FASD Across the Lifespan." The IHE recommended ten FASD prevention measures, including "A higher proportion of prevention resources should be targeted at high risk populations. The Parent-Child Assistance Program (PCAP) has shown great success. Canadian programs based on the PCAP model should be encouraged" (Institute of Health Economics and the Alberta Government, 2009). The purpose of this chapter is to describe the PCAP model, to examine how PCAP strategies might help prevent the birth of alcohol-exposed children in the next generation, and to discuss policy considerations.

11.3
Background

PCAP began in 1991 at the University of Washington (Seattle, Washington) as a federally funded research demonstration designed to test the efficacy of an intensive, three-year home visitation/case management model with high-risk mothers and their children. The primary aim of the model was to prevent subsequent alcohol- and drug-exposed births among birth mothers who abused alcohol and/or drugs during an index pregnancy. Research findings demonstrated the model's efficacy (Ernst *et al.*, 1999; Grant *et al.*, 2003, 2005a), and the Washington State Legislature subsequently funded PCAP sites in counties throughout Washington. The model has been replicated at numerous other locations in the United States and Canada.

A woman is eligible to enroll in PCAP if she is characterized by one of these two profiles:

- Pregnant or up to six months postpartum; *and* has abused alcohol and/or drugs heavily during the pregnancy; *and* is ineffectively engaged with community service providers.

- Has previously delivered a child who has a medical diagnosis of FASD; *and* continues to drink heavily; *and* is capable of becoming pregnant.

Mothers who enroll in PCAP exemplify the intergenerational nature of familial substance abuse and dysfunction; they were themselves the neglected and abused children in our communities just a decade or two ago. For example, among 817 women currently enrolled in the Washington PCAP, the typical client was born to substance-abusing parents (92%), she was physically and/or sexually abused as a child (71%), ran away from home as a child (67%), and did not complete high school (49%). She is now in her late 20s, and her life circumstances are grim: she is unmarried (91%), and is beaten by her partner (76%), is homeless or lives in temporary housing (24%), has a history of incarceration (75%), and is on public assistance (69%). She does not use family planning on a regular basis (91%), and now has an average of 2.7 children, half of whom are not in her custody (Grant and Ernst, 2009).

11.4
The PCAP Intervention

PCAP is an intensive home visitation case management model that offers mothers direction and support over three years. The model incorporates fundamental and well-known characteristics of effective case management: it is individually tailored, promotes the competency of the client, uses a relational approach to deliver services, it is family-centered, community-based, and multidisciplinary.

Well-trained and closely-supervised PCAP case managers have a caseload of 15 to 16 families each, and begin their work with clients during pregnancy or within six months after the birth of an index child. The intervention is not delivered according to a specific model of behavioral intervention. Instead, it is based on three theoretical constructs that inform the therapeutic approach with clients, and that give shape to day-to-day case management activities. These three constructs are relational theory, stages-of-change, and harm reduction.

11.4.1
Relational Theory

Relational theory underscores the importance of interpersonal relationships to women as they grow, develop, and define themselves (Miller, 1991; Surrey, 1991). A related concept, the therapeutic alliance, generally refers to the process through which a mental health professional builds rapport and engagement with a patient in order to help that person achieve desired change (Orlinsky, Ronnestad, and Willutski, 2004). This dynamic has been studied by researchers and practitioners in other disciplines, who have found that positive interpersonal relationships: (i) are critical to successful outcomes among women with substance abuse disorders who are in intervention, treatment, and recovery settings (Amaro and Hardy-Fanta, 1995; Finkelstein, 1993); (ii) may determine the extent of patient compliance and retention in an intervention (Barnard *et al.*, 1998); and (iii) may be more important to improvement than concrete services received (Pharis and Levin, 1991).

The PCAP model puts these concepts of relational theory and therapeutic alliance into practice by offering personalized, knowledgeable and compassionate case management for three years, a period of time long enough for the process of gradual and realistic change to occur among women who have overwhelming problems and few personal or social resources. In addition, many of the PCAP case managers have themselves overcome difficult personal or familial life circumstances similar to those experienced by their clients. They can better understand clients because of this shared history and culture, and they are able to gain access and build rapport with clients who might otherwise be unapproachable. PCAP case managers, who sometimes are termed "advocates" or "mentors," are able to inspire hope from a realistic perspective because they have achieved long-term goals in recovery, school, employment or parenting, and can provide credible guidance toward change.

11.4.2
Stages-of-Change

The PCAP model incorporates stages-of-change theory, and practices motivational interviewing (MI) strategies, and the case managers are trained accordingly. This approach recognizes that clients will be at different stages of readiness for change; motivation is a process, not a characteristic or personality trait; and client ambivalence about change is understandable and should be expected (Miller and Rollnick, 1991; Prochaska and DiClemente, 1986; Rollnick and Bell, 1991). Motivation for change occurs within the context of interpersonal relationships, and the quality of a PCAP case manager's relationship with her client, as well as her MI skills and style, can strongly influence whether a client resists or makes changes. In practice, at home visits with the client, the PCAP worker listens closely and respectfully, accepts the client's situation, trusts in the client's perception and judgment about her own life, and encourages the client to explore both the positive and negative aspects of her own problem behavior.

Stages-of-change theory dovetails with the constructs of self-efficacy (Olds et al. 1997; Sherman, Sanders, and Yearde, 1998). Self-efficacy is the belief in one's ability to actually behave in ways that will produce desired changes; expectations about self-efficacy (and the potential for change) are influenced most strongly by a person's own past accomplishments (Bandura, 1977). PCAP case managers put these ideas into practice consistently throughout the three-year intervention by helping clients to: (i) pinpoint specific areas in which they want or need to make changes; (ii) identify corresponding meaningful and realistic goals; (iii) identify the incremental ("baby") steps required to reach each of those goals; and (iv) evaluate progress and re-establish goals and steps every two to four months. Initially, the client and the case manager each take responsibility for some of the steps; for example, the case manager may obtain local bus schedules, and they will both research childcare options; the client will attend outpatient treatment groups at least three times per week. It is critical that some of the steps—no matter how small—are attainable by the client during the two- to four-month period, because it is only as she observes herself accomplishing behaviors she desires that her sense of self-efficacy will develop (Grant et al., 1997). This process allows for a client's gradual transition from an initial dependence on the case manager's assistance and support, to interdependence as they work together to accomplish steps toward goals, to independence as the client begins to trust in herself as a worthwhile and capable person, and learns (to the extent she is able) the skills necessary to manage a healthy, positive life.

11.4.3
Harm Reduction

The framework of the PCAP intervention is influenced by harm reduction theory. Harm reduction is based on the assumption that alcohol and drug addiction and the associated risks can be placed along a continuum, with the goal being to help a client move along this continuum from excess to moderation, and ultimately to abstinence, in order to reduce the harmful consequences of the habit (Marlatt and Tapert, 1993). In this view, "... any steps toward decreased risk are steps in the right direction." (Marlatt, Somers, and Tapert, 1993). In practice, case managers focus attention not simply on reducing alcohol and drug use, but also on reducing other risk behaviors and addressing the health and social well-being of the clients and their children. For example, an important PCAP goal is to reduce the incidence of future alcohol-exposed births. While not every client will be able to become abstinent from alcohol, harm can be reduced by encouraging the woman to use effective family planning methods to avoid becoming pregnant if she is still drinking.

11.5
PCAP: A Two-Pronged Intervention

11.5.1
Between the Client and the Case Manager

11.5.1.1 Establishing Trust
Case managers and clients begin by getting to know each other and establishing the trust that will enable them to work closely together for three years. This bonding process sometimes takes months for clients whose lifelong experiences of abuse and abandonment have taught them not to trust anybody. During early home visits the case manager identifies client, family, and community strengths that the client might draw on throughout the intervention as she works towards recovery. At the same time, the case manager addresses immediate problems such as obtaining diapers for the newborn or locating temporary housing – activities that demonstrate from the beginning that the case manager cares and can be trusted to follow through.

11.5.1.2 Working with the Family
Case managers work within the context of the client's personal network, and attempt to establish rapport with the children, the husband or partner, members

of the extended family, and friends. Everyone in this network of relationships is involved in some way with the client's substance abuse and related problems. Family and friends may have a powerful influence over the woman, and they will all be affected by changes that the woman makes as she attempts to break long-established behavioral patterns. Gaining the family's trust is a preliminary step that then allows the case manager access and the opportunity to communicate with this important group throughout the intervention.

11.5.1.3 Role-Modeling

Role-modeling and teaching basic life skills are critical strategies for PCAP case managers. PCAP clients rarely have a mental template for what "normal" adult life or parenting might look like, and their bleak backgrounds have done little to prepare them for adult responsibilities. In addition, they may have cognitive impairments as a result of their own prenatal exposure to alcohol/drugs and years of substance abuse. PCAP clients typically have poor emotion regulation skills, and may respond to problems and disappointments with angry outbursts or by withdrawing. PCAP case managers role model positive interperson behaviors that are likely to elicit support; they help clients organize and articulate their thoughts, and practice these skills with the clients. Case managers find that the most effective teaching techniques are those that are explicit, hands-on, concrete, and experiential. This approach is particularly important when working with clients who have cognitive impairments or FASD (Grant et al., 2004).

11.5.2
Between the Client and the Community Service Providers

PCAP does not offer direct services, such as substance abuse treatment or healthcare, but instead connects clients with appropriate services and providers in the client's community. The PCAP case manager acts as a communications liaison within this network, arranging meetings or conference calls to bring members of a client's provider network together, with the client present whenever possible. The case manager facilitates the development of a service plan that gives voice to the client's needs, creates realistic expectations, and addresses concerns of service providers. She then helps the client follow through with the plan.

In practical terms, this strategy means that a client in recovery will not be faced with trying to comply with a court's stipulation that she attend outpatient treatment five mornings a week in one part of town, while the housing authority assigns her to an apartment in a neighborhood requiring her to make two bus transfers to get to treatment, and while child welfare grants her reunification with her two children with no contingencies for childcare (a realistic scenario). When a woman learns to speak up in a way that demonstrates respect for herself and others and, with her case manager as guide and advocate, most providers will listen, recognize the realities of the client's circumstances, and help develop a plan that will help her succeed in her recovery, rather than set her up for another failure.

11.6
Preventing Alcohol- and/or Drug-Exposed Births

A basic formula for preventing alcohol- and/or drug-exposed births is to either motivate women to stop drinking before and during pregnancy, or to help those women who cannot stop drinking to avoid becoming pregnant.

11.6.1
Substance Abuse Treatment

Concern for their children and fear of losing their housing are the most common barriers to women entering and staying in inpatient treatment, and for these reasons most PCAP clients attend outpatient rather than inpatient residential treatment. Gender-specific, women-only treatment programs are preferable because, in these settings, a client is usually more comfortable discussing difficult life circumstances associated with the onset of her substance abuse, for example, childhood sexual abuse or domestic violence.

During and after treatment, the PCAP case manager and her client continue to discuss very frankly the risk for relapse, including stressors and relapse triggers, and relapse-prevention strategies. Clients are most likely to succeed in recovery if they are active in support groups or meetings that meet their needs. For example, is the meeting place accessible and are times convenient? And/or, is childcare available?

Clients are not asked to leave PCAP because of noncompliance, setbacks, or relapse. Instead, they are taught to learn from their mistakes and are inspired to believe in their potential to succeed. This PCAP policy has resulted in clients' ability to overcome shame after relapse, to contact the case manager quickly, to resume recovery (or treatment), and repair the damage done. Case managers use client relapse experiences to help them examine events that triggered the setback, and to develop resiliency strategies. When a client is able to rebound from a relapse event, she develops self-efficacy as she observes herself overcoming a crisis, coping, and moving on.

> "There were times when I felt like I was going to relapse and my advocate would be there for me, and she'd keep checking on me and I'd get through it. I've learned so much about myself and being responsible again and being a good mother. It was all what she taught me – she changed my life for me."
> PCAP Client Comment

11.6.2
Family Planning

PCAP is not a pregnancy-prevention program *per se*; rather, the goal is to help women prevent future alcohol- and drug-exposed births. Many clients who achieve a stable and sober lifestyle in PCAP choose to become pregnant because, for the first time, they may be able to have a healthy pregnancy and the opportunity to raise their own child. Case managers help clients understand that "family planning" means exactly that – planning a pregnancy to occur at an optimal time.

Case managers connect clients with family planning clinics or healthcare providers who provide physical examinations, identify potential contraindications for specific birth control methods, and determine the safest and most appropriate method for the woman. Introducing family planning concepts, educating, motivating, and helping a client obtain a method takes time and may involve setbacks, missed appointments, birth control side effects or failure, or subsequent unintended pregnancy. Understanding this can reduce case manager frustration. The important points are to be sensitive to the client's perspective and cultural and religious beliefs, to follow-up, and to be persistent in meeting client and program goals. Case managers use motivational interviewing strategies, and as they continue goal-setting exercises with clients throughout the intervention, they explore how having another child might affect the client's ability to achieve her goals. The process is gradual and thoughtful, so that any decisions are longlasting, grounded in the client's own belief system, and based on her individual choices.

11.7
PCAP Outcomes

The PCAP model includes a rigorous evaluation component. Clinical supervisors are trained to interview clients at PCAP intake and at three-year exit, by using the Addiction Severity Index (ASI), fifth edition (McLellan *et al.*, 1992). The ASI is a widely used standardized interview instrument that demonstrates good reliability and validity, and assesses seven potential problem areas: medical status; employment and support; drug use; alcohol use; legal status; family/social status; and psychiatric status. In 1997, PCAP researchers developed supplemental questions for pregnant and postpartum women with regard to childhood history, alcohol and drug use during pregnancy, and service utilization.

The following are the PCAP three-year outcome data from 433 clients who exited PCAP from July 2005 through June 2009 (Grant and Ernst, 2009). At PCAP exit:

- During the three-year program, 13% had a subsequent alcohol- or drug-exposed birth.

- At exit, 81% were at reduced risk for delivering another exposed child, either because they had been abstinent from alcohol and drugs for at least six months, or were using a family planning method on a consistent basis.

- At exit, 90% had completed inpatient or outpatient treatment (or were in progress); 66% were currently abstinent from alcohol and drugs for at least one month; 48% for at least six months; and 41% for at least one year.

- At exit, 65% were using a method of family planning on a regular basis; 52% were using regular Depo-Provera injections, Norplant implant, intrauterine device, or had obtained a tubal ligation.

- While 69% of the women received welfare income at enrollment, only 45% received welfare income at exit (mean US$458 per month). While 3% of the women were employed at enrollment, 34% were receiving income from employment at exit (mean US$935 per month). At PCAP exit, employment was the main source of income for 28% of clients.

- At exit, 79% (279/353) of target children were living either with their biological mother (64%) or with other family members, including the biological father (15%); 8% had been legally adopted; 11% were in foster care; and 1% were in care of Tribal courts or Tribal child welfare.

- While only 36% were in stable housing at enrollment, 71% were in permanent, stable housing at exit.

11.8
PCAP Cost Effectiveness

At two PCAP replication sites (Seattle and Tacoma), it was found that among the 78 clients who had been binge drinkers during the index pregnancy, at exit 65% were no longer at present risk of having another alcohol- or drug-exposed pregnancy, either because they were using a reliable contraceptive method (31%), or they had been abstinent from alcohol/drugs for at least six months (23%), or both (11%) (Grant *et al.*, 2005a). Based on state subsequent birth rates it was assumed that, without PCAP intervention, about 30% ($n = 23$) of these 78 drinking mothers would have delivered another highly exposed child. Instead, that number was reduced by 65%, preventing approximately 15 exposed births. The incidence of FAS is estimated at 4.7% to 21% among heavy drinkers (Abel, 1995; Barr and Streissguth, 2001; Majewski, 1993), hence, it has been estimated conservatively that PCAP prevented at least one and up to three new cases of FAS.

PCAP costs approximately US$15 000 per client for the three-year program, including intervention, administration, and evaluation. The estimated average lifetime cost for an individual with FAS is at least $1.5 million (Harwood, Fountain, and Livermore, 1998; Rice, 1993). If PCAP were to prevent a single new case of FAS, the estimated lifetime cost savings would be equivalent to the cost of the PCAP intervention for 102 women.

11.9
PCAP Intervention with Women who Themselves Have FASD

While neuropsychological deficits and other adverse outcomes associated with prenatal alcohol exposure have been well documented for over 30 years, interventions for adults with FASD have not been systematically developed and evaluated. In 1999, PCAP expanded its eligibility criteria to enroll a sample of women with FASD; subsequently, in 2001 a 12-month community intervention pilot study was conducted to examine more specifically how these women could be helped within the existing framework of PCAP (Grant et al., 2004). A total of 19 women who either had FASD ($n = 11$) or were prenatally exposed to heavy alcohol and had suspected FASD ($n = 8$), were enrolled into the pilot study. Their average age was 22 years, most were unmarried (84%) and poorly educated (47% had a 9th grade education or less), and almost all had been physically or sexually abused as children (94%). All reported having many unmet basic service needs, and the quality of life and levels of psychiatric distress among these women were found to be very poor (Grant et al., 2005b).

The neuropsychological and cognitive deficits of the PCAP mothers with FASD required the implementation of specific strategies to increase their connection to community services, and to improve the quality of services they were receiving. The pilot community intervention delivered the standard PCAP model enhanced in two ways: (i) by modifying PCAP in order to accommodate the special cognitive needs of clients with FASD; and (ii) by educating community service providers *about* FASD so that they could better accommodate the clients (Grant et al., 2004, 2007).

11.10
Policy Recommendations: Collaborative Approaches for Preventing Alcohol-Exposed Pregnancies

The reduction and prevention of alcohol-exposed pregnancies requires many approaches. Indeed, those countries currently addressing these problems might first consider the steps that have been found useful in the US and in Washington State, involving many types of public health actions (Streissguth and Grant, 2010).

Government warnings regarding "no safe level of alcohol use during pregnancy" are a good first step. At the national level, the bottle-labeling of all alcohol-containing beverages may be an important strategy, as well as point-of-purchase warning signs at all local venues where alcohol is sold in any form. Grant et al. (2009) recently reported that between 1989 and 2004, while rates of any alcohol use *before* pregnancy remained stable in western Washington (slightly >40%), rates of any alcohol use *during* pregnancy decreased from 30% between 1989 and 1991 to 12% between 2002 and 2004. US national rates decreased similarly (to 10.2% in 2002) (Centers for Disease Control and Prevention 2004). These findings suggest that public health messages about the potential risks to the fetus of maternal drinking during pregnancy have been effective. Indeed, most women who drink alcohol stop doing so when they recognize that they are pregnant.

An essential component of evidence-based research demonstration programs is the ultimate incorporation of successful strategies into the ongoing public health programming of a country. Positive outcomes can be achieved when governments develop policies that complement effective community programs. Grant et al. (2005) have described how PCAP study outcomes were improved by a number of public policies and programs geared to pregnant and parenting substance-abusing women. For example, between 1991 and 2003 the Washington State Division of Alcohol and Substance Abuse (DASA) almost tripled (from 55 to 149) the number of gender-specific inpatient residential treatment beds for pregnant and postpartum women. The availability of these specialized treatment facilities had a dramatically positive impact on PCAP treatment and abstinence outcomes during this same time period (Grant et al., 2005a).

Other areas in which to commit national funds for research and policy implementation include increasing funding for addressing specific aspects of women's alcohol problems, for example, developing detoxification protocols and practices for pregnant women; and developing guidelines for routine prenatal screening for alcohol use and abuse.

Enlisting the support of professional organizations can also be extremely important. At the national level, the American Medical Association (American Medical Association, 1999) endorsed universal alcohol screening for all patients over the age of 14 years. The American Academy of Pediatrics (American Academy of Pediatrics, 2000) endorsed the Surgeon General's recommendation for abstinence during pregnancy and additionally recommends federal legislation requiring the inclusion of health and safety messages in all print and broadcast alcohol advertisements.

It is recommended that physicians routinely should screen *every* woman of childbearing age about her alcohol use and her risk for becoming pregnant (Grant et al., 2009). In 2004, the American College of Obstetricians and Gynecologists (American College of Obstetricians and Gynecologists (ACOG) Committee on Ethics, 2004) outlined the ethical rationale for using a consolidated alcohol protocol, including universal screening, brief intervention, and referral to treatment. Brief alcohol screening instruments that can help healthcare providers to identify problem drinkers include the CAGE, TWEAK, AUDIT (Alcohol Use Disorders Identification Test), and T-ACE instruments (Ewing, 1984; Russell, 1994; Saunders et al., 1993; Sokol, Martier, and Ager, 1989). New codes have been established by the American Medical Association that allow physicians to screen patients for alcohol problems, to deliver behavioral interventions, and to report these activities to health insurance programs.

Noting that approximately one-half of all pregnancies in the US are unintended (Mohllajee et al., 2007), it is recommended that women who are not pregnant should be questioned about their use of contraception, and educated about the potential risks of frequent or binge drinking at conception and throughout pregnancy (Grant et al., 2009). Screening for binge alcohol use among nonpregnant women may also identify those who are underusing contraception and who may benefit from a more aggressive counseling regarding the risks for pregnancy and methods that do not require daily use. They may also benefit from having plan B (emergency contraceptive backup method) supplies available. Women who are pregnant and who drink should be advised to stop drinking, because research has not identified a universally safe level of alcohol use during pregnancy.

The coordination of efforts with parent organizations and advocacy organizations can be a powerful and effective strategy for raising public awareness, for providing hope and help to families of people with FASD, and for developing and raising funds for much-needed programs (Streissguth and Grant, 2010). One such advocacy organization with international connections is the National Organization on Fetal Alcohol Syndrome (NOFAS).

Clearly, comprehensive prevention efforts are ultimately more cost-effective for countries than managing the costs to society, families and children of raising future generations of alcohol-affected individuals.

References

Abel, E.L. (1995) An update on incidence of FAS: FAS is not an equal opportunity birth defect. *Neurotoxicol. Teratol.*, **17** (4), 437–443.

Amaro, A. and Hardy-Fanta, C. (1995) Gender relations in addiction and recovery. *J. Psychoactive Drugs.*, **27**, 325–337.

American Academy of Pediatrics (2000) Committee on substance abuse and committee on children with disabilities. Fetal alcohol syndrome and alcohol-related neurodevelopmental disorders. *Pediatrics*, **106** (2 Pt 1), 358–361.

American College of Obstetricians and Gynecologists (ACOG) Committee on Ethics (2004) Committee opinion No. 294. At-risk drinking and illicit drug use: ethical issues in obstetric and gynecologic practice. *Obstet. Gynecol.*, **103** (5 Pt 1), 1021.

American Medical Association (AMA) (1999) Summaries and recommendations of council on scientific affairs reports [Online]. AMA Interim Meeting, p. 22, http://www.ama-assn.org/ama1/pub/upload/mm/443/csai-99.pdf.

Bandura, A. (1977) Self-efficacy: toward a unifying theory of behavioral change. *Psychol. Rev.*, **84** (2), 191–215.

Barnard, K.E., Magyary, D., Sumner, G., Booth, C.L., Mitchell, S.K., and Spieker, S. (1998) Prevention of parenting alterations for women with low social support. *Psychiatry*, **51**, 248–253.

Barr, H.M. and Streissguth, A.P. (2001) Identifying maternal self-reported alcohol use associated with Fetal Alcohol Spectrum Disorders. *Alcohol. Clin. Exp. Res.*, **25** (2), 283–287.

Barry, K.L., Caetano, R., Chang, G., DeJoseph, M.C., Miller, L.A., O'Connor, M.J., et al. (March 2009) *Reducing Alcohol-Exposed Pregnancies: A Report of the National Task Force on Fetal Alcohol Syndrome and Fetal Alcohol Effect*, Centers for Disease Control and Prevention, Atlanta, GA.

Centers for Disease Control and Prevention (CDC) (2004) Alcohol consumption among women who are pregnant or who might become pregnant—United States, 2002. [Online]. *Morb. Mortal. Wkly Rep.*, **53** (50), 1178–1181, http://www.cdc.gov/mmwr/preview/mmwrhtml/mm5350a4.htm.

Coles, C.D., Platzman, K.A., Smith, I., James, M.E., and Falek, A. (1992) Effects of cocaine and alcohol use in pregnancy on neonatal growth and neurobehavioral status. *Neurotoxicol. Teratol.*, **14** (1), 22–23.

Ernst, C.C., Grant, T.M., Streissguth, A.P., and Sampson, P.D. (1999) Intervention with high-risk alcohol and drug-abusing mothers: II. 3-year findings from the Seattle model of paraprofessional advocacy. *J. Community Psychol.*, **27** (1), 19–38.

Ewing, J.A. (1984) Detecting alcoholism: the CAGE questionnaire. *JAMA*, **252** (14), 1905–1907.

Finkelstein, N. (1993) Treatment programming for alcohol and drug-dependent pregnant women. *Int. J. Addict.*, **28** (13), 1275–1309.

Grant, T.M. and Ernst, C.C. (2009) 2007–2009 biennial report from the University of Washington Parent Child Assistance Program to the Washington State Department of Social and Health Services, Division of Behavioral Recovery and Health.

Grant, T.M., Ernst, C.C., McAuliff, S., and Streissguth, A.P. (1997) The difference game: an assessment tool and intervention strategy for facilitating change in high-risk clients. *Fam. Soc.*, **78** (4), 429–432.

Grant, T., Ernst, C.C., Pagalilauan, G., and Streissguth, A.P. (2003) Post-program follow-up effects of paraprofessional intervention with high-risk women who abused alcohol and drugs during pregnancy. *J. Community Psychol.*, **31** (3), 211–222.

Grant, T., Huggins, J., Connor, P., Pedersen, J., Whitney, N., and Streissguth, A. (2004) A pilot community intervention for young women with Fetal Alcohol Spectrum Disorders. *Community Ment. Health J.*, **40** (6), 499–511.

Grant, T., Ernst, C., Streissguth, A., and Stark, K. (2005a) Preventing alcohol and drug exposed births in Washington State: intervention findings from three Parent–Child Assistance Program sites. *Am. J. Drug Alcohol Abuse*, **31** (3), 471–490.

Grant, T., Huggins, J., Connor, P., and Streissguth, A. (2005b) Quality of life and psychosocial profile among young women with Fetal Alcohol Spectrum Disorders. *Mental Health Aspects Dev. Disabil.*, **8** (2), 33–39.

Grant, T.M., Youngblood Pedersen, J., Whitney, N., and Ernst, E. (2007) The role of therapeutic intervention with substance abusing mothers: preventing FASD in the next generation, in *Attention Deficit Hyperactivity Disorder and Fetal Alcohol Spectrum Disorders: the Diagnostic, Natural History and Therapeutic Issues through the Lifespan* (ed. K. O'Malley), Nova Science Publishers, Inc, Hauppauge, NY, pp. 69–93.

Grant, T.M., Huggins, J.E., Sampson, P.D., Ernst, C.C., Barr, H.M., and Streissguth, A.P. (2009) Alcohol use before and during pregnancy in Western Washington, 1989–2004: implications for the prevention of Fetal Alcohol Spectrum Disorders. *Am. J. Obstet. Gynecol.*, **200** (3), 278e1–8.

Harwood, H., Fountain, D., and Livermore, G. (1998) *The Economic Costs of Alcohol and Drug Abuse in the United States 1992, Report prepared for the National Institute on Drug Abuse and National Institute on Alcohol Abuse and Alcoholism.* NIH publication No. 98-4327. National Institutes of Health, Rockville, MD.

Institute of Health Economics and the Alberta Government (2009) Consensus Statement on Fetal Alcohol Spectrum Disorder (FASD)–Across the Lifespan. [Online]. [24 screens], http://www.ihe.ca/publications/library/2009 (accessed 24 December 2009).

Jacobson, J.L., Jacobson, S.W., and Sokol, R.J. (1994a) Effects of prenatal exposure to alcohol, smoking and illicit drugs on postpartum somatic growth. *Alcohol. Clin. Exp. Res.*, **18** (2), 317–323.

Jacobson, S.W., Jacobson, J.L., and Sokol RJ. (1994b) Effects of fetal alcohol exposure on infant reaction time. *Alcohol. Clin. Exp. Res.*, **18** (5), 1125–1132.

McLellan, A.T., Kushner, H., Metzger, D., Peters, R., Smith, I., Grissom, G., et al. (1992) The fifth edition of the addiction severity index. *J. Subst. Abuse Treat.*, **9** (3), 199–213.

Majewski, F. (1993) Alcohol embryopathy: experience in 200 patients. *Dev. Brain Dysfunct.*, **6**, 248–265.

Marlatt, G.A. and Tapert, S.F. (1993) Harm reduction: reducing the risks of addictive behaviors, in *Addictive Behaviors across the Lifespan* (eds J.S. Baer, G.A. Marlatt, and R. McMahon), Sage Publications, Newbury Park, CA, pp. 243–273.

Marlatt, G.A., Somers, J.M., and Tapert, S.F. (1993) Harm reduction: application to alcohol abuse problems. *NIDA Res. Monogr.*, **137**, 147–166.

Miller, J.B. (1991) The development of women's sense of self, in *Women's Growth in Connection* (eds J.D. Jordan, A.G. Kaplan, J.B. Miller, I.P. Stiver, and J.L. Surrey), Guilford Press, New York, pp. 11–26.

Miller, W.R. and Rollnick, S. (1991) *Motivational Interviewing: Preparing People* to Change Addictive Behavior, Guilford Press, New York.

Mohllajee, A.P., Curtis, K.M., Morrow, B., and Marchbanks, P.A. (2007) Pregnancy intention and its relationship to birth and maternal outcomes. Obstet. Gynecol., **109** (3), 678–686.

Office of Applied Studies (2002) Results from the 2001 National Household Survey on Drug Abuse: Volume 1: Summary of National Findings. [Online], U.S. Department of Health and Human Services, SAMHSA, Rockville, MD, http://www.oas.samhsa.gov/nhsda/2k1nhsda/PDF/cover.pdf.

Olds, D.L., Eckenrode, J., Henderson, C.R. Jr, Kitzman, H., Powers, J., Cole, R., et al. (1997) Long-term effects of home visitation on maternal life course and child abuse and neglect. JAMA, **278** (8), 637–643.

Orlinsky, D.E., Ronnestad, M.H., and Willutski, U. (2004) Fifty years of psychotherapy process–outcome research: continuity and change, Handbook of Psychotherapy and Behaviour Change, 5th edn (ed. M.J. Lambert), John Wiley & Sons, Inc., New York, pp. 207–291.

Pharis, M.E. and Levin, V.S. (1991) "A person to talk to who really cared": high-risk mothers' evaluations of services in an intensive intervention research program. Child Welfare, **70** (3), 307–320.

Prochaska, J.O. and DiClemente, C.C. (1986) Toward a comprehensive model of change, in Addictive Behaviors: Processes of Change (eds W.R. Miller and N. Heather), Plenum Press, New York, pp. 3–27.

Rice, D.P. (1993) The economic costs of alcohol abuse and dependence: 1990. Alcohol Health Res. World, **17** (1), 10–11.

Rollnick, S. and Bell, A. (1991) Brief motivational interviewing for use by the nonspecialist, in Motivational Interviewing: Preparing People to Change Addictive Behavior (eds W.R. Miller and S. Rollnick), Guilford Press, New York, pp. 203–213.

Russell, M. (1994) New assessment tools for drinking in pregnancy: T-ACE, TWEAK, and others. Alcohol. Health. Res. World., **18** (1), 55–61.

Saunders, J.B., Aasland, O.G., Babor, T.F., de la Fuente, J.R., and Grant, M. (1993) Development of the Alcohol Use Disorders Identification Test (AUDIT): WHO collaborative project on early detection of persons with harmful alcohol consumption–II. Addiction, **88** (6), 791–804.

Sherman, B.R., Sanders, L.M., and Yearde, J. (1998) Role-modeling healthy behavior: peer counseling for pregnant and postpartum women in recovery. Womens Health Issues, **8** (4), 230–238.

Sokol, R.J., Martier, S.S., and Ager, J.W. (1989) The T-ACE questions: practical prenatal detection of risk-drinking. Am. J. Obstet. Gynecol., **160** (4), 863–868.

Stratton, K., Howe, C., and Battaglia, F. (eds) (1996) Fetal Alcohol Syndrome: Diagnosis, Epidemiology, Prevention, and Treatment, Institute of Medicine, National Academy Press, Washington, D.C..

Streissguth, A.P. and Grant, T.M. (2010) Prenatal and postnatal intervention strategies for alcohol–abusing mothers in pregnancy, in Drugs in Pregnancy–the Price for the Child: Exposure to Foetal Teratogens and Long-Term Neurodevelopmental Outcome (eds D. Preece and E. Riley), MacKeith Press, London, England.

Surrey, J.L. (1991) The "self-in-relation": a theory of women's development, in Women's Growth in Connection (eds J.D. Jordan, A.G. Kaplan, J.B. Miller, I.P. Stiver, and J.L. Surrey), Guilford Press, New York, pp. 51–66.

12
FASD in the Perspective of Primary Healthcare
June Bergman

Primary healthcare is the provision by clinicians of integrated, accessible healthcare services that address a large majority of personal healthcare needs. It includes the prevention of disease; the early identification of disease and risk factors; and the diagnosis, treatment and follow-up of various health conditions. As the first point of contact with the healthcare system, primary care also opens doors to secondary and tertiary care. Of particular value is the primary care practitioner's ability to develop relationships with individuals over time, and to care for the whole person in the context of family and community. This is reflected in the principles of family medicine as outlined by the College of Family Physicians of Canada.[1] The World Health Organization's broader definition of primary healthcare includes aspects of social determinants of health and other community resources and community capacity, as defined at ALMA ATA in the early 1960s.[2]

It is known from the work of Barbara Starfield[3] and others that a good, strong, primary care system results in improved health of the community, and is highly cost-effective. Effective primary care shifts the emphasis from "illness care" to true "healthcare": people are empowered to stay healthy, families are supported to meet the healthcare needs of family members, and only thereafter is access to the healthcare system necessary. International studies show that the strengthening of a country's primary care system is associated with improved population health outcomes and lower rates of all-cause mortality and premature mortality. Furthermore, increased availability of primary care results in higher patient satisfaction and reduced aggregate healthcare spending. Countries such as New Zealand, Finland, Denmark, Norway and the United States (Kaiser Permanente) have demonstrated significant improvements in the equity, efficiency, effectiveness and responsiveness of their health systems. In light of these considerations, there has been a worldwide resurgence of interest in strengthening primary care.

1) http://www.cfpc.ca/English/cfpc/about%20us/principles/default.asp?s=1
2) http://www.who.int/hpr/NPH/docs/declaration_almaata.pdf
3) http://iis-db.stanford.edu/staff/4001/Barbara_Starfield-CV.pdf

12.1
Primary Care Approaches to FASD

Fetal alcohol spectrum disorder (FASD) is a major concern in the current environment. Those who suffer from this syndrome are affected for life, and it creates a need for significant care and support, both for those individuals and for their families. Since 90% of all contacts with the healthcare system are in primary care, the primary care clinic is the opportune setting in which to provide interventions for patients at risk for FASD. Many organizations in Canada have created guidelines to identify and treat individuals who have FASD or are at risk of drinking during pregnancy. All are agreed that prevention is the area in which major efforts should be focused. The Canadian Task Force on the Periodic Health Exam has, for example, recommended case finding, counseling and follow-up as effective means of managing problem drinking.

In 1997, the Canadian Pediatric Association led a coalition that included the College of Family Physicians and sixteen other signatories who came together to create a framework for prevention of FASD. The framework divides prevention into primary, secondary, and tertiary levels:

- *Primary prevention* includes actions that would avert FASD before it occurs, such as disseminating information about the dangers of drinking during pregnancy.

- *Secondary prevention* includes actions that identify persons at risk, such as screening and early intervention services for pregnant women and women of childbearing potential at risk for having a child with FASD.

- *Tertiary prevention* includes actions that prevent recurrence and attempt to lessen the cognitive, behavioral and social impact of FASD. Examples would be programs for children with FASD and their caregivers, as well as treatment for women who already have one FASD child and plan to have more children.[4]

The document articulates the following specific roles for primary care physicians: (i) to be involved in screening for alcohol use in pregnant women, referral for diagnosis, follow-up and linking patients to the community resources; and (ii) to serve as a catalyst for the prevention of continued exposure and subsequent affected siblings, referrals for diagnostic assessment, and links to services and resources to improve outcome.

More recently, the Toward Optimized Practice (TOP) group in Alberta has developed guidelines for both the prevention and treatment of FASD [5),6]. The prevention guidelines, last revised in 2005, mirror the framework of the broader coalition, but are specifically directed to the primary care community. In the TOP

---

4) http://www.cps.ca/english/statements/fn/cps96-01.htm
5) http://topalbertadoctors.org/informed_practice/cpgs/fasd_diagnosis.html
6) http://www.topalbertadoctors.org/informed_practice/cpgs/fasd_prevention.html

guidelines, primary prevention is the abstinence from alcohol during pregnancy. The goal is that no fetuses are exposed to alcohol, thus eliminating the problems of FAS before they develop. Secondary prevention seeks to reduce the duration and severity of maternal drinking by identifying the women at risk. Tertiary prevention is aimed at mitigating the harm to affected individuals and their families through reducing the complications, impairments and disabilities caused by FASD. Tertiary prevention also includes activities that prevent recurrence of the condition in subsequent children.

The TOP guidelines outline an approach to prevention that is easily integrated into primary care practice and provides clear decision points for action. The interventions at each stage are focused on asking the right questions and offering support to individuals. The primary provider is asked to "engage in education regarding FAS and the adverse effects of alcohol on the fetus; be aware of and use promotional materials in offices and as handouts for patients; be aware of and access community resources; and discuss and enhance access to contraceptive strategies with all women and their partners."

Questions that should be asked of every woman of childbearing age who comes into a primary care office are:

1) Do you use alcohol?
2) Has alcohol use ever caused a problem for you or any family member?
3) Do you use other drugs/substances (illicit, prescription or over-the-counter).[7]

A positive answer to any of these questions will lead the inquirer to the next suggested set of questions:

1) In a typical week, on how many days do you drink?
2) On those days, how many drinks are usual?[8]

Based on the answers to these question, a standardized questionnaire is suggested. The following T-Ace questionnaire has long been used in primary care environments as it is quick and easy to administer.

The T-ACE Questionnaire (Sokol, Martier, and Ager, 1989)
Tolerance – How many drinks does it take to make you feel high?

Score 2 for more than 2 drinks. Score 0 for 2 drinks or less
Score 1 point for each YES answer to the following:

Annoyance – Have people annoyed you by criticizing your drinking?
Cut down – Have you felt you ought to cut down on your drinking?
Eye opener – Have you ever had a drink first thing in the morning to steady your nerves or get rid of a hangover?

High risk score = 2 or more points.

[End of questionnaire.]

7) http://www.topalbertadoctors.org/informed_practice/cpgs/fasd_prevention.html
8) http://www.topalbertadoctors.org/informed_practice/cpgs/fasd_prevention.html

The identification of high-risk drinking behaviors leads to a brief intervention that could consist of any of the following:

- Providing personalized feedback to the patient, such as advising the patient to stop or reduce drinking.
- Assisting the patient by providing with material to facilitate change.
- Referring the patient to appropriate resources. If the identified risk is substantial, immediate referral to a specialized resource is suggested, along with ongoing support and follow-up.

12.2
Barriers to Screening

What are the issues that stand in the way of providing screening and referral services in a primary care environment? Very few studies have been conducted concerning the barriers to screening specifically for FASD; however, the literature on prevention and screening in general divides these into provider issues, system issues, and patient issues.

Provider issues include a lack of training; another is a lack of time, often due to business models of patient care that leave little time for screening. Clinicians often know *what* they could do, but do not have the ability to do it at that particular time. A third provider issue is uncertainty about where to get help if screening is done and a problem is found. A fourth barrier is personal discomfort in dealing with the issues.

System issues are shortages of resources. These include trained counselors and other service providers who have the appropriate skills to help individuals identified with FASD, programs to treat the often complex issues that are present, and access to these resources where people live, including in rural and isolated areas. Once a patient is identified as having FASD, there is an immediate need for support and for a process of mentoring the patient into the appropriate programs. These resources need to be available and closely linked to the primary care system.

Patient issues include feeling stigmatized by the diagnosis and failing to comply with treatment because of a lack of understanding, or an uncertainty that it is the correct treatment. The more that specialized resources can be integrated into primary care, the less stigmatizing it is for individuals. The system is organized based on diagnoses, because that works for us as healthcare professionals, but patients have a different paradigm based on their identity as individuals, within their family and community. Finding the balance between specialized resources that address a disease and broad-based individual care focused on the family and community in which the individual lives is a difficult issue for the future.

12.3
Impact of Healthcare Reform

During the past five to ten years, all jurisdictions in Canada have invested in primary healthcare reform. Currently, shared-care models that demonstrate partnership between primary care and specialty care, and between healthcare and other areas such as social services, are being developed throughout the country.[9] Each province has set goals and objectives to be achieved, and all include improved access, improved coordination of care, and an enhanced comprehensiveness of service at the primary care level. They also include, in many cases, defined and measurable outcomes, and the integration of primary care with the broader system. Programs now exist to support primary care in making the changes to business and clinical processes that are required to achieve higher quality and consistency of care. In Alberta, for example, TOP provides mentoring in the area of quality improvement, and AIM (Access Improvement Measures) provides help with improving access to services. Other provinces and territories are initiating similar support programs. The direction of these changes is generally acknowledged to be successful, although it is too early for true outcome measurement.

Reforms in primary care have affected the previous system of independent physician offices in the following major ways. A stronger population focus has been developed which builds on the personal relationship that has always been central to primary care. Members have been organized into networks of primary care physicians working in multidisciplinary teams that vary with the needs of the communities they serve. There is an emphasis on maintaining a continuity of relationship with patients, and on ensuring that individuals can remain within their communities for healthcare services. Information technology, which provides a means of sharing information and improving the continuity of patient support, is an important underpinning of each model.

In the new team-based, shared-care models, primary care is improving its capacity to partner with the rest of the healthcare system. As a result, a broader, more comprehensive skill-set is available at the primary care level. Individual physicians and their patients have access to the skills of pharmacists, social workers, psychologists, dieticians, chronic disease nurses, social services, nurse practitioners, and other caregivers. These individuals provide services in traditional ways, but also work in teams to provide new and innovative programs, such as individual life skills counseling and more effective liaison within the broader health system.

The new team-based models will provide an enhanced ability for primary care to do more of the recommended interventions for FASD prevention. This is in part because of the availability in team-based clinics of more professional hours, more skill-sets and services, and improved information technology support. The prevention of FASD would require that the protocols for periodic health assessments include questions for screening. Recently, a TOP program developed a

[9] http://www.hc-sc.gc.ca/hcs-sss/pubs/acces/2002-htf-fass-prim/over-aper-eng.php

reminder sheet that is integrated into an electronic medical record. This reminder sheet includes 10 preventive maneuvers that should be carried out on all individuals. The use of an integrated reminder sheet has reduced the number of omissions (not offering a test that needs to be offered) by 70%. Once an individual or family is identified as being at risk, the breadth of skills now offered at primary care provides the ability to be aware of and partner with more community resources, and thus to support the individual in finding appropriate programs.

In summary, research has measured and proven the inherent value of a strong primary healthcare system that provides services in the areas of prevention, treatment, rehabilitation, and health maintenance. Today, primary care is being enhanced through reforms that improve the ability to provide services in a manner that is accessible, comprehensive, continuous, and integrative. As primary care physicians, we have the opportunity to perform anticipatory counseling, early identification, and primary, secondary, and tertiary prevention. Our new shared-care models provide a greater ability to refer patients to appropriate services, and to bring those services into the community. Longitudinal relationships with patients enable long-term monitoring to be conducted, and to support both individuals and families within the community. Fetal alcohol disease prevention is one part of a bigger picture of improving health among communities at all levels, through partnerships both inside and outside of the healthcare system.

Reference

Sokol, R., Martier, S., and Ager, J. (1989) The T-ACE questions: practical prenatal detection of risk drinking. *Am. J. Obstet. Gynecol.*, **160**, 863–871.

Part Four
FASD and the Legal System

13
The Manitoba FASD Youth Justice Program: Addressing Criminal Justice Issues

Mary Kate Harvie, Sally E.A. Longstaffe, and Albert E. Chudley

> The horrible thing about all legal officials, even the best, about all judges, magistrates, barristers, detectives and policemen, is not that they are wicked (some of them are good), not that they are stupid (several of them are quite intelligent), it is simply that they have gotten used to it. Strictly they do not see the prisoner in the dock; all they see is the usual man in the usual place. They do not see the awful court of judgment; they only see their own workshop.
>
> (G. K. Chesterton – "Tremendous Trifles")

13.1
Introduction

A basic premise of the criminal justice system is that offenders have the ability to learn from their own mistakes and from the mistakes of others. Two central principles of sentencing set out in *Criminal Code section 718* are "specific deterrence" (imposing the appropriate sentence to deter the individual offender) and "general deterrence" (imposing a sentence designed to deter others from committing like offences). But what happens if it turns out that a group of individuals, suffering from organic brain damage, are challenged or even unable to learn from their own mistakes or from the mistakes of others? How should the Court apply these basic sentencing principles to these offenders, recognizing the need to address public safety, to maintain public confidence in the criminal justice system and to attempt to prevent offending behavior in these and other offenders?

The Manitoba Fetal Alcohol Spectrum Disorder (FASD) Youth Justice Program was initiated as a pilot project in September of 2004, with a view to answering these and other questions, to gathering data, and to develop recommendations for the Court and support in the community for those FASD-affected youth involved in the criminal justice system.

Fetal Alcohol Spectrum Disorder–Management and Policy Perspectives of FASD. Edited by Edward P. Riley, Sterling Clarren, Joanne Weinberg, and Egon Jonsson
Copyright © 2011 WILEY-VCH Verlag GmbH & Co. KGaA, Weinheim
ISBN: 978-3-527-32839-0

13.2
The Legislative Context

Legislators have long recognized that youthful offenders ought to be dealt with differently than those offenders involved in the adult system. When Parliament of Canada passed *The Young Offenders Act* (*YOA*), this legislation replaced the previous *Juvenile Delinquent's Act* with the intention of taking a new approach to youth crime. The declaration of principles set out in *Section 3* of *The Young Offenders Act* included the following:

- While young persons should not in all instances be held accountable in the same manner or suffer the same consequences for their behavior as adults, young persons who commit offences should nonetheless bear responsibility for their contraventions.

- Society must, although it has the responsibility to take reasonable measures to prevent criminal conduct by young persons, be afforded the necessary protection from illegal behavior.

- Young persons who commit offences require supervision, discipline and control, but because of their state of dependency and level of development and maturity they also have special needs and require guidance and assistance.

It was recognized that background information, including some medical information, could assist the Court in determining an appropriate disposition for youthful offenders. The YOA contained provisions under section 13 for the preparation of medical and psychological reports for purposes such as applications to transfer a youth to adult court, to determine fitness to stand trial on account of insanity, or for making or reviewing a disposition under the Act.

On April 1, 2003 the *YOA* was replaced by the *Youth Criminal Justice Act* (YCJA). The YCJA purported to introduce a new regime for youth justice in Canada. The Declaration of Principles of the YCJA is set out in section 3, and included the following:

3. (1) The following principles apply in this Act:

    a)   the youth criminal justice system is intended to

        i)   prevent crime by addressing the circumstances underlying a young person's offending behavior;

        ii)   rehabilitate young persons who commit offences and reintegrate them into society; and

        iii)   ensure that a young person is subject to meaningful consequences for his or her offense in order to promote the long-term protection of the public.

The legislation took a somewhat different approach to youth-sentencing principles, eliminating general deterrence as a sentencing principle for young persons, emphasizing "timeliness" of youth proceedings and emphasized dispositions that include "meaningful consequences," that ensure "fair and proportionate accountability." The declaration of principles in the YCJA identified the need to prevent crime by "... addressing the circumstances underlying a young person's offending behavior," and that "accountability" be made meaningful "... for the individual young person given his or her needs and level of development and, where appropriate, involve the parents, the extended family, the community and social and other agencies in the young person's rehabilitation and reintegration."

Once again, the legislation authorized the Court to order medical or psychological reports in certain circumstances. Section 34 of the YCJA outlined that such reports can be ordered at any stage of the proceeding where:

i) the court has reasonable grounds to believe that the young person may be suffering from a physical or mental illness or disorder, a psychological disorder, an emotional disturbance, a learning disability or a mental disability.

Many reports and other information available under the YOA continue to be provided to the Court under the YCJA. While the YOA contemplated the preparation of "predisposition reports," the YCJA mandates the preparation of pre-sentence reports when custody is being considered. The preparation of forensic reports by qualified psychologists continues to be available under the YCJA. Other collateral information, such as relevant medical information, can be provided to the Court, usually at the discretion of defense counsel.

These provisions for young offenders differ significantly and are far more expansive than those for adults contained within the Criminal Code of Canada. In adult proceedings, Courts can order the preparation of a pre-sentence report, which will contain relevant background information. Under the Criminal Code, other than the provisions which allow the Court to order an assessment to determine if the accused is fit to stand trial or not criminally responsible, there is no general provision comparable to much wider YCJA section 34 provisions, which enables the Court to order an assessment to determine if the accused is suffering from a "... physical or mental illness or disorder, a psychological disorder, an emotional disturbance, a learning disability or a mental disability."

It is important to remember that both YCJA and Criminal Code proceedings are subject to principles enshrined in the Canadian Charter of Rights and Freedoms (the "Charter"). Section 15 of the Charter provides that:

> Every individual is equal before and under the law and has the right to the equal protection and the equal benefit of the law without discrimination and, in particular, without discrimination based on race, national or ethnic origin, colour, religion, sex, age or mental or physical disability.

13.3
The Information Gap

Despite the legislative provisions allowing the Court to order certain reports for youthful offenders, it was suspected that there were significant gaps in the information being provided to the Court as it related to the issue of FASD. While the actual prevalence of individuals with FASD within the criminal justice was statistically unknown in Manitoba and Canada, there was evidence to suggest a disproportionately high representation. In their 1996 Report entitled "Understanding the Occurrence of Secondary Disabilities in Clients With Fetal Alcohol Syndrome (FAS) and Fetal Alcohol Effects (FAE)," A. Streissguth and colleagues found the following results in a survey of individuals with an FASD diagnosis:

- Trouble with the law experienced by 60% of those aged 12 and over.
- Confinement (inpatient treatment for mental health, alcohol/drug problems, or incarceration for crime), experienced by 60% of those age 12 and over.
- Inappropriate sexual behavior was reported in 45% of those age 12 and over.
- Alcohol/drug problems were experienced by 30% of individuals age 12 and over.
- Disrupted school experience (suspension or expulsion or drop out), was experienced by 43% of children of school age (Streissguth et al., 2006).

Other studies raised concerns about the impact of FASD on youth in the care of child welfare agencies. Research findings from a study by British Columbia's Representative for Children and Youth indicated that 44% of youth receiving services from the Ministry of Children and Family Development end up facing criminal charges, and that 36% of children in care are incarcerated. The Representative has further found that children in care are becoming involved in the youth criminal justice system at a younger age, and are involved in the justice system for a longer time. The study also found that 70% of the children in care involved in the justice system have special needs, such as learning disabilities, mental health issues, and FASD (Culbert, 2008).

Local studies identified similar concerns. A 2005 Manitoba report suggested that 17% of Manitoba children in care were either diagnosed (11.3%) or suspected of being directly affected by FASD (Fuchs et al., 2005).

While it was suspected that FASD was a significant problem for those involved in the criminal justice system, little medical or psychological information was being provided to the Court to confirm these suspicions. In part, this was due to a lack of diagnostic resources within the community. Until recently, other than privately funded assessments, there were few diagnostic services in Manitoba for individuals over 12 years of age (which, coincidentally, is the age at which one can be charged criminally). For those involved in the justice system, an appropriate diagnosis was difficult – if not impossible – to obtain.

The lack of confirmatory clinical information was also due to a lack of recognition and knowledge about FASD on the part of those involved in the criminal justice system. Information was provided to the Court on a sporadic basis, and usually as a result of an FASD assessment conducted when the accused was an infant or toddler. These assessments often provided limited up-to-date information about the individual's functional challenges. In general, the impact of FASD went unrecognized by lawyers, probation officers, other justice participants and the Court.

13.3 The Information Gap

The criminal justice system is highly structured and language-based. Information is provided to the Court within the confines of the adversarial system, although the crown and defense counsel have usually discussed their respective positions on a case in advance of the matter appearing before the Court. Busy dockets run quickly; courtrooms are often crowded, with lawyers, accused, reporters, and members of the public coming and going. Many Court orders are reduced to writing for the accused.

As awareness of the many issues around FASD increased, the deficiencies of the criminal justice system for those affected youth and adults were highlighted. An undiagnosed accused who may have had difficulty processing a Judge's decision (made in a noisy Courtroom environment) may also have challenges with the written word and with short-term memory. Such an accused, who failed to follow the Court order, was often mislabeled as a result of his/her noncompliance. As an example, consider the characterizations in the following extract from a pre-sentence report prepared prior to the youth being diagnosed pFAS:

> D related that he had been going to Court for such a long time and has not missed many court dates. D admitted that he is lazy and always sleeps in and "must have missed some". When confronted about his noncompliance with disposition (ISSP) and undertaking (bail) he stated, "I was stupid– especially this summer". D did not appear to have a good understanding of time of time frames when the offences were committed, even after it was explained to him. He seemed confused and questioned, "Was that before I was in MYC the last time or after I was in that other time?" While D did not appear remorseful for his actions he vowed to get back on track. D does not appear to comprehend the seriousness of not complying with court imposed conditions and could not explain why he continues to be noncompliant.

At the time of his Court appearance, D was 16 years old and had a criminal record which included break-ins and failing to comply with Court orders. The probation officer expressed the opinion that D was a candidate for a community sentence, although the officer also expressed considerable concern about D's inability to follow through with Court orders for an extended period of time.

The FASD assessment conducted on D revealed a different picture. D came from a family of five children; his two brothers were both deceased. D lived with both of his parents, who had complained and sought assistance for D's impulsivity, lack of focus, and his noncompliant behavior at home and school, but no services had been provided. D was reported as having "huge gaps" in his learning profile due to sporadic attendance at school and behavioral issues. He self-reported the use of marijuana to "calm me down."

Testing revealed that D's brain domain functions that were seriously impaired included language comprehension, memory, and academic achievement. He also had significant impairment in the verbal and performance categories of IQ testing, and demonstrated difficulties with attention and executive functioning. He did demonstrate some strength in visual processing.

The findings and comments of the probation officer in the pre-sentence report which related to D's performance were not uncommon. The negative self-image that flowed from the inability to comply with Court orders was interpreted as evidence of a negative attitude. Rather than recognizing that D's actions flowed from his deficits, the probation officer concluded that he was not "remorseful" and did not "... comprehend the seriousness of not complying with court-imposed conditions" and "... could not explain why he continues to be noncompliant". Clearly, a new interpretation of D's behavior and a different approach to sentencing was needed to assist D in avoiding further criminal activity, and to comply with further Court orders.

Justice system participants wondered, but were unable to determine, how many individuals might be affected by FASD but were, as yet, undiagnosed. Concerns were expressed about the criminalization of individuals who lack the capacity to follow Court orders. Concern was also expressed with respect to the ease with which these individuals were led into gang activity and other criminal organizations, as well other individuals who were presenting as victims of crime.

13.4
The Manitoba FASD Youth Justice Program

The FASD Youth Justice Program began as a collaborative initiative, spearheaded by the Court with the assistance of a variety of players who had an interest in making available Court-ordered FASD assessments. The Steering Committee included members of the Provincial Court of Manitoba, crown attorneys, defense counsel, probation and corrections officials, members of Courts administration and a representative of the Winnipeg Police Service. After a series of meetings, seed funding was applied for and provided in September 2004 by the Federal Government's Youth Justice Renewal Fund for an 18-month period. Staffing was provided on an "in kind" basis for coordinators and community facilitators. The Manitoba Clinic for Alcohol and Drug Exposed Children (formerly "CADEC" and now the "Manitoba FASD Center") provided the medical services. Psychological assessments were made available through the Manitoba Adolescent Treatment Center ("MATC").

The Steering Committee identified the following as the goals of the program:

1) To assess youth involved in the Justice System who may be FASD affected.
2) To provide recommendations to the Court for appropriate dispositions consistent with the sentencing principles of the Youth Criminal Justice Act.
3) To build capacity within the family and the community while enhancing government and non-government FASD supports and services.
4) To implement meaningful multidisciplinary interventions and reintegration plans with supports for youth affected by FASD, and their family.

In October 2006, the program was adopted by the Province of Manitoba and is now fully funded by the Justice Department.

13.5 Screening

The Steering Committee recognized early that the demands on the Program might easily exceed its capacity. Criteria for acceptance into the program were established in an effort to focus the work of the program and to control the number of youth eligible to participate, recognizing that in doing so, certain youth would be left out of the process. In short, the Program sought to identify and provide access to an FASD diagnosis to those accused who may have "slipped through the cracks," and whose criminal behavior was therefore being misconstrued due to a lack of medical information.

Some of the indications or "red flags" that would trigger a referral for diagnosis (where a history of confirmed prenatal exposure to alcohol is present) include:

- Repeated "Fail to Comply"
- Lacking empathy
- Poor school experiences
- Difficulties within the institution:
  - Following expectations
  - Poor peer interactions
  - Academics
- Unable to connect their actions with consequences
- Does not seem affected by past punishments
- Crime committed may be of opportunity rather than planned
- Crimes that involve risky behavior for little gain
- Gang involvement
- Superficial relationships/friends

The Program accepts accused who meet the following criteria:

1) The accused young person must be between 12–18 years of age.
2) The accused young person must not yet have disposed of their criminal charges.
3) The young person must be residing in Winnipeg (or in our expansion location in The Pas, Manitoba) or planning to reside in either of those locations.
4) Both the young person and their parent or guardian must consent to participation in the program.
5) There must be confirmation of significant prenatal alcohol exposure prior to the youth being accepted.
6) The youth must not have been previously diagnosed as FASD.

The preliminary screening process takes a very short period of time. Referrals for preliminary screening have come from a wide range of players within the justice system. Those identified as potentially eligible for an FASD assessment are usually prescreened within a week of their being identified. It is also a requirement that they have not already been subject to an FASD assessment. It is a concerning discovery that some youth who are before the Courts, many with an extensive criminal record, have already been diagnosed FASD, and that the information is either not being provided or sought by their counsel, or provided to the Court.

13.6
The Preassessment Period

Once a youth has completed the prescreening process, the matter appears in court and an FASD assessment order is made pursuant to section 34 of the YCJA. The coordinators then begin the process of preparing for the assessment. The assessment information collected by the coordinators includes:

- Inventory for Client and Agency Planning (ICAP);
- Behavior-Rating Inventory of Executive Functioning (BRIEF);
- Conners Scale (behavior rating);
- Short sensory profile and the Clinic for Alcohol- and Drug-Exposed Children/Manitoba FASD Center intake form (Winnipeg Health Sciences Center).

Simultaneous to the intake process is the *psychological assessment*, which is completed on those youth who are in custody at the correctional institute by a Youth Forensic Services Psychologist through the MATC. This involves an assessment of the domains of the brain, including cognitive and language skills, academic achievement, attention, memory, executive functioning, and social adaptive skills.

The coordinators also spend a significant amount of time with the young person and their family or caregivers. This includes attending to the family residence, interviewing and obtaining background information about the young person, while at the same time developing a level of trust with the family members or caregivers. The bond which develops is extremely important for the long-term success of the program, ensuring that a "team" approach is taken when supports are eventually established for the young person.

13.7
Medical Assessment

Clinics occur monthly, with participation of physicians from the Manitoba FASD Center. The preclinic data gathering is provided in advance to the physicians. Every effort is made to assist the parents and/or caregivers in attending on assessment day. For those youth in custody, the assessments are conducted at the Manitoba Youth Center ("MYC"). Due to limitations in resources, only two assessments are conducted each month.

At the clinic, the physicians interview the youth and the youth's caregivers for any further background information. The diagnostic team uses the Canadian Guidelines for FASD diagnosis (Chudley et al., 2005). The physicians have introduced innovative tools for diagnosis, including routine facial photoanalysis to augment the clinical examination of the youth in the clinic. They confer after a review of all preclinic materials and the psychological testing, and discuss the findings arising from the medical history, the physical, neurological and genetic (dysmorphology) examinations, along with the photos and the growth assessments. Physicians look not only for subjective and objective evidence of findings consistent with an FASD diagnosis, but also for any other medical, psychiatric or other problem which could explain the findings or be a result of the persisting FASD condition.

The results of the clinic are shared with the youth and their caregivers with the assistance of visual printed tools, which have been developed to assist in explaining the relevant strengths and affected brain domains to the youth and family. The results are conveyed in a manner which is nonjudgmental, the goal being to identify specific challenges that the youth may be facing, in order to empower the youth and their caregivers to seek a new approach to meet those challenges.

Other recent innovations have included the addition of assessments of resilience for each youth, with a hope of understanding factors that might improve the advice given to youth and caregivers and to predict better outcomes.

The challenges of the program for the healthcare professionals have been diverse, but fortunately they have been answered in part by the development of a strong trust between team members who have worked hard to strengthen communication across systems that operate with different rules, mandates, and cultures. It has been necessary to determine which decisions belong to which system, and how to respect and at times modify each system. The physicians have had to work in the space provided by MYC that was not designed for clinical activities. The team has been successful in developing a report format that is acceptable to all participants, and has also worked hard to meet the timelines of the court system in situations where there may be only varying degrees of readiness by the family and youth to deal with the issues of diagnosis and its implications for their relationships.

Ultimately, the physicians and coordinators deserve credit for any successes demonstrated by this program. They have ensured that the assessments are conducted in such a way as to provide respect, privacy, support and practical suggestions within a setting where the youth is safe and supervised. Coordination with a hospital-connected youth clinic has facilitated a post-diagnosis follow-up for youth receiving long-term medications. A community development facilitator is an integral part of the team, working with the community supports to ensure the appropriate and practical application of the information contained with the report.

13.8
The Doctor's Report and Its Use

The report prepared by the physicians contains significant detail about the youth's background, the mother's prenatal history respecting alcohol and drug exposure, and the youth's developmental and medical history. Also included is information about school history, psychological information, a sensory profile, and the results of a physical and genetic examination.

It is important to note that the recommendations which are contained at the conclusion of the report outline the youth's strengths and deficits, and propose a multidisciplinary and multisystemic approach in responding to the youth's needs. These reports are completed from a medical perspective, and not from the justice perspective. The physicians do not interview the youth about their criminal activity, and the report does not make specific, justice-based conclusions or recommendations. The recommendations can be shared with other medical professions, and are shared with the youth and his/her caregivers at the time of the assessment.

13.9
Sentencing Conferences

Once the assessment report is completed, the accused can proceed directly to a sentencing hearing. This is usually the situation where the coordinators have been successful in establishing the necessary community supports for the youth. Where it appears that there are gaps in the community plan, a sentencing conference can be convened to assist the Court in developing a plan for the youth. *YCJA section 41 states:*

Recommendation of conference

41. When a youth justice court finds a young person guilty of an offense, the court may convene or cause to be convened a conference under section 19 for recommendations to the court on an appropriate youth sentence.

In order for this conference to be successful, it is important to have in attendance as many of the supporting agencies, organizations or individuals that are involved in the young person's life. This provides each of the players with an opportunity to learn more about the young person's diagnosis, and to discuss as a group the development of a community plan. It is a mechanism which also assists the youth in better understanding the conditions and expectations that may be placed upon them. Finally, it allows the Court to explore whether traditional sentencing conditions are helpful or realistic, given the findings with respect to the individual youth.

From a practical point of view, the Court can consider whether a youth who may have memory difficulties, may have executive functioning challenges, or may not be able to do something as simple as reading, writing or telling the time, can understand and follow specific conditions, such as reporting to a probation officer or following curfew conditions. That is not to say that in some instances the young person may not require significant supervision. However, in order for this supervision to be meaningful and for the imposed conditions to be realistic, all of the players need to consider the impact that the conditions will have on the young person's life, and how they can work together to ensure any Court-ordered conditions are followed. The recommendations made by the sentencing conference participants can be considered by the Judge at the actual sentencing hearing.

13.10
The Sentencing Process

Regardless of whether or not a sentencing conference has been held, the judge who sentences the youth is provided with all information, including the assessment, in advance of Court. The coordinators make themselves available during the sentencing to answer any questions that the Court may have with respect to the report, including any proposed community plan.

The YCJA mandates the Court to impose "meaningful consequences" on youthful offenders through "fair and proportionate accountability" which is "... meaningful for the individual young person given his/her need and level of development, and where appropriate, involve parents, the extended family, the community and social or other agencies in the young person's rehabilitation and reintegration". All of the forgoing is applied in such a manner so as "... to promote the long-term protection of the public."

The information provided assists the Court in better understanding the offender, and thereby properly applying the provisions of the YCJA. Under this program, youth are not diverted out of the Court system, but are sentenced for the offences they commit. However, the information serves to fill the information gap which existed previously, with a view to assisting the Court in imposing "meaningful consequences," as mandated by the YCJA. Establishing a network of supports to assist the youth upon their return to the community helps to ensure long-term compliance with Court orders and, ultimately, to reduce recidivism.

13.11
The Statistical Outcomes

A statistical review of those youth involved with the FASD Youth Justice Program confirms the initial concerns which gave rise to the program. As of March 2010, the program has received 385 referrals, from which 110 assessments have been completed and 73 youth diagnosed [nine with pFAS and 64 with Alcohol-Related Neurodevelopmental Disorder (ARND)]. The significant and concerning number of ARND assessments speaks to the hidden nature of the disability.

Of the 73 youth diagnosed, 23 had their biological mothers attend with them to the assessment, and four had their biological fathers attend. Of the remaining youth, 32 had other support staff present (Child and Family Services worker, probation officer, Group Home worker, or Adoptive/Foster parents). The remaining 14 had attended alone, without the support of a significant caregiver.

A review from a social history perspective indicated that 41 of the 73 youth were in the care of Child and Family Services; 32 youth resided with their birth families (14 with their biological mother, six with their biological father, five with both birth parents, five with grandparents, and two with aunts and uncles). Of concern was that the remaining 26 youth were of no fixed address at the time of the assessment. Of the 72 youth, 10 were teen parents, 45 were reported to have been in a gang, 55 had issues with alcohol and drugs, and 11 had been sexually exploited.

An analysis of the psychiatric/psychological history of the 72 youth indicated the following:

- only six were in the average IQ range (90–109)
- 17 were in low average IQ range (80–89)
- 27 were in borderline IQ range (70–79)
- 22 were in mentally deficient IQ range (<70)
- 39 had severe verbal mediation deficits.

The assessments outlined the following areas of severe deficits:

- Executive functioning ($n = 58$)
- Academic function ($n = 55$)
- Attention ($n = 53$)
- Memory ($n = 47$)
- Communication ($n = 34$)
- Sensory function ($n = 22$)

13.12
One-Day Snap-Shot of Age of Majority Youth (28 February 2010)

One of the limitations of the FASD Youth Justice Program is that its focuses almost exclusively on youthful offenders. As noted earlier, this is due in part to the limitations of the *Criminal Code* as compared to the YCJA, the latter of which enables the Court to order an FASD assessment. However, those diagnosed through the program are followed and support is provided by the coordinators beyond the offender's eighteenth birthday. A "one-day snap-shot" of those individuals reveals the following:

- 37 of the 72 youth diagnosed were aged 18 or over
- 33 of the 37 have qualified for long-term adult supports
- 9 were in supported living homes
- 16 of the 37 have never been in custody as an adult
- 27 of the 37 were in the community
- 2 were deceased

13.13
Other Initiatives

13.13.1
"This is Me"

An interdisciplinary team in Manitoba worked together to create "This is Me – A Tool for Learning About and Working with People Affected by FASD." It is an interactive package, including a CD-ROM and interactive guide designed to illustrate typical behaviors of people affected by FASD, and to identify strategies to assist caregivers and those affected in dealing with those behaviors. The gender-neutral, race-neutral animated character "Me" is affected by FASD, and is shown in a common range of social circumstances. All scenes are depicted visually, with no verbalization throughout the video.

The project was originally inspired by the need to provide those in the criminal justice system with a tool to better understand, work with, and communicate with those people affected by FASD. It is now used both as a teaching tool for those working with FASD-affected youth and with the youth themselves in better understanding their own deficits and strengths. For more information, see: www.mefasd.com.

13.13.2
This is Me Life Books

The FASD Youth Justice Program has begun using "This is Me Life Books" as a creative way for youth to better understand themselves and their disability. The staff had been using "This is Me" as a graphic aid in debriefing with the youth after their assessment. It became apparent in those sessions that the youth were responding in a concrete way to the scenarios, and that an individualized approach expanding on those scenarios was needed.

In creating their own "Life Book," the young person is allowed to identify their preferred learning style, their interests, learning goals, and strengths. They also work to identify triggers or stressors, and to provide strategies and supports to avoid or respond to those stressors. The stories and experiences are told from the youth's point of view, and their words and personal style are reflected throughout the book. With assistance, the youth makes all of the design choices, including fonts, colors stories, and pictures. The book is designed in such a way as to allow the youth to add to it on a continual basis.

This program has been funded through the Department of Justice Youth Justice Fund: "Understanding Youth with FASD and Making Accommodation Project," and will be evaluated in March 2011. A study group has been identified and an assessment is underway to determine if use of the books assists in increasing in understanding of the disability; identifying strengths; increasing confidence levels; increasing awareness of sensory issues; increasing awareness of triggers; and increasing awareness of possible adaptations and interventions. The study will also assess if there a decrease in number and or severity of incidents, outbursts, and consequences.

13.13.3
The Icons Project

The interdisciplinary team in Manitoba became aware that some youth who were diagnosed with FASD were struggling to understand many of the language-based Court orders that were imposed upon them. The team worked with a group of graphic arts specialists to develop a series of "icons" to which justice officials can refer when reviewing and discussing court-ordered conditions. These icons assist those for whom a visual representation is easier to comprehend than a written explanation. They are included on some Court-related documents, but are available to all probation officers in the form of large laminated posters that can be used and copied when working with clients. Extra details related to each of the conditions can be added below the icon, such as the name of an individual(s) who is/are part of a "no contact or communication" order.

The Manitoba probation officers have been very diligent in using all tools available to them, and the "icons" are no exception. The probation officers were asked to provide feedback and to make suggestions for any improvements they thought would assist this project. It was in this context that the following e-mail was received from a senior probation officer, relating his experience:

> GOOD AFTERNOON! I just had to contact both of you to relate my experience earlier this afternoon when I met with one of my clients. He is not diagnosed FASD though yesterday in Court his mother admitted on the record she drank during the first month of her pregnancy. This young man is cognitively impaired and though 20 cannot read or write and has serious difficulties remembering.
>
> Today we sat down and went through both his old Youth Order and his new Adult one using the Icons. It was an absolute **EUREKA** moment for him. I would read him the condition and he would pick out the matching picture usually with no problem. Once we had all the Icons selected, I had him read them back to me in his own words and then I would write a word or two under each Icon.
>
> He was just so proud of himself and stated, "I understand my conditions for the first time". He was heading home to put them on his fridge. It actually brought a tear to this old goat's eye.
>
> TO SAY THE LEAST IT MADE MY DAY!!!!!!!

Depicted below are some examples of the icons used to explain conditions in bail or probation orders:

From left to right: No Weapons
Abstain from Alcohol and Drugs
No contact or communication

13.13.4
Youth Accommodation Counsel

Legal Aid Manitoba has commenced a two-year project that endeavors to accommodate the needs of Legal Aid's youth clients who are affected by FASD, recognizing that these youth require accommodation within the justice system. Legal Aid Manitoba's "FASD Youth Accommodation Counsel" provides a lawyer with extensive knowledge, training and experience with FASD-affected youth and adults, who is dedicated to improving service and outcomes for youth clients. Concerns identified by Legal Aid Manitoba respecting this client group include: (i) the risk of wrongful convictions; (ii) the imposition of inappropriate sentences, which can lead to criminalization through noncompliance and breach charges; (iii) denial of bail due to inappropriate case planning; and (iv) a denial of access to government services.

Legal Aid Manitoba identifies the need for effective advocacy for these clients, which requires an understanding of the unique needs of these young clients so as to ensure effective lawyer–client communication. Accommodation counsel seeks to ensure that the client understands the court process, receives appropriate referrals to community services and support, and that the court understands the client's unique needs, circumstances and abilities, thereby contributing to appropriate dispositions that reflect and accommodate the client's unique needs, circumstances, and abilities.

In summary, whilst Manitoba has taken a number of steps to accommodate the special needs of youth involved in the Justice system, it is recognized that there is clearly much work still to be done.

13.14
Strengths and Challenges

The fact that FASD assessments are now available to youth involved in the justice system represents a positive development. Innovations that have as their goal the modification of the criminal justice system so as to make it more understandable are laudable. The team building and the increase in communication between systems that have traditionally been separate has resulted in an improvement for those who rely on those systems. The dedication of the FASD coordinators and Community Development workers is unfailing, yet much more is needed.

To date, the work in Manitoba has centered largely on those who reside in Winnipeg. Much needs still to be done to expand the accessibility of the diagnostic and supports that have flowed through the FASD Youth Justice Program to those living in other centers and the other regions of the province. In addition, the work of the program has been limited to two youth per month, and only to those youth who have not yet disposed of their charges. Many youth are turned away or choose not to participate, as they or their counsel do not wish to postpone dealing with their charges until after an assessment can be completed. Given the "timeliness" provisions of the YCJA, such a position is understandable from a criminal–legal point of view, and demonstrates the tension between the criminal justice/youth justice system and lifelong disability and long-term planning that a FASD diagnosis represents.

Recommendation: More services need to be provided for those offenders seeking access to ensure rapid access to an FASD diagnosis. Programs such as the FASD Youth Justice Program need to be available on a province-wide basis.

As its name denotes, the FASD Youth Justice Program is one which focuses almost exclusively on young offenders. Ironically, some significant successes have been seen with those offenders who have been transitioned into supported adult living environments. However, such supports are of limited availability, and are only in place for adults who meet certain criteria, some of which are IQ-based, and others of which are dependent upon the individual having a confirmed FASD diagnosis. It is interesting to note that with respect to the former, IQ results are often deceiving and are viewed as a poor tool in determining the support needs of a person diagnosed with FASD. With respect to the latter, few FASD diagnostic services are available for adults within the community, and the Criminal Code of Canada does not contain provisions to allow the Court to order such a diagnosis.

Recommendation: Eliminate the use of IQ testing alone as a basis for determining eligibility for supported housing and other community-based programming.

Recommendation: Amend the Criminal Code of Canada to allow Courts to order FASD and other medical assessments when considered appropriate for identified Court purposes.

Many FASD-affected youth thrive when living in a supported, nurturing environment but, conversely, many fail terribly when they live an unstructured life. Ironically, those FASD youth who require a structured school attendance and would benefit greatly from it, cannot conform to the traditional demands of the school environment. Similarly, the need for a safe living environment is paramount; these youth, many of whom have either dropped out or have been kicked out of school, are living in the most dangerous and undesirable parts of the city, and are vulnerable to gang recruitment. Many who have difficulty linking cause with effect, and may have trouble recognizing the high-risk nature of their activities, can become involved in a criminal lifestyle which is difficult to reverse. Equally disturbing, they too are often victimized. The need for secure and safe housing is crucial to seeing an end to this cycle of crime and victimization.

Recommendation: An increase in all community-based supports, including housing for FASD-affected youth and adults. Supports need to be based on the needs of the individual, and may require varying levels of supervision and control. Such housing needs to be included in safe neighborhoods and be paired with other community supports, such as school and recreational activities.

References

Chudley, A.E., Conry, J., Cook, J.L., Loock, C., Rosales, T., and LeBlanc, N. (2005) Fetal alcohol spectrum disorder: Canadian guidelines for diagnosis. *Can. Med. Assoc. J.*, **172** (Suppl. 5), S1–S21.

Culbert, L. (2008) For kids in B.C. care, jail a more likely future than graduation. Vancouver: *Vancouver Sun*, Friday, 27 June 2008, http://www.canada.com/vancouversun/news/story.html?id=d4176365-77fd-49b4-bd36-85962a55e571 (accessed 9 November 2008).

Fuchs, D., Burnside, L., Marchenski, S., and Mudry, A. (2005) *Children with Disabilities Receiving Services from Child Welfare Agencies in Manitoba*, Centre of Excellence for Child Welfare.

Streissguth, A.P., Bar, H.M., Kogan, J., and Bookstein, F.L. (1996) *Understanding the Occurrence of Secondary Disabilities in Clients With Fetal Alcohol Syndrome (FAS) and Fetal Alcohol Effects (FAE)*, Final Report to the Centers for Disease Control and Prevention (CDC), University of Washington, Fetal Alcohol & Drug Unit, Seattle, WA.

14
Understanding FASD: Disability and Social Supports for Adult Offenders

E. Sharon Brintnell, Patricia G. Bailey, Anjili Sawhney, and Laura Kreftin

Fetal Alcohol Spectrum Disorder (FASD) can affect people of all backgrounds, ethnicities, religious persuasions, and socioeconomic groups, and represents a significant factor in predicting involvement in the correctional system. The following discussion is focused on the issues facing the corrections system and offenders or former offenders with FASD, as well as issues facing society on the release of people with FASD back into the community. The strong correlation between the disability, unmet multifaceted needs and incarceration of people with FASD, along with the stigma and hard realities of incarceration, necessitate a thoughtful perspective to improve the chances of success for this population.

Adults with FASD experience a range of disabilities that contribute to the likelihood that they will enter the correctional system. These include the primary brain deficits of cognitive and behavioral problems, gullibility, inhibition and poor judgment; secondary disabilities, such as substance abuse and mental illness; and deficits in the social determinants of health, such as poverty and racism. However, the judicial and correctional systems are generally not prepared either to identify FASD, or to address the disabilities of FASD among the offender population. Screening and assessment for the spectrum in these systems is limited, and at present there is no system in place to screen for FASD in adult prisons in Canada (Chapman, 2008). Although there are overtures in some jurisdictions to address the latter deficit, the needs of adult offenders with FASD are currently not adequately met in the corrections system (Chapman, 2008; Miller, 2005; Boland, Chudley, and Grant, 2002). It is important that the correctional system recognizes and identifies affected individuals, because people with FASD are vulnerable and their disabilities contribute to their getting into trouble with the law (Moore and Green, 2004; Streissguth *et al.*, 2004). Moreover, research has indicated that identifying FASD at any age improves options for interventions and helps to reframe problematic behaviors (Malbin, 2004).

Fetal Alcohol Spectrum Disorder–Management and Policy Perspectives of FASD. Edited by Edward P. Riley, Sterling Clarren, Joanne Weinberg, and Egon Jonsson
Copyright © 2011 WILEY-VCH Verlag GmbH & Co. KGaA, Weinheim
ISBN: 978-3-527-32839-0

14.1
Fetal Alcohol Spectrum Disorder (FASD) is a Disability

The term "disability" has a variety of definitions. For the purposes of discussing the interrelationship between FASD, social services and incarceration, disability occurs when individuals have activity limitations that create barriers to their participation in society. In the International Classification of Functioning, Disability and Health (ICF), developed by the World Health Organization (WHO), three broad health domains are noted: body functions and structures; activities or the execution of actions; and participation, or life involvement (Cieza et al., 2009). Using this international framework, it is appropriate to view FASD as a brain dysfunction that causes impairments (Olson et al. 2009) in all three main areas of functionality.

Fetal alcohol exposure (FAE) results in a continuum of impairments that cause affected people to be functionally disabled in a variety of ways. There are both primary disabilities and secondary disabilities associated with FASD. *Primary disabilities* are those that are directly caused by exposure to alcohol before birth, while *secondary disabilities* arise out of the interplay between primary disabilities, psychosocial factors, and environmental influences. Other factors that contribute to the barriers experienced by people with FASD are deficits in the social determinants of health.

Incarcerated people with FASD are an especially vulnerable subgroup of the affected population. Offenders may never have received a diagnosis, and they commonly have a pattern of repeated trouble with the law. People with extensive criminal histories are frequently considered high-risk offenders, and factors contributing to their criminality, including brain damage caused by FASD, may not be taken into account by the justice system. The incidence of primary and secondary disabilities caused by the prenatal exposure to alcohol is very high in offenders with FASD, and interventions to ameliorate adverse outcomes generally have not been employed (Fast and Conry, 2004). The interrelationship between the primary and secondary disabilities of FASD can lead to behavior that results in incarceration, and as a result affected people will require the same societal accommodation and supports that are available to people with other disabilities. Offenders with FASD are disabled. FASD can be viewed as a disability with societal challenges and needs for support similar to those of other chronic impairments that result from injury, such as spinal cord injuries.

14.1.1
Primary Disabilities Associated with FASD

The effects of prenatal alcohol exposure (PAE) include damage to brain structures that results in deficits in cognition, development, and behavior. PAE can also cause separate medical conditions, such as seizure disorders, skeletal problems, cardiovascular disease, dental problems, and reactions to medications (Paley and O'Connor, 2009). All of these conditions influence an individual's ability to function and participate fully in society.

In particular, the diminished cognitive function in FASD results in major deficits in an individual's ability to meet the demands of daily life. Poor short-term memory can result in information and instructions being quickly forgotten; and while long-term memory may be unaffected, information storage is often disorganized and therefore information becomes difficult to retrieve (Grant et al., 2004). Effective executive function (EF), or the integration of basic cognitive processes, is often lacking in people with FASD. Deficiencies in EF affect multiple areas of functioning, including work and school performance, social interactions, parenting, daily living skills, and ability to plan, organize and learn from mistakes (Grant et al., 2004; Connor et al., 2000). The scores on intelligence tests of people with FASD may or may not be abnormal, but many people with FASD are not able to perform at the level predicted by their IQ scores (Malbin, 2004; Kodituwakku, 2009; Fast and Conry, 2009; Chudley et al., 2007; Wass, Persutte, and Hobbins, 2001). These factors, combined with insufficient inhibition and poor cause-and-effect reasoning, can lead to life-long difficulties in adapting to social expectations and functioning socially (Rasmussen et al., 2008).

14.1.2
Secondary Disabilities Associated with FASD

Secondary disabilities are conditions, behaviors or situations that develop after birth. Secondary disabilities associated with FASD include mental illness (depression, anxiety and psychoses), substance abuse, restlessness, trouble at school, homelessness, and unemployment (Paley and O'Connor, 2009; Chudley et al., 2007). Secondary disabilities cause significant upheaval in people's lives. Trouble at school can lead to delinquency and then to trouble with the law, which can in turn cause difficulty in attaining and retaining employment and lead to further encounters with the legal system. According to Streissguth et al. (2004), adolescents and adults with FASD have a 60% risk of getting into trouble with the law. Incarceration can result in a loss of housing and possessions, and association with other criminals can lead to stigma and victimization.

These problems are also referred to as "neurobehavioral." Malbin (2004) states that the observable behavioral effects of FASD need to be recognized as the result of the physical changes in the brain and their impact on brain processing. Dysfunction in behavior includes poor judgment, poor impulse control, conduct problems, poor problem-solving skills, learning problems, fine motor skills deficits, hyperreactivity to stress, sexual promiscuity, resistance to change, difficulties in forming lasting relationships, gullibility, victimization, and an inability to understand or conform to social norms (Paley and O'Connor, 2009; Chudley et al., 2007; Rasmussen et al., 2008; Aragon et al., 2008). Some of these behaviors develop over time as protective reactions to feelings of frustration (Malbin, 2004). Aragon et al. (2008), indicate a link between a lack of services due to undiagnosed FASD and the development of secondary disabilities in adolescence and adulthood. Other research suggests that the early evaluation and identification of exposed people is vital to anticipating the confounding occurrence of secondary disabilities (Schonfeld, Mattson, and Riley, 2005). Among offenders with FASD, secondary disabilities are a key aspect of their involvement with the justice system.

14.1.3
The Social Determinants of Health and FASD

Housing, sanitation, nutritious food, health care, employment, access to services, justice and human rights are vital for everyone, and are known as the "social determinants of health." These are the economic and social conditions in which people are born, grow up, live and age, and the wider economic, social and political systems established to deal with illness (World Health Organization, 2008). Research indicates that social determinants of health have a greater influence on an individual than behavioral risk factors.

Social determinants of health influence both the incidence of FASD and the outcomes of people affected by it. There are a number of factors that lead to FAE. The mother's alcohol consumption during pregnancy is the direct cause; however, women do not live their lives in a vacuum. Trauma, poverty, inadequate nutrition and housing, gender inequity, racism, addiction, abusive relationships and other conditions affect their health, personal options and behavior. Children born with FASD in such environments face serious environmental and social adversities that place them at risk for adverse outcomes (Streissguth et al., 2004; Rasmussen et al., 2008). Adults with FASD have frequently grown up in unstable environments and have experienced difficulty in meeting their basic needs; many have been separated from biological parents and lived in multiple foster homes (Olson et al., 2009). They often have experienced learning problems and unemployment, and may have ended up in jail (Streissguth et al., 2004). Studies indicate that FASD-affected children have lower health-related quality of life than children who have survived cancers and other significant physical disabilities (Stade et al., 2006). High-quality caregiving and a stable home environment are important factors in successful outcomes for people with FASD, yet positive and stable family environments are not typical for this population (Olson et al., 2009). The negative outcomes associated with FASD are thus related to the very high rates of environmental risk factors and the inadequacy of family resources for affected children (Olson et al., 2009). Offenders with FASD represent the output of an environment that is characteristically rated low on social determinants of health.

The principal health risks such as trauma, physical and sexual abuse, racism, poverty and housing instability are rarely addressed in FASD prevention (Reid, Greaves, and Poole, 2008). In order to improve the health of people with FASD and lower the prevalence of the spectrum, the contributing social influences on alcohol use need to be attended to (Gearing, McNeill, and Lozier, 2005). All of these factors are important to address, as they contribute to sustaining a disadvantaged life situation. When added to the secondary disabilities such as substance abuse, unemployment and incarceration, social and environmental risk factors are increased. Attending to the social determinants of health should help to reduce not only the occurrence of complex conditions but also the societal costs associated with FASD.

14.1.4
Human Rights and FASD

FASD is best considered within a broader disability context wherein participation in society is accommodated and functionality is managed through service provisions, much as would occur for someone with a progressive degenerative disorder. Canadian law, as detailed in the Charter of Rights and Freedoms, recognizes that individuals are equal under the law and entitled to equal protection and benefit, without discrimination based on age, race or ethnicity, religion, skin color, gender, mental or physical disability (Canada, Charter of Rights and Freedoms, 1982). The purpose of equality rights is to respect the dignity, worth and value of all Canadians, and to ensure that laws and government action are based on circumstances and needs, and not on negative stereotypes. One of the roles of government is to facilitate equal access to services and supports for citizens. The impairments caused by FASD necessitate that affected individuals receive the supports they need to participate equally in society.

14.1.5
Incarceration and FASD

The prevalence of FASD is difficult to assess. Current estimates of the percentage of the overall population that is affected range from 2% to 5% in North America and Western Europe, with the spectrum generally under-recognized (Malbin, 2004; Fast and Conry, 2004; May et al., 2009). There is a higher incidence of FASD in certain regions and groups. Typically, rural, isolated and remote areas have a higher incidence, but the precise numbers are unknown (Bohjanen, Humphrey, and Ryan, 2009). Canadian FASD research suggests that rates are also higher in some Aboriginal communities (Boland et al., 1998). In particular, the prevalence of FASD is significantly higher among incarcerated people in Canada, and an estimated 10-fold greater than among the general population (MacPherson and Chudley, 2007).

The problems that people with FASD have in adhering to social norms of behavior can result in their being considered socially deviant instead of disabled. Children with FASD are viewed as victims of the birth defects caused by alcohol exposure (Donohue, 2008). Adults with FASD in the correctional system are regarded in an entirely different way, as deviants who have violated social norms (Donohue, 2008). Research indicates that the disabilities caused by FASD become de-medicalized as the affected person develops into an adult (Donohue, 2008; Golden, 1999). The term "de-medicalization" refers to affected people coming to be regarded as in charge of their lives and perpetrating social wrongs through choosing to display challenging behaviors, rather than as people with physical and mental limitations caused by damage to the brain (Golden, 1999). This change in perspective results in adults with FASD being considered fully responsible and then being relegated to the criminal justice system (Donohue, 2008). However, the child with FASD inevitably matures into the adult with FASD, because the brain deficits caused by PAE do not resolve or lessen as the individual ages. As a result, the adult with FASD is as much a victim of PAE as the child. The change in social attitude, from child as victim to adult as deviant, poses special challenges to an adult's ability to live in, and participate equally in, society.

14.2
Correctional Environment in Canada for Adults with FASD

In 2007, there were 268 533 adult offenders in federal, provincial or territorial custody in Canada. Of these, 154 768 were held in remand centers and 90 831 were sentenced (Government of Canada, 2008). While the precise number of adults in the correctional system who are affected by FASD is unknown, the percentage is higher than in the general population. Research indicates that 60% of people with FASD get into trouble with the law (Streissguth et al., 2004). Current prevalence estimates for FASD are at approximately 2–5% of the population for North American and Western European countries (May et al., 2009). However, Canadian research denotes that there is an estimated 10-fold greater incidence of FASD in the correctional population (MacPherson and Chudley, 2007) than in the general population. If this estimate is correct, it means that a substantial proportion of the correctional population may have FASD. These figures are difficult to substantiate, however, and the results must be interpreted with caution. Nevertheless, the recent literature suggests that the percentage of adults in Canadian correctional systems who are affected by FASD is significant. Brain deficits related to judgment, learning and problem-solving, along with gullibility, make it more likely that a person with FASD will be caught by legal forces during an illegal act. People with FASD are also known to have falsely confessed to crimes, including murder, in efforts to appease authority figures (Fast and Conry, 2004; Fast and Conry 2009). The brain impairments that result in efforts to please others have ramifications for offenders with FASD on multiple levels of the legal system.

Those people whose brain functions are impaired by PAE are not currently considered intellectually disabled by the corrections system. Unlike people with other mental disabilities, adults with FASD are treated as people without disabilities. This lack of fit between the demands of the system and the abilities of the individual has a negative influence on the chance for rehabilitation in the corrections environment for adults with FASD. One of the most significant events in recent correctional policy is the shift from a strictly punishment-based model to one that incorporates offender rehabilitation (Ward and Marshall, 2007; McFarlane, 2006; Canada, Criminal Code of Canada, RSC, 1985). In order for people with FASD to have opportunities to be rehabilitated, adaptations for their disabilities must be recognized as necessary and imposed by the system.

A central factor to consider regarding the incarceration of people with FASD is the rights of disabled people. A fundamental principle of the Canadian Criminal

14.2 Correctional Environment in Canada for Adults with FASD

Code is that sentencing is proportionate to the gravity of the offense and the degree of responsibility of the offender (Canada, Criminal Code of Canada, RSC, 1985). Both, researchers and government agencies acknowledge that the legal rights of people with disabilities in corrections facilities must be legally protected (Fast and Conry, 2004; Public Health Agency of Canada, 2008; Green, 2006; Brown, 2004; Cox, 2005). In the case of people with FASD, the ability of the person to comprehend the nature of the violation, the legal processes, and the consequences of their behavior may be severely limited, even if such limitations are not recognized at sentencing.

Offenders with FASD frequently encounter problems during incarceration. Prison culture is well known to be hierarchical, with a strong emphasis on the personal status of inmates. For adults with FASD, prisons are very "macho" environments, where being stigmatized as stupid or "retarded" has dangerous consequences. Victimization during incarceration, which includes psychological and physical abuse or assault (Miller, 2005; Malbin, 2004; Fast and Conry, 2004; Mitten, 2003), continues a cycle of trauma and abuse that is commonly experienced by people with FASD. People with FASD are likely to be used as scapegoats by other offenders in prison, subjected to the negative influence of peers, and exposed to recruitment by professional criminals (Malbin, 2004; Fast and Conry, 2004; Mitten, 2003).

Due to the potentially high percentage of FASD-affected inmates in the Canadian correctional system, and the extent of their cognitive and behavioral deficits, the establishment of units to deal with FASD-affected adults in Canadian prisons is essential (Boland *et al.*, 1998; Bell, Trevethan, and Allegri, 2004). Moreover, there remains a widespread ignorance of the invisible disabilities of FASD in the criminal justice system, and what that signifies for affected people (Chapman, 2008; Miller, 2005; Boland, Chudley, and Grant, 2002; Moore and Green, 2004; Burd *et al.*, 2003). Their problematic behavior is linked to their disability, and in order for correctional interventions to be effective, holistic environmental adaptations to the disability must be addressed (Malbin, 2004).

Acknowledging the realities of FASD-affected people in the Canadian correctional system will allow the development of targeted and more effective services. However, the development of successful interventions first requires diagnosis (Malbin, 2004; Chudley *et al.*, 2007). The research recommends that, for correctional institutions to deal with this population, it is vital that offenders be screened for FASD at intake into the system (Chapman, 2008; Boland, Chudley, and Grant, 2002; Burd, Martsolf, and Juelson, 2004; Chudley *et al.*, 2005; Astley, 2004; Clarren and Sebaldt, 2009). An FASD diagnosis allows officials to identify affected people so that potential relationships within the correction system which could be beneficial for the individual are strengthened (Chapman, 2008). A diagnosis also helps to ensure that people with FASD have access to programs and diagnostic services in the community after release. The FASD research clearly indicates that it is important to respond to the needs of people with FASD while they are in custody (Streissguth *et al.*, 2004).

14.2.1
Treatment Programs

The brain deficits in offenders with FASD, combined with the secondary disabilities associated with the spectrum, cause problems in functioning that make traditional prison treatment programs ineffective. It is an expensive use of scarce resources to have people enrolled in programs that cannot be effective. Moreover, a lack of response to programs by people with FASD can lead to their not being able to access prison programs because they have a history of not being helped by such programs (Fast and Conry, 2009; Chudley et al., 2007; Boland et al., 1998; Burd et al., 2003). The literature calls for corrections to do a better job with offenders with FASD. One way to accomplish this goal is to effect a change in offenders' behavioral skills and social circumstances before they are released into the community. The research calls for programs in prisons to address affected people's abilities (Chapman, 2008; Debolt, 2009; Alberta Health Services, 2009; SAMHSA, 2007). In addition, the prison programs need to be client-centered in order to assess each individual's strengths and weaknesses and to develop appropriate skills training and treatment. In order to attend to the different needs and functional abilities of people with FASD, existing programs in the corrections system require restructuring. Notably, this can be accomplished through focusing on meaningful outcomes, such as building adaptive skills.

The rehabilitation of offenders is an evaluative and capability-building process (Ward and Marshall, 2007). Current research indicates that rehabilitation programs, and the release plans developed for individuals, need to include constructive conceptions of positive lives (Ward, 2002). It is important to assess individuals' life histories to better understand their psychological dispositions and vulnerabilities, as well as the internal and external factors that may prevent them from meeting their primary needs (Ward, 2002). The multiple and comprehensive requirements of people with FASD in the correctional system require interdisciplinary teams to deliver programs (Evans and Brewis, 2008; Egger, Binns, and Rossner, 2009). The research on PAE conducted in schools indicates that a structured environment which is heavily oriented around order and routine is beneficial for people with FASD; developing integrated programs in the corrections system could help to address these requirements for adults (Chapman, 2008; Bell, Trevethan, and Allegri, 2004).

14.2.2
Recidivism and Alternative Sentencing

Numerous studies have noted the high rates of recidivism among offenders with FASD (Boland, Chudley, and Grant, 2002; Malbin, 2004; Fast and Conry, 2004; Chudley et al., 2007; Mitten, 2003). Recidivism among FASD-affected people occurs for many reasons. The brain structural deficits that cause difficulty with school and socializing continue to cause problems in the correctional system. People with FASD have difficulty adhering to parole conditions, which is a primary reason for the recidivism problems for this group (Fast and Conry, 2004; Chudley et al., 2007). Moreover, research indicates a link between a lack of community living skills and high rates of recidivism (Eggers et al., 2006; Lindstedt et al., 2004). For offenders with mental illnesses, a lack of community services after release can exacerbate their illnesses and contribute to recidivism (Sneed et al., 2006).

Considering the primary and secondary disabilities of people with FASD, a strong argument can be made for addressing recidivism among offenders with FASD through purposeful sentencing, improving skills, and attending to other issues such as mental illness. Research into chronic offenders acknowledges that high-needs offenders, such as people with FASD, are more likely to return to prison (Government of Canada, 2007), and that receiving services to meet those needs may act to reduce recidivism. However, this research also suggests that the criminal law paradigm is a less than appropriate way to deal with disabled offenders (Government of Canada, 2007).

Incarceration is based on principles of deterrence and rehabilitation, along with denouncing unlawful conduct. Incarceration is designed to deter people from breaking laws, and rehabilitation is intended to return offenders to the community in a better state than before the crime. These principles assume that offenders have the capacity to understand the nature and consequences of unlawful behavior and to enact changes in their behavior and personal circumstances. The brain damage caused by FASD makes deterrence from future crimes and rehabilitation unlikely, if not impossible (Malbin, 2004). Alternatives to incarceration for people with FASD, such as diversions, conditional sentences and sentencing circles, need to be considered (Fast and Conry, 2004; Fast and Conry 2009; Roach and Bailey, 2009; Mitten, 2004). The literature calls for environmental accommodations for affected people within the correctional system, much as would be considered standard for persons with other disabilities (Malbin, 2004). Chudley et al. (2007) have recommended alternative sentences and parole conditions that consider the disability caused by FASD.

Currently, the justice system in Canada may consider FASD as either a mitigating or an aggravating factor in sentencing offenders. An analysis of Canadian case law indicates a lack of consistency in the approach to offenders with FASD (Roach and Bailey, 2009; Justice Canada, 2009). In certain cases, offenders were found unfit to stand trial, some were sentenced as youths to adult correctional facilities, while others were labeled as dangerous offenders (Roach and Bailey, 2009; Justice Canada, 2009). FASD may be considered or mentioned by the court, but the brain impairments are frequently not given substantial weight during sentencing (FASD Ontario Justice Committee, 2007).

For offenders with FASD, it is essential for the justice system to focus on achieving long-term positive outcomes. Alternatives to incarceration are recommended in the literature for offenders with FASD who are not a risk to the public (Fast and Conry, 2009; Mitten, 2003; Debolt, 2009; Roach and Bailey, 2009). Suggestions for incarceration alternatives include holistic community-based programs that address both substance-abuse problems and mental health issues (Mitten, 2003). According to the research, the most purposeful sentences for people with FASD may be those that effect change in the person's life or social circumstances, rather than strictly aiming to change the person's behavior (Fast and Conry, 2009). Another factor to consider in the sentencing of FASD-affected offenders is their susceptibility to victimization while in custody (Chudley et al., 2007; Jones, 2007). Incarceration is not appropriate for many affected people because of the brain deficits caused by PAE, which impair learning and reasoning and influence behavior. Inmates with mental illnesses may accumulate disciplinary sanctions during incarceration, and thereby reduce their opportunities for parole (Baillargeon et al., 2009). In addition, the negative social networks extant in prison increase the probability of re-offending through post-release affiliations.

Reducing recidivism by offenders with FASD will require adjustments within the correctional system. Canadian government research acknowledges that incarceration is an inefficient and ineffective way of addressing recidivism in chronic offenders for whom prison is not a deterrent (Government of Canada, 2007). This research recommends that, for chronic offenders with cognitive impairments, investing resources in health and social systems may yield better results than repeated processing through the criminal justice system (Government of Canada, 2007). Chudley et al. (2007) have called for the allocation of appropriate resources within social services to improve the outcome and quality of life for affected people. Addressing the lack of fit between the abilities of people with FASD and release requirements will also reduce the costs of recidivism associated with FASD, and improve outcomes (Alberta, Government, 2007–2008; Every et al., 2000; Canada Government, 2009; Canada, Government, 2007).

The criminal justice system is currently limited in its ability to impose the type of sentencing that would be beneficial to people with FASD. Court jurisdictions were not intended to recommend the community-based holistic supports and services that help people with FASD live effectively in the community (Roach and Bailey, 2009).

14.2.3
Release Planning

Statutory release from custody is governed by a federal law that allows offenders who are not considered dangerous to serve the last one-third of their sentence in the community on parole to the Correctional Service of Canada (CSC). Parole is granted to offenders to facilitate rehabilitation and ease transition into the community. The criteria for granting parole includes the risk to society, the prisoner's post-release plan, criminal record, behavior in prison, and information provided by psychiatrists/psychologists, police, victims, and family. The CSC is responsible for preparing offenders for consideration by the National Parole Board. Upon release, prior offenders are required to report to a parole supervisor and to adhere to a variety of conditions, including curfews, restrictions on travel, movement and behavior, as well as prohibitions on alcohol consumption and associating with certain individuals. Should the conditions of release be violated, the offender may be returned to custody (Correctional Service of Canada, 2008).

Before an offender is released from custody, the individual must agree to a correctional plan, which outlines the procedures to maintain a law-abiding lifestyle in the community. The plan details restrictions on movement and commitments to participate in employment and programs. Each plan is individualized to the person's needs and focuses on specific issues, such as job training and substance abuse. Successfully re-engaging prior offenders in the community requires adequate supervision and effective community programs (Correctional Service of Canada, 2008). The inherent limitations faced by people with FASD can result in significant barriers to reintegration into the community for this group.

Community reintegration programs for offenders with brain damage, including those with FASD, are vital to success. However, the research identifies a general need for community reintegration programs in the corrections systems (Eggers et al., 2006), and a lack of continuity and program consistency between correctional sites and community settings. This can result in people with FASD "falling through the cracks" in service provision if bridging and transitions between the two environments are not in place.

Prior to the release of an offender with FASD into the community, planning for that eventuality is critical. Both pre-release and post-release plans for affected individuals need to be developed (Eggers et al., 2006; Sneed et al., 2006). Any treatment needs of individuals, such as for mental illness and substance abuse, need to be started while the person is in custody and then continued in the community upon release (Magaletta et al., 2009). Establishing contacts with a team of service providers in the community is therefore essential for affected people before release (Eggers et al., 2006). It is also important that probation and parole orders be interpreted to offenders with FASD. Otherwise, their learning disabilities may preclude adherence to the guidelines and increase their risk of re-offending (Fast and Conry, 2004). Research recommends that affected people receive comprehensive but simply written and meaningful discharge planning before being released into the community (Debolt, 2009).

Housing for offenders with FASD upon release appears to be a key to succeeding in the community. Stable housing for adults with FASD upon release from incarceration will also improve the delivery of necessary services. The correctional system has an opportunity to maximize the chance of successful reintegration for offenders with FASD by tailoring services and supports, programs, sentences and probation or parole orders to reflect the needs and functionality of the individual.

14.2.4
Correctional System Needs

The challenges facing the corrections system in addressing the functional disabilities caused by conditions such as FASD are increasingly being recognized by both researchers and governments. The complexity is in altering the existing system in such a way as to enable the needs of FASD-affected offenders to be successfully met within the corrections context, and then extending these services into the community upon the person's release. The FASD literature calls for purposeful sentencing that is focused on outcomes, and effecting changes in the behaviors and skills of this population while they are in custody.

A recent formal evaluation of federal Canadian correctional programs revealed that targeting the specific needs of offenders is both relevant and effective (Correctional Service of Canada, 2009). Offenders who participated in correctional service programs, such as for substance abuse, exhibited changes in behavior and were more likely to be granted a discretionary release than those who did not participate in programs (Correctional Service of Canada, 2009). Program participation was also associated with a reduction in re-admissions into the correctional system. However, the evaluation also revealed that the correctional system is significantly lacking in its ability to deliver programs to offenders with learning deficits, cognitive disabilities, and mental disorders. The report formally recommended developing a strategy to address the programming needs of such offenders (Correctional Service of Canada, 2009).

Studies indicate a need to develop a clear approach to FASD in the criminal justice system, and to develop clear practice guidelines (Cox, Clairmont, and Cox, 2008). FASD research calls for specific units to be established within the correctional system that are environmentally sensitive to the offenders' behavioral profile (Fast and Conry, 2009; Chudley *et al.*, 2007; Boland *et al.*, 1998; Bell, Trevethan, and Allegri, 2004; Burd *et al.*, 2003). Founding FASD units would enable the corrections system to address the cognitive and behavioral impairments of offenders with FASD. Training corrections personnel about FASD and its effects is also vital since, unless they have an awareness of the condition, such personnel are unlikely to recognize that potentially difficult behaviors are not deliberately chosen but are, rather, a result of brain impairment. Corrections system staff require both a knowledge of FASD impairments and the skills to interact effectively with offenders with FASD (Fast and Conry, 2009; Chudley *et al.*, 2007; Boland *et al.*, 1998; Bell, Trevethan, and Allegri, 2004; Burd *et al.*, 2003). In these ways, the neurobehaviors caused by the impairment can be addressed through modifying the social and physical environment experienced by affected people.

In addition to training for corrections personnel, programming is needed that targets meaningful and relevant roles for FASD-affected offenders after release. Corrections is an environment that can be viewed as both a barrier and a facilitator to participating in society. Because people do get released, it can play a vital role in preparing people to re-enter society (Canada, Criminal Code of Canada, RSC, 1985). Certain aspects of correction culture – structure, repetition, habit – are positive environmental supports that can enhance this pre-release preparation. Working with the FASD-affected population in prisons is an opportunity to connect with individuals who might otherwise "fall through the cracks" in the system.

14.3
Interventions and Social Supports for Adults with FASD after Release

Viewing FASD as a disability under the WHO International Classification of Functioning (ICF) framework allows us to see FASD as a condition that can benefit from many of the same strategies and principles used in developing services for other disability groups, such as those with acquired brain injury or spinal cord injury. In the case of FASD, there is some research indicating that the disabilities (both primary and secondary) may not be in a stable state (Moore and Green, 2004; Paley and O'Connor, 2009; Kodituwakku, 2009; Rasmussen et al., 2008; McGee et al., 2008). (This point will be discussed later, because it has implications for the design of a community service network.) Drawing on the prevention model (Caplan and Grunebaum, 1967), secondary prevention (which is sometimes called "treatment") and tertiary prevention (which can be viewed as "rehabilitation") are the focus of the activities both within corrections system and in the community.

Appropriate interventions and accommodations are vital for people with FASD if they are to remain in the community after being released from the correctional system. One major issue here is to establish seamless and continuous support from the correctional system to the community. Providing continuity in services requires that the correctional system establishes methods and policies to interact with community partners who will help individuals function and access services in the outside world. The limited knowledge regarding adults with FASD, and the challenges that they face in the community, dictates that interventions be specifically tailored to their individual profiles, but still be within the principles of general social service offerings.

14.3.1
Client-Centered Lifelong Multisectoral Supports

A client-centered approach recognizes that the specific characteristics and circumstances of individuals must influence service need and service delivery. The clients are the best persons to describe their experiences of reality, and as a result it behooves service providers to spend the necessary time to learn about the client's life experiences (Law, 1998). An essential element of client-centered practice is facilitating clients in solving their problems (Law, 1998). In the client-centered approach, clients may be involved in making choices about their rehabilitation, rather than having pre-established external decisions imposed on them. The approach allows for flexibility in program delivery and recognizes the particular requirements of the individual.

A client-centered approach has multiple advantages in delivering interventions for people with FASD, because of the highly variable nature of the brain deficits associated with the disorder. The needs of individuals with FASD depend in part on the extent of brain damage by PAE and on the secondary disabilities. Interventions that focus on the individual will be most able to address specific requirements (Chudley et al., 2007; O'Connor and Paley, 2009; Bertrand, 2009). The literature recommends the establishment of multidisciplinary teams to work with community partners to maximize the interventions for individuals, based on their needs and abilities (Grant et al., 2004; Chudley et al., 2007). By tailoring interventions to the needs of the individual, personally beneficial programs and treatments can be provided (Lindstedt et al., 2004; O'Connor and Paley, 2009).

The disabilities caused by FASD do not self-correct over time, and the FASD research indicates that lifelong interventions are necessary for affected people (Paley and O'Connor, 2009; Grant et al., 2004; Chudley et al., 2007; O'Connor and Paley, 2009; Bertrand, 2009). As a result, people with FASD need multisectoral coordinated services that are accessible across their lifespan (Grant et al., 2004). The involvement of multiple service providers in comprehensive interventions has a variety of benefits. Service providers generally follow different mandates to deliver services, which can result in a "silo" approach. Affected people can be lost to follow-up and monitoring if left alone to negotiate the support system. A comprehensive approach that offers coordinated care across multiple systems, including corrections, is important to maximize success for people with FASD (Paley and O'Connor, 2009). Furthermore, traumatic brain injury research reveals that rehabilitation involving an interdisciplinary team leads to improved function and independence (Evans and Brewis, 2008). Successful initiatives are multidisciplinary and multisectoral, and also involve partnerships with community agencies to provide services (Olson et al., 2009; Brown, 2004).

There are multiple barriers to accessing community services for adults with FASD. Affected individuals may not realize they are in need of formal support, and therefore do not seek it out (Debolt, 2009). In addition, they may not possess sufficient cognitive faculties to negotiate separate application processes, and adhere to the variety of system requirements developed for people without brain damage. For offenders with FASD, other impediments include a lack of coordination between the correctional system and programs available in the community (Hartwell and Orr, 1999). The literature identifies a need for community programs for newly released offenders, in part due to the prevalence of waiting lists for entry into treatment programs in communities (Brown, 2004). Further, establishing connections between the services that people receive in the correctional system and in the community will assist in informing service providers about the specific needs and goals of individual offenders (Magaletta et al., 2009).

14.3.2
Employment and Housing

For adults with FASD, an additional impediment to successful community establishment after release from prison is trouble finding employment and other forms of support (Brown, 2004). Other barriers include deficient work skills, gaps in employment records, difficulty accessing transportation to employment or community services, and the availability of health care, child care, and medication (Brown, 2004; Magaletta et al., 2009; Fonfield-Ayinla, 2009).

The availability of stable housing for newly released offenders with FASD is also crucial to success in the community. Homelessness worsens secondary disabilities, such as mental illness and substance abuse, even in people without the brain deficits caused by FASD (Fonfield-Ayinla, 2009; Zlotnick, 2009; Shand, 2004). Moreover, offenders may be released into unfamiliar communities, and with few financial resources (Brown, 2004). It is well known that limited financial resources have a negative influence on the ability to find and maintain a stable home. As a result, developing a comprehensive approach to intervention across multiple systems of care, including stable housing, is strongly advocated for people with FASD (Paley and O'Connor, 2009; Grant et al., 2004; Chudley et al., 2007; Bohjanen, Humphrey, and Ryan, 2009).

Secure and stable housing has multiple benefits for adults with FASD, and acts as a cornerstone to success in other programs. Housing can help people with FASD to protect against, cope with, and minimize secondary disabilities, such as mental illness (Burd et al., 2003; Brinda, 2006). Stability in housing also assists them in adhering to treatment goals (Debolt, 2009), and reduces recidivism in offenders with mental illnesses (Lindstedt et al., 2004; Case et al., 2009). The difficulties of affected individuals in tracking finances, controlling impulses and following rules necessitates support to ensure stability in housing, such as access to a case worker who can assess the person's needs and ensure that they are met (Brinda, 2006).

The variable and particular effects of FASD on individuals require that programs cover a wide range of housing options appropriate to the range of abilities and functionality, including specialized support and more independent programs. For a newly released offender with FASD, housing within the community is preferred over placement in halfway houses, due to the negative influence of other criminals on affected individuals (Brinda, 2006). Following a client-centered model for providing housing to newly released offenders with FASD is beneficial. A comprehensive umbrella program to provide housing for people with FASD has been recommended in recent research. Such a program would have the capacity to maintain housing stability while accommodating the changing needs of individuals.

14.3.3
Training and Programs

Social and vocational skills training is important for people with FASD. Adults with FASD may have significant social and vocational skill deficits because the impairments caused by PAE do not improve over time, and may even intensify (Moore and Green, 2004; Paley and O'Connor, 2009; McGee et al., 2008). Moreover, even if individuals have received social skills training during their incarceration, repetitive training is vital due to problems in short-term memory and information storage and retrieval (Grant et al., 2004).

The deficits in social problem-solving skills in adults with FASD also require specific interventions. Social problem-solving is affected by impairments in working memory, initiating and planning tasks, and organization and monitoring behavior. Interventions aimed at providing appropriate training and adequate academic and social support can improve the individuals' skills (McGee et al., 2008). Research into interventions in anger management and social skills programs for adults with FASD after release from prison suggests that the management of high-risk situations can be successful with appropriate interventions (Brinda, 2006).

Integrated treatment programs to address mental illness and substance abuse for people with FASD are important. The literature clearly indicates that the secondary disabilities of mental illness and substance abuse are common in people with FASD. Furthermore, mentally ill offenders are at risk of social isolation on release, which is known to worsen substance abuse-related problems (Hartwell and Orr, 1999). Considering that many offenders with FASD are affected by mental illness and addictions, linking treatment for these conditions will help to reduce feelings of isolation. Research into the experiences of mentally ill offenders strongly advocates for an integrated mental health and substance abuse treatment to improve outcomes (Magaletta et al., 2009; Roskes and Feldman, 1999).

Treating mental illnesses or substance abuse in people with FASD requires certain accommodations. Treatment approaches that focus on changing behaviors that are symptoms of FASD disability are inappropriate and ineffective (Malbin, 2004). Research indicates that, like people who have received a traumatic brain injury, those with FASD can be more sensitive to the medications used to treat mental illnesses, and may react to such medications in unexpected ways (Paley and O'Connor, 2009; Fast and Conry, 2009; O'Connor and Paley, 2009). People with FASD may also be resistant to medications and psychosocial therapy (O'Connor and Paley, 2009; Hellemans et al., 2008). These atypical reactions to typical interventions require that service providers monitor individuals closely in order to ensure that medications are prescribed and administered effectively, and that any negative side effects are minimized.

Interventions provided for people with FASD after release from corrections must be tailored to address the needs that are specific to people of their age and gender. Youth require access to education, vocational and life skills training, social skills training and to stable housing. Adaptations to the learning problems associated with FASD require environments of reduced stimulation, the use of visual schedules, repeated instructions, and positive behavioral support (Bohjanen, Humphrey, and Ryan, 2009). Women require access to birth control and child care, as well as certainty in release planning to aid in re-establishing relationships with their children (Pedlar et al., 2008). It is also important to screen for alcohol use in women during pregnancy (SAMHSA, 2007). The social and life skills training for women and men with FASD should reflect particular life circumstances, gender roles, culture and behaviors, such as anger management and parenting skills.

14.3.4
External Executive Function Support

People with FASD require the assistance of designated people to act as transitional navigators or advocates, mentors, advocates, and trustees. The problems that affected people have in learning, reasoning, judgment and adaptive skills frequently result in their becoming lost in the system, without assistance. They may have a limited insight into their lack of abilities, and over-represent their capabilities to themselves and others (Chudley et al., 2007). The FASD research recommends that an "external brain" in the form of formal caregivers and advocates be established to help affected individuals adapt, function, and meet their social needs (Chapman, 2008; Chudley et al., 2007; Boulding, 2007; Kellerman, 2003). Access to stable and funded contacts for service providers helps to improve adherence to recommendations and retain participants in programs (Grant et al., 2004; Debolt, 2009). Assistance in financial management is also necessary, and affected people should benefit from the establishment of trustees to manage their personal finances (Chudley et al., 2007). The experience of disability can be minimized with integrated approaches to optimize the person's capacities, strengthen their access to available resources, and improve their interaction with the environment (Stucki and Celio, 2007).

Accommodating the disabilities caused by FASD requires whole system modifications to the person's social and political environment. The ability of affected people to function in society can be improved through coordinated multisectoral targeted interventions, and the establishment of an "external brain" composed of caretakers, advocates, and trustees. Alterations in the environment to support and enhance the functionality of the person disabled by FASD need to be made at the community level, through programs and services provided by governments and community agencies.

One essential aspect of providing interventions to adults with FASD is to follow individuals through the system. When multiple systems of care provide services, it is vital to establish a process to follow the individual across services, to communicate between providers, and to ensure that the individual does not fall through any gaps in the services. Research indicates that, without follow-up, individuals can become lost in the system (SAMHSA, 2007; Hartwell and Orr, 1999). Recent FASD-associated literature calls for a full continuum of services to be available to affected individuals across the lifespan (Malbin, 2004; Olson et al., 2009; Paley and O'Connor, 2009; Fast and Conry, 2009; Chudley et al., 2007). Developing official procedures to follow affected people from the correctional system through the services supplied by multiple providers is critical.

14.3.5
FASD Costs

The economic costs of FASD are important to consider when evaluating the benefits of multisectoral assisted support across the lifespan for adult offenders with FASD. The overall costs of FASD are difficult to assess, due to the variety of factors that influence individuals. Furthermore, researchers acknowledge that the estimates of costs are minimum values, due to the unavailability of many types of data on individuals (Fast and Conry, 2009; Thanh and Jonsson, 2009; Stade et al., 2009). However, there are several basic categories of cost, including direct short-term and long-term costs of FASD to society, correctional system costs, and costs of homelessness.

Evaluations of the general costs of FASD for Canada and for Alberta indicate considerable expenditures at provincial and federal levels. In Canada, the cost of FASD from birth to 53 years of age is CA$5.3 billion (at 2007 price levels) (Stade et al., 2009). (For updated figures on the cost of FASD, see Chapter 4.) Over half of these costs are attributed to education and healthcare for affected children. This estimate does not include the costs of incarceration, or the cost of lost productivity among adults. The costs in Alberta of FASD amount to about CA$400 million annually in long-term costs, and about CA$143 million annually in short-term costs such as healthcare, education and the justice system (based on 2009 price levels) (Thanh and Jonsson, 2009).

Although the correctional system costs are not exclusive to people with FASD, the corrections system includes many with FASD. The cost of the corrections program in Alberta is CA$106.31 per person per day (CA$38 696 per year), while the federal corrections cost is CA$260.10 per person per day (CA$94 676 per year) (Alberta, Government 2008–2009). Despite these significant expenditures, the figures do not include all of the costs of FASD to the legal system, as there are also costs associated with the crime itself, policing, and court appearances (Fast and Conry, 2009). Boland, Chudley, and Grant (2002) have advocated that the costs associated with adult offenders with FASD be reduced by identifying affected individuals, so that the system could accommodate their disabilities.

The costs of homelessness are also significant. For example, in 2008 there were approximately 11 000 homeless people in the province of Alberta, while the provincial government spends annually an estimated CA$ 100 000 per person to deliver programs and services to chronically homeless people, such as emergency medical services. The cost of shifting from managing homelessness to ending homelessness in Alberta would save an estimated CA$ 7 billion over a 10-year period (The Alberta Secretariat for Action on Homelessness, 2008). Clearly, the avoidance of homelessness in people with FASD would lead to a major reduction in associated costs.

The extent of costs to society from FASD is substantial, even without an accurate breakdown of all the costs associated with the spectrum. The figures indicate, unmistakably, that to assist affected individuals in avoiding the correctional system and homelessness would not only greatly help the person but also have significant and lasting economic benefits to society.

14.3.6
Developmental Disability Assistance

The FASD research recognizes that a significant impediment to improving the community participation of people with FASD is the narrow criterion for disability assistance, which includes developmental disabilities (DD) and financial assistance (Streissguth et al., 2004; Grant et al., 2004; Bertrand, 2009). Chudley et al. (2007) have identified the need for access to developmental disability assistance for adults with FASD. Moreover, DD services can help in secondary prevention efforts. Burd et al., (2003) have identified the receipt of developmental disability services as a protective factor for avoiding the secondary disabilities associated with FASD. The literature clearly indicates the lasting and meaningful solutions that could be developed through providing access to developmental disability assistance to people with FASD, thus improving the quality of their lives. These supports may be just enough to keep some of these adults out of justice systems.

Alberta has developed a vision and innovative approaches to FASD in its ten-year strategic plan. There are a number of important ongoing projects in the province, and recently programs have been funded to address the needs of adults with FASD, one of which is Corrections and Connections to the Community (3C). The program has three components: evaluation, transition, and follow-up. Evaluation occurs within the correctional site, while transition begins with programming and relationship-building prior to release, and then continues into the community. The project also has an assertive follow-up phase.

Existing funded programs should be an exceptional source of data to expand the current knowledge of FASD. The annual reports that they submit to funders illustrate the importance of reporting requirements that go beyond accountability, and actually provide important information.

The above strategies are all reasonable and currently in existence in many areas of the community. The challenge is access to support for those people who are not in the greater urban areas and, more specifically, those returning to remote geographical areas with limited health and social services.

14.4 Policy Considerations for Adults with FASD

1) Comprehensive diagnostic capacity. In order to access services and supports to assist people with FASD to be contributing members of society, it is necessary to have access to a diagnosis. Diagnostic services for people who may be affected by FASD are needed for children, youth, and adults in all jurisdictions. The usual diagnostic clinics should also seek to evaluate the functional performance of individuals.

2) Seamless and equitable services across the lifespan. The brain impairments caused by FASD are lifelong. Thus, services that transcend the traditional "silo" approach, are multisectoral and connect the separate social, health and corrections systems are vital if people with FASD are to have access. It is important for these services to be available, regardless of the affected person's chronological age.

3) Expand Persons with Developmental Disabilities (PDD) legislation to incorporate the functional disabilities of FASD. Legislation that governs developmental disabilities assistance needs to be expanded to include adults with FASD. The legislation which guides eligibility and available services to people with developmental disabilities varies by province. Currently, the

leading provinces in Canada for providing inclusive disability support are Manitoba, Ontario, Prince Edward Island, and Saskatchewan. Alberta, British Columbia, the Northwest Territories and Quebec continue to link access to developmental disability support to narrow eligibility criteria: people with an IQ two standard deviations below the norm.

4) Transitions from child to adult services be pre-planned and allow for wraparound services, including follow-up, housing, and supported employment.

5) Sustainably funded services based on functional needs.

6) Ongoing life skills and socialization assistance.

7) FASD prevention efforts to target the social determinants of health.

8) Alternative sentences for offenders with FASD whenever possible.

9) Pre-release and post-release plans for offenders with FASD.

10) Enhance correctional environment to reflect the needs and functionality of offenders with FASD.

11) Reduce reconnecting with justice and correctional systems through provision of training, programs, and ongoing assertive supports.

12) Create a safe and supporting community-based "virtual world." Such a world has destinations, activity programs and shelters in which staff and volunteers understand and can manage the neurobehavioral presentation and profile of individuals with FASD. This world is interconnected and linked to resources and case management options across disparate service sectors.

References

Alberta Health Services (2009) *FASD Adult Assessment and Diagnosis Discussion Paper*, Alberta's Fetal Alcohol Spectrum Disorder Cross-Ministry Committee.

Alberta, Government (2007–2008) Annual Report Solicitor General and Public Security.

Alberta, Government (2008–2009) Annual Report Solicitor General and Public Security.

Aragon, A., Kalberg, W., Buckley, D., Barela-Scott, L., Tabachnick, B., and May, P. (2008) Neuropsychological study of FASD in a sample of American Indian children: processing simple versus complex information. *Alcohol. Clin. Exp. Res.*, 32 (12), 2136–2148.

Astley, S. (2004) Fetal Alcohol Syndrome prevention in Washington State: evidence of success. *Paediatr. Perinat. Epidemiol.*, 18, 344–351.

Baillargeon, J., Williams, B., Mellow, J., Harzke, A., Hoge, S., Baillargeon, G., and Greifinger, R. (2009) Parole revocation among prison inmates with psychiatric and substance use disorders. *Psychiatr. Serv.*, 60 (11), 1516–1521.

Bell, A., Trevethan, S., and Allegri, N. (2004) A needs assessment of federal Aboriginal women offenders. Correctional Service Canada Research Branch Report 156.

Bertrand, J. (2009) Interventions for children with fetal alcohol spectrum disorders (FASDs): overview of findings for five innovative research projects. *Res. Dev. Disabil.*, 30, 986–1006.

Bohjanen, S., Humphrey, M., and Ryan, S. (2009) Left behind: lack of research-based interventions for children and youth with fetal alcohol spectrum disorders. *Rural Spec. Educ. Q.*, 28 (2), 32–38.

Boland, F., Burrill, R., Duwyn, M., and Karp, J. (1998) *Fetal Alcohol Syndrome: Implications for Correctional Service*, Correctional Service Canada Research Branch, Corporate Development.

Boland, F., Chudley, A., and Grant, B. (2002) The Challenge of Fetal Alcohol syndrome in adult offender populations. *Correctional Service Canada, Forum Correct. Res.*, 14 (3), 61–64.

Boulding, D. (2007) Fetal Alcohol Spectrum Disorder and the Law: Some Practical Steps for Parents and Legal System Professionals. Available at: http://davidboulding.com/pdfs/21.pdf (accessed 29 June 2009).

Brinda, A. (2006) Housing for adults with FASD: towards development of a comprehensive program. Masters of Environmental Design thesis. University of Calgary, Calgary, AB.

Brown, J. (2004) Challenges facing Canadian federal offenders newly released to the community: a concept map. *J. Offender Rehabil.*, 39 (1), 19–35.

Burd, L., Selfridge, R., Klug, M., and Juelson, T. (2003) Fetal Alcohol Syndrome in the Canadian Corrections system. *J. FAS Int.*, 1, e14.

Burd, L., Martsolf, J., and Juelson, T. (2004) Fetal Alcohol Spectrum Disorder in the corrections system: potential screening strategies. *J. FAS. Int.*, 2, e1.

Canada, Charter of Rights and Freedoms 1982, Section 15 (1).

Canada, Criminal Code of Canada. R.S.C. 1985, c. C-46.

Canada, Government (2007) *National Roundtable on the Development of A Canadian Economic Model for the Impact of FASD*, Public Health Agency of Canada, Ottawa, ON. Available at: http://www.ndss-snsd.gc.ca/fasd-etcaf/publications-eng.php (accessed 20 November 2009).

Canada, Government (2008) The Changing Profile of Adults in Custody 2006/2007. Statistics Canada Juristat, http://www.statcan.gc.ca/pub/85-002-x/2008010/article/10732-eng.pdf (accessed 30 December 2009).

Canada, Government (2009) *Evaluation Report: Correctional Services Canada Correctional Programs*, Evaluation Branch Performance Assurance Sector.

Caplan, G. and Grunebaum, H. (1967) Perspectives on primary prevention:

a review. *Arch. Gen. Psychiatry*, **17**, 331–346.

Case, B., Steadman, H., Dupuis, S., and Morris, L. (2009) Who succeeds in jail diversion programs for persons with mental illness? A multi-site study. *Behav. Sci. Law*, **27**, 661–674.

Chapman J. (2008) Fetal Alcohol Spectrum Disorder (FASD) and the criminal justice system: an exploratory look at current treatment practices. Master of Arts Thesis. Simon Fraser University, School of Criminology, Burnaby, BC

Chudley, A., Conroy, J., Cook, J., Loock, C., Rosales, T., and LeBlanc, N. (2005) Fetal alcohol spectrum disorder: Canadian guidelines for diagnosis. *Can. Med. Assoc. J.*, **172** (5 Suppl.), S1–S21.

Chudley, A., Kilgour, A., Cranston, M., and Edwards, M. (2007) Challenges of diagnosis in fetal alcohol syndrome and fetal alcohol spectrum disorder in the adult. *Am. J. Med. Genet. C Semin. Med. Genet.*, **145C**, 261–272.

Cieza, A., Hilfiker, R., Chatterji, S., Kostanjsek, N., Ustun, B., and Stucki, G. (2009) The International Classification of Functioning, Disability and Health could be used to measure functioning. *J. Clin. Epidemiol.*, **62**, 899–911.

Clarren, S. and Sebaldt, R. (2009) Comprehensive Fetal Alcohol Spectrum Disorder Clinical Assessment Form. Confidential draft, The Canada Northwest FASD Research Network.

Connor, P., Sampson, P., Bookstein, F., Barr, H., and Streissguth, A. (2000) Direct and indirect effects of prenatal alcohol damage on executive function. *Dev. Neuropsychol.*, **18** (3), 331–354.

Correctional Service of Canada (2008) National Parole Board. http://www.csc-scc.gc.ca/text/organi/role-eng.shtml (accessed 30 December 2009).

Correctional Service of Canada (2009) The net Federal fiscal benefit of CSC programming. Report No. 208.

Cox, L. (2005) *FASD – Developing an Appropriate Response in the Justice System*, Nogemag Healing Lodge, New Brunswick.

Cox, L., Clairmont, D., and Cox, S. (2008) Knowledge and attitudes of criminal justice professionals in relation to fetal alcohol spectrum disorder. *Can. J. Clin. Pharmacol.*, **15** (2), e306–e313.

Debolt, D. (2009) *Creating Touchstones Support to Adults with Fetal Alcohol Spectrum Disorder*, Lakeland Centre for FASD, Alberta FASD Service Network.

Donohue, E. (2008) What once was sick is now bad: the shift from pathologized victim to deviant identity for those diagnosed with fetal alcohol spectrum disorder. Master of Arts thesis. Carleton University, Ottawa, ON.

Egger, G.J., Binns, A.F., and Rossner, S.R. (2009) The emergence of 'lifestyle medicine' as a structured approach for management of a chronic disease. *Med. J. Aust.*, **190** (3), 143–146.

Eggers, M., Munoz, J.P., Sciulli, J., and Hickerson Crist, P.A. (2006) The community reintegration project: occupational therapy at work in a county jail. *Occup. Ther. Health Care*, **20** (1), 17–37.

Evans, L. and Brewis, C. (2008) The efficacy of community-based rehabilitation programmes for adults with TBI. *Int. J. Ther. Rehabil.*, **15** (10), 446–458.

Every, N.R., Hochman, J., Becker, R., Kopecky, S., and Cannon, C. (2000) Critical pathways: a review. *Circulation*, **101**, 461–465.

FASD Ontario Justice Committee (2007) FASD and the Justice System, http://fasdjustice.on.ca/cases/summary-of-legal-literature.html (accessed 3 February 2010).

Fast, D. and Conry, J. (2004) The challenge of fetal alcohol syndrome in the criminal legal system. *Addict. Biol.*, **9**, 161–166.

Fast, D. and Conry, J. (2009) Fetal alcohol spectrum disorders and the criminal justice system. *Dev. Disabil. Res. Rev.*, **15**, 250–257.

Fonfield-Ayinla, G. (2009) Commentary: a consumer perspective on parenting while homeless. *Am. J. Orthopsychiatry*, **79** (3), 299–300.

Gearing, R., McNeill, T., and Lozier, F. (2005) Father involvement and fetal alcohol spectrum disorder: developing best practices. *J. FAS Int.*, **3**, e14–e25.

Golden, J. (1999) An argument that goes back to the womb: the demedicalization of fetal alcohol syndrome, 1973–1992. *J. Soc. Hist.*, **33** (2), 269–298.

Government of Canada (2007) Federal Provincial Territorial [FPT] Working Group on Chronic Offenders, Final Report.

Grant, T., Huggins, J., Connor, P., Pedersen, J., Whitney, N., and Streissguth, A. (2004) A pilot community intervention for young women with fetal alcohol spectrum disorders. *Community Ment. Health J.*, **40** (6), 499–511.

Green, M. (2006) A judicial perspective. Fetal alcohol syndrome disorders Symposium for Justice Professionals.

Hartwell, S.W. and Orr, K. (1999) The Massachusetts Forensic Transition Program for mentally ill offenders re-entering the community. *Psychiatr. Serv.*, **50**, 1220–1222.

Hellemans, K., Verma, P., Yoon, E., Yu, W., and Weinberg, J. (2008) Prenatal alcohol exposure increases vulnerability to stress and anxiety-like disorders in adulthood. *Ann. N. Y. Acad. Sci.*, **1144**, 154–175.

Jones, J. (2007) Persons with intellectual disabilities in the criminal justice system: review of issues. *Int. J. Offender Ther. Comp. Criminol.*, **51**, 723–733.

Justice Canada (2009) Highlights from Canadian Caselaw. Paths to Justice–Research in Brief, Department of Justice, http://www.justice.gc.ca/eng/pi/rs/rep-rap/2009/rb09/p2.html (accessed 25 January 2010).

Kellerman, T. (2003) External Brain. http://come-over.to/FAS/externalbrain.htm (accessed 29 June 2009).

Kodituwakku, P. (2009) Neurocognitive profile in children with fetal alcohol syndrome disorders. *Dev. Disabil. Res. Rev.*, **15**, 218–224.

Law, M. (1998) *Client Centered Occupational Therapy*, SLACK Inc., Thorofare.

Lindstedt, H., Soderlund, A., Stalenheim, G., and Sjoden, P. (2004) Mentally disordered offenders' abilities in occupational performance and social participation. *Scand. J. Occup. Ther.*, **11**, 118–127.

McFarlane, A. (2006) *Exploration of Persons with Developmental Disabilities (PDD) Services in Alberta for People with Fetal Alcohol Spectrum Disorder (FASD)*, Lakeland Centre for FASD, Alberta FASD Service Network.

McGee, C., Fryer, S., Bjorkquist, O., Mattson, S., and Riley, E. (2008) Deficits in social problem solving in adolescents with prenatal exposure to alcohol. *Am. J. Drug Alcohol Abuse*, **34**, 423–431.

MacPherson, P. and Chudley, A.E. (2007) FASD in a Correctional Population: Preliminary Results from an Incidence Study. Addictions Research Centre, Correctional Service Canada. Powerpoint presentation http://events.onlinebroadcasting.com/fas/090707/ppts/correctional.ppt (accessed 29 June 2009).

Magaletta, P., Diamond, P., Faust, E., Daggett, D., and Camp, S. (2009) Estimating the mental illness component of service need in corrections: results from the Mental Health Prevalence Project. *Crim. Justice Behav.*, **36**, 229–244.

Malbin, D.V. (2004) Fetal Alcohol Spectrum Disorder (FASD) and the role of family court judges in improving outcomes for children and families. *Juv. Fam. Court. J.*, Spring **2004**, 53–63.

May, P., Gossage, J., Kalberg, W., Robinson, L., Buckley, D., Manning, M., and Hoyme, H. (2009) Prevalence and epidemiologic characteristics of FASD from various research methods with an emphasis on recent in-school studies. *Dev. Disabil. Res. Rev.*, **15**, 176–192.

Miller, A. (2005) *Fetal Alcohol Spectrum Disorder and the Incarceration of Métis and Other Aboriginals*, Vancouver Métis Community Association. Available at: http://www.walkbravelyforward.com/FASD.pdf (accessed 30 November 2009).

Mitten, H.R. (2003) *Barriers to Implementing Holistic, Community-Based Treatment for Offenders with Fetal Alcohol Conditions*, University of Saskatchewan Department of Law, Saskatoon, SK.

Mitten, R. (2004) Section 9: fetal alcohol spectrum disorders and the justice system. First Nations and Metis Justice Reform Commission Final Report, volume 2.

Moore, T. and Green, M. (2004) Fetal Alcohol Spectrum Disorder (FASD): a need for closer examination by the criminal justice system. *Clin. Rep.*, **19**, 99–108.

O'Connor, M. and Paley, B. (2009) Psychiatric conditions associated with prenatal alcohol exposure. *Dev. Disabil. Res. Rev.*, **15**, 225–234.

Olson, H., Oti, R., Gelo, J., and Beck, S. (2009) 'Family Matters:' Fetal Alcohol Spectrum Disorders and the family. *Dev. Disabil. Res. Rev.*, **15**, 235–249.

Paley, B. and O'Connor, M. (2009) Intervention for individuals with fetal alcohol spectrum disorders: treatment approaches and case management. *Dev. Disabil. Res. Rev.*, **15**, 258–267.

Pedlar, A., Arai, S., Yuen, F., and Fortune, D. (2008) *Uncertain Futures: Women Leaving Prison and Re-Entering Community*, University of Waterloo, Department of Recreation and Leisure Studies, Waterloo, ON.

Public Health Agency of Canada (2008) *Federal Government Actions on Fetal Alcohol Spectrum Disorder (FASD)*, Public Health Agency of Canada.

Rasmussen, C., Andrew, G., Zwaigenbaum, L., and Tough, S. (2008) Neurobehavioural outcomes of children with fetal alcohol spectrum disorders: a Canadian perspective. *Paediatr. Child Health*, **13** (3), 185–191.

Reid, C., Greaves, L., and Poole, N. (2008) Good, bad, thwarted or addicted? Discourses of substance-using mothers. *Crit. Soc. Policy*, **28**, 211–234.

Roach, K. and Bailey, A. (2009) The relevance of fetal alcohol spectrum disorder in Canadian Criminal law from investigation to sentencing. *UBC Law Rev.*, **41**, 1–58.

Roskes, E. and Feldman, R. (1999) A collaborative community-based treatment program for offenders with mental illness. *Psychiatr. Serv.*, **50**, 1614–1619.

SAMHSA (2007) *Task 6: Identifying Promising FASD Practices: Review and Assessment Report*, Substance Abuse and Mental Health Services Administration, Fetal Alcohol Spectrum Disorders Center of Excellence, National Institutes of Health, U.S. Department of Health and Human Services.

Schonfeld, A., Mattson, S., and Riley, E. (2005) Moral maturity and delinquency after prenatal alcohol exposure. *J. Stud. Alcohol*, **July**, 545–554.

Shand, A. (2004) The lived experience of men residing in an inner city shelter for the homeless. Master of Science thesis. University of Alberta, Edmonton, AB.

Sneed, A., Koch, D.S., Estes, H., and Quinn, J. (2006) Employment and psychosocial outcomes for offenders with mental illness. *Int. J. Psychosoc. Rehabil.*, **10** (2), 103–112.

Stade, B., Stevens, B., Ungar, W., Beyene, J., and Koren, G. (2006) Health-related quality of life of Canadian children and youth prenatally exposed to alcohol. *Health Qual. Life Outcomes*, **4**, 81–91.

Stade, B., Ali, A., Bennett, D., Campbell, D., Johnston, M., Lens, C., Tran, S., and Koren, G. (2009) The burden of prenatal exposure to alcohol: revised measurement of cost. *Can. J. Clin. Pharmacol.*, **16** (1), e91–e102.

Streissguth, P., Bookstein, F., Barr, H., Sampson, P., O'Malley, K., and Young, J. (2004) Risk factors for adverse life outcomes in Fetal Alcohol Syndrome and Fetal Alcohol Effects. *Dev. Behav. Pediatr.*, **25** (4), 228–238.

Stucki, G. and Celio, M. (2007) Developing human functioning and rehabilitation research Part II: interdisciplinary university centers and national and regional collaboration networks. *J. Rehabil. Med.*, **39** (4), 334–342.

Thanh, N.X. and Jonsson, E. (2009) Costs of Fetal Alcohol Spectrum Disorder in Alberta, Canada. *Can. J. Clin. Pharmacol.*, **16** (1), 80–90.

The Alberta Secretariat for Action on Homelessness (2008) A Plan for Alberta: Ending Homelessness in 10 Years. http://www.housing.alberta.ca/documents/PlanForAB_Secretariat_final.pdf (accessed 7 July 2009).

Ward, T. (2002) Good lives and the rehabilitation of offenders: promises and problems. *Aggress. Violent Behav.*, **7**, 513–528.

Ward, T. and Marshall, B. (2007) Narrative identity and offender rehabilitation. *Int. J. Offender Ther. Comp. Criminol.*, **51** (3), 279–297.

Wass, T., Persutte, W., and Hobbins, J. (2001) The impact of prenatal alcohol exposure on frontal cortex development in utero. *Am. J. Obstet. Gynecol.*, **185**, 737–742.

World Health Organization (2008) Social Determinants of Health: Key Concepts, http://www.who.int/social_determinants/thecommission/finalreport/key_concepts/en/index.html (accessed 1 December 2009).

Zlotnick, C. (2009) What research tells us about the intersecting streams of homelessness and foster care. *Am. J. Orthopsychiatry*, **79** (3), 319–325.

15
Policy Development in FASD for Individuals and Families Across the Lifespan

Dorothy Badry and Aileen Wight Felske

15.1
Introduction

Fetal Alcohol Spectrum Disorder (FASD) is a complex, multidimensional disability creating a vexing social problem in terms of government response. Different policy responses are required for different sectors of government in answer to concerns for children, adolescents, adults and birth mothers who are often involved in one or several service delivery systems. Supporting individuals with FASD and their families across the lifespan has implications for policy development and coordination in the areas of health, children and family services, education, adult social welfare, and justice (Public Health Agency of Canada, 2006). In 2009, the Institute of Health Economics (IHE, 2009) consensus conference on FASD in Edmonton, Alberta offered an opportunity to focus on this critical social issue.

The importance of FASD as a public issue requires visibility at levels beyond professional disciplines such as medicine, social work and psychology, which are intricately linked to FASD through diagnostic and human service supports. The need to involve disciplines such as economics and social/healthcare planning takes FASD beyond a personal concern, and highlights public concern over this growing problem.

This chapter examines the critical need to develop a policy and practice that is consistent with the needs of individuals, families and communities and governments responding to, and struggling with the multilevel complexity of FASD.

The development of policy for support and intervention services being delivered to individuals and families living with FASD should have certain values embedded within its policies:

- To treat all Canadians with fairness and equity.
- To promote equality of opportunity for all Canadians.
- To value family, kinship and community as partners in state support and interventions.
- To support – not blame – the birth family if they are to continue their participation in interventions, and offer support to family members.

- To offer culturally appropriate supports and interventions.
- Geography should not be a barrier to timely responses in relation to health and social problems.

A human rights framework of equality supports the family and individual's rights to a range of inclusive support services driven by social policy. Policy needs to be clearly established for pre- and postdiagnosis of FASD across various intervening disciplines in relation to roles and functions, so as to minimize the risk of duplicating services.

Although FASD has been suggested to be a preventable disability, moral quandaries exist in Canadian society with regards to the role of the state in monitoring and intervening in individual behavior related to alcohol consumption. In theory, many social problems are preventable, yet FASD is differentiated due to the stigma associated with the disability related to the specific cause – that of alcohol consumption during pregnancy at levels that are serious enough to cause damage to the developing fetus. The problem of FASD represents a serious conflict between personal behavior and consequences for the community. Highlighting this point places the issue into a policy context and helps in recognizing that the cause of FASD is not simply the use of alcohol during pregnancy, but rather can be attributed to the complex social histories that precede the birth of a child with FASD (Badry, 2008; Poole, 2003). The connection between alcohol consumption during pregnancy and the subsequent FASD brain injury is not disputed; however, the state's ability to influence and/or intervene is unclear due to a lack of Canadian legislation in this area (Burgoyne, 2006). In the US, the Substance Abuse and Mental Health Services Administration (SAMHSA) Fetal Alcohol Spectrum Disorders (FASD) Center for Excellence has reported that 199 specific policies related to FASD have been introduced, and in some cases were passed in the house (Fetal Alcohol Spectrum Disorders Legislation Report, 2009). The driver for the multitude of American policy is often related to the criminalization of substance-using women, which can lead to incarceration while pregnant. Meanwhile, Canada offers a different response to alcohol use during pregnancy that is largely related to harm reduction (Rutman *et al.*, 2000).

Warren and Hewitt (2009) have traced the roots of the use of alcohol both medicinally and historically, while presenting the conundrum that inconsistent responses still exist within the medical community in relation to the use of alcohol during pregnancy. A clash between philosophies, human rights and personal freedoms, as noted by Warren and Hewitt, are reflective of the challenging moral climate surrounding FASD, despite solid science that has clearly established the linkage between *in utero* alcohol use and consequent disabilities in children (Jones *et al.*, 1973; Lemoine *et al.*, 1968; Lemoine, 2003). In order to consider a lifespan policy framework, this chapter will examine the topics of birth, childhood, adolescence/teenage years, adulthood, a disability paradigm for FASD, and cultural fairness in order to support a life trajectory policy model, while appreciating and considering FASD from a disability lens. Finally, policy and research recommendations are made.

> **Box 15.1**
>
> The CAN Northwest FASD Network Research Action team on women's health has representatives from British Columbia, Alberta, Northwest Territories, Saskatchewan and Manitoba. It is critical that intervention take place early and prior to birth or multiple births of children with FASD. The CAN Northwest FASD Network Research Action Team (NAT) on women's health has developed a mothering policy framework (2009) drawing on the work of co-lead Nancy Poole, (Executive Director, BC Center of Excellence in Women's Health) in collaboration with members of the NAT.

15.2 Birth

Every day, babies with FASD are born across Canada. The "Y" in the road may be a simple metaphor representative of FASD as a social problem. At the apex of the "Y," one road is reached that the mother will take, and one road that the child will take. But, what happens after birth? What factors come into play that result in the separation of mother and child? The provision of care for those with FASD is a trigger for discussions that should focus on health promotion, and not blaming women (Network Action Team on FASD Prevention from a Women's Health Determinants Perspective, CanNorthwest FASD Research Network, 2007) (see Box 15.1). A social justice perspective in the interest of children might restrict alcohol use during pregnancy. Another framework, of human rights, would support personal freedoms while leaning towards doing the least harm. From a feminist perspective, restricting personal freedoms such as alcohol consumption during pregnancy would be counter to the rights of women, although such action is more acceptable from a harm-reduction perspective (Poole, 2003; Armstrong, 1998, 2003). These points simply highlight the constraints related to the development of public policy in relation to FASD. Despite the arguments in relation to the freedom to choose to ingest alcohol or drugs during pregnancy, some studies support the concern that women do not intentionally set out to cause harm (Grant et al., 2003; Poole 2003 Badry, 2008; Boyd, 2004; Rutman et al., 2000; Boyd and Marcellus, 2007).

15.3 Childhood

Childhood is a critical arena for policy development, as children with FASD have high needs for resources within education and family support services. The existence of bio-health fragilities (which are endemic in FASD), alongside invisible neurological disabilities, leads to cascading vulnerabilities for the person with

FASD in the social world. The term "cascading vulnerabilities" is used by electrical engineers, and suggests that if one part of the system goes out, then the rest will follow in a cascading pattern (Badry, 2009). Children with FASD have existing predisposition and vulnerabilities to attention defection disorder and hyperactivity, drug and alcohol abuse and psychiatric disorders, and are also at risk for abuse and neglect over their lifespan.

The *active treatment hospital*—where diagnosis and early intervention so often originate—has a responsibility to build a trusting relationship with families, and to prepare the bridge between the health sector, children's services and education. Early intervention should follow a diagnosis, and timely referrals to early intervention programs are critical. While birth families may be involved in the support of their child with an FASD diagnosis, the reality is that many find the difficulties in their own lives, which are related to poverty, alcohol abuse, histories of trauma, and housing instability, overwhelm the possibility of taking on exceptional parenting. Through child welfare intervention, individuals representing the state—foster families—quickly become engaged in the parenting process. Aronson (1997) has suggested that:

> "... even though early fostering did not appear to eliminate the harmful effects of exposure to alcohol *in utero*, foster care seems to be the most favorable alternative for children whose biological mothers, continue to abuse alcohol and experience severe personal psychological problems. Children prenatally exposed to alcohol who remain in biological families ..."remain at continued risk" (1997, p. 24).

In Canada, governments provide services to children and families, generally under an umbrella of provincial or territorial child welfare. While this has resulted in a "broken road" for Canadians who wish to relocate, the differing systems from province to province does provide the researcher with a good lens with which to examine difference models. At present, only the Yukon has directly addressed the issue of FASD in its legislation (Children's Act Revision—Prevention/Early Intervention Policy Forum Paper Yukon, 2003). Alberta has developed an FASD 10-Year Strategic Plan published in 2008, and suggests that all government ministries include priorities regarding FASD within their operational plans (Alberta Government FASD 10-Year Strategic Plan 2008) and also established an FASD Cross Ministry Committee (see Box 15.2). One report that specifically calls for public policy development in relation to social determinants related to healthy pregnancy hails from British Columbia. This report not only reviews strategic initiatives outlined in 2003, but also outlines future plans and strategies strongly related to women's health in the following policy document: Understanding Fetal Alcohol Spectrum Disorder; Building on Strengths: A Provincial Plan for British Columbia (2008–2018).

Education is a key issue in childhood, and one that requires special attention in the school system in response to FASD. The educational needs of children with FASD are highly individual and, as a group, they are different from the majority of children receiving special education. Children, adolescents and adults exhibit a

> **Box 15.2**
>
> The FASD Cross Ministry Committee of the Government of Alberta has incorporated the following service areas in government in the 10 year strategic plan aimed at responding to and preventing FASD:
>
> Alberta Children and Youth Services – co-chair/administrative lead
> Alberta Health and Wellness – co-chair
> - Alberta Health Services/AADAC
> - Alberta Health Services/Alberta Mental Health Board
>
> Alberta Aboriginal Relations
> Alberta Advanced Education and Technology
> Alberta Seniors and Community Supports
> Alberta Education
> Alberta Employment and Immigration
> Alberta Justice and Attorney General
> Alberta Solicitor General and Public Security
> - Alberta Gaming and Liquor Commission
>
> Health Canada
>
> - Public Health Agency of Canada
> - First Nations and Inuit Health Branch
>
> This Cross Ministry Committee represents an outstanding example of various government ministries, provincially and federally and engages community stakeholders in sharing and exchanging resources and knowledge to develop relevant policy and practice. (Source: http://www.fasd-cmc.alberta.ca/home/index.cfm; http://www.child.alberta.ca).

range of co-occurring disabilities such as attention deficit and hyperactivity, behavioral disorders and a range of intellectual functioning (Clark et al., 2004). Public educational policy is a provincial concern, and Alberta has developed a specific curriculum for teachers educating children with FASD that has been updated on a regular basis as part of special education curriculum, and can be retrieved online (Alberta Education, 2004). The unique needs of children with FASD and other disabilities require the partnership of educators with a range of services drawn from medical and social policies that can then provide intervention and support in the schools. In the interest of supporting children with complex needs, and dealing with problems such as bullying, attendance, and serious behavioral concerns, the Government of Alberta has partnered with academic and community agencies (Mount Royal University; Teaching & Learning Center, University of Calgary; Society for the Treatment of Autism; LEAD Foundation and Media Learning Systems) to launch the Positive Behavior Supports for Children Website in

2010 (http://www.pbsc.info/). Resources of this nature tap into the world-wide-web, and can be a viable mechanism for the sharing of knowledge and resources relating to supporting children with complex needs.

15.4
Adolescence/Teenage Years

Adolescence is considered a very challenging time for youth with FASD, and also for biological, foster and adoptive families as well as other caregivers. Streissguth *et al.* (2004) offer the following portrait of adolescence. Clinical descriptions of patients with Fetal Alcohol Syndrome (FAS) and other alcohol-related disabilities suggest major problems with adaptive behavior (Streissguth *et al.*, 2004). Five operationally defined adverse outcomes and 18 associated risk/protective factors were examined using a Life History Interview with knowledgeable informants of 415 patients with FAS or other alcohol-related effects (median age 14 years; range 6–51 years; median IQ 86; range 29–126). Of the 415 participants in this study, 80% of the youth were not raised by their biological mothers. The lack of connection to their parents for youth with FASD is a critical finding in relation to life trajectory issues. For adolescents and adults, the lifespan prevalence was 61% for "disrupted school experiences," 60% for "trouble with the law," 50% for "confinement" (in detention, jail, prison, or a psychiatric or alcohol/drug inpatient setting), 49% for "inappropriate sexual behaviors on repeated occasions," and 35% for "alcohol/drug problems." Although, Streissguth and colleagues have identified daunting problems, the receipt of a diagnosis of FASD at an early age helps to mediate against some of the identified adverse life outcomes through establishing supports early in life.

As teenage years occur, involvement in the youth justice system, and as an adult in the federal system of justice, is a reality for many young people with FASD (Youth Criminal Justice Act of Canada, 2002). It is the specific neurological challenges of teenagers with FASD that lend to their involvement in criminal activity, in direct relation to their vulnerability in effectively responding to peer pressure and fears of rejection (Fast and Conry, 2000). It is during the teen years that critical breakdowns in service delivery occur, and youth "fall through the cracks." Recent studies have predicted dire life events for many young people with FASD (Gough and Fuchs, 2008), with eight key areas having been identified for youth requiring intervention to experience a smoother transition to adulthood: relationships; education; housing; life skills; identity; youth engagement; emotional healing; and financial support. The first seven of these areas are referred to as "pillars," while the final area – financial support – is considered to be the foundation on which all of those pillars are built. Each of these areas is interrelated and needs to work collaboratively so as to create the necessary supports for a successful transition to adulthood (Reid and Dudding, 2006). Often, this does not happen, and the new young adult moves from young offenders services to the adult justice system. The policy bridge from teen years to adulthood is not well established, and it is critical that all jurisdictions including provinces and territories define protocols for the transition of youth with FASD to the adult system.

15.5 Adulthood

In adulthood, it appears for policy purposes that services delivery and response to FASD has fallen into the realm of disability supports, although this is not always explicitly clear (Canadian Disability Agenda, 2003). These services and supports, which are delivered provincially and thus vary across the nation, have developed during the past two decades. Historical services with segregated institutions, and parental advocacy have resulted in the closure of most large institutions. Within the community, a range of social policies support adults who are unable to secure employment because of their disability; an example of this is Alberta's Assured Income for the Severely Handicapped (AISH) (http://www.seniors.gov.ab.ca/AISH/). Attention has also been focused on employment opportunities and supports, both federally funded and provincially based. Disability supports offers cost-shared technological devices related to specific disabilities, while health and well-being supports are delivered as home care for individuals, though this is often temporary in nature. Overall, this complex system maintains a minimal flexibility for what can be the differing needs of an individual with FASD who does not necessarily qualify for services, but has serious difficulties with memory and stability in terms of day-to-day functioning.

In a society and a system where expectations of independence influence social program development, FASD does not fit in smoothly. The criteria that qualify individuals for support are dependent upon stringent guidelines and the will of policy makers. For example, the current policy for AISH in Alberta employs criteria that consider a disability to be severe and prohibitive to earning a living. One of the challenges of this approach is that it is subjectively open to interpretation, particularly in response to determining eligibility for young adults with FASD. Even then, the policy response will differ between jurisdictions, such that no common national framework exists for adult support in relation to FASD. A forthcoming (2010) report on the Assessment and Diagnosis of FASD among Adults: A National and International Systematic Review (Badry and Bradshaw, 2010) from the Public Health Agency of Canada should offer a current review of practice and policy issues alongside an inventory of related research. The determination of eligibility and receipt of disability support for individuals with FASD are not only often time-consuming but also highly dependent on the voices of advocates to state the case for need on behalf of the young adult who has failed to receive such support on their own.

Mental health policy, social policy in relation to mental health services to persons with FASD, requires ongoing development. The mental health system has a key role with individuals struggling with social problems associated with FASD, which often manifest as co-occurring disorders. Youth who leave the child welfare system are highly likely to become involved with the adult judicial system. Individuals with FASD require lifelong managed or supportive care that is often unavailable when they transition to adulthood. Although desirable, this response inevitably fails individuals with FASD, as interventions must be purposeful in relation to FASD that prevents unique and challenging parameters that are not consistently understood throughout the service delivery system. Mental health is often a reflection of social health, and this is an important consideration in policy development. A national framework for FASD training does exist, under Alcohol and Drug Strategies (Canadian Center on Substance Abuse, 2009). However, although Public Health has been very strong in its initiatives, no federal minister has yet championed FASD as a cause in relation to developing federal policy.

The mental health systems, in being provincially based, are left with a key role in the case of individuals struggling with co-occurring disorders associated with FASD. Ironically, professional practice models are often based on client-driven goals towards independence, and do not appreciate the need for managed care. Policies developed for persons with FASD should examine and consider the studies on interdependent decision-making that were initiated by the Canadian Association on Community Living (http://www.cacl.ca/). It is necessary to have dialogues around the constructs of independence and interdependence in professional circles, as the subjective interpretation of independence may result in a misguided denial of service to those individuals who cannot advocate well for themselves, nor manage well in the absence of such services. Implications for the rehabilitation goal of the justice system are challenging. Disability services cannot fully respond to the unique needs of persons living with an FASD brain injury, who have a range of biological effects that cannot be ameliorated but can be supported effectively through the application of consistent supports related to day-to-day living and the mediation of social problems.

Vitale Cox, Clairmont, and Cox (2008) have identified the challenges within the justice system in responding to FASD, and have advised that further education within this system is required in order to respond to the unique problems of those individuals who, as offenders, become incarcerated. Clearly, there are gaps in the individual's ability to learn from past experience that leads to a cycle of reoffending and re-incarceration. This information may present a clue that a diagnostic process may be important for the individual, and the court has the capacity to order such as assessment. Surprisingly, Vitale Cox, Clairmont, and Cox (2008) identified a major issue relevant to a serious policy gap. In a review of court cases up to 2006, New Brunswick, Quebec, Nova Scotia, Prince Edward Island and Nunavut made no mention of FASD. Yet, the issue of FASD and the law was identified as a serious public issue in the leading-edge book, *Fetal Alcohol Syndrome and the Criminal Justice System* (Conry and Fast, 2000). The lack of a common Canadian policy and justice framework remains a concern for those whose lives become entangled with the law, and whose disabilities put them at risk in this system.

15.6
A Disability Paradigm for FASD

A new paradigm for treating individuals and families living with FASD needs to be developed that considers and appreciates the discovery of scientifically based medical research, and its poor fit in current provincial and federal policy. Social science research has not kept pace with biomedical research in this area (Badry, 2008). Responses to FASD that are structurally related to economic rationalism and managerialism do not work in the policy framework required to respond to FASD as a lifespan issue. Rather, the situation must be examined through a new lens, that acknowledges the similarity of FASD to other disabilities while appreciating novel differences, such as multilayered problems and the complex intergenerational social histories of birth mothers. Just as the FASD community begins to examine the supports based in community, the fiscal restraints within many provincial governments may limit these possibilities. For those individuals living with FASD, enhanced supports across their lifespan are required for a positive experience of community living. Indeed, without these supports, history threatens to repeat itself, with institutionalization (most likely in the criminal justice system) as the service model for individuals with FASD (Conry and Fast, 2000). In the arenas of policy for families of children with FASD, early intervention, child welfare, education, adult living, mental health and justice are all components of a lifespan management policy model that must be developed. A multilateral cross- and inter-disciplinary collaborative approach which offers a 24-7 delivery model is distinct, expensive, and is required for individuals with FASD.

Disability supports are delivered provincially, and vary across the nation. In the 2004 report, *Inclusion of People with Disabilities*, the major dimensions of inclusion were: disability support; skills development and training; employment; income; health and well-being; and the capacity of the disability community. In their National Report Card, 2008, for Canadians living with intellectual disabilities, the Canadian Association for Community Living (CACL) has established the following agenda, which could easily be considered relevant and applied to individuals with FASD:

1) To achieve equality rights and recognition.
2) To close institutions and assure a home in the community.
3) To secure child rights and needed supports.
4) To ensure that families have needed supports.
5) To achieve Inclusive Education.
6) To secure the right and access to disability supports.
7) To establish safe and inclusive communities.
8) To eradicate poverty for people with intellectual disabilities and their families.
9) To achieve employment equality.
10) To make a global impact on inclusion of intellectual disabilities.
    (Source: www.cacl.ca/)

Whilst professional practice models generically demand client-driven responses, individuals with FASD may need assistance in many aspects of decision-making.

Decisions related to independent living and financial management are particularly important supports. An interdependent model of decision-making needs to be used (CACL). Expectations of independence parameters challenge social practice while not appreciating the explicit need for managed care. Although desirable, this response inevitably fails individuals with FASD, as interventions must be purposeful in, and aimed at, the prevention of unique and challenging problems that are not consistently understood throughout the service delivery system. Young adults with FASD who have a range of biological effects that cannot be ameliorated can be supported effectively through the application of consistent supports related to day-to-day living and mediating social problems. Overall, the Disability Services are not able to fully respond to the unique needs of persons living with FASD, a brain injury with a range of biological effects that cannot be fully ameliorated.

Intensive supports that are life-enhancing can mediate against the vulnerabilities of FASD. In response to plans that extend beyond current policy allowance and case management, practice standards – particularly those allotted to hours per case being inadequate due to budgetary issues – require examination. For example, when taken as a model, it uses harm-reduction initiatives, strengthening supports during periods of increased vulnerability, and thus reduces the cost of any extraordinary support that is needed when things go wrong.

15.7
Cultural Fairness

The concerns of Aboriginal communities in relation to FASD and differing policy responses are not clearly addressed. For example, what about the over-representation of aboriginal children in care? Although FASD is not a uniquely Aboriginal issue, the need exists to develop a policy that is culturally relevant and specific, and addresses the needs of any community where FASD is identified. Policy and programs that are culturally defined are limited; those children who are Aboriginal, who have FASD and are in care, reflect (in part) a history of intergenerational trauma related to the residential school era. Bastien (2004) acknowledges the need to respect cultural protocols in relation to Aboriginal concerns. Pacey (2009), in association with the National Collaborating Center for Aboriginal Health, published a report entitled *Fetal Alcohol Syndrome and Fetal Alcohol Spectrum Disorder among Aboriginal Peoples: A review of prevalence*. This pivotal paper reports the following:

> "Other issues relating to FAS/FASD beyond prevalence have not been extensively explored. There has been silence, for instance, on what the long-term effects of FAS have been on Aboriginal communities as individuals with the syndrome age. There is no clear sense of how historically-deep the syndrome may be, and thus First Nations reserves may have been dealing with its unobserved consequences for decades." (p. 23)

It is critical, Pacey (2009) observes, that assumptions related to prevalence are not made without actual evidence to substantiate any claims of higher levels of FASD in Aboriginal communities, as general prevalence rates among non-Aboriginal populations have not been established. Reports originating from Aboriginal researchers and agencies could help inform public policy responses that are culturally sensitive. Federal organizations, such as the First Nations Inuit and Aboriginal Health Branch of Health Canada, also have a role in informing policy and practice (www.he-sc.gc.ca).

15.8
Life Trajectory Policy Model

A life trajectory policy model presents multiple challenges. First, the model must be lifelong, from birth to death. Second, for Canada this is a national challenge as different provinces and territories offer differing supports. Third, this is a cross-ministry challenge, even within a provincial system that occasionally adopts justice policy from the federal level (Fuchs et al., 2005). Thus, there is a need to consider the communication links between various departments of government, and also to consider the sharing of resources as a basis for a federal intervention strategy. Finally, cross-professional work is critical in the endeavor towards supporting individuals with FASD across the lifespan, and this construct leads to the identification of a need for a consistent cross-disciplinary professional educational agenda. For example, the Canadian Center on Substance Abuse (CCSA) has identified a training database alongside various relevant curriculum in an effort to standardize different levels of training (this work is in progress and requires further development). A broad cross-disciplinary framework recognizes that individuals with FASD require program support in many areas of their life, and supports the expansion in scope, and the development of intervention models designed to effectively meet these needs.

Policy in relation to FASD is at an evolutionary stage. From a holistic framework, policy must address children, adolescents and adults (see Box 15.3). Problems related to FASD have received differential responses in provinces and territories of Canada, due to varying resource availabilities, including funding infrastructures. One of the most problematic issues relevant to policy development is that of prevalence. Fuchs et al., (2009) suggested that an increased knowledge about the economic costs associated with FASD could positively influence the creation and development of public policy in response to the problem of prenatal alcohol exposure and consequent lifelong disabilities for children in care with FASD, and have offered a review of the complexities associated with estimating costs that might be representative of the "tip of the iceberg." The aim of this study was to examine the costs associated with health, education and child care for child welfare agencies within Manitoba, and to offer an informed perspective to support knowledge development related to the high costs associated with children with a diagnosis of FASD who were in care of the state. The only population that lends itself more easily to identification for the purpose of prevalence are children in the care of various children's services authorities, as no major Canadian prevalence studies have been conducted (due to systemic constraints) that have not supported adult diagnoses on a large scale. The diagnostic framework in Canada, which is driven primarily through the CanNorthwest FASD Research NAT on Diagnosis, supports a consistent diagnostic framework across various jurisdictions, based on the Canadian model for FASD diagnosis, and has established Canadian diagnostic guidelines (Chudley et al., 2005).

15.9
Conclusions

The infrastructure to respond to FASD as a disability with access to community living supports from childhood to adulthood exists in pockets across Canada. Knowledge, assessment, diagnosis and follow-up supports are evolving, but have not yet been fully established, and this supports the need to maintain a focus on FASD as a critical national issue for intervention/prevention. The policy issues presented by FASD are not limited to Canada. Indeed, globalization has facilitated the sharing of both knowledge and difficulties across different cultures. One example of such knowledge sharing was the International Circumpolar Conference on Health (2009) held in Yellowknife, NWT, where several members of the CanNorthwest FASD Research Network presented papers related to FASD. It is through educational conferences and courses of study with specific competencies related to understanding FASD, that all Canadians can be supported.

Box 15.3 Policy and Research Recommendations

- Clear diagnostic definitions are needed in social policy across the country, and across policy sectors.
- The development of discipline-specific standards of practice are required in response to FASD in psychology, social work, and other human services.
- Develop alcohol education modules, starting in elementary school, and continue through to high school and perhaps even to post-secondary education.
- Inform all people at all ages of the risks of alcohol consumption during pregnancy.
- Develop a framework on various programs that offer education to those on the front line to those in program administration.
- Engage a new model of lifespan trajectory planning at the outset of entry into the child welfare system or alternatives.

- Develop a policy that maximizes support and improves the quality of life for children with FASD who live with challenges over their entire life, while professionals move on to new cases, particularly in child welfare.

- No time limits on policy for FASD, as it is a lifespan diagnosis/issue that requires ongoing support at all ages.

- Examine, critically, those policies that require individuals to show improvement to maintain support services, and reduce contingency policies that can result in an individual losing funding due to stability, rather than maintaining funding and necessary supports.

- Create a channel of informational research that opens the door to best practice under a federal umbrella.

- All provinces and territories should have a children's advocate that is responsive to the needs of children and families.

- Develop a unilateral model of understanding FASD from a social paradigm that is consistent and easily applied with respect to specific geographic needs.

- Develop child welfare research on policy and practice that supports longitudinal cases studies; also develop an infrastructure in research policy frameworks that offer funding for longitudinal research.

- Provide research funding to examine differences in child and youth transition experiences, across provinces and territories.

- Develop models of inter-jurisdiction cooperation between federal/provincial, urban/rural, and between social services, education judicial and health authorities and their governance, in order to increase the sharing of best-practice programs.

- The need exists to examine and develop policy research on Mental Health's systemic ability to respond to FASD.

- Develop a current best-practices inventory that is gender-based, considers the needs of children and families and systemic/policy responses – both nationally and internationally.

References

Alberta's Assured Income for the Severely Handicapped (AISH), http://www.seniors.alberta.ca/aish/tipsheets/eligibility.pdf; http://www.ccsa.ca/Eng/Priorities/NationalFramework/Pages/default.aspx.

Alberta Cross Ministry Committee, http://www.child.alberta.ca/home/877.cfm (accessed 11 November 2009).

Alberta Government FASD 10 Year Strategic Plan (2008) http://www.fasd-cmc.alberta.ca/home/documents/

FASD_10yr_plan_FINAL.pdf (accessed 7 February 2010).

Alberta Education (2004) Teaching Students with Fetal Alcohol Disorder, http://education.alberta.ca/media/377037/fasd.pdf (accessed 8 February 2010).

Armstrong, E. (1998) Diagnosing moral disorder: the discovery and evolution of Fetal Alcohol Syndrome. *Soc. Sci. Med.*, **47**, 2025–2042.

Armstrong, E. (2003) *Conceiving Risk, Bearing Responsibility: Fetal Alcohol Syndrome & the Diagnosis of Moral Disorder*, Johns Hopkins University Press, Baltimore, MD.

Aronson, M. (1997) Children of alcoholic mothers: results from Goteborg, Sweden, in *The Challenge of Fetal Alcohol Syndrome: Overcoming Secondary Disabilities* (eds A.K. Streissguth and J. Kanter), University of Washington Press, Seattle, WA, pp. 15–24.

Badry, D. (2008) Becoming a Birth Mother of a Child with Fetal Alcohol Syndrome. Unpublished Doctoral Dissertation, University of Calgary, Canada.

Badry, D. (2009) Fetal alcohol spectrum disorder standards: supporting children in the care of children's services. *First People Child Fam. Rev.*, **4** (1), 47–56, http://www.fncfcs.com/pubs/vol4num1/Badry_pp47.pdf (accessed 7 February 2010).

Badry, D. and Bradshaw, C. (2010) *Assessment and Diagnosis of FASD among Adults: A National and International Systematic Review*, Public Health Agency of Canada.

Bastien, B. (2004) *Blackfoot Ways of Knowing: The Worldview of the Siksikaitsitapi*, University of Calgary Press, Calgary, AB.

Boyd, S. (2004) *From Witches to Crack Moms: Women, Drug Law and Policy*, Carolina Academic Press, Durham, NC.

Boyd, S. and Marcellus, L. (2007) *With Child: Substance Use during Pregnancy. A Woman Centred Approach*, Fernwood Publishing, Halifax, NS.

British Columbia FASD Ten Year Plan, http://www.mcf.gov.bc.ca/fasd/pdf/FASD_TenYearPlan_WEB.pdf (accessed 22 January 2010).

Burgoyne, B. (2006) *What We Have Learned: Key FASD Awareness Campaigns*, Public Health Agency of Canada, http://www.phac-aspc.gc.ca/publicat/fasd-ac-etcaf-cs/pdf/fasd-ac-etcaf-cs_e.pdf (accessed 7 February 2010).

Canadian Association on Community Living, http://www.cacl.ca/english/index.asp (accessed 8 February 2010).

Canadian Disability Agenda (2003) http://www.hrsdc.gc.ca/eng/disability_issues/goc_agenda/index.shtml (accessed 8 February 2010).

Canadian Centre on Substance Abuse, http://www.ccsa.ca/ENG/PRIORITIES/FASD/Pages/default.aspx (accessed 7 February 2010).

Children's Act Revision – Prevention/Early Intervention Policy Forum Paper Yukon (2005) http://www.yukonchildrensact.ca/downloads/policyforum/Prevention-Intervention.pdf (accessed 7 February 2010).

Chudley, A.E., Conroy, J., Cook, J.L., Loock, C., Rosales, T., and Leblanc, N. (2005) Fetal Alcohol Spectrum Disorder: Canadian guidelines for diagnosis. *Can. Med. Assoc. J.*, **172**, S1–S21.

Clark, E., Lutke, J., Minnes, P., and Ouellette-Kuntz, H. (2004) Secondary disabilities among adults with fetal alcohol spectrum disorder in British Columbia. *J. FAS Int.*, **2**, e13, http://www.motherisk.org/JFAS_documents/Secondary_Disabilities_Adults.pdf (accessed 11 February 2010).

Conry, J., and Fast, D. (2000) *Fetal Alcohol Syndrome and the Criminal Justice System*, The Law Foundation of British Columbia, Vancouver, BC.

Fast, D.K. and Conry, J. (2004) The challenge of fetal alcohol syndrome in the criminal Legal system. *Addict. Biol.*, **9** (2), 161–166.

Fetal Alcohol Spectrum Disorders Legislation Report (October 2009) http://fasdcenter.samhsa.gov/documents/FASD_Related_Legislation_11_12_2009.pdf (accessed 29 January 2010).

First Nations Inuit and Aboriginal Health Branch of Health Canada, http://www.hc-sc.gc.ca/fniah-spnia/finance/

agree-accord/prog/index-eng.php (accessed 11 February 2010).

Fuchs, D., Burnside, L., Marchenski, S., and Mudry, A. (2005) *Children with Disabilities Receiving Services from Child Welfare Agencies in Manitoba*, Centre of Excellence for Child Welfare, Winnipeg, MB, http://www.cecw-cepb.ca/publications/577.

Fuchs, D., Burnside, L., De Riviere, L., Brownell, M., Marchenski, S., Mudry, A., and Dahl, M. (2009) Economic Impact of Children in Care with FASD and Parental Alcohol Issues Phase 2: Costs and Service Utilization of Health Care, Special Education, and Child Care, http://www.cecw-cepb.ca/publications/1146 (accessed 10 February 2009).

Gough, P. and Fuchs, D. (2008) *Transitions Out-of-Care: Youth with FASD in Manitoba*, Centre of Excellence for Child Welfare, Toronto, ON, http://www.cecw-cepb.ca/sites/default/files/publications/en/FASDTransitions67E.pdf (accessed 7 February 2009).

Grant, T., Ernst, C., Pagalilauan, G., and Streissguth, A. (2003) Post program follow-up effects of paraprofessional intervention with high-risk women who abused alcohol and drugs during pregnancy. *J. Community Psychol.*, **31** (3), 211–222.

Institute of Health Economics (IHE) Consensus Conference on Fetal Alcohol Spectrum Disorder Across the Lifespan (2009) http://www.buksa.com/fasd/ (accessed 7 February 2010).

Jones, K., Smith, D., Ulleland, C., and Streissguth, A. (1973) Pattern of malformation in offspring of chronic alcoholic mothers. *Lancet*, 1267–1271.

Lemoine, P. (2003) The history of alcoholic fetopathies. *J. FAS Int.*, **1**, e2, http://www.motherisk.org/JFAS_documents/History_Alcoholic_Fetopathies.pdf (accessed 7 February 2010).

Lemoine, P., Harousseau, H., Borteyru, J.-P., et al. (1968) Les enfants de parents alcooliques: Anomalies observées. A propos de 127 cas (Children of alcoholic parents: Abnormalities observed in 127 cases). *Ouest Med.*, **21**, 476–482.

Mount Royal University Positive Behavior Supports for Children website 2010, http://www.pbsc.info/ (accessed 11 February 2010).

Network Action Team on FASD Prevention from a Women's Health Determinants Perspective, CanNorthwest FASD Research Network (2007) *Coalescing on Women and Substance Use: Linking Research, Practice and Policy Information Sheet: Women-Centred Approaches to the Prevention of FASD: Barriers to Accessing Support for Pregnant Women and Mothers with Substance Use Problems*, Network Action Team on FASD Prevention from a Women's Health Determinants Perspective, CanNorthwest FASD Research Network. Available at: http://www.coalescing-vc.org/virtualLearning/community3/documents/MotheringandSubstanceUse-InfoSheet1.pdf.

Canadian Association for Community Living (2008) National Report Card 2008: Inclusions of Canadians with Intellectual Disabilities, http://www.cacl.ca/english/documents/ReportCards/2008ReportCard_Nov26.pdf (accessed 7 February 2010).

Pacey, M. (2009) *Fetal Alcohol Syndrome and Fetal Alcohol Spectrum Disorder among Aboriginal Peoples: A Review of Prevalence*, National Collaborating Centre for Aboriginal Health, Prince George, BC, Canada. www.nccah.ca (accessed 11 February 2010).

Poole, N. (2003) *Mother and Child Reunion: Preventing Fetal Alcohol Spectrum Disorder by Promoting Women's Health*, BC Centre for Excellence in Women's Health, Vancouver, BC.

Public Health Agency of Canada (2006) Summary Report: National Thematic Workshop on FASD, http://www.phac-aspc.gc.ca/publicat/fasd-ntw-etcaf-atn/index-eng.php (accessed 7 February 2010).

Reid, C. and Dudding, P. (2006) *Building A Future Together: Issues and Outcomes for Transition-Aged Youth*, Centre of Excellence for Child Welfare League of Canada, http://www.cecw-cepb.ca/children-youth-care/transitions-adulthood (accessed 11 February 2010).

Rutman, D., Callahan, M., Lundquist, A., Jackson, S., and Field, B. (2000) *Substance*

Use and Pregnancy: Conceiving Women in the Policy Making Process, Status of Women Canada, Ottawa, ON.

Streissguth, A.P., Bookstein, F., Barr, H., Sampson, P., O'Malley, K., and Young, J. (2004) Risk factors for adverse life outcomes in fetal alcohol syndrome and fetal alcohol effects. *J. Dev. Behav. Pediatr.*, **25** (4), 228–238.

Understanding Fetal Alcohol Spectrum Disorder: Building on Strengths: A Provincial Plan for British Columbia (2008–2018), http://www.mcf.gov.bc.ca/fasd/pdf/FASD_TenYearPlan_WEB.pdf (accessed 22 January 2010).

Vitale Cox, L., Clairmont, D., and Cox, S. (2008) Knowledge and attitudes of criminal justice professionals in relation to Fetal Alcohol Spectrum Disorder. *Can. J. Clin. Pharmacol.*, **15** (2), e306–e313, http://www.cjcp.ca/pubmed.php?issueId=129 (accessed 7 February 2010).

Warren, K., and Hewitt, B. (2009) Fetal alcohol spectrum disorders: when science, medicine, public policy and laws collide. *Dev. Disabil. Res. Rev.*, **15**, 170–175. doi: 10.1002/ddrr.71, www.interscience.wiley.com (accessed 2 February 2010).

Youth Criminal Justice Act of Canada 2002, http://www.justice.gc.ca/eng/pi/yj-jj/ycja-lsjpa/back-hist.html (accessed 8 February 2010).

16
The Impact of FASD: Children with FASD Involved with the Manitoba Child Welfare System

Linda Burnside, Don Fuchs, Shelagh Marchenski, Andria Mudry, Linda De Riviere*, Marni Brownell, and Matthew Dahl*

16.1
Introduction

Fetal Alcohol Spectrum Disorder (FASD) encompasses a range of conditions that are caused by maternal alcohol consumption during pregnancy, and which has lifelong implications for the affected person, the family, and society in general. Although considered to be a preventable condition (Zevenbergen and Ferraro, 2001), the adverse effects of the maternal consumption of alcohol have been noted throughout history, and were first described as a pattern of disabling effects under the term "Fetal Alcohol Syndrome" (FAS) during the early 1970s (Overhoser, 1990). Although no national statistics are currently available regarding the rates of FASD in Canada, the incidence of the condition in Manitoba has been estimated as ranging between 7.2 per 1000 live births (Williams, Obaido, and McGee, 1999) and 101 per 1000 live births (Square, 1997).

Because of the range of effects that result from prenatal alcohol exposure, the diagnosis of FASD can be complex (Chudley et al., 2005; Hay, 1999; Wattendorf and Muenke, 2005; Zevenbergen and Ferraro, 2001). Indicators include physical characteristics (such as distinct facial features and inhibited growth), neurodevelopmental problems (such as impaired fine motor skills), and behavioral and cognitive difficulties that are inconsistent with developmental level (such as learning difficulties, poor impulse control, or problems in memory, attention or judgment), often in conjunction with a confirmation of maternal alcohol use. Although the diagnosis is often most easily made between the ages of four and fourteen years (Lupton, Burd, and Harwood, 2004), an early diagnosis and intervention is strongly recommended so as to ameliorate the negative effects of FASD, through the provision of cognitive stimulation, speech and language therapy, educational supports, and other interventions (Sonnander, 2000).

The effects of FASD are manifested throughout the individual's lifespan (Streissguth et al., 1999; Zevenbergen and Ferraro, 2001). Infants who have been

*Contributed to Chapter 16.6 only.

Fetal Alcohol Spectrum Disorder–Management and Policy Perspectives of FASD. Edited by Edward P. Riley, Sterling Clarren, Joanne Weinberg, and Egon Jonsson
Copyright © 2011 WILEY-VCH Verlag GmbH & Co. KGaA, Weinheim
ISBN: 978-3-527-32839-0

exposed to alcohol *in utero* may show decreased arousal, sleeping problems, irritability, and feeding difficulties. Difficulties with speech, language development, and attention span are often also identified in preschool years. Poor attention, impulsivity, and hyperactivity often persist throughout childhood and adolescence, leading to behavioral issues that arise in school settings, and which only exacerbate the academic challenges that stem from learning disabilities and other cognitive impairments related to FASD. These academic and social difficulties often contribute to low self-esteem, to conduct problems, and to delinquent behaviors in adolescence. As adults, individuals with FASD are vulnerable to mental health problems, conflict with the law, alcohol and drug issues, and problems with employment (Streissguth *et al.*, 1996).

In addition to the personal implications for a person affected with FASD, the societal impact of FASD is profound (Lupton, Burd, and Harwood, 2004). Individuals with FASD often require high levels of medical care, residential services, special education supports, adult vocational services, and other social services throughout their lifetimes. The increased risk for deleterious outcomes in adulthood as a result of FASD (i.e., unemployment, homelessness, poverty, criminal activity, incarceration, and mental health problems) all have a social cost in terms of the support services, organizational structures, and associated financial costs that must be provided to respond to the needs of this vulnerable population (Lupton, Burd, and Harwood, 2004).

During the past decade, the professional literature has begun to detail more clearly the relationship between parental substance use and involvement with the child welfare system, through the identification of three affected populations: (i) the families involved with child welfare due to parental substance abuse; (ii) the children who are victims of maltreatment as a result of parental substance abuse; and (iii) children who come into care due to their prenatal exposure to alcohol (Young, Boles, and Otero, 2007). Early studies based on child welfare data acquired during the 1990s estimated the percentage of families affected by parental substance abuse and involved with child welfare to range from 40% to 80% (Besinger *et al.*, 1999; Curtis and McCullough, 1993; Department of Health and Human Services, 1999; Dore, Doris, and Wright, 1995; McNichol and Tash, 2001; Semidei, Radel, and Nolan, 2001; Young, Gardner, and Dennis, 1998). With parental substance abuse and its relationship to child abuse and neglect being one of the major reasons for the involvement of the child welfare system with families (Barth, 2001; Bartholet, 1999), it is not surprising that many children with FASD come into out-of-home care, often on a permanent basis (Jones, 1999).

In their study of children in care with disabilities in child and family services agencies in Manitoba, Fuchs *et al.* (2005) found that 11% of the total number of children in care on 1st September 2004 were diagnosed with FASD. Further, a considerable number of children were suspected of having the condition, as they were in the process of being tested for FASD or were receiving services consistent with the diagnosis. Combining those children who were currently being tested for FASD with those who had already been diagnosed, inferred that 17% of all children in care were affected by the condition. Yet, given the challenges associated with diagnosing FASD, this is considered to be a conservative estimate.

This concerning proportion of children in care with FASD in Manitoba inspired further research into the context and circumstances of these children's lives. In this chapter, a summary is provided of five studies of children with FASD in care of child welfare agencies in Manitoba, conducted by the present authors between 2005 and 2009 through a joint initiative between the Faculty of Social Work at the University of Manitoba and the Manitoba Department of Family Services. The study was conducted under the auspices of the Prairie Child Welfare Consortium, and funded by the Public Health Agency of Canada through the Center of Excellence for Child Welfare. Additional partnerships with the University of Winnipeg and the Manitoba Center for Health Policy were instrumental to completion of the fifth research project on the economic impact of FASD on children in care in Manitoba.

16.2
Study One: Children in Care with Disabilities

The initial exploratory study by Fuchs *et al.* (2005) was intended to advance the understanding of the nature and scope of disabilities affecting children in child welfare care in Manitoba. Children with disabilities were defined as those children in whom the ability to participate in age-appropriate activities of daily living was compromised by limitations in one or more areas of functioning. This definition was broad enough to include children with congenital conditions, complex medical needs, chronic psychological or mental health concerns, Fetal Alcohol Spectrum Disorder (FASD) and/or learning difficulties.

One-third ($n = 1869$) of the children in care in Manitoba in this study were found to have a disability. First Nations children comprised just over two-thirds (68.7%) of children with disabilities; their representation in the disability population approximated their representation in the overall child in care population. Most children with disabilities were permanent wards (69%), but a significant proportion (13%) was in care under a Voluntary Placement Agreement (VPA). The proportion of permanent wards was somewhat greater among First Nations children. The most frequently cited reasons for children with disabilities coming into care were related to the conduct or condition of their parents. Children in care under a VPA were the exception. Approximately one-half of those children were in care for reasons related to the conduct or condition of the child. Most children (75%) were placed in foster homes, and only 2% required hospital or residential care at the time of the study. The proportion of children requiring more intensive care was greater among those under a VPA (41%) than among those who were permanent wards (16%).

Specific disabilities were organized within six main categories: intellectual; mental health; medical; physical; sensory; and learning. The most common disabilities were intellectual, which affected 75.1% of the children with disabilities, and mental health, which impacted 45.8%. More than half of the children had more than one type of disability (58.1%), while the most common combination of disabilities was intellectual disabilities and mental health conditions. FASD was diagnosed in one-third of children with disabilities (34.2%), or 11% of all children in care. Children with a mental health diagnosis were almost always (95%) given a diagnosis that fell into the Attention Deficit/Disruptive Behavior Disorders group. Attention Deficit Disorders were the most frequently diagnosed (73%); FASD and Attention Deficit Hyperactivity Disorder (ADHD) were coincident in 39.1% of children with an FASD diagnosis. The remaining disability types affected smaller proportions of children with disabilities: medical disabilities 22%; physical disabilities 18%; sensory disabilities 5%; and diagnosed learning disabilities 3%.

The majority of disabilities (51%) resulted from an unknown cause. The second most common cause of disability was substance abuse, which affected 34% of the disability population. Further, substance abuse was a suspected cause for an additional 17% of the children with disabilities.

The overwhelming majority of children with FASD (89%) were in permanent agency care, with 61% placed due to conduct or conditions of the parent. By comparison, among the non-FASD population 54% of children were in care for reasons related to parents. Children with FASD had limited contact with parents; typically, 46% had no contact with parents, but for those maintaining contact with parents 12% had regular contact (monthly or more), 4% had regular contact (but less than monthly), and 17% had irregular contact. By way of comparison, in the non-FASD group 24% had no contact with parents, and 24% had regular monthly or more contact. Regular but less than monthly contact involved 4% children in the non-FASD group, and an additional 20% had irregular contact.

The results of this study revealed the concerning proportion of children in care in Manitoba with FASD, and signaled some key differences in their characteristics and experiences compared to other children in care with disabilities. Examining these differences became the focus of the second study. The emerging recognition of the impact that FASD was having on the child welfare system also became evident, and furthered the importance of research in this area.

16.3
Study Two: The Trajectory of Care for Children with FASD

Because of the significant proportion of Manitoba children in care identified with a diagnosis of FASD, understanding the relationship between this population and child welfare agencies is particularly important. The second study (Fuchs *et al.*, 2007) utilized the identified population of children in care in Manitoba from the previous project, from which random samples were created: (i) children diagnosed with FASD; (ii) children with a disability that was not FASD; and (iii) children with no disability. The children were grouped according to their legal status on 1st December 2005, and their records of child welfare care were analyzed in terms of changes in legal status over time (a variable that traces a child's general path of care in child welfare) and placement histories (a variable that identifies changes in placement). Although VPAs were frequently used for other groups of children, they were seldom used for children with FASD. Therefore, due to the majority of children with FASD who were permanent wards, this summary will focus on the trajectory of care for those children with a permanent ward legal status. It is important to note that the diagnosis of FASD was generally made after the child's admission to care, and therefore did not play a direct role in the first admission to care.

The data showed clearly that children with FASD had come into care at least a year younger than any other group of children, at a mean age of 2.5 years. These children became permanent wards more quickly than those with no disability or with a non-FASD disability, on average by age 4.6 years. This mean age of becoming a permanent ward was two years younger than children with no disabilities, and three years younger than children with other disabilities.

Further, at just over two years on average between their first legal status and becoming a permanent ward, children with FASD became permanent wards almost a year sooner than other children who became permanent wards. To an extent, this is an expected outcome, as provincial legislation requires child welfare agencies to pursue permanent plans, including permanent orders of guardianship, more quickly when children enter care in the preschool years than with children who are admitted at an older age, to ensure that the normative need for stable and consistent caregivers in early childhood is adequately met.

Because the age of children is a factor that influences the length of their total time in care, a more accurate comparison of the relative time spent in care was made by comparing the children's time in care as a proportion of their age. Children with FASD in the sample were found to have spent over 70% of their lives as a proportion of their age in care, compared to approximately 60% of their lives for the other two permanent ward groups.

Placements followed a similar pattern. Permanent wards with FASD spent the most time in child welfare agency placements (7 years), in comparison to permanent wards with other disabilities (6 years), and permanent wards with no disabilities (5 years). Possibly as a consequence of being in care longer, they had a higher number of placements than other permanent wards. The mean length of their placements was also greater than that of other permanent wards. There was no evidence that children with FASD were any more disadvantaged in terms of placement changes than other children in care.

Both, the legal and placement histories confirmed that permanent wards with FASD spend on average close to three-quarters of their lives in the care of an agency – about 15% more than any other children who are permanent wards. This makes them more reliant than any other group of children examined in this study on the parenthood of the state, which in turn creates a range of implications for service delivery, policy and prevention for child welfare agencies, and also for society.

Social workers must be aware of the possibility that children for whom they are providing service may be alcohol-affected. Because children with FASD come into care earlier and spend more of their life in placement, workers must recognize the critical role of lifelong planning for their needs. There are some services available in the community for children with FASD and/or their families. Workers need to know about the availability of services in their region, and be able to advocate for FASD-related services for both children in care and their caregivers.

Similarly, expertise related to FASD is critical for foster parents and other direct service providers, who must be prepared to manage the unique needs of children with this condition. When recognizing the long-term placement needs of these children, foster parents must be able to make a long-term commitment to their care. They must also be aware of the additional stresses that may result from caring for children with FASD, and develop some reliable stress management strategies.

The long-term nature of providing child welfare care to children with FASD makes it clear that it is not enough to plan for the needs of children only while they are in care. Every year, increasing numbers of children who have been identified with FASD will be transitioning out of care and into the community. The shift to independence is difficult for all permanent wards. The move to independence for persons with FASD is further complicated by the nature of their disability. They are often not eligible for services related to cognitive impairments because their level of intellectual functioning is above the eligibility criteria. There are few adult services directly related to FASD. As adults, their disability tends to be invisible, but their behavior can present many challenges. Long-term planning for children with FASD needs to include special attention to their transition into adulthood.

To better understand the needs of youth with FASD who are approaching adulthood, the next study by the authors focused on the experiences of permanent wards with FASD at age of majority.

16.4
Study Three: Youth with FASD Leaving Care

The age of leaving home has increased steadily over the past decade. Today, it is commonplace for youth to remain economically and emotionally dependent on their parents until well into their twenties (Reid and Dudding, 2006). The termination of child welfare support at the age of majority serves to increase the risks of an already disadvantaged group. Considerable research on children leaving care has documented their increased risk for failing to complete high school, living in poverty, homelessness, early pregnancy, and mental health issues; for youth in care with disability, the risks are even more daunting. Moreover, when that disability is FASD, the transition to adulthood becomes further complicated by an adult service system that does little to recognize the impact of an invisible disability.

There is limited literature documenting the transition of those who experience both disability and foster care, especially when that disability is FASD. Knowing from previous research the proportion of children in care with FASD in Manitoba and that they will eventually reach age of majority, this third study by Fuchs *et al.* (2008a) involved a review of the administrative database records for a small sample of youth with FASD to examine the process of transition planning to adulthood that had occurred in relation to their legal and placement histories, including decisions to extend care beyond age of majority. A sample of 27 youth with FASD was identified.

A detailed review of the legal and placement histories of the youth in the sample was undertaken. As would be anticipated from previous research on legal status and placement history for children with FASD (Fuchs et al., 2007), many of the individuals in this group ($n = 8$; 30%) became permanent wards before the age of three years, and an additional 12 (for a total of 44%) were permanent wards by the time they were in Grade 1. It is interesting to note that the next most frequently noted age for becoming a permanent ward was between 11 and 14 years.

Considering all possible changes in placement recorded in the child welfare database (including short-term respite stays, hospitalizations, incarceration in a youth detention facility, and being on the run from one's placement) gives an indication of the degree of disruption that some children and youth experience. Using this figure, the average number of placements in this group was 14.1 (range 1 to 55), while the mean length of placement was 39.9 months (range 1 to 182 months).

Particular interest was paid to the longest period of placement, when the longest placement occurred in a child's placement history, and the age of the child at the time their longest placement was broken. Information on a longest placement was contained in 22 files. In those cases, the length of the longest placement ranged from a low of 1.8 years to a high of 15.2 years. The mean length of the longest placement was 6.5 years. The mean age at the end of the longest placement was 15.3 years. Most frequently, the longest placement was the first recorded placement.

Although the length of placement is limited by the age at which a child enters care, it is clear that the large proportion of children with FASD who become permanent wards in their preschool years is not matched by a large proportion of children with placements lasting in excess of 10 years. The age at which the longest placements end is informative. Of 21 youth for whom information was available, six (29%) were in their longest placement at age of majority. However, eight youth (38%) experienced their longest placement in early adolescence (ages 11–14) and five youth's longest placement (24%) ended in mid-adolescence (ages 15–17). It is clearly demonstrated that adolescence poses a threat to the endurance of placements.

Once the longest placement is interrupted, some youth experienced many short placements. While a substantial proportion of the group ($n = 10$; 48%) experienced relatively stable placement histories with one or fewer placements following their longest placement, the majority were moved more than twice after the termination of their longest placement, and many youth experienced multiple moves during the time they needed to be preparing for independence.

In addition to an examination of placements, transitional planning (TP) was reviewed. The TP designation indicates that an extension of care (beyond the 18th birthday) has been granted, a provision available for permanent wards only. In the

FASD sample, nine youth were designated as TP prior to their file closing. Six (67%) of the group had their care extended for less than one year. The individual whose care was extended the longest had his/her file closed at age 21; in that case, the extension was related to the necessity of significant medical procedures rather than what would usually be considered support for the transition to independence. Despite this information about extensions of care, there were no details available regarding the quality or nature of transitional planning. It is not clear from the electronic administration records which elements are incorporated into plans, or even how often transitional planning occurs, although such information may have been recorded in paper files.

Although there were some limitations in data availability, for those with records, the mean number of workers was 5.7 (range 2 to 15 workers). Because these histories are likely to be incomplete, however, these would be considered to be conservative estimates of the number of workers involved with each child.

The stability of residential placements was clearly an issue for this group of children in care. While stability in placements is always important, as children age, placement stability has increasing impacts on educational continuity. Without placement stability in later adolescence, the process of transitional planning is also made more difficult. What this study demonstrated was that this group of children with FASD tended to have stable placements in their early years, but faced increasing instability as they entered adolescence—a time critical for both their education and transition planning.

This later instability also reduced the likelihood of establishing enduring relationships with foster parents or teachers, reducing the pool of possible adults who might serve as the advocate/mentor that has been characterized as important to successful transition. Workers who might have filled the role of mentor appeared to be even more changeable than placements.

16.5
Study Four: The Cost of Child Welfare Care for Children with FASD

The emerging evidence of the detrimental impact of FASD on children in terms of their admission to child welfare care, their trajectories of care while involved with the child welfare system, and their experiences when leaving care at age of majority led to questions about the costs incurred to provide for their care needs. Few data exist on the specific costs of FASD, and those studies that do exist use different definitions and measurement criteria (Abel and Sokol, 1987; Harwood and Napolitano, 1985; Rice, Kelman, and Miller, 1991; Harwood, Fountain, and Livermore, 1998). In particular, very little research has been conducted on the cost incurred by child welfare to care for children with FASD. Knowledge of the costs associated with children with FASD in care of child welfare agencies can lead to a more informed process for addressing the needs of those children, and also provide an increased impetus for efforts to reduce the incidence of FASD. The fourth study (Fuchs et al., 2008b) built on the known population of children in care with FASD in Manitoba, and examined the cost of providing child welfare services for calendar year 2006, for a sample of 400 permanent wards who were in care for every day of that year.

Three main categories of costs were examined by researchers: basic maintenance; special rate/special needs; and exceptional circumstances. The cost for the 400 children in the random sample for 2006 was $3 124 600 for basic maintenance, $6 074 974 for special rate/special needs, and $230 752 for exceptional circumstances. This totaled $9 504 094, with an average of $23 760 for the year or $65 per day per child. When the total financial costs were examined by age group, children aged 11–15 years had the highest average cost per child for 2006 at $26 021 or $71 per day per child; they also had the highest total cost of $4 865 910 for 2006. Children aged 16+ years had the next highest average at $24 742 for 2006 or $68 per day per child, with a total cost of $1 781 404 for 2006. The 6–10-year-old age group had a yearly average of $20 633 and a daily average of $57 per child, with a total of $2 496 616 for 2006. The youngest children had the lowest total cost for 2006 at $360 165 and the lowest averages, with $18 008 for 2006 or $49 per day per child.

Of particular interest were the costs incurred in the special rate/special needs category. This type of funding is available to cover costs that exceed basic maintenance, or were not intended to be covered by basic maintenance. This includes both increases in the per diem (through a fee for service) and one-time-only expenses such as respite, therapy, initial clothing, age of majority, home visits, medical, and other special expenses. While basic maintenance is paid automatically, special rate/special needs funding must be requested, supported, and justified by social work staff, and approved through internal agency procedures. Of the 400 children in the sample, 389 (97%) had some cost recorded in this category, for a grand total of $6 074 974 in 2006. Costs recorded for the year ranged from $166 to $172 135 per child, with an average cost of $15 617 for 2006 or $43 per day per child. The average special rate/special needs costs were also found to increase with age of the child.

The average daily financial cost per child reported by this study was compared to the cost reported by the Child Protection Branch of Manitoba Family Services for all children in care for a similar time period. While the children from this financial record were of all legal statuses and included the children of our sample, it was of interest to note the average special rate cost per day for each group. This study found a daily special rate average of $43 per child for children who are permanent wards with diagnosed FASD. This is higher than the daily special rate average of $35 per day reported by the Child Protection Branch for all children in care over a similar time period.

At an average cost of $65 per day, children in care diagnosed with FASD deserve attention. As basic maintenance rates are standard for all children, the increased costs of care are primarily a reflection of special rate/special needs funding. It is clear from examination of actual expenditures that the fee-for-service portion of special rate/special needs is the most important driver of the total cost of services for this population. It is also evident that this cost varies directly with the age of the child; that is, as the age of children increases, so too does the average fee for service cost. Previous research (Fuchs et al., 2007) demonstrated the earlier admission of children with FASD to child welfare care and their increased likelihood of becoming permanent wards and spending the majority of their lives in care. Therefore, not only are the daily special rate costs higher for this group of children, but those costs are also extended over a lengthier period of time. Further, in examining the placement histories of youth with FASD leaving care, Fuchs et al. (2008a) found a tendency for an initial breakdown in placement when children reached the 11–14-year age group, the group that had the highest median costs in the cost of child welfare care study (Fuchs et al., 2008b).

In estimating the cost of providing services to children with FASD, a significant oversight occurs if it is not recognized that many children do not generate a direct cost – not because they do not need service, but because no service is available to them. This is especially true for children who live on reserves and who are disadvantaged by distance, isolation, and federal funding structures. This summary of direct expenses for a particular sample of children does not include a variety of indirect costs that are incurred on behalf of every child who enters the child welfare system, or the costs that are associated with efforts to prevent children entering the system. For example, the cost of prevention services and the cost of maintaining a trained staff of agency workers have not been added. Costs incurred by other systems, such as education, medical and dental costs not covered by child welfare agencies on behalf of these children, also were not included in this study. These themes, however, were the focus of the fifth study by the authors.

16.6
Study Five: Economic Impact of FASD for Children in Care

As has been demonstrated in the preceding discussion, children with a diagnosis of FASD present child welfare agencies with an array of complex and variable needs, as a consequence of a range of detrimental health outcomes. Given the significant proportion of FASD-affected children in care, as well as the nature of their needs, it is imperative to understand the service demands of this population to agencies, governments, and communities. Although research into the economic impact of FASD is in its infancy, the recognition that FASD is a contributing factor to large social and economic costs is helping to advance research on this important issue.

The fifth study examined the broader economic impact of FASD involving children in child welfare care, including the costs of health care, special education, and subsidized child care services (Fuchs et al., 2009). For the purposes of comparison, a sample group of children in care with no FASD diagnosis, but who were involved with the child welfare system due to parental alcohol abuse, was created. In addition, a random sample of children from the general population was created using a random matching methodology. The children in this group were not in care and not involved with an agency due to parental alcohol misuse. The monetary value of health, education, and child care service utilization were estimated for the various comparison groups. This study contributes to a broader national strategy to develop a Canadian model for calculating the economic impact of FASD.

During the 1980s and 1990s, some of the earlier economic costing exercises calculated the lifetime cost for each child affected by FAS typically, which was found to be in the range of $596 000 in 1980 (Harwood and Napolitano, 1985) to $1.4–1.5 million less than a decade later (Lupton, Burd, and Hardwood, 2004; Klug and Burd, 2003; Manitoba Child and Youth Secretariat, 1997; Thanh and Jonsson, 2009). Frequently, studies calculate the annual aggregated costs to a province, state, or a nation incurred on behalf of FASD-affected individuals.

More recently, these figures have been revised to $2.0 million (in 2002 $US) and $2.8 million (in 2008 $Can). Fiscal costs comprise approximately 80% of the total costs, and the balance of 20% is allocated to productivity losses (Lupton, Burd, and Harwood, 2004; Thanh and Jonsson, 2009). The Harwood and Napolitano (1985) figure of $596 000 in 1980 has recently been adjusted to $2 774 400 (in 2008 $Can) (Thanh and Jonsson, 2009). It should be noted that some of the costs cited in this section are undiscounted–and, thus, overstated–because they do not account for the time value of money. Nevertheless, the literature reveals that the fiscal and societal impact of FASD is strikingly large.

The most comprehensive Canadian-based study to date, which was conducted by Stade et al. (2006), measured both societal and individual costs in 2003 dollars. The study findings showed that medical and education costs comprised approximately 63% of the total costs. Productivity losses made up 8% of the costs, although if the children had been aged over 21 years these costs would be much higher. Based on a prevalence rate of 3 in 1000 people, the estimated cost of FASD-affected children up to age 21 is $344 208 000 (in 2003 $Can). The annual average cost for all children combined totaled $14 342.

These costing exercises have informed the knowledge gap around this significant health and social issue. Many costs are thought to be the "tip of the iceberg" in terms of the true costs and, as result of the substantial estimations of the cost of care, a common theme in the literature is the need for primary prevention. Primary prevention costs are thought to be lower than the discounted lifetime cost for an alcohol-affected child (Lupton, Burd, and Harwood, 2004; Klug and Burd, 2003; Stade et al., 2006; Fuchs et al., 2007, 2008b).

To gather information on the costs of services delivered outside of the child welfare system, it was necessary to determine what additional services were being provided to children who were either in care or involved with a child welfare agency, and at what cost those other services were being delivered. In order to do this, the services of the Manitoba Centre for Health Policy (MCHP) were utilized. The MCHP is a research centre of excellence that develops and maintains a comprehensive population-based data repository from the Province of Manitoba. All data found in the repository are derived from administrative records kept by different government departments to deliver health and social services.

Through the child welfare database in Manitoba, three sample groups were created: (i) permanent wards diagnosed with FASD (FASD-PW); (ii) children in temporary care whose parents presented with alcohol as a primary issue (PA); and

(iii) permanent wards whose parents presented with alcohol as a primary issue (PA-PW). The intent in including parental alcohol as a factor was to determine if there were any differences between those children diagnosed with FASD and those who did not have a diagnosis (who may or may not have been prenatally alcohol-exposed), although both sets of children had been affected by parental alcohol in some manner. The fourth group was identified through the Clinic for Alcohol and Drug Exposed Children (CADEC, now known as the Manitoba FASD Centre), and included children who had been diagnosed with FASD but were not currently found on the child welfare database. The final group of children was a sample from the MCHP data repository of the Manitoba population of children who received healthcare services in 2006. To create the general population group a sample based on a 4:1 matched cohort for the population was used. More specifically, four children were selected for every one child from the three children in care groups. A random matching method was used.

A summary of the population groups and their sample sizes is listed as follows. The FASD-PW, PA, and PA-PW children will be referred to as the children in care or CIC children in the discussion:

- **FASD-PW:** Children who had been diagnosed with FASD and were permanent wards of child welfare in 2006 ($n = 603$).

- **PA:** Children whose parents presented with alcohol as a primary issue, who were in care under a temporary order of guardianship or a voluntary placement arrangement for some period of time in 2006 ($n = 587$).

- **PA-PW:** Children whose parents presented with alcohol as a primary issue at a child welfare agency, who are permanent wards in 2006 ($n = 51$).

- **FASD-CADEC:** Children who had been diagnosed by CADEC as having FASD, and who were not noted on CFSIS as children in care in 2006 ($n = 119$).

- **General Population group:** Children who were selected on the basis of a random matching methodology ($n = 4964$).

For 1360 children who were FASD-affected, or for whom parental alcohol was an issue [the Children in Care (CIC) groups and FASD-CADEC], the costs of hospital and physician visits, plus prescription drugs, totaled $1 388 642 in 2006, compared to a representative sample of 4964 children in the general population with total costs of $1 993 849. Considering the entire sample of children for whom costs were tallied in this study ($n = 6324$), the FASD-affected and parental alcohol children comprised 21.5% of the total children; however, their hospital (inpatient and outpatient), physician, and drug costs made up 41.1% of the total costs.

The data in Figure 16.1 show that the FASD-PW and PA groups comprised the majority of the average total costs for hospital visits (inpatient and outpatient), physician services, and prescription drugs. Both cost estimates, as well as FASD-CADEC, were statistically different with respect to the General Population (Gen Pop) group. The average total costs for the FASD-PW group were 3.5-fold higher than for the General Population. In other words, an additional $1001 in healthcare

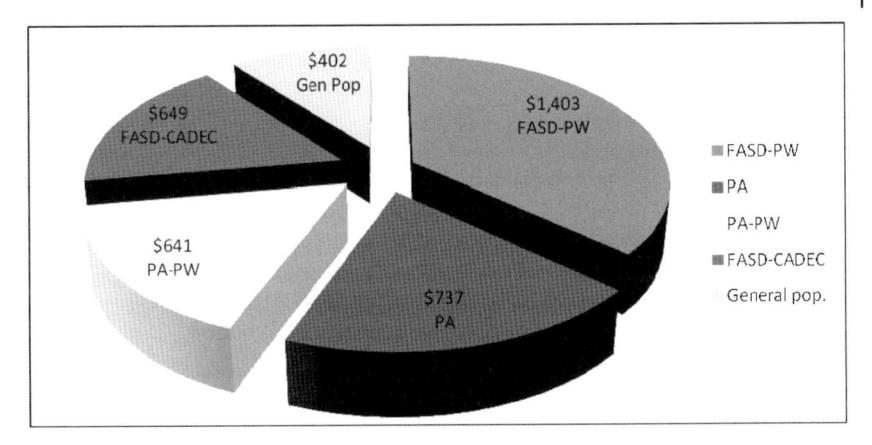

Figure 16.1 Average costs per child—hospitalizations (inpatient and day procedures), physician visits, and prescription drugs—by group (2006).

costs was incurred each year for every child who was FASD-affected and a permanent ward, compared to the General Population group.

This finding was anticipated, since Klug and Burd (2003) also found that health costs in North Dakota for FAS-affected children were 5.7-fold the cost of medical care for children who did not have FAS (annual averages of $2842 versus $500, respectively). It is known that children with a diagnosis of FAS have more severe health problems (Stade et al., 2006), which could explain an average cost which is 5.7-fold the health costs of the general population in the Klug and Burd study compared to 3.5-fold for FASD-PW children in the current study. The PA group's average costs were 1.8-fold higher than the General Population group.

The estimate of the percentage of PA-PW children using prescription drugs was comparable to that of the General Population, and the difference between the two estimates was not statistically significant. However, the small PA-PW sample size may yield unreliable estimates, making any comparison or interpretation rather difficult. The data in Figure 16.2 show that the FASD-PW children consumed the largest share of average total expenditures on prescription drugs, followed by the FASD-CADEC group, and the PA and PA-PW children. Prescription medications were one health expenditure category where a substantial deviation between the costs of healthcare provided to FASD-PW and PA groups of children was observed.

Table 16.1 reports a three-way cross-tabulation—group, gender, and age category—of the mean number of prescriptions per child in 2006. FASD-PW children were prescribed medications at 5-fold the rate of children in the General Population sample. Likewise, the PA and FASD-CADEC children were prescribed medications at 2.5-fold the rate of children in the General Population. The differences of the estimates were statistically significant although, here again, the PA-PW estimates may not be reliable due to a small sample size. However, the difference of estimates compared to the General Population group was not statistically significant.

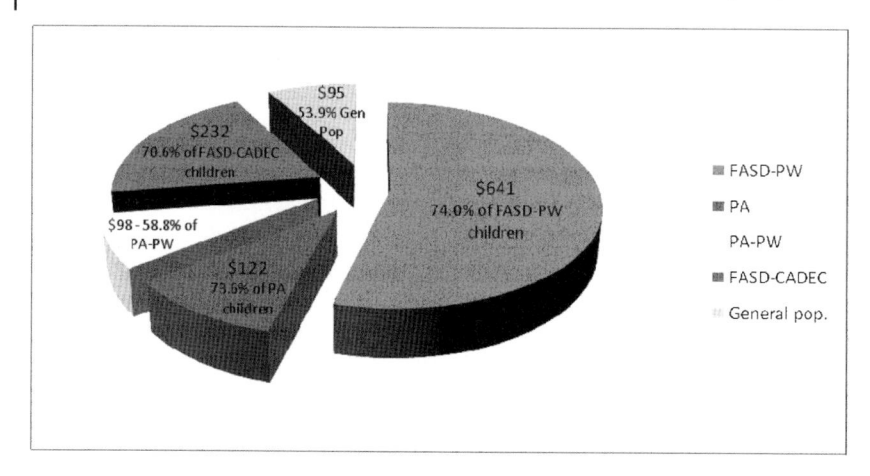

Figure 16.2 Average costs of prescriptions per child and percentage of children using prescription medications by group (2006).

Table 16.1 Mean number of prescriptions per child by age category, gender, and group.

Group	Mean no. of prescriptions	Age 0–5 years Male	Age 0–5 years Female	Age 6–10 years Male	Age 6–10 years Female	Age 11–15 years Male	Age 11–15 years Female	Age 16+ years Male	Age 16+ years Female
FASD-PW	12.1[a]	4.2	6.2[a]	11.7[a]	13.7[a]	15.0[a]	8.7[a]	13.6[a]	12.4[a]
PA	6.1[a],[b]	6.5[a]	6.9[a]	3.2[a],[b]	6.6[a],[b]	7.0[a],[b]	5.8[a]	4.1[b]	8.7
PA-PW	4.5[b]	4.7	10.7	2.7[b]	3.0[b]	3.4[b]	6.9[a]	1.3[b]	4.3
FASD-CADEC	6.3[a],[b]	4.0	4.3	5.6[a],[b]	4.9[b]	7.2[a],[b]	7.9[a]	4.2[b]	12.5
General pop.	2.4	3.2	2.5	1.9	2.0	2.3	2.0	2.2	4.5

a) Statistically significant difference with respect to the General Population group.
b) Statistically significant difference with respect to the FASD-PW group.

In the younger age categories (0–5 and 6–10 years) of the FASD-PW group, females had a higher average number of prescriptions compared to males, whereas this situation was reversed in the two highest age categories (11–15 and 16+ years). FASD-PW males aged 11 years and above received 6.0–6.5-fold the medication compared to males of a similar age in the General Population group. The average number of prescriptions issued to FASD-PW males was increased 3.5-fold as they aged, typically from a mean of 4.2 in the youngest age category to a high of 15.0 in the 11–15-year category. This substantial increase was not observed to the same degree in the other groups of children, although the number of prescriptions was doubled from 6.2 to 12.4 as the FASD-PW females aged, and tripled for the FASD-CADEC females, from 4.3 to 12.5. Notably, in all age categories, females had a higher mean number of prescriptions in the PA-PW group compared to males.

Table 16.2 Average costs (in $) of prescriptions per child by age category, gender, and group (2006).

Group	Average costs per child in the group ($)	Age 0–5 years Male	Age 0–5 years Female	Age 6–10 years Male	Age 6–10 years Female	Age 11–15 years Male	Age 11–15 years Female	Age 16+ years Male	Age 16+ years Female
FASD-PW	641[a]	77	212[a]	602[a]	722[a]	914[a]	460[a]	672[a]	475[a]
PA	122[a,b]	117	96[a]	60[b]	142[a,b]	243[b]	142[b]	116[b]	153
PA-PW	98[b]	72	150	61[b]	30[b]	51[b]	184	16[b]	149
FASD-CADEC	232[a,b]	76	53	202[a,b]	105[b]	313[b]	398[a]	98[b]	367
General pop.	95	95	50	65	62	173	80	93	139

a) Statistically significant difference with respect to the General Population group.
b) Statistically significant difference with respect to the FASD-PW group.

For most groups of children, the costs will either increase or remain constant from the youngest (0–5 years) to the oldest (16+ years) age categories. The only deviation from this pattern was the PA-PW group, and this was attributed to a small sample size yielding unreliable estimates. The average costs for each group, age category, and gender are listed in Table 16.2.

Among those children who were FASD-affected, or for whom parental alcohol was an issue (CIC and FASD-CADEC), 72.9% were users of prescription drugs, whereas only 53.9% of children in the General Population had a prescription in 2006. The average costs were much lower in the General Population sample, since approximately half of the children were prescribed a medication, compared to almost three-quarters of the children in the FASD-PW, PA, and FASD-CADEC groups.

The General Population group was prescribed central nervous system drugs less frequently; in fact, approximately 20% of these children were prescribed a drug which fell into this classification, compared to more than 60% of the FASD-affected children and more than 25% of those in the parental alcohol groups. The top three drug classifications were consistent across the groups: drugs related to the nervous system and respiratory conditions, as well as general anti-infectives for systemic use. As the FASD-PW children who were prescribed medication aged, the percentage of prescriptions for nervous system drugs they received was increased (from 30.4% in the youngest age category to 77.4% for children aged 16+ years). Yet, the opposite situation occurred for the other two most commonly prescribed drugs, namely respiratory system drugs or general anti-infectives. Genitourinary system and sex hormone prescriptions were used by youth aged 11–15 years and 16+ years.

Potentially, there are a few explanations for the high rate of prescribed nervous system medications as the children became older in the FASD-PW group. First, this might be the result of the issues that led to admission to care, the trauma and attachment disruption often associated with coming into care, as well as the experience of spending most of their life in care. Second, it could be the result of a co-occurring condition, such as ADHD. Also, when an individual has been prescribed such nervous system medications, there may be a high probability of their long-term use, possibly into and through adulthood.

In addition to an examination of health data, this study examined the usage and costs of educational supports for the study sample groups. A much higher percentage of FASD-affected children accessed special education services and corresponding funding, compared to PA children or the General Population group. The provision of special education funding is often based on functioning or the spectrum of need, and not necessarily on a diagnosis of FASD. Although the differences in these estimates were statistically significant, only 50.2% of FASD-PW children and 45.6% of FASD-CADEC children accessed special education funding during 2006. The average cost of education funding for FASD-PW children was 3.4-fold that incurred for children in the General Population, and 2.7-fold higher than for the FASD-CADEC children. While the FASD-affected or parental alcohol children comprised 20.4% of the total children enrolled in school, their aggregated costs comprised 38.2% of the total education costs.

A lower percentage of FASD-affected children was still enrolled in school after age 15 years when compared to the General Population or PA groups. The FASD-PW children had lower graduation rates, and had a lower chance of having completed eight or more credits in Grade 9. This finding was congruent with other studies which tracked school outcomes for youth in care in general, where the deleterious impact of growing up in care has been well documented (Courtney and Dworsky, 2005; Kufeldt, 2003; Merdinger et al., 2005). In a trajectory of care study, Fuchs et al. (2007) also found that many of the youth with FASD had a placement break down after age 12 years, which further disrupted school attendance.

The FASD-PW, PA, PA-PW and FASD-CADEC children all had much higher school retention rates compared to the General Population group. Unfortunately, individual data were not available on the specific reasons for retention, nor on follow-up supports in order to prevent future retention. Neither was it possible to obtain retention rates for each grade (these would have assisted greatly in deciphering whether or not retention occurred earlier on in the child's schooling). For example, it is not unusual for children to be held back in kindergarten or grade one at the parent's request. There was no indication for the children who had been retained as to whether this was as a consequence of being late starters, as opposed to being retained in the late primary school or junior high years. Guevremont, Roos, and Brownell (2007) found that school retention has a high predictive power in terms of school dropout rates. In fact, there is heightened (8-fold) risk of school withdrawal if a child has been retained more than once, and 3-fold if held back once.

These outcomes raise the probability that alcohol-affected individuals will have a reduced lifetime participation and employment rates in the labor market, as well as lower earnings. The fiscal impact of these educational outcomes is known to be a higher reliance on social services, including housing subsidies and income assistance, for the duration of their lives. Whilst such costs are excluded from the current study, it is known that children of problem drinkers may have a higher probability of experiencing detrimental outcomes in the labor market, for example, unemployment and lower wages (Balsa, 2008).

Finally, this study examined the cost of subsidized day care. Subsidized child care funding is available to families in Manitoba with qualifying reasons, such as medical or special needs, or low family income, as well as to children in care of a child welfare agency. The cost of nonsubsidized child care and the rate of accessing day care outside of subsidized child care services were not available. As expected, subsidized child care was concentrated in the two youngest age categories. Compared to the General Population, the FASD-PW children had a much higher likelihood of accessing subsidized day care funding. The difference of the FASD-PW estimates compared to the General Population group was statistically significant.

The PA group of children, with a higher proportion of preschoolers, was also more likely to access subsidized child care. For example, 15.1% of PA children were in subsidized day care, which was statistically different from 5.6% of children in the General Population group. Further, a higher percentage of PA children accessed subsidized child care compared to the FASD-PW group, and the difference was statistically significant. However, of the 41 FASD-PW children aged 0–5 years, almost half were in subsidized child care, while only one-quarter of the 251 preschoolers in the PA group were in subsidized child care. It is quite possible that being a permanent ward helps the child to settle into a routine, which could include day care and a stable placement, whereas the PA group may still be in flux within the child welfare system, having more recently come into care, not yet being in a stable placement, or in an emergency placement where day care supports have not yet been arranged.

The average costs of subsidized child care funding for FASD-PW children was more than double the costs incurred for children in the General Population group. Although the FASD-PW comprised 10.2% of children accessing subsidized child care, their aggregated costs comprised 19.5% of the total costs of subsidized child care.

To summarize, these research studies found that FASD-affected children who were permanent wards used more services in all categories that were examined in 2006. Moreover, the services used were more costly compared to a random sample of children in the General Population. Indeed, compared to the General Population the empirical evidence revealed a similar finding of higher costs and service utilization for the PA group of children for whom problematic parental drinking was an identified issue at a child welfare agency. In the categories of physician costs and hospitalizations, the findings revealed that FASD-PW children had the highest utilization and costs, followed by the PA group of children. However, the FASD-CADEC group was second to the FASD-PW children with regards to prescription medications and special education costs. Several of the differences in the estimates between the FASD-PW and PA groups of children were not statistically significant, which implies that many of the estimates did not yield true and reliable differences between these two groups of children.

One thesis of this study was that it is short-sighted only to investigate how FASD – the most extreme impact of parental alcohol abuse – affects children, since FASD is but one identifiable outcome of prenatal substance abuse. Children who have been exposed to alcohol prenatally, as well as those who have been affected by parental alcohol abuse postnatally, may also be at risk of deleterious outcomes, even if they do not have FASD.

Until the current study was conducted, there had been a significant research gap in costing the adverse health and education outcomes for children with no diagnosed FASD who were affected by their parents' misuse of alcohol. This analysis reveals an equally detrimental impact in some cost areas for those children without an FASD diagnosis, but for whom parental alcohol misuse is a risk factor that contributed to their being a child in care. Further, given that children with FASD were generally found by Fuchs *et al.* (2007) to be admitted to care mainly for reasons related to parental conditions or conduct (including substance abuse), and not related to the child's diagnosis of FASD, it is important to consider the impact of both prenatal and/or postnatal parental alcohol abuse on children.

16.7
Conclusions

It is widely recognized that children with a diagnosis of FASD present child welfare agencies, as well as the health and education sectors, with an array of complex and variable needs, as a consequence of a range of detrimental health outcomes. These observable facts are corroborated by the empirical findings in the various studies described in this chapter.

Given the significant proportion of FASD-affected children in care, as well as the nature of their needs, it is imperative to understand the service demands of this population to agencies, governments, and communities. Meeting the needs of children with FASD in care presents a number of challenges. Four particular challenges are associated with providing care for children in this group: the length of time they are in care; their special developmental needs; their needs as they transition out of care; and the number of affected children that will continue to enter the system and require care. Each of these challenges is a factor that increases the demand on resources available for children in care in Manitoba.

It is well-recognized, and has been widely discussed, that children who are diagnosed with FASD enter care at an earlier age, tend to become permanent wards, and spend a greater proportion of their lives in care. These factors place additional responsibility on child welfare agencies, which must assume the role of substitute parents for the majority of the individual's childhood years. Children in care are known to be disadvantaged, and face additional risks to their successful adaptation to adulthood. These risks must be addressed by the agency representing the caring of the community. Likewise, the duty of a responsible substitute parent carries a fiscal commitment.

In addition, the needs of this group of youth suggest that their care should be extended to the limit that is possible under current legislation. Although this would increase the cost of care for this group, that cost could be balanced by them remaining in school longer and being mentored beyond adolescence. An investment made earlier in the life of the individual could be used to maximize their adult independence, thus reducing the potential cost to adult health, income, and judicial systems.

In addition to systemic responses to the needs of children in care with FASD, interventions are required at a child-specific level. FASD is a complex constellation of symptoms that varies from individual to individual. Managing the condition necessitates effective assessment and planning. Workers and care providers need to be aware of the nature of the condition, its specific impact on the children in their care, the best practices in managing children with FASD, and the protocol for transitioning out-of-care for children with disabilities. Training specific to the challenges related to FASD is recommended for foster parents, social workers, and other service providers who deliver care services to children with FASD. In addition, workers and foster parents must be prepared with the knowledge and skill to advocate for appropriate service provision from the medical system, the education system, the disability service sector and the providers of adult services. Each of these strategies would contribute to the ability of a child with FASD to successfully navigate from care to the adult world.

Finally, closing the gap in knowledge around the economic costs of FASD-affected individuals can help to make better public policy, which partly focuses on preventing children from being born with significant FASD-related disabilities in future. An emphasis on primary prevention would effectively expand the public policy approach to FASD in Canada, as well as creating efficiencies in the allocation of scarce resources.

References

Abel, E.L. and Sokol, R.J. (1987) Incidence of fetal alcohol syndrome and economic impact of FAS-related anomalies. *Drug Alcohol Depend.*, **19** (1), 51–70.

Balsa, A.I. (2008) Parental problem drinking and adult children's labor market outcomes. *J. Human Resources*, **43** (2), 454–486.

Barth, R.P. (2001) Research outcomes of prenatal substance exposure and the need to review policies and procedures regarding child abuse reporting. *Child Welfare*, **80** (2), 275–296.

Bartholet, E. (1999) *Nobody's Children: Abuse and Neglect, Foster Drift, and the Adoption Alternative*, Beacon Press, Boston, MA.

Besinger, B.A., Garland, A.F., Litrownik, A.J., and Landsverk, J.A. (1999) Caregiver substance abuse among maltreated children placed in out-of-home care. *Child Welfare*, **78** (2), 221–239.

Chudley, A.E., Conry, J., Cook, J.L., Loock, C., Rosales, T., and LeBlanc, N. (2005) Fetal alcohol spectrum disorder: Canadian guidelines for diagnosis. *Can. Med. Assoc. J.*, **172**, 1–21.

Courtney, M. and Dworsky, A. (2005) *Midwest Evaluation of the Adult Functioning of Former Foster Youth: Outcomes at Age 19*, Chapin Hall, Chicago, IL.

Curtis, P.A. and McCullough, C. (1993) The impact of alcohol and other drugs on the child welfare system. *Child Welfare*, **72**, 533–542.

Department of Health and Human Services (1999) *Blending Perspectives and Building Common Ground: A Report to Congress on Substance Abuse and Child Protection*. USA.

Dore, M., Doris, J.M., and Wright, P. (1995) Identifying substance abuse in maltreating families: a child welfare challenge. *Child Abuse Negl.*, **19**, 531–543.

Fuchs, D., Burnside, L., Marchenski, S., and Mudry, A. (2005) *Children with Disabilities Receiving Services from Child Welfare Agencies in Manitoba*, Centre of Excellence for Child Welfare, Ottawa, http://www.cecw-cepb.ca/sites/default/files/publications/en/DisabilitiesManitobaFinal.pdf

Fuchs, D., Burnside, L., Marchenski, S., and Mudry, A. (2007) *Children with FASD: Involved with the Manitoba Child Welfare System*, Centre of Excellence for Child Welfare, Ottawa, http://www.cecw-cepb.ca/sites/default/files/publications/en/FASD_Final_Report.pdf

Fuchs, D., Burnside, L., Marchenski, S., and Mudry, A. (2008a) *Transition out-of-care: Issues for youth with FASD*, Centre of Excellence for Child Welfare, Ottawa. Available at: http://www.cecw-cepb.ca/publications/626

Fuchs, D., Burnside, L., Marchenski, S., Mudry, A., and De Riviere, L. (2008b) *Economic Impact of Children in Care with FASD, Phase 1: Cost of Children in Care with FASD in Manitoba*, Centre of Excellence for Child Welfare, Ottawa, http://www.cecw-cepb.ca/sites/default/files/publications/en/FASD_Economic_Impact.pdf

Fuchs, D., Burnside, L., De Riviere, L., Brownell, M., Marchenski, S., Mudry, A., and Dahl, M. (2009) *Economic Impact of Children in Care with FASD and Parental Alcohol Issues Phase 2: Costs and Service Utilization of Health Care, Special Education, and Child Care*, Centre of Excellence for Child Welfare, Ottawa, http://www.cecw-cepb.ca/sites/default/files/publications/en/FASD_Economic_Impact_Phase2.pdf

Guevremont, A., Roos, N.P., and Brownell, M. (2007) Predictors and consequences of grade retention: examining data from Manitoba, Canada. *Can. J. School Psychol.*, **22** (1), 50–67.

Harwood, J.H. and Napolitano, D.M. (1985) Economic implications of the Fetal Alcohol Syndrome. *Alcohol Health Res. World*, **10** (1), 38–43.

Harwood, J.H., Fountain, D.M., and Livermore, G. (1998) *The Economic Costs of Drug and Alcohol Abuse in the United States, 1992*, National Institute on Drug Abuse and National Institute on Alcohol Abuse and Alcoholism, Washington, DC.

Hay, M. (1999) A practical roadmap for the imperfect but practical-minded clinician, in *Fetal Alcohol Syndrome/Effect: Developing A Community Response* (eds J. Turpin and G. Schmidt), Fernwood, Halifax, NS, pp. 26–43.

Jones, K. (1999) The ecology of FAS/E: developing an interdisciplinary approach to intervention with alcohol-affected children and their families, in *Fetal Alcohol Syndrome/Effect: Developing A Community Response* (eds J. Turpin and G. Schmidt), Fernwood, Halifax, NS, pp. 80–87.

Klug, M.G. and Burd, L. (2003) Fetal alcohol syndrome: annual and cumulative cost savings. *Neurotoxicol. Teratol.*, **25** (6), 763–765.

Kufeldt, K. (2003) Graduates of guardianship care: outcomes in early adulthood, in *Child Welfare: Connecting Research, Policy, and Practice* (eds K. Kufeldt and B. McKenzie), Wilfrid Laurier University Press, Waterloo, ON, pp. 203–216.

Lupton, C., Burd, L., and Harwood, R. (2004) Cost of fetal alcohol spectrum disorders. *Am. J. Med. Genet. C.*, **127C**, 42–50.

McNichol, T. and Tash, C. (2001) Parental substance abuse and the development of children in family foster care. *Child Welfare*, **80** (2), 239–256.

Manitoba Child and Youth Secretariat (1997) *Strategy Considerations for Developing Services for Children and Youth*, Manitoba Children and Youth Secretariat.

Merdinger, J., Hines, A., Osterling, K.L., and Wyatt, P. (2005) Pathways to college for former foster youth: understanding factors that contribute to educational success. *Child Welfare*, **84** (6), 867–896.

Overhoser, J.C. (1990) Fetal alcohol syndrome: a review of the disorder. *J. Contemp. Psychother.*, **20** (3), 163–176.

Reid, C. and Dudding, P. (2006) *Building A Future Together: Issues and Outcomes for Transition-Aged Youth*, Centre of Excellence for Child Welfare, Ottawa, ON.

Rice, D.P., Kelman, S., and Miller, L.S. (1991) *The Economic Costs of Alcohol and Drug Abuse and Mental Illness: 1985*, DHHS Publication No (ADM) 90-1694. U.S. Department of Health and Human Service, Rockville, MD.

Semidei, J., Radel, L.F., and Nolan, C. (2001) Substance abuse and child welfare: clear linkages and promising responses. *Child Welfare*, **80** (2), 109–128.

Sonnander, K. (2000) Early identification of children with developmental disabilities. *Acta Paediatr.*, **434**, 17–23.

Square, D. (1997) Fetal alcohol syndrome epidemic on Manitoba reserve. *Can. Med. Assoc. J.*, **157** (1), 59–60.

Stade, B., Ungar, W.J., Stevens, B., Beyenne, J., and Koren, G. (2006) The burden of prenatal exposure to alcohol: measurement of cost. *J. FAS Int.*, **4** (e5), 1–14.

Streissguth, A.P., Barr, H.M., Koga, J., and Bookstein, F.L. (1996) *Understanding the Occurrence of Secondary Disabilities in Clients with FAS and FAE*, University of Washington Fetal Alcohol and Drug Unit, Seattle, WA.

Streissguth, A.P., Barr, H.M., Bookstein, F.L., Sampson, P.D., and Olson, H.C. (1999) The long-term neurocognitive consequences of prenatal alcohol exposure: a 14-year study. *Psychol. Sci.*, **10** (3), 186–190.

Thanh, N.X. and Jonsson, E. (2009) Costs of fetal alcohol spectrum disorder in Alberta, Canada. *Can. J. Clin. Pharmacol.*, **16** (1), e80–e90.

Wattendorf, D.J. and Muenke, M. (2005) Fetal alcohol spectrum disorders. *Am. Fam. Physician*, **72** (2), 279–285.

Williams, R.J., Obaido, F.S., and McGee, J.M. (1999) Incidence of fetal alcohol syndrome in northeastern Manitoba. *Can. J. Public Health*, **90** (3), 192–194.

Young, N., Gardner, S., and Dennis, K. (1998) *Responding to Alcohol and Other Problems in Child Welfare*, CWLA Press, Washington, DC.

Young, N.K., Boles, S.M., and Otero, C. (2007) Parental substance use disorders and child maltreatment: overlaps, gaps and opportunities. *Child Maltreat.*, **12** (2), 137–149.

Zevenbergen, A.A. and Ferraro, F.R. (2001) Assessment and treatment of fetal alcohol syndrome in children and adolescents. *J. Dev. Phys. Disabil.*, **13** (2), 123–136.

17
British Columbia's Key Worker and Parent Support Program: Evaluation Highlights and Implications for Practice and Policy

Deborah Rutman, Carol Hubberstey, and Sharon Hume

17.1
Introduction

This chapter provides an overview of the Key Worker and Parent Support program, which was implemented province-wide in British Columbia, Canada, as of 2006. A comprehensive formative and summative evaluation of the program ran from 2006–2009, conducted by the present authors[1]. Selected evaluation findings are presented, along with discussion highlighting several emerging promising practices in working with families with a child or youth with Fetal Alcohol Spectrum Disorder (FASD).

17.2
Background

In 2003, the province of British Columbia released the first comprehensive plan in Canada related to FASD entitled *Fetal Alcohol Spectrum Disorder: A Strategic Plan for British Columbia*. In 2005, funding was provided to implement a cross-ministry initiative to support children and youth with FASD and their families. As part of this initiative, the Ministry of Health established new screening, assessment and diagnosis teams throughout the province; the Ministry of Education created a provincial FASD outreach program designed to support teachers of students with FASD through sharing FASD-related research, resources, and successful practices; and the Ministry of Children and Family Development (MCFD) established the Key Worker and Parent Support program to support families of children and youth (birth to 19 years of age) with FASD.

The overall goal of the Key Worker and Parent Support program is to maintain and enhance the stability of families with children or youth with FASD and other similar complex neurodevelopmental conditions in order to improve the children's

---

1) Interim and Final Evaluation Reports on the Key Worker and Parent Support program are available via the Ministry for Children and Family Development website: http://www.mcf.gov.bc.ca/fasd/kw_evaluation.htm

Fetal Alcohol Spectrum Disorder–Management and Policy Perspectives of FASD. Edited by Edward P. Riley, Sterling Clarren, Joanne Weinberg, and Egon Jonsson
Copyright © 2011 WILEY-VCH Verlag GmbH & Co. KGaA, Weinheim
ISBN: 978-3-527-32839-0

long-term outcomes. The objectives of the evaluation were to: (i) identify the program's strengths and challenges; (ii) learn what difference the program made for families and communities; and (iii) identify areas for improvement.

17.3
Program Model and Components

The Key Worker and Parent Support program is based on research and practice evidence from four areas: (i) research in FASD, such as that conducted by Dr Ann Streissguth and her colleagues (Streissguth *et al.*, 1996, 2004); (ii) by Dr Heather Carmichael Olson (Carmichael Olson, 2007) and by Diane Malbin; (iii) research in the disabilities field; and (iv) community-based, FASD-related parent support projects that were operating in BC at the time the program began. The program is rooted in an understanding of FASD as a brain-based physical disability with behavioral symptoms, and was influenced by Diane Malbin's conceptual framework (Malbin, 2002, 2004), which emphasizes the need for appropriate environmental accommodations that take into account the brain-based nature of the disability. The model is premised on the notion that when parents, caregivers, professionals and other community members are knowledgeable about FASD, and have a common understanding of the disability, they will reinterpret the behavioral symptoms associated with FASD and shift their expectations of the child/youth accordingly. When this happens, appropriate environmental accommodations and supports then can be identified and put into place, resulting in additional positive outcomes for families.

In the BC program, the Key Worker acts as a facilitator with the broad role of engaging families, service providers, and community members in learning about FASD and helping them to access and/or devise environmental accommodations that are appropriate for the child and supportive for the family. The core components of the role are to provide FASD-related information and education to families and to community professionals, and to provide support and assistance to families and community partners in accessing appropriate services and/or other resources. Key Workers may be involved in: (i) integrated service/care planning; (ii) bringing an FASD-lens to discussions; and (iii) (re)-interpreting a child's behaviors, strengths and needs in light of this understanding. Additional components of the BC Key Worker program include: (i) clinical supervision by a qualified professional; (ii) access to consultation with an FASD expert for MCFD regions and the contracted Key Worker agencies on an ongoing, as-needed basis; (iii) participation in provincial training specifically for Key Workers and their supervisors; and (iv) a regional service delivery model whereby each MCFD region contracts with agencies to deliver the Key Worker program.

In 2008, Provincial Practice Standards were developed to guide and support the Key Worker program. At the same time, given MCFD's regionalized operational structure, wherein services are organized, managed, and delivered in each of five regions, regional variation exists in the delivery of the Key Worker program.

17.4
Literature

As noted previously, the conceptual framework articulated by Malbin (2004) was influential in the development of the Key Worker program. According to Malbin's framework, because FASD is a neurobehavioral disability resulting in impairments in memory, cause and effect reasoning, impulsivity, and abstract thinking (among other sensory and cognitive problems), standard learning and cognitive behavioral approaches to intervention with the child are contraindicated. Instead, both Malbin's Oregon-based Fetal Alcohol Syndrome Consultation Education and Training, Inc. (FASCETS) model and the BC Key Worker and Parent Support program model emphasize the importance of everyone developing a common understanding of FASD and its behavioral symptoms. This common understanding then results in a shift in expectations for the child/youth, and ultimately in the implementation of strategies that take into account the ways in which each individual child with FASD thinks, learns, and experiences the world.

At the same time, recent intervention studies in the US have shown that children can benefit from direct and explicit skills-based training, while their parents improved their knowledge of FASD and of related parenting issues when they were provided with FASD-specific education and training[2] (Bertrand, 2009). Two of the five studies reviewed focused chiefly on direct parent training as the primary intervention, while three studies employed direct and explicit instruction of children with FASD. Equally as important, these results were achieved by providing existing practitioners and community services such as counseling, special education, and therapy services, with FASD specific knowledge and skills.

At a system level, an FASD "intervention" introduced in Alberta similarly aimed to improve outcomes for children. Responding to concerns that children with FASD living in government care were not being adequately served by existing practice standards, the Alberta Children's Services created enhanced practice standards that were "... designed to provide early identification of FASD through screening and referral; planning for specific needs of children based on assessment/diagnosis; and increased training about FASD for caseworkers, foster parents and community-based program staff" (Badry, 2009, p. 53). Research and evaluation following the implementation of these FASD-related enhanced practice standards found: (i) significant decreases in the number of placement changes over time; (ii) fewer disruptions and school absences among the children; and (iii) an improved ability by caseworkers and foster parents to meet the needs of the children they were working with when they received specific training in FASD (Badry and Pelech, 2005).

---

2) One study provided a 12-week training course aimed at improving children's social skills functioning; a second study was aimed at improving behavioral and mathematical functioning through a psycho-educational program (Behavioral Regulation Training) designed to improve readiness to learn – with acquisition of math skills as the focus; and a third study used a neurocognitive habilitation model that incorporated instruction on self-regulatory strategies and skills to improve executive functioning.

Common themes in both the US and Canadian studies were that: parents and caregivers need to be included as part of the intervention or practice; parents and caregivers benefit from provision of FASD-related education and training; practitioners and community partners alike benefit from provision of support and training in FASD; and, it is possible to make a difference for families by enhancing knowledge about FASD within already existing community resources (e.g., school system, healthcare providers).

Finally, literature from the disabilities field also influenced development of the BC Key Worker and Parent Support program. Of particular note were results from an evaluation of a British Key Worker program for families with children with disabilities (Greco and Sloper, 2003; Liabo et al., 2001). In the British model, the Key Worker assists families by providing information; identifying and addressing the needs of all family members; providing emotional and practical support as required; assisting families in their dealings with agencies; and acting as an advocate as required (Greco et al., 2005; Greco and Sloper, 2003). Key Workers were also regarded as a trusted source of information for families, community resources, and schools, as well as an ally to families. Evaluations of the British program revealed that Key Workers helped to increase cooperation between all parties, which resulted in increased access to resources for parents and families (Webb et al., 2008).

17.5
Evaluation Methods

The three-year formative and summative evaluation of the BC Key Worker and Parent Support program employed both qualitative and quantitative methods of data collection. Likewise, the evaluation involved both province-wide data collection from all Key Worker agencies, and community-based data collection from a sample of several Key Worker agencies in each of five MCFD regions.

The province-wide data were collected via multiple methods: (i) Annual Agency Questionnaires, which were collected from all Key Worker agencies and which focused on a variety of components of the formative evaluation, including program administration issues such as funding issues, staffing, training, supervision, program activities, and program strengths and challenges; (ii) monthly output data submitted by all agencies; and (iii) Intake and Exit Questionnaires for Parents/Caregivers.

The Intake Evaluation Questionnaire for Parents/Caregivers and the Exit Questionnaire for Parents/Caregivers consisted of several standardized instruments that were modified to meet the needs of the study, and questionnaire items that were developed specifically for the Key Worker evaluation; a detailed description of the sources of the questionnaire items can be found in the Final Summative Evaluation Report (Hume et al., 2009b).

Parents and caregivers were asked to complete the Intake Questionnaire at or near intake, and the Exit Questionnaire when they ended their involvement with the program. A total of 394 parents or caregivers completed the Intake Questionnaire, and 115 parents or caregivers completed the Exit Questionnaire; 81 were "matched pairs," in that respondents completed both the Intake and Exit Questionnaires.

In addition to the province-wide data, semi-structured qualitative interviews were conducted in a total of 21 communities; 125 parents/caregivers, 115 community service providers, and approximately 60 program staff and agency managers were thus engaged in the evaluation.

An important feature of the evaluation is that it was conducted simultaneously with provincial implementation of the Key Worker and Parent Support Services program. Thus, there was a reliance on agencies contracted to provide Key Worker services to comply with the evaluation and assist in the gathering of evaluation data. In addition, the parents' and caregivers' participation in the evaluation was voluntary.

Staff at the contracted agencies had a diverse range of experience with, and knowledge about, comprehensive evaluation processes and methods, which led to some challenges for the evaluation. For example, not all agencies submitted Intake and/or Exit Questionnaires, leading to questions as to whether the sample of parents/caregivers who completed the questionnaires were representative of the total population receiving services. In addition, several Key Workers did not administer the Intake Questionnaire, which was intended to provide baseline data, until several weeks or months after the parents/caregivers had joined the program. As a result, some of the intake data cannot be considered to be true baseline data.

Furthermore, data collection challenges resulted in modifications to the Annual/Exit Questionnaire in order to shorten its length and simplify the language. Thus, it was not possible to compare directly all items with those on the Intake Evaluation Questionnaire. Lastly, there was a low number of matched Intake and Annual/Exit Questionnaires ($n = 81$), which means that the quantitative questionnaire findings may not be generalizable to all parents and caregivers receiving Key Worker services.

17.6
Formative Evaluation Findings

Fifty-six diverse community agencies, including 29 multi-service agencies, 13 Aboriginal agencies, 12 child development centers, one community college and one local MCFD office, were contracted to provide Key Worker services across MCFD's five regions in the province.

Between April 2006 and March 2009, over 1800 families were referred to the program; moreover, intakes and referrals more than doubled over the program's first three years. In addition, output findings revealed that the program was reaching a range of families (Table 17.1). In Year 3, the largest percentage of families accessing the program were birth parents (34%), followed by foster parents (23%), adoptive parents (20%), and grandparents (15%). The increase in the percentage

Table 17.1 Parent's/caregiver's relationship to the child (i.e., type of families receiving Key Worker services).

Family type	Year 2 (2007/2008)	Year 3 (2008/2009)
Birth parent	23%	34%
Foster parent	32%	23%
Adoptive parent	20%	20%
Grandparent	16%	15%

Source: Parent/Caregiver Intake and/or Exit Questionnaires.

of birth parents accessing the program from the second to third year was noteworthy, as was the high number of grandparents in all years. Both groups face unique challenges in accessing programs. Birth parents report feeling shamed and stigmatized in their communities, and also report fear that their use of services maybe perceived by authorities as indication of their lack of parenting capacity (Poole and Isaac, 2001; Rutman et al., 2000). Grandparents who are caring for their grandchildren report feeling isolated, out of synch with their peers, and out of the loop in terms of knowing what resources are available to them (Whittington et al., 2007).

Additional demographic information about the parents/caregivers accessing the program included:

- 20% of parents/caregivers reported annual earnings of less than $20 000; by contrast, provincially 8% of the BC population reports annual earnings of less than $20 000 (BC Stats, 2006)
- Approximately 33% of parents/caregivers accessing the program were headed by one parent; by contrast, 22% of families in BC were headed by one parent (BC Stats, 2006)
- 68% of parents/caregivers were Caucasian; 17% were Aboriginal, and 15% were South Asian, Central/South American, African, American or "other."

Demographic information about the children/youth of the families served by the program included:

- Two-thirds of children/youth were males.
- 60% of children/youth were aged between 0–10 years; however, the breakdown of child/youth intakes by age varied substantially by region, especially the percentage of intakes that were young children (age 0–6 years) and older youth (16–19 years). One region consistently had a relatively high number of older youth intakes (i.e., 20% of the child/youth intakes were 16–19 years old), many of whom were no longer living with their families.
- 46% of the children/youth were of Aboriginal origin.

Source: Key Worker and Parent Support program agencies' monthly output data.

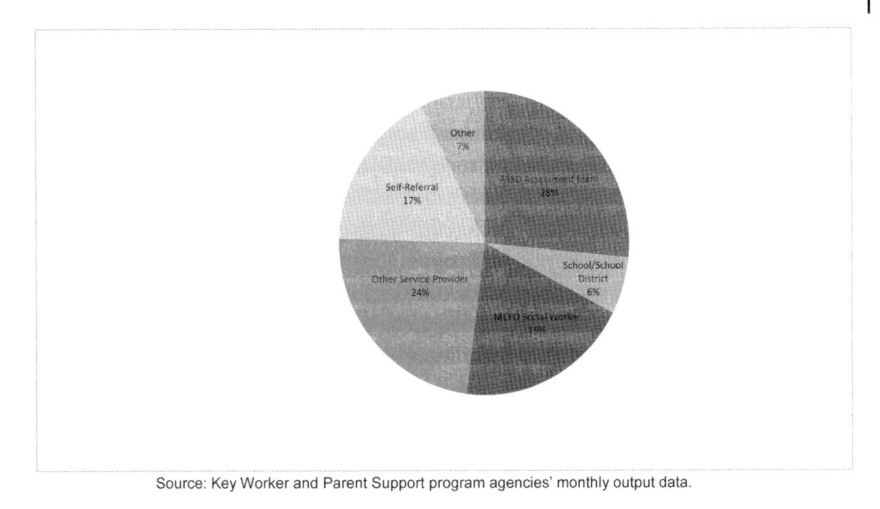

Figure 17.1 Referral sources. Source: Key Worker and Parent Support program agencies' monthly output data.

As shown in Figure 17.1, families learned about and/or were referred to the Key Worker and Family Support program through a variety of sources (e.g., community service providers, Ministry social workers, and via regional FASD assessment teams), and 17% of the parents/caregivers were self-referred to the program. At the same time, there were substantial regional differences in referral source. For example, in the North, the primary referral source was the Health Authority assessment team, which was in keeping with that region's service model for the Key Worker program (i.e., having the Key Workers serve as the Intake Worker on the assessment teams).

17.6.1
Activities and Role of the Key Worker

The activities of the Key Worker are both family- and community-focused. Key Workers provide both groups with information and education about FASD and other relevant topics, and information is shared in both informal, unstructured ways (e.g., kitchen table conversations with family members or "water-cooler" chats with colleagues), as well as through more formal information-sharing sessions (e.g., in-service seminars with colleagues in the same organization, community-based FASD workshops, parent education workshops, and so forth). In their information sharing, Key Workers focused on discussing FASD as a neurodevelopmental disability that was most often invisible; they also emphasized the need for appropriate accommodations, and strategies for ways of caring for and/ or working with the child or youth. Key Workers stressed the importance of adjusting everyone's (e.g., parents', caregivers', school personnel's, other community members') expectations for the child/youth, which often was key to reducing stress for the family and ultimately reducing secondary behaviors associated with FASD. In the words of the evaluation participants:

> "The Key Worker helped the foster parent set appropriate boundaries and rules – the child needs more supervision than the foster parent anticipated. The child is 14 but because of dysmaturity is much younger, and the foster parent did not understand this at first". (Community service provider)

> "The Key Worker helped change my expectations of my daughter – helped me to reframe them. I am more realistic about them and about my daughter". (Parent)

> "I found that having the meaning of FASD spelled out into 'developmental age' is helpful. It changes how we (as teachers) work with these students. We work with students as a three year old rather than a seven year old. Intellectually we know this, but it's important to be reminded and to see it in practice". (Community service provider)

Key Workers also provide families with practical and emotional support, as directed by families' needs and self-identified goals. For example, Key Workers have supported families by: (i) attending school-based or other service-related meetings with them; (ii) providing support through the FASD assessment process; (iii) assisting families to access funding support or various types of services or resources (e.g., summer camps); and (iv) providing emotional support and/or someone to talk with. In describing the support component of their role, Key Workers have stated:

> "I support families by helping brainstorming ideas, linking them to services, attending meetings if they want me and being available for crises". (Key Worker)

> "I explain to the parents or caregivers what the various professionals do, about the assessment tools, what co-morbid conditions are – it gives the families more comfort when they go into the assessment". (Key Worker)

> "I spend a lot of time with different professionals, for example, a new worker at the Boys and Girls Club who knows nothing about FASD. I will meet with her over coffee to provide information and support about working with youth with FASD". (Key Worker)

Parents/caregivers, service providers and Key Workers also noted that a large component of the role was to provide families – and community partners – with a voice in order to advocate for and facilitate families linking with available services, and to help ensure that there would be follow-up and accountability in terms of individualized service or educational plans.

> "Families need a voice. Sometimes I just go with parents and don't say anything – mainly with the school district, but with MCFD too. I attend meetings to support parents with social workers." (Key Worker)

"I attend the IEP (Individualized Educational Plan) meetings as a guest of the parents. ... I get the parents/social worker to provide consent to gather information and then I go to the school and say, "What can we do?" "Is there going to be an IEP?" Initially it felt adversarial, but it is less so now. Teachers were afraid I was seeking a bunch of accommodations that were impossible to do. What seems to happen instead is that my presence seems to help hold everyone accountable to the IEP, including me. The more people who can be involved, the better".

Over time, there has been a growing clarity about and consensus in understanding of the role and activities of the Key Worker. For example, during the first year of the program, a number of Key Worker agencies expressed some confusion – as did community partners – regarding the role of the Key Worker. Specifically, some agencies grappled with whether the Key Worker role was a case manager role and, if not, how Key Workers would be working with community partners and families. By Years 2 and 3, however, the Key Worker role became increasingly clear to all involved – that is, to Key Workers themselves, their agency managers, families, and community partners. The role is not that of case manager or case coordinator, but rather to provide families with information, support and connections to existing community resources, while similarly providing community partners with information, education and linkages. That there is clarity in the Key Worker role is positive and important, since clarity in worker role has been associated with positive outcomes for service users (Greco et al., 2005; Greco and Sloper, 2003).

In addition, over the first three years of program implementation, the shifts in the Key Worker role were minor. The primary way in which the activities of the Key Worker shifted in Year 3, relative to previous years, was that a stronger emphasis was given to the Key Worker's community-focused activities. Key Workers operated in collaboration with community partners to organize and facilitate/deliver professional training, community-based workshops and learning opportunities about FASD and other relevant topics. Key Workers also took a leadership role in community awareness activities around FASD (e.g., International FASD Day). In addition, Key Workers were increasingly involved in inter-agency committees focusing on youth and families with multiple and/or complex needs.

17.6.2
Regional and Provincial Supports for Key Workers

As mentioned above, the Key Worker program was developed at a provincial level but administered regionally by the five MCFD regions throughout BC. From the outset, the MCFD Provincial Office stipulated some minimum requirements that the contracted agencies were required to fulfill with regard to Key Worker qualifications – for example, that Key Workers had a Bachelors degree in social work, child and youth care or psychology and extensive knowledge of FASD, and that they had access to clinical supervision from graduate level health or social work professionals. These requirements were based on the British Key Worker evaluation (Greco et al., 2005), which found that having these standards supported positive outcomes for families and children. Despite challenges to providing clinical supervision at the level the Ministry had anticipated (which were due to lack of qualified professionals in small communities and insufficient funding in agency contracts), the Key Workers did receive regular supervision from their manager, or through team meetings, peer supervision, and/or regular staff meetings.

Regional Coordination was another source of support and networking for Key Workers, who were often the only person in their agency or community working as a Key Worker. Regional Coordinator positions were created in two regions when the Key Worker program was first implemented. Based on recommendations from the Formative Evaluation Report (Hume et al., 2009a), the other regions followed suit and created regional coordination positions or processes. The Regional Coordinators hosted monthly or bi-monthly conference calls for the Key Workers in their region to promote information sharing, case consultation, and education. In some regions they developed and coordinated regional training in FASD and acted as a resource and clearing-house for information about FASD. In addition, the Regional Coordinators worked to strengthen the linkages and smooth the referral process both to the Assessment Teams, and from the Assessment Teams to the Key Workers.

Key Workers reported that the conference calls with other Key Workers in the region were an important source of information, support and networking, and contributed to the quality of their work life.

> "The support and networking and information exchange between everyone is great and a factor in me staying in the position".
>
> "Support from the Regional Coordinator has been key–essential. She brings all the information together from the Health Assessment Network, from other Key Workers in the region, from the evaluation team. She is the 'eyes and ears' of the program".
>
> "Key Workers modeled good practice for each other. When [we] heard about a Key Worker doing some activity or trying new strategy it inspired the rest of us to try".

Provincial Key Worker training was developed centrally and offered to all the Key Workers and their supervisors in the province. The training–with Level 1 and Level 2 each being run as three-day sessions–offered participants a foundation for understanding FASD as a neurodevelopmental disability, and a framework for working with families and communities to develop effective strategies and accommodations for children with FASD. Level 2 provided more in-depth discussion about strategies and working effectively with communities. Both levels of the training provided information about the FASD assessment process. In addition, because the Key Worker role was a new one in the province, the training helped new Key Workers and their supervisors to gain a more concrete understanding of their roles and responsibilities with children, families and their communities. Evaluations of the training were very positive. In the words of one participant:

"[I gained an] increased understanding of the multidisciplinary composition of the assessment teams, the objectives of the assessment process, and strategies for interpreting and implementing the (assessment) reports' recommendations".

Unfortunately, with the downturn in the economy in the fall 2008, the training sessions were canceled. For new Key Workers, even those experienced in working with children and families, the lack of training was troubling. One new worker stated that:

"I am really feeling at sea without training. I don't know how to address challenging behaviours or what to present for public information/education sessions. I have limited information and practical experience and no real introduction as to the [Key Worker] role, only a job description".

17.6.3
Parents', Caregivers' and Community Partners' Perceptions of the Program

Parents and caregivers were very positive about the Key Worker program. Indeed, 91% of respondents to the Exit Evaluation Questionnaire said they were satisfied or very satisfied with the program. In addition, the majority of Exit Questionnaire respondents (71%) were satisfied with the frequency that they saw their Key Worker and the amount of time spent together, although about one-third indicated that they wanted to have more contact with their Key Worker.

What parents/caregivers said they liked most about the program was that it was *family-centered and family-directed*. In other words, families' needs set the agenda and were the starting point for Key Worker services. Many parents found this a refreshing departure from their previous experiences with services.

"She helps me to stand back and decide what are the priorities. Every one of the specialists has his or her own view, but [the Key Worker] helps me identify what is important for me and [my children]". (Parent)

Parents and caregivers perceived Key Workers as being *flexible, non-judgmental, responsive, respectful,* and *knowledgeable,* and valued these aspects of the program. Moreover, the program overall was viewed being very accessible, and this was important to families:

"It makes a difference that the Key Worker comes to my house. Accessibility is huge. I'm not sure I would have followed through to make an appointment without that – I am a single parent, and I work part time and am going to school, so flexibility is very important".

From the perspective of Key Workers, having a high degree of autonomy and flexibility enabled them to meet families' needs better; flexibility in how they carried out the role gave Key Workers the ability to spend as much time as needed when meeting with parents and/or their family members, and to decide where and how they would meet (for example, by telephone, via e-mail, in families' homes, and occasionally in the Key Workers' offices).

In addition, parents and caregivers valued the emotional and practical support they received from Key Workers. Along these lines, many parents and caregivers said that what distinguished the Key Worker program from others was that they, as parents/caregivers, were the focus of the service. As these two parents stated:

> "The Key Worker is a support to the family, not to the child, which was what I needed".

> "The Key Worker has come to my home and spent time with me one to one. The focus was on me. She was willing to come to my home when I was in crisis. We talked about stuff that was getting in the way of parenting. She helped give me perspective and focus on the positive. I could cry, be angry and be honest. We laughed together".

Community partners were likewise very positive about the role and impact of the Key Worker. Indeed, the themes in relation to community partners mirrored those of parents/caregivers: Key Workers were perceived as being knowledgeable about FASD and the families they worked with, accessible and available to consult with service providers/practitioners, and were a link to other service sectors.

> "The Key Worker was like buffer between my client and me. As a child protection social worker, I hold a lot of power, so the Key Worker really helps buffer communication between social workers and clients whose backs would often be up because of power dynamics".

Very few parents or caregivers identified anything that they did not like about the Key Worker and Parent Support program. The relatively few comments that were made came from families that had only had minimal time with their Key Worker and expressed feelings that not much, by way of support, had yet occurred.

17.6.4
Program Challenges

BC's Key Worker and Parent Support program, like all new programs, experienced some implementation issues during its first three years. Initially, these were concerned with clarifying the Key Workers' role and function, and developing consistency in service delivery approach across the province. Challenges to the program noted at the end of the three-year evaluation were cited as:

- In Year 3, the absence of Key Worker training that had been instrumental in helping to provide orientation and a conceptual framework for new staff and furthering skill development for existing Key Workers.

- Key Worker turnover, which in some areas was quite high. Findings in Year 3 of the evaluation linked Key Worker turnover to a combination of lack of training opportunities, low pay, and high expectations for the position.
- Program eligibility criteria, which were perceived by some Key Worker agencies and community partners to be unclear. In response to this, the program eligibility criteria were clarified by MCFD's Provincial Office such that only families with children and youth diagnosed with FASD or with a constellation of needs and challenges similar to those of children with FASD, were accepted for service.
- Potential for program drift resulting from pressures within multi-service agencies to deliver the Key Worker program along the lines of, and more in keeping with, other one-to-one intervention programs.

17.7
Summative Evaluation Findings

Summative outcomes for program participants (parents and caregivers) fell into four broad categories: (i) parents/caregivers increased their knowledge about FASD; (ii) parents/caregivers made changes to their parenting behaviors and responses; (iii) parents/caregivers felt supported; and (iv) parents/caregivers increased their access to services and resources. Qualitative findings produced the strongest evidence that program outcomes were being achieved. These outcomes are summarized below.

17.7.1
Increased Knowledge about FASD

Parents and caregivers consistently reported that they enhanced their knowledge and understanding of FASD as a brain-based disability; moreover, they attributed their heightened knowledge to their contact with a Key Worker or Key Worker-facilitated parent education groups. For example:

> "It helps to know that (my son) can't do things as opposed to he won't do things. I had a real problem with repetition. I thought: 'Why don't you get this? I have told you so many times.' Now I understand that I need to repeat things. That is how he learns. The part that is hard to accept is that he will always need to be reminded".

In addition and as a related outcome, findings revealed that parents and caregivers increased their understanding of their own child(ren)'s needs, strengths and capabilities, increased their understanding of the FASD assessment process and the assessment's recommendations, and increased their confidence in their parenting.

17.7.2
Shifts in Parenting Strategies and Responses

Parents and caregivers reported that, as a result of their enhanced understanding of FASD as a neurodevelopmental disability, they modified aspects of their parenting practices. Typically, they became more reflective and curious about the root causes of their child's behavior, more patient and not engaging in power struggles, made accommodations in the physical environment at home, and became more assertive with professionals. The following statements from parents and caregivers underscore these themes.

> "The program has given me tools and knowledge, which led to changes in my parenting style. Things are calmer around the house now because the Key Worker showed me that when son says 'I don't know' that he means it. I am way more understanding of my son. We used to think that when he said, 'I don't know' that he was just trying to avoid responsibility. It relieves frustration as a parent to have this understanding".

> "I have learned the ability to step back from intensity of our situation and think about what he may be processing and understanding".

> "My son kept running away–ALL the time. It is hard to say what would trigger him running away, but since I backed off in terms of my expectations, he has stopped running away".

> "I now know from the diagnosis that she has intellectual limitations. Before, she would not finish tasks around the house. So now, we have to simplify things for her and only do one thing at a time, like taking a shower and changing clothes".

17.7.3
Feeling Supported

Quantitative and qualitative findings from both the Exit Questionnaires and face-to-face community-based interviews with parents and caregivers over the course of the three-year evaluation revealed that parents and caregivers reported feeling less isolated, less stressed, better supported emotionally and practically, and having better "peace of mind" as a result of their involvement with the Key Worker and Parent Support program. The following comments from parents and caregivers illustrate the impacts of Key Worker support for parents and caregivers:

> "The Key Worker has been supportive, resourceful, helpful. It is huge for me to have someone like her available–knowing that she is there for support. She is so available through phone or email, and is flexible".

> "The Key Worker made a huge difference. I had no one to talk to. I felt very isolated and guilty. He helped fill in the answers and helped

us feel good about our parenting. He connected us with a couple of other parents".

"The program has allowed us to think ahead – we are not in crisis mode all the time. I can think and plan ahead and anticipate what is coming. I'm not perfect and don't have to have all the answers. It takes considerable burden off my shoulders".

17.7.4
Increased Access to Services and Resources

Parents and caregivers reported being better equipped to navigate various service systems following their involvement with the program. Numerous parents and caregivers specifically credited Key Workers with helping them to navigate the school system and facilitating their interactions with school personnel:

"The Key Worker helps take up some of the load/stress. I wouldn't have had time to do the research for new resources, so having the Key Worker to do that is such a help".

"The Key Worker has helped me access things I couldn't on my own. This helps my foster child, which makes me happier".

"I have been able to get special consideration for my child's needs in the school. Lots of recommendations that came from the Key Worker and the FASD assessment – the school is doing these. My son has support now. Before, the school would call and want me to pick him up. Now that the school knows I am working with a Key Worker, they are more supportive. I get support from people I didn't have before".

In terms of outcomes for community partners, qualitative findings over the three-year evaluation provided consistent evidence of several key outcomes; community service providers reported improving their understanding of FASD as a neurodevelopmental disability and increasingly sought out information regarding relevant accommodation strategies and interventions through the support of Key Workers. Moreover, community service providers reported that they were beginning to change their own practices based on their knowledge and understanding of FASD.

17.8
Discussion

Considerable research was undertaken by BC government staff prior to and in preparation for the implementation of the Key Worker program. Nonetheless, the BC program did not replicate an already established program or model that followed specific procedures laid out in a program manual. The program was also not piloted in a handful of sites prior to full-scale implementation. Instead, MCFD chose to launch the program province-wide from the outset. In addition, given MCFD's regionalized approach to service delivery, and the fact that the program was operated from several different types of service delivery agents/agencies, there was flexibility in development of the model and in delivery of Key Worker services.

As such, the model for the BC Key Worker program was consolidated during implementation. Province-wide Key Worker training, regional coordination, development of practice standards for the program, and formative and summative evaluation findings contributed to the ongoing evolution and consolidation of the Key Worker program during the first three years of implementation.

To date, a primary strength of the Key Worker and Parent Support program has been in its ability to be sensitive to and build capacity within families and communities. In this regard, the program utilizes a community development approach, increasing community knowledge and reaching many more people than typical intervention programs targeted to individuals.

17.8.1
Promising Practices

Based on the evaluation study, a number of promising practices emerged in relation to working with families with a child or youth with FASD. Foremost among these were the program's use of a *family-centered* model of practice and a *relational approach*, which recognizes that growth and change occur as a result of social interactions that are grounded in non-judgmental, supportive, and trust-based relationships (Umlah and Grant, 2003). In this regard, parents/caregivers spoke many times of their relief at finally feeling understood and not blamed for the child's shortcomings or behaviors. This non-judgmental approach was also key to being able to work with and gain the trust of communities and/or parents who were potentially struggling with guilt and shame associated with the effects of prenatal exposure to alcohol.

Additional promising practices identified were:

- *Holistic/ecological approach*, whereby the Key Worker works with parents/caregivers, siblings, extended family and even neighbors. This approach was well suited to working with Aboriginal families because it focused on the family rather than an individual within the family and acknowledged the importance of community ties.

- *Flexibility* as to where, when, and for how long Key Workers were able to work with families.

- *Collaborative approach* in working with community partners.

- *Thinking beyond "services,"* whereby the Key Worker pays attention to local contexts and opportunities and to the child's and family's need to be part of the larger community

The findings from both the formative and summative evaluations highlighted the connection between the Key Workers' provision of information and support, and parental and caregiver knowledge about FASD. With a better grasp of FASD as a neurodevelopmental disability, many parents and caregivers made important shifts in their own thinking and behavior. Their expectations of their child's behavior shifted to be more in keeping with the child's developmental age rather than his/her chronological age. Moreover, parents and caregivers began to think more about environmental antecedents to behavior, and about incorporating relevant environmental accommodations that would prevent certain problem behaviors. These shifts also led to parents and caregivers having increased confidence in their own parenting ability, which contributed to increased confidence in their ability to advocate on behalf of their child(ren).

Importantly, parents and caregivers credited the Key Worker role for promoting positive outcomes for families. They saw the Key Worker as source of information and resources and as a knowledge-related change agent. Both, parents and caregivers also reported that, in addition to a reduction in stress, they experienced improvements in their quality of life, especially in relation to peace of mind, and emotional and mental health. They also reported that Key Workers played a vital role in connecting them to social supports as well as community programs and services. These positive outcomes contribute to child and family resiliency; support stable, long-term placements for children with FASD; and have the potential over the longer term to contribute to the reduction of in secondary disabilities (Bertrand, 2009; Streissguth *et al.*, 1996).

17.9
Policy Considerations

Other jurisdictions can learn from the BC experience and the promising practices that emerged from the comprehensive evaluation of the Key Worker program (Hume *et al.*, 2009a, 2009b). First and foremost, policy makers are urged to consider funding the creation and ongoing implementation (i.e., not just short-term "project" funding) of a Key Worker-type of program in jurisdictions across Canada.

The BC Key Worker program is founded on research evidence and a clear conceptual framework related to FASD as a brain-based disability that responds to environmental accommodations. The focus of the Key Worker program has been to work with families and communities, rather than to implement specific child-focused interventions. Among the program's strengths are its flexibility, its holistic approach and its eligibility criteria, which have enabled all families (foster, adoptive and birth, including those who may have FASD themselves) with a child suspected of having FASD to access the program and obtain support, even prior to obtaining a FASD assessment or diagnosis. For policy makers, a key message is the importance of developing eligibility criteria that facilitate families' access to the program, as well as developing practice guidelines and standards that recognize the value of having flexibility in Key Workers' activities and in how, where, and when they work with families.

The formative evaluation showed that the success of the Key Worker program was linked to the existence of a number of program components and supports at a local, regional and provincial level. First, Key Workers must be skilled practitioners with extensive knowledge of FASD and community development. Training, clinical supervision, regional coordination, networking and information-sharing opportunities, and program practice standards contribute to creating a knowledgeable and stable workforce as well as facilitating quality assurance. Without these in place, there is a risk that the integrity of the Key Worker program will drift toward one-to-one intervention programs, such as family counseling or child-focused interventions. As a key message for policy makers, it is important to emphasize the provision of funding for annual or semi-annual training of all Key Workers. Policy makers and program developers are also urged to ensure that the supervisory component of the program is adequately resourced.

Furthermore, there is a need to develop expertise and a shared understanding of FASD as a neurodevelopmental disability among agency supervisors, managers and community service providers, as well as the Key Workers. Thus, annual provincial training should be provided not only to the Key Workers, but also to their supervisors, agency managers, and community members. Along these lines, information about FASD should form part of a mandatory curriculum in the professional training of community-based service providers in all health, education, and social service sectors.

The program's emerging promising practices–and in particular its use of a family-centered and a relational approach, as well as Key Workers' flexibility–have implications for program delivery, and thus for policy. Specifically, program planners and policy makers need to recognize that these approaches can only occur when Key Workers are permitted to have adequate time to develop trusting relationships with families, and to have families identify and prioritize their needs. Moreover, as families' needs and priorities change, so too can the nature of the work and the intensity with which the Key Worker engages with the family. From the families' perspectives, the BC Key Worker was not a short-term or time-limited service. A consideration for policy makers is that families living with a child with FASD will have ongoing needs, albeit needs and issues that ebb and flow over time. While a timeframe for service may be targeted, it is important that families believe they can access support through the program when need arises. Thus, allocating resources to enable the Key Worker program–and other programs of a similar nature–to have the capacity to utilize relational and family-centered approaches is critical to these programs' success.

17.10
Conclusions

The BC Key Worker and Parent Support program is a province-wide program that was established in 2005. The program model, which was grounded in a relational approach and a conceptual framework of FASD as a brain-based disability that responds to appropriate environmental accommodations, evolved and was refined during implementation. The summative and formative evaluations showed that the program was successful in making a difference for families. Using a relational approach the Key Worker and Parent Support program has been successful at building capacity in families to support children and youth with FASD.

References

Badry, D. (2009) Fetal Alcohol Spectrum Disorder standards: supporting children in the care of Children's Services. *First Peoples Child Fam. Rev.*, 4 (1), 47–56.

Badry, D. and Pelech, W. (2005) *Fetal Alcohol Spectrum Disorder Practice Standards Evaluation Project*, Centre for Social Worker Research and Development, School of Social Work, University of Calgary, Calgary.

BC Stats (2006) *Census Fast Facts: Children in BC Families*, Ministry of Labour and Citizens' Services, Victoria, BC, http://www.bcstats.gov.bc.ca/data/cen06/facts/cff0603.pdf (accessed 8 October 2009).

Bertrand, J. (2009) Interventions for children with Fetal Alcohol Spectrum Disorders (FASDs): overview of findings for five innovative research projects. *Res. Dev. Disabil.*, 30 (5), 986–1006.

Carmichael Olson, H. (2007) Fetal Alcohol Spectrum Disorders (FASDs): Meeting the Challenge of Intervention – or – How the Families Moving Forward Program Was Developed and Why! Powerpoint presentation, http://depts.washington.edu/fmffasd/FMFPublications.html (accessed 2 October 2009).

Greco, V. and Sloper, P. (2003) *Care Co-Ordination and Key Worker Schemes for Disabled Children: Results of A UK Wide Survey*, Social Policy Research Unit, University of York, York, UK.

Greco, V., Sloper, P., Webb, R., and Beecham, J. (2005) An Exploration of Different Models of Multi-Agency Partnerships in Key Worker Services for Disabled Children: Effectiveness and Costs, Research Report 656, Department for Education and Skills, University of York, York, UK.

Hume, S., Hubberstey, C., Rutman, D., Lentz, T., and Van Bibber, M. (2009a) *Final Formative Evaluation Report: Key Worker and Parent Support Program*, Ministry of Children and Family Development, British Columbia.

Hume, S., Hubberstey, C., Rutman, D., Lentz, T., and Van Bibber, M. (2009b) *Final Summative Evaluation Report: Key Worker and Parent Support Program*, Ministry of Children and Family Development, British Columbia.

Liabo, K., Newman, T., Stephens, J., and Lowe, K. (2001) A review of key worker systems for children with disabilities and development of information guides for parents, children and professionals: Summary, http://www.barnardos.org.uk/resources (accessed 10 November 2006).

Malbin, D. (2002) *Trying Differently Rather Than Harder*, 2nd edn FASCETS, Portland, OR.

Malbin, D. (2004) Fetal Alcohol Spectrum Disorder (FASD) and the role of Family Court Judges in improving outcomes for children and families. *Juv. Fam. Court J.*, 55, 53–63.

Poole, N. and Isaac, B. (2001) *Apprehensions: Barriers to Treatment for Substance-Using Mothers*, Centre of Excellence for Women's Health, Vancouver, BC.

Rutman, D., Callahan, M., Lundquist, A., Jackson, S., and Field, B. (2000) *Substance Use and Pregnancy: Conceiving Women in the Policy Making Process*, Status of Women Canada, Ottawa.

Streissguth, A., Barr, H., Kogan, J., and Bookstein, F. (1996) Understanding the Occurrence of Secondary Disabilities in Clients with Fetal Alcohol Syndrome (FAS) and Fetal Alcohol Effects (FAE). Final Report to the Centers for

Disease Control and Prevention (CDC), Technical Report No. 96-06, University of Washington, Fetal Alcohol & Drug Unit, Seattle.

Streissguth, A., Bookstein, F., Barr, H., Sampson, P., O'Malley, K., and Young, J. (2004) Risk factors for adverse life outcomes in Fetal Alcohol Syndrome and Fetal Alcohol Effects. *J. Dev. Behav. Pediatr.*, **25** (4), 228–238.

Umlah, C. and Grant, T. (2003) Intervening to prevent prenatal alcohol and drug exposure: the Manitoba experience replicating a paraprofessional model. *Manitoba J. Child Welfare*, **2**, 1–12.

Webb, R., Greco, V., Sloper, P., and Beecham, J. (2008) Key workers and schools: Meeting the needs of children and young people with disabilities. *Eur. J. Spec. Needs Educ.*, **23** (3), 189–205.

Whittington, B., Brown, L., MacKenzie, P., Pearson, T., Burns, D., and Gracey, M. (2007) *Supporting Grandparents Raising Grandchildren Resource Booklet: A Work in Progress*, 2nd edn, University of Victoria.

18
FASD and Education Policy: Issues and Directions
Elizabeth Bredberg

18.1
Introduction

Recognition and analysis of the learning and behavioral characteristics of Fetal Alcohol Syndrome (FAS) began to develop almost immediately on recognition, during the early 1970s, of the existence of the physical abnormalities, growth retardation and central nervous system damage associated with the condition (Jones and Smith, 1973; Jones *et al.*, 1973). Since then, knowledge of the impact of the neurological damage that ensues from prenatal alcohol exposure has expanded substantially. In their recent review of the implications of the findings of neurobehavioral research into what is now termed Fetal Alcohol Spectrum Disorder (FASD), Kalberg and Buckley (2006) described the complexity and diversity of the effects of alcohol exposure on learning and behavior, and the implications of that complexity for instructional planning for the individual child.

In addition to the challenges that the breadth of effect within FASD poses to the development of instructional practice for affected students, there remains to date minimal research into the efficacy of educational interventions for this population, and even less that is classroom-based. Although there is an emerging base of diagnostic capacity in North America, debate persists about diagnostic procedures, and access to diagnosis is limited. It is clear that many children with full FAS, and many more with FASD, remain undiagnosed (May *et al.*, 2009). Indeed, the most recent estimates of FASD prevalence suggest a value of between 2% and 5% of the school-aged population of North America and some countries of Western Europe (2006).

Thus, the makers of education policy are confronted by a daunting combination, of a spectrum disorder with a complex and varied range of effects, limited research-based information about instructional planning, and a very high prevalence in the general population. Moreover, the education community of practice has not always bought into the implications of the diagnostic findings of FASD (Ryan and Ferguson, 2006). Kalberg and Buckley (May *et al.*, 2009) have stressed the importance of integrating diagnosis and functional assessment for educational planning, but (in part due to a lack of access to diagnosis) it appears that has yet to become

Fetal Alcohol Spectrum Disorder–Management and Policy Perspectives of FASD. Edited by Edward P. Riley, Sterling Clarren, Joanne Weinberg, and Egon Jonsson
Copyright © 2011 WILEY-VCH Verlag GmbH & Co. KGaA, Weinheim
ISBN: 978-3-527-32839-0

common practice. There is, therefore, no body of common practice for evaluation, nor is there any consensus of what are realistic educational goals for prenatally alcohol-affected students. These challenges, whilst inherent in the population to be supported, are increasingly recognized at times of limited economic resources, when education authorities may face cuts in funding that not only threaten established programs but also reduce receptivity to programming for newly identified needs and for practices that have not been established as effective. If the population of students with FASD is to receive equitable and effective service delivery, then their distinctive characteristics must be recognized in education policy. As matters stand today, many elements of established policy can work against the interest of affected students, and this equally true of policy for special education and the general student population. In this chapter, some of these policies are addressed, and some alternative suggestions offered that will enable students within the FAS spectrum to benefit from education.

18.2
Where Do Students with FASD Fit Into the Education System?

A fundamental grasp of education policy in general is essential in order to address policy issues concerning any subpopulation in a realistic way. Turnbull (2008), when writing of policy contexts for special education for students with learning disability (LDs) in the US, used a "nested doll" image, working outward from policy that directly addresses LDs and extending to employment issues, social security issues and criminal justice issues. This is a useful approach, since educational goals must be understood as reflections of current social values.

There is growing evidence of a shift within educational policy and practice, from a model that treats education as an at least nominally egalitarian enterprise to a neoliberal model supporting a "...narrow, market-based conception of education, skill and talent" (Brown and Tannock, 2009). Individual skill-sets are seen as a commodity to be offered on a global market, a perspective that removes the role of education from a social service provided by governments to their citizens, to a contribution to a global corporate meritocracy. Although this perspective has yet to be made explicit in mission statements of education systems, the rhetoric of the "knowledge economy" is increasingly evident in the discussion of education policy (Ball, 1998; Levin, 1998).

Reconsiderations and discussions of services for students with special learning needs are occurring concurrently with this shift within policy discussion (Turnbull, 2008; Alberta Education, 2009). Although inclusion has been the predominant model of special education throughout North America and Western Europe for the past two decades, inevitably, if the goals of the general education system are under reconsideration, so too will be the elements within it. Thus, the discussion of inclusion seems to be departing from a recognition of the diverse ("exceptional") learning needs of specific populations and their support within education systems, to a more homogeneous diversity. A more individualized model can replace the necessity of recognizing and addressing exceptionality and the etiologic factors that may drive it.

This shift in perspective risks over-generalizing from observed behavior and function in an attempt to move away from a "medicalized" or "categorical" approach to services (Alberta Education, 2009; Gibbard, 2009). Without a grasp of the underlying causes of behavior, however, programming may not address the real needs of a student. This becomes especially true among students with FASD, whose behavior and learning styles may superficially resemble those of their non-disabled classmates but who, because of their neurological characteristics, may require quite distinctive instruction and support.

Public education systems, regardless of their response to shifts in the social mandate of education, retain a number of key policy elements that affect students with FASD. These elements include eligibility for enrollment, curriculum and programming, accountability, school discipline, and the qualifications set for education professionals. Likewise, inter-agency collaboration (e.g., with healthcare, social services, and corrections) is often an area of policy discussion. All of these policy areas begin by assuming developmental and behavioral norms of their students. Special education policy often addresses exceptions to the policies listed above, on the basis of student characteristics.

18.2.1
Eligibility

Typically, student eligibility for school enrollment is based upon age and residence within a specific catchment area. Schools or school districts are responsible for the education of students who have reached a certain age (often 5 years within North American schools) until they reach an upper limit. This upper limit often has two tiers—one after which students may leave school (or be excluded), and one after which students are no longer eligible for enrollment in public education. These age boundaries often fail to meet the needs of students with developmental exceptionalities, including those with FASD. The upper limit presupposes a level of maturation that no longer requires schooling. Both boundaries presuppose that the years within them are the years at which a student can optimally benefit from instruction. The developmental unevenness of students with FASD often indicates that the duration of their schooling could usefully be extended. Students with FASD do not learn at the same rate as their nondisabled classmates, and may need a longer period of time to benefit from the content of public instruction. They are, however, often excluded from schooling at the age of 18, with the expectation that their education is "complete," and that they can move into an adult role in society.

18.2.2
Exclusion and Discipline

In many instances, moreover, students are excluded at the younger eligibility boundary, because they are regarded as unable to benefit from instruction, or as unwilling to cooperate with school systems. Again, this application of an age norm to a developmentally immature youth is based on an expectation of capacity and responsibility established for a nondisabled population. It also assumes that the excluded student has the same capacity to function outside of the supports of an education system as a nondisabled student of the same chronological age. The same system that has a policy against expulsion of a child aged six will expel a youth aged 16, even though he or she may be, in many functional respects, on a par with the six-year-old.

Suspension and other school disciplinary measures, as first noted in 1997 (Streissguth et al., 1997), remain part of the experience of many students with FASD. A follow-up to the 1997 study (Streissguth et al., 2004) noted that, among a sample of 415 affected subjects, 60% had experienced or were experiencing negative school experiences, identified as either expulsion or suspension, or had dropped out.

18.2.3
Accountability: Curriculum and Learning Outcomes

Curriculum and learning requirements for each grade and for school completion are typically set by government departments of education, with varying degrees of flexibility afforded to school districts concerning the way in which they are met. Because these requirements are essential elements of any accountability framework, schools are often under scrutiny with respect to the percentage of their students demonstrating the successful achievement of learning goals for their grade. Evaluation often includes system-wide standardized achievement testing of all students within a selection of grade levels.

Students with some learning exceptionalities may be exempted from this testing, and from instructional programming driven by the standardized learning expectations of their grade. Exemptions are typically based on cognitive abilities, as indicated by IQ. As noted by Kodiwakku in a recent review of neurocognitive abilities of children with FASD, although the IQ-values of the population tend to be in the borderline to low average range, other neurocognitive capacity is often impaired (Kodiwakku, 2009). The use of IQ may, therefore, be an poor indicator to determine educational programming for students with FASD, and individualized learning plans based on a detailed neuropsychological assessment and an evaluation of classroom and adaptive skills is indicated for the population. Regardless of IQ, functional skills will be an essential component of the learning outcomes for students with FASD (Kalberg and Buckley, 2006).

18.3
Students with FASD within Special Education Systems

Instructional programming for students with FASD has been slow to develop and, as noted earlier, research-based interventions are very rare (Bohjanen, Humphrey, and Ryan, 2009). This situation is, in all likelihood, exacerbated by the diversity of neurocognitive profiles among the population, and by the relatively small percentage of affected students who have a valid diagnosis of the condition. The placement of students varies widely, and is not infrequently based on behavioral phenotypes rather than on any underlying etiology.

Students with FASD who have identified intellectual disabilities may be among those with the most appropriate programming (Streissguth et al., 2004; Kodi-wakku, 2009). Students with IQs in borderline and average ranges have been frequently categorized as presenting with behavior or mental health disorders as their primary disability, with programming that has not addressed their learning needs nor resolved their behaviors. Underlying receptive language disabilities are frequently overlooked or mistaken for oppositional behavior. Sensory issues among students with FASD have been identified, and several programs – most notably those within the Winnipeg School Division – have been developed with environmental modifications to address sensory issues. Although these have been shown to be successful in reducing behavioral difficulties and supporting student learning, rigorous research-based evaluations of their educational efficacy have yet to be conducted.

18.4
Education Professionals and FASD

Education professionals, whether teachers, administrators or classroom assistants, recognize the challenges they face in supporting students with FASD, whether in a regular classroom or a resource room (Bohjanen, Humphrey, and Ryan, 2009). Despite that recognition, the observed poor educational outcomes common to affected students, and the prevalence figures of the condition, there remains minimal pre-service instruction about students with FASD in most teacher education programs. A survey of faculties of education in North America revealed no programs that specifically addressed FASD, either in teacher education or in more advanced programs. Textbooks used in teacher pre-service education typically devoted between one half-page to five pages to the condition.

Many school districts offer teachers in-service instruction about FASD. However, these vary in their quality, and even the best seldom leave teachers or administrators with a sense of preparedness to support their students (Ryan and Ferguson, 2006; Bohjanen, Humphrey, and Ryan, 2009). Although several Canadian ministries of education have attempted to address this gap by preparing manuals about students with FASD, and joining in inter-ministerial initiatives, teachers generally remain under-informed and under-prepared to address the complexities posed by this student population.

The dearth of research-based instructional interventions does, admittedly, pose a limitation to the development of teacher education regarding FASD. In addition, education systems remain shy of the "medical model" of special education (Alberta Education, 2009), and often perceive a FASD diagnosis as an exclusively medical diagnosis, rather than as a functional diagnosis. There is, however, an emerging body of practice and expertise, and typically much education practice in other areas has become established without the substantiation of research (Levin, 2009). In addition, the integration of multidisciplinary diagnostic findings, including the neurocognitive characteristics of a student, with functional classroom assessment as described in Kalberg and Buckley (2006), seems a promising model for the development of informed and effective programming for students with FASD. There is, therefore, a substantial body of material that should form part of the instruction of new teachers, preparing them to support affected students.

18.5
Inter-Agency and Community Supports for Students with FASD

In addition to the physiologically situated effects of prenatal alcohol exposure, many children and youth with FASD experience post-natal adversities in their home lives. Many find themselves in foster placements, and among these many present with behavioral challenges that result in placement breakdowns and numerous relocations. This lack of stability, whatever its cause, has implications both for their mental health and for their schooling. Although an effective collaboration between guardianship agencies and schools represents a challenged area, there is potential to enhance the stability in the lives and schooling of students outside of the classroom. Community supports that specifically support parents and caregivers of children with FASD can play an important role in improving students' school outcomes.

Students with FASD are often described as benefitting from learning skills in the specific setting in which they are to be used. The transfer from school-based skills to their application in community and employment settings is less spontaneous than it is for the nondisabled student (Streissguth, 1997). Collaboration between future employers or community service providers and school professionals can facilitate this skill transfer for students. Despite an apparent cognitive capacity to meet established learning outcomes, students with FASD may benefit more from a modified instructional program, with a focus on "real-world" skills in a community setting, than on the analog skills that are often the focus of classroom instruction. The successful achievement of this learning will require collaboration between the school and the community, and also an awareness on the part of educators of potential placements within the community.

18.6
Policy indications

Eligibility criteria should be based on functional ability rather than on age or IQ. An exclusive reliance on age or IQ for admission and for provision of educational services is inappropriate for students with the developmental unevenness typical of FASD. Functional ability such as that described by Kalberg and Buckley (2006), situated in the classroom and community, are potentially effective guides for decision making.

Functional assessments must be sensitive to the culture and community of the child or youth assessed.
To date, most standardized functional assessments are developed in urban settings. Competence in culturally appropriate assessment is an essential for meaningful assessment, as is familiarity with community goals and norms. For rural and remote areas, these skills can include wildlife awareness and competence around farm animals. These may be more telling about a child's adaptive ability than an ability to function in a neighborhood supermarket or knowing a home address (which may simply be a post office box number).

The recognition of cultural strengths, both in Aboriginal and recent immigrant populations, must be appreciated.

Programming should be based on an integration of diagnostic assessment data (such as areas of executive functioning, sensory issues, and language abilities) and on functional assessment.
Programming should address and build on a student's abilities, as well as recognizing their disabilities. Strengths should be recognized and developed, *and their exercise should never be made contingent on meeting remedial goals or behavioral norms.* A role as a contributing member of a school community should be developed at an early stage in all children, rather than merely rendering them objects of remediation.

Educational goals should be directed towards students becoming contributing, rather than independent, members of society.
FASD is a long-term condition, and many of its disabling effects will remain with students for life. It will be far more valuable and realistic to recognize and develop areas in which students can make a contribution and/or hold down employment than to strive towards an unattainable independence. Areas of disability should not be made impediments to the exercise of skills. For example, students with a poor sense of time and poor travel skills should be supported in these areas, in order to enable them to reach a place of employment training on time.

Disciplinary measures should take into account the fact that students' working memory deficits may make consequences ineffective deterrents to inappropriate behavior.
Pro-active measures will be a more effective means of addressing behavior issues. If intervention is unavoidable for safety reasons, it must be recognized as a temporary strategy rather than a solution, and the likelihood that such intervention will not reduce future occurrences of a behavior must be recognized.

Disciplinary measures such as suspension and expulsion should be eliminated.
Excluding a student with FASD from the classroom is an ineffectual way of addressing inappropriate behavior. In many instances, the behavior is likely to be

a response to stimuli or demands within the classroom that the student finds aversive. Removing the student temporarily will not address this problem. Both, expulsion and suspension put students at risk of involvement with undesirable and dangerous community elements that they are even less able to recognize and avoid than their age peers.

Functional behavior analysis and positive behavior support should be applied in lieu of punishment.
The learning environment should be seen as extending beyond sensory stimuli. The behavior of an entire school community should be shaped into an environment in which all students, and not just those with FASD, can learn and function in a healthy and positive way.

Initial baselines for skills of students with FASD should be recorded and growth and development monitored, through Individual Education Plans (IEPs) and through appropriate classroom-based assessment.
The actual learning trajectories of students with FASDs in effective learning conditions have yet to be observed or clearly understood. Recording learning and progress can inform this understanding and enable the evaluation of both programming and student success.

University faculties of education and teacher education programs should develop courses in FASD, and programs of research into instructional effectiveness. Instruction about FASD should be obligatory for any teacher seeking qualifications in special education.
To date, minimal research into the instruction of students with FASD has been conducted at faculties of education. It is the responsibility of the education system to educate all students, and the role of universities should include research into the effective instruction of this population. School districts should collaborate with universities to provide informed and rigorous in-service instruction about FASD, and to act as settings for program evaluation.

Education and healthcare systems should collaborate to provide optimally informed diagnosis, assessment and support for students with FASD.
The value of a multidisciplinary assessment in instructional programming is often overlooked in favor of the categorization of the student for funding purposes. Although this omission is not in the student's interest, diagnosticians and other healthcare professionals involved with children and youth with FASD must also recognize the value of informed observation by educational professionals. Classroom observation can inform diagnosis in meaningful ways, particularly when the channels of communication between the two professions are open to respectful information sharing.

Input from parents and caregivers should be sought and respected by education professionals.
Cultural issues, social issues, and difference in goal perception are but a few of the impediments to meaningful collaboration between parents and teachers. The

very real stressors involved in parenting a child or youth with FASD should be acknowledged by school systems, and additional challenges such as expecting a parent to be ready to intervene in behavioral episodes should be recognized as unrealistic and, in many cases, punitive.

The diversity inherent in the spectrum nature of the condition should be reflected in educational programming for students with FASD.
Not all students with FASD will benefit from similar programming, nor can similar outcomes and expectations be derived from a diagnosis. Again, functional assessment and diagnostic findings can serve to shape decisions regarding optimal programming and placement and meaningful evaluation of program and student achievement.

18.7
Conclusions

Although the understanding of the nature of this complex and challenging condition has moved forward during the past few decades, education policy has not kept pace with that progress. The "instructional nihilism" that existed early in the history of the diagnosis must not continue to influence programming; nor can programming simply be based on interventions that are effective with other populations. The distinctiveness – but also the potential – of students with FASD must be recognized and addressed in all education systems and in all communities. Failure to do so will be at the cost, not merely of students with FASD and their families, but of contemporary society.

References

Alberta Education (2009) *Setting the Direction Framework*, Alberta Education.

Ball, S.J. (1998) Big policies/small world: an introduction to international perspective in education policy. *Comp. Educ.*, **34**, 119–130.

Bohjanen, S., Humphrey, M., and Ryan, S.M. (2009) Left behind: lack of research-based interventions for children and youth with fetal alcohol spectrum disorders. *Rural Spec. Educ. Q.*, **28**, 32–37.

Brown, P. and Tannock, S. (2009) Education, meritocracy, and the global war for talent. *J. Educ. Policy*, **24**, 377–392.

Gibbard, W.B. (2009) Extent and impact on child development. Conference presentation, Consensus Development Conference on Fetal Alcohol Spectrum Disorder – Across the Lifespan, October 7–9, 2009, Edmonton, Alberta. Power-point, http://www.buksa.com/fasd//docs/004-Gibbard.pdf (accessed 23 January, 2010).

Jones, K.L. and Smith, D.W. (1973) Recognition of the fetal alcohol syndrome in early infancy. *Lancet*, **2**, 999–2001.

Jones, K.L., Smith, D.W., Ullelean, C.H., and Streissguth, A.P. (1973) Pattern of malformation in offspring of chronic alcohol mothers. *Lancet*, **1**, 1267–1271.

Kalberg, W.O. and Buckley, D. (2006) Educational planning for children with fetal alcohol syndrome. *Ann. Ist. Super. Sanità*, **42** (1), 58–66.

Kodiwakku, P.W. (2009) Neurocognitive profile in children with fetal alcohol spectrum disorders. *Dev. Disabil. Res. Rev.*, **15**, 218–224.

Levin, B. (1998) An epidemic of education policy: (what) can we learn from each other? *Comp. Educ.*, **34**, 131–141.

Levin, B. (2009) Build a relationship between research and practice. *Phi Delta Kappa*, **90**, 528–529.

May, P.A., Gossage, J.P., Kalberg, W.O., Robinson, L.K., Buckley, D., Manning, M., and Hoyme, H.E. (2009) Prevalence and epidemiologic characteristics of FASD from various research methods with an emphasis on recent in-school studies. *Dev. Disabil. Res. Rev.*, **15**, 176–192.

Ryan, S. and Ferguson, D.L (2006) On, yet under, the radar: students with fetal alcohol syndrome disorder. *Except. Child.*, **72**, 363–379.

Streissguth, A.P. (1997) *Fetal Alcohol Syndrome: A Guide for Families and Communities*. Paul Brookes Publishing, Baltimore, London.

Streissguth, A.P., Barr, H., Kogan, J., and Bookstein, F. (1997) Primary and secondary disabilities in fetal alcohol syndrome, in *The Challenge of Fetal Alcohol Syndrome: Overcoming Secondary Disabilities* (eds A. Streissguth and J. Kanter), University of Washington Press, Washington, pp. 25–39.

Streissguth, A.P., Bookstein, F.L., Barr, H.M., Sampson, P.D., O'Malley, K., and Kogan Young, J. (2004) Risk factors for adverse life outcomes in fetal alcohol syndrome and fetal alcohol effects. *Dev. Behav. Pediatr.*, **25**, 228–238.

Turnbull, H.R. (2008) Today's policy contexts for special education and students with specific learning disabilities. *Learn. Disabil. Q.*, **32**, 3–9.

19
Shifting Responsibility from the Individual to the Community
Audrey McFarlane

> Never doubt that a small group of thoughtful, committed citizens can change the world; indeed, it's the only thing that ever has.
>
> *Margaret Mead*

19.1
Introduction

Serving individuals with Fetal Alcohol Spectrum Disorder (FASD) may require a shift in thinking and in how such services are delivered in communities. The Lakeland Center for Fetal Alcohol Spectrum Disorder (LCFASD) is one example of a community collaboration and community development model prepared to meet the needs of individuals with FASD and their families. This chapter describes the development, services, successes and challenges of the LCFASD, together with several policy recommendations based on the sixteen years of community development and service to individuals with FASD.

19.2
Why Do We Need to Make a Shift?

Individual responsibility is a key cornerstone of Canadian society. If you do harm in society, you must take responsibility for your actions; if you make a mistake at work, you are expected to take responsibility; if you mess up as a parent, you may lose your responsibility in that role. Society expects individuals to follow its rules, and if you don't you will be punished. This is based on a belief that we are each able to be responsible citizens if we try hard enough; however, when individuals with undetected cognitive disabilities are held to this kind of thinking, it frustrates society that our punishments are not having the desired behavioral changes.

Part of this individual responsibility belief is grounded in the Canadian (or North American) goal for independence. We want people to be independent to live their lives, and North American Governments often make this a goal for its citizens. We measure success in society by how hard we work, how much money we make, how many things we have, how many exotic trips we take, and sometimes by how much we give to others. These are seen as independent activities or achievements. But if we examine our lives, we do not do any of these things alone: our partners plan the trips that we go on; we only work as hard as those around us will allow; we rely on our families and friends to help us when we are in a difficult situation; and we seek out professional supports when we need them. We are not independent people – we are *interdependent* people who rely on each other for almost everything.

People with Fetal Alcohol Spectrum Disorder (FASD) have varying degrees of cognitive damage that often requires them to have a range of supports in order to function well in society. It is hoped that supports at an early age will reduce the outcomes for many youth and adults with FASD (Malbin, 1993). Adults with FASD often face homelessness, children that they have difficulty in raising, overrepresentation in the legal system, and alcohol and drug abuse (Clark *et al.*, 2004). Today, however, systems are not coordinated together, but rather expect those needing to access such services to be capable of navigating the agencies and programs without assistance, and often to be independent to prove they can benefit from certain services. Yet, when well-intentioned agencies and well-meaning staff follow these procedures, it can become system abuse (Booth, McConnell, and Booth, 2006) for individuals with FASD.

To make a shift in thinking about individuals as needing an interdependent existence in order to be productive citizens would begin to better support individuals with FASD in the community.

19.3
Examples of Individual's Situations

The first of these examples is true, and permission has been granted for its use in this chapter.

Example 1: A young boy was first diagnosed by the Lakeland Center for FASD (LCFASD) when he was aged 10 years, at which time he had a stable and loving foster parent in his First Nations community. The LCFASD post-diagnostic outreach worker has had minimal involvement with the family because the skills of the foster parent were meeting the boy's needs. However, when the teenager's foster mother died unexpectedly just before his 18th birthday, the LCFASD was contacted to provide the Children Services department with information on what services they might consider applying for. The Children Services felt they had the situation under control, and support was not required from the LCFASD. However, about a year later the lad appeared at a LCFASD office and asked for help. He was homeless (living in a old abandoned car); people were looking for him to beat him up; he was actively using drugs when available; he had no money; and any extended family had told him he was not welcome. Thankfully, he was not in trouble with the law. It should be noted that this young man had an IQ of 51. So, what to do? The On-Reserve Children Services said he was aged over 18 and not their responsibility because he had not signed an extended agreement; the provincial Children Services said that it was On-Reserve's jurisdiction; the On-Reserve financial services said that they had given him funds earlier in the month, he could get no further funds, and if he was now living off reserve it was not their jurisdiction; provincial Employment and Immigration would support him once he had an address, but there were no men's shelters or homeless places in rural communities; the extended family was unwilling to help; Provincial Persons with Developmental Disabilities said that they would not open a file as he was still considered On-Reserve, but they could not provide a date when he would be considered Off-Reserve. Now, this situation was frustrating and downright maddening: how is a young man of IQ 51 expected to maneuver this kind of system, independently?

This situation highlights the system breakdowns that are experienced by individuals with FASD who are also from First Nations communities. The frustrations in jurisdictional protectionism paired with individuals who do not understand waiting for services to begin, or money to flow, makes service delivery very complicated. Thus, many of these individuals continue to couch-surf and move about, making service engagement even more challenging.

Although this second example involves an adolescent with FASD, it is not an exact story but rather a repeated theme in serving youth with FASD.

Example 2: A young girl who was diagnosed with an FASD when aged 10 or 11 years lives with a long-term foster family who are committed to her. School has managed her cognitive deficits with a modified program of less homework than her peers, and there have been occasional behavioral difficulties involving following peers, or being provoked to do something. When in grade 7 or 8, aged about 13 years, the girl becomes belligerent and listens only to her friends, who have unhealthy interests such as smoking and experimenting with alcohol. This negative attitude becomes progressively more normal, the foster parents seek respite supports, but likely there are minimal opportunities for them. They also seek counseling supports for the girl, to help her make better decisions. The counseling sessions become a reason to develop a crisis so that she has something to talk about with her therapist. Her behavior continues to escalate; the school is unwilling to modify her academics as they believe that she can do the work if she applies herself, because her IQ is in the normal range. The school will be unwilling to provide a social coach for the teen, and appear to allow the situation to worsen until they are able to expel her. The teen will run away a few times, or does not come home for an evening. Children Services will remove the child and the family feels blamed for this behavior. One or two attempts will be made to reunite the family and the child without supports; these are unsuccessful. The teen is then placed in a variety of placements where the negative attitude continues to escalate, but now as a coping mechanism. As the girl approaches legal age, Children Services becomes less committed to her and she runs away to the city and begins working on the streets. She will return in a year or so to the community, pregnant or with a child, who may have been exposed to alcohol and drugs. It should be noted that this situation also happens to young men.

This scenario highlights the repeated system failures of expectations of young people with FASD, and the lack of thoughtful community resources. These situations can be avoided by understanding the disability and working together to plan services.

19.4
One Model of Community of Care

There is not one model that will work for all communities, nor will it meet all the needs of individuals with FASD. One model that has worked well in Northeastern Alberta is the LCFASD, which has linked diagnosis, post-diagnostic supports and prevention with high-risk women by using a community-based delivery model.

The LCFASD is a unique rural, not-for-profit organization that is based on the principles of interdependence, lifelong support, harm reduction, prevention of problems, and respect. This rural area is located in north eastern Alberta, nestled up against the Saskatchewan border and about three-and-a-half hours from Edmonton. The geographical area has a travel distance of about a three-hour drive across the region, and serves almost 100 000 people; it consists of one small city (Cold Lake), 25 small towns, seven First Nations communities, four Métis[1] settlements, and one military base. This organization was a result of hard work and dedication by an FAS committee who believed that, in order for systems to change to meet the needs of individuals with FASD, communities needed access to diagnostic/assessment services and support for families to advocate for their needs after diagnosis.

19.4.1
Diagnosis

The LCFASD has developed a community-based model of service delivery (McFarlane and Rajani, 2007) in order for individuals exposed prenatally to alcohol to access diagnostic and assessment services in this rural area. The diagnosis and assessment for children is completed by a multidisciplinary team that is comprised of professionals and clinicians from the community: Speech Language Pathologist; Occupational Therapist; Mental Health Therapist; Addiction Counselor; Public Health Nurse; Teacher; Social Worker; and Cultural Liaison. These individuals are given one or two days per month from their community agencies to

1) Métis settlements are located across the northern part of Alberta. The Métis are the descendants of European fur traders and Indian women who emerged as a distinct group on the Prairies towards the early part of the nineteenth century.

participate on the diagnostic teams, with their salary being provided by their agencies. From outside the community, the Neuropsychologist and Pediatricians travel from Edmonton to participate on clinic days, and are provided an honorarium or fee for service. The Lakeland area does not have these specialties. The LCFASD provides a Clinic Coordinator and post-diagnostic support workers who participate in the clinic, and receives funding for these positions from a variety of government departments.

The adult diagnostic clinics were developed in 2002 and function in the same manner as the children's diagnostic clinics, but include from the community: Physician; Mental Health Therapist; Justice Representative; Disability Services Representative; Career Counselor; Addictions Counselor; and Cultural Liaison. From outside the community, a Neuropsychologist and a Psychiatrist are provided a fee for their services. The LCFASD provides the clinic coordinator and post-diagnostic outreach support workers.

The LCFASD is responsible to support the intakes and the referrals, to ready the families for clinic, to collect supporting documentation, determine alcohol exposure, and provide long-term post-diagnostic outreach support and file management. The post-diagnostic support workers assist families and individuals in understanding the disability, following through on the diagnostic team recommendations, developing support systems that meet their needs, and ongoing case management. The post-diagnostic support is available for as long as it is required by the family or the individual.

As of early 2010, the traveling diagnostic/assessment teams have seen 350 individuals in the Lakeland area. The LCFASD has current capacity to see four children per month, and one adult per month. The LCFASD diagnostic/assessment teams follow the Canadian Guidelines (Chudley et al., 2005).

The LCFASD has experienced very few "no-shows" at clinics, as families are well-supported prior to clinic as part of the intake process. Initially, most of the children seen in the clinic were part of the Children Services department, and today about 30% of the children are still within a biological home, with referrals coming from schools, doctors, community nurses, Children Services, and parents.

The benefits to the partnering agencies such as Health Services, Children Services, School authorities, and community agencies are many, including:

- having a local FASD diagnostic/assessment service to refer clients to;
- building an FASD expert in their agency;
- increasing staff retention in rural areas with collaborative work;
- the agency staff feel less overwhelmed by working on complex cases together;
- collegial relations for complex cases outside the FASD team are increased;
- there is a cross-sector understanding of roles; and
- the identification of needed system changes is made from within partnering agencies.

The benefits to the family and individuals include:

- access to a local services;
- no or minimal travel to services, as the diagnostic/assessment teams travel to the communities in the Lakeland rather than being static in one locale;
- partnering agencies are invested in serving individuals with FASD;
- recommendations from the diagnostic teams are relevant and appropriate, as they know the services that are available and the informal community supports;
- pre- and post-diagnostic supports encourages families to seek services;
- team members are often known to the families or individuals thus reducing anxieties; and lifelong outreach supports are available.

The benefits to the community include:

- a greater awareness of the problems associated with alcohol and pregnancy due to the large number of community members involved in diagnosis;
- a greater understanding and awareness of the needs of individuals with FASD as follow-up outreach workers assist each family in accessing the services they need and how they need them, such as school programming;
- a service that one can make referrals to; and
- a sense of ownership of the services because of the wide range of community involvement.

The challenge with this type of model is to maintain a trained team of people, as rural areas tend to experience a large amount of professional turnover. In order to address this issue, the LCFASD provides annual Diagnostic team training for new and back-up team members. Back-up team members are identified for all professionals, and fill in for holidays or illness; they may also fulfill the role when a position becomes vacant, in order that canceled clinics are minimized. Another challenge is the extensive travel required of team members to move around the region, but more so for the outreach workers who have large travel expenses. Notably, the travel costs are supplemented by the LCFASD-fundraised dollars, as government rarely covers the full costs of infrastructure to programs.

19.4.2
Prevention

Diagnosis is linked to prevention. Through the process of diagnosing children, mothers are found who may still be in the cycle of drinking and having babies. Each time the diagnostic clinic finds such a mother she is supported by a Mentor, who works long-term and intensively in an outreach capacity to support the women to be alcohol- or pregnancy-free. Referrals can also come from other sources in the community. The Parent–Child Assistant Program (PCAP) (Grant,

Ernst, and Streissguth, 1996) program, which the LCFASD refers to as "Mothers-to-be Mentorship," is the most intensive form of prevention by supporting women over a three-year period in an effort to get their lives back on track. This program is very integrated into each of the communities in the Lakeland, with sub-offices in three locations plus the main office, as the program needs to work very closely with community partners to assist the women in moving to a healthier lifestyle. Currently, the LCFASD has the capacity to serve 40 women.

The circle back to diagnosis may occur if the women herself, or her children, need to have an FASD assessment.

19.4.3
Intervention

Diagnosis/assessment and mentorship to pregnant substance using women are termed interventions. The LCFASD has identified post-diagnostic outreach supports as ethical and critical to the diagnostic process, as well as the link to mentorship programming for the women. The post-diagnostic outreach services work with the families to follow through on the recommendations of the clinic team, assist in linking to additional community resources, advocate for needed services, rebuild bridges with community agencies, assist adults in meeting basic needs, and listening to families. This case management model shows promise in linking all the services together over a long period of time. However, the more outreach supports provided, the more gaps in services are identified. When the current community services are unable to meet these needs, the LCFASD works to develop new service models to demonstrate effective service delivery that others can duplicate or expand upon.

Specific programs developed by the LCFASD include a summer camp that is specifically designed for children with FASD. These camps have been a huge success and, with children and families returning each year, the agency finds itself needing to build a larger facility. The LCFASD has recently launched a transition program to guide youth with FASD to adult services. This new program is promising to reduce the gaps in services and maintain stability into adulthood. Also newly launched is an employment coordination program to assist adults with FASD who are stable and wanting to work or volunteer to enter this phase.

The LCFASD provides training and education opportunities to anyone who is interested, numerous awareness campaigns, information sharing and community development in FASD around the country. In a recent review of training completed by the LCFASD it was discovered that over 10 000 people within the Lakeland area had received training from the LCFASD.

19.4.4
Outcomes

Does this model improve outcomes for the individuals with FASD? In two independent evaluations (Stonehocker, 2004, 2007) of the services, the LCFASD families and individuals have indicated that they feel supported and in a better position with the diagnosis than without it (Stonehocker, 2004). The following example highlights the effectiveness of this model:

Example 3: A mother, with the encouragement of Child and Family Services, sought diagnostic services from the LCFASD for her teenage daughter who was experiencing extreme behavioral problems. The LCFASD provided diagnostic/ assessment services, and the daughter received a diagnosis of Alcohol-Related Neurodevelopmental Disorder. The post-diagnostic outreach services soon discovered that the mother was pregnant with her 26th pregnancy, and was still in the cycle of binge drinking. The mother was offered the mentorship program, and she readily accepted the help. Since having this baby, with the support of the Mentor, she has had permanent birth control. The baby, who is now 5 years old, is waiting to be seen by the diagnostic team. The teenage daughter found herself in an unstable relationship and became pregnant, but did not use substances and had a healthy baby; however, she still receives extensive supports from the LCFASD as she has developed mental health issues and needs an advocate to access services. The daughter has stable birth control. The mother's other five living children that are not in her care have been referred for diagnostic services and supports. The mother, who continues to struggle with binge drinking, has been employed with the support of the LCFASD. A whole family has been supported by this model. It is great when families are identified when the mother is on her first or second pregnancy, rather than the 26th.

19.5
History

How did a model such as the LCFASD come about in rural Alberta? A dedicated group of Lakeland professionals began the Lakeland FAS committee in 1994 after a conference in the area on FAS/Fetal Alcohol Exposure (FAE). This group of professionals focused on raising the awareness of prenatal alcohol exposure, and of the disability of FAS, the aim being to work towards a better understanding of the needs of individuals with FASD and the services that would support them. The committee continued to gain momentum and interest until a belief was finally formed that community action was not enough to change systems or service delivery, unless a diagnosis was provided. From this point, the committee worked to develop a diagnostic service and launched the first clinic in 2000, after a team had been trained at the University of Washington.

The Committee had no legal structure, no money, and no mandate from anyone, other than the energy and dedication to make a diagnostic service work. Along with this desire for diagnostic services was a commitment to provide post-diagnostic supports and outreach to substance-using pregnant women. In 2001, the Committee was successful in securing funding for two projects with provincial and national funders, and in 2003 moved to become a not-for-profit society and charity of Canada.

The originating committee utilized community economic development practices to build the FASD services and supports needed. The use of social capital is the practice of utilizing personal and professional networks to move a particular issue forward (Lotz and Macintyre, 2003). The committee used this method to spread the information about FASD, to expand a knowledge base, and to encourage participation in the committee work. A strategic decision of which professional groups needed to be involved were targeted, thus further building the capacity. Both, politicians and community leaders were engaged in the same manner to highlight the need for services and to give voice to the needs. The principles of Community Economic Development (CED) include equity, participation, community building, cooperation and collaboration, self-reliance and community control, integration, interdependence, capacity building, diversity, and appropriate indicators (Simon Fraser University Community Economic Development Centre, 1997). These principles were the key elements in moving the issues of individuals with FASD forward in the Lakeland area. CED holds many keys to understanding the engagement and development of new community services, particularly those of a difficult nature. The LCFASD continues to utilize these principles today to sustain and grow needed services.

19.6
Future

In 2010, the LCFASD will celebrate 10 years of FASD diagnostic services. None of this would have been possible to build if the communities served had not been prepared, involved, and committed. As noted, CED principles were utilized by the Lakeland FASD Committee over a six-year (1994–2000) period of readiness. In 2010–2011, the LCFASD will undergo an extensive independent evaluation of the entire model, so as to offer the center some feedback for future decision making and to provide developing agencies or communities with a model option.

In development are:

- plans for residential supports for youth with FASD who have the highest risk behaviors and are currently not well supported;
- an emergency residence for adults with FASD who find themselves homeless and unable to access any other supports;
- a women's health center to meet all medical and social needs; and
- a residential home for young women who are pregnant, using substances and homeless.

In most of these new developments, the program models have been developed and capital dollars are being sought to bring these models to completion.

19.7
Policy Considerations

Although the LCFASD works hard to bring about change for all the individuals it is involved with, it does not always meet with success, usually due to system gaps or failures. The following policy recommendations would greatly assist in making a difference in the lives of individuals with FASD and to build communities of care:

1) *Funding* to establish programs and supports to individuals and families is important. Along with funding, *time* is needed to prepare the community to receive these services, while they develop the service models that will best serve the purpose. On many occasions, community services and projects have been seen to fail because they have no foundation on which to grow.

2) On/Off First Nations inadequacies need to be addressed. It cannot simply be on or off reserve status – people move about, are homeless, or couch-surfing – it is almost impossible to find them supports when they are in this unstable situation. The back and forth of whose jurisdiction it is, is not helpful to individuals with FASD. To be clear, FASD is not an Aboriginal issue – prenatal alcohol exposure includes all cultural groups (Koren and Nulman, 2006). It is the present author's experience that Aboriginal communities have been most prepared to address the issues related to FASD; given the long history of alcohol-related problems, it is not a difficult leap that this use affects the babies.

3) In developing policy, plan to serve the most complex of individuals with FASD, such as the young man described above in Example 1. Meeting the needs of the most complex will inform systems to meet the needs of those that are less complex. There is a tendency to meet the easiest needs first, while somehow never getting around to meet the most complex of society (Elliott, 2005).

4) Recommend payment models for professionals that reward the work that is needed to serve those with FASD (Elliott, 2005). It takes a lot of time to support and keep stable those individuals with FASD; professionals such as doctors and psychiatrists are not supported to take this time.

5) Significant cost savings could be found in reducing the number of individuals with FASD who enter the legal system. Catching them on a first offense and redirecting them into community services has proven to reduce recidivism in these individuals (Waage and Debolt, not dated). The present author's experience has been that individuals may be charged for a minor offense, such as stealing cup hooks at Canadian Tire. Yet, because they do not appear at court they receive a warrant for their arrest; then they do not know how to get a lawyer and so arrive in court without one; the process just keeps

going, to end up on probation. So, the systems have spent thousands of dollars in court time for a person who has stolen cup hooks. A couple of outreach workers instead would be much more cost-effective!

6) A policy to address housing and residential programming is critical to ensuring stability and productivity for youth and adults with FASD. Having a continuum of residential options for youth and adults with FASD will reduce overall costs to the human service systems and develop more productive individuals.

7) Emergency placements in rural areas is important, to stabilize individuals and keep them safe while their needs are assessed as they are linked to services. Women shelters are an excellent support to women, but men are often left to find accommodations on their own, leaving them disconnected from services.

8) In-house facility addiction programs that target pregnant homeless women. The number of these is growing, and many are themselves FASD and having children who have been exposed prenatally to alcohol. Women who are using substances are unable to access many women shelters.

9) As all of these services are built, or system changes made to better meet the needs of individuals with FASD, a skilled work force will be needed. Advanced education needs specific programs and information integrated into current human services, education, and medical programs. Specific college programming needs to be established to utilize skilled front-line workers.

10) A focus on research to assist governments in making good policy decisions is critical. The field of research in FASD services is just developing, and needs continual support in order to better understand best practices and to support the work of new initiatives.

11) The stigma associated with FASD is ours, as society, not the women's. There is a need to move beyond this to build the supports within current systems, and new systems to better meet their needs, and the needs of their children.

19.8 Conclusions

Policies that are thoughtful of promoting interdependent opportunities for individuals with FASD, and provide communities with CED strategies will result in long-term sustainable, positive outcomes. Successful examples such as the LCFASD are important to examine in order to understand and build upon this success. However, it is also important to explore new ideas and give opportunities to communities who have developed creative solutions to meet the needs of individuals with FASD.

"There are risks and costs to a program of action. But they are far less than the long-range risks and costs of comfortable inaction."

John F. Kennedy

Acknowledgments

The author would like to thank especially the Lakeland Center for FASD for providing the opportunity to carry out the research for this chapter. These data are also used in her research essay, *Fetal Alcohol Spectrum Disorder: Is there a Community Economic Development Solution?*, for the MBA(CED) Program, Cape Breton University, 2009.

References

Astley, S. and Clarren, S. (1999) *Diagnostic Guide for Fetal Alcohol Syndrome and Related Conditions: The 4 Digit Code*, University of Washington, Seattle.

Booth, T., McConnell, D., and Booth, W. (2006) Temporal discrimination and parents with learning difficulties in the UK child protection system. *Br. J. Soc. Work*, **36** (6), 997–1015.

Chudley, A.E., Conry, J., Cook, J.L., Loock, C., Tosales, T., and LeBlanc, N. (2005) Fetal alcohol spectrum disorder: Canadian guidelines for diagnosis. *Can. Med. Assoc. J.*, **172** (5), 1–21.

Clark, E., Lutke, J., Minnes, P., and Ouellette-Knutz, H. (2004) Secondary disabilities among adults with fetal alcohol spectrum disorder in British Columbia. *J. Fetal Alcohol Syndr. Int.*, **2** (e13), 9–11.

Elliott, D. (2005) No Simple Solutions for Complex Needs. Canadian Public Policy. V XXXI Special supplement on Mental Health reform in the 21st Century, August 2005, pp. 53–58. Available at: http://economics.ca/cpp/en/special2005.php

Grant, T., Ernst, C., and Streissguth, A. (1996) An intervention with high risk mothers who abuse alcohol and drugs: Seattle Advocacy model. *Am. J. Public Health*, **86** (12), 1816–1817.

Koren, G. and Nulman, I. (2006) *The Motherisk Guide to Diagnosing Fetal Alcohol Spectrum Disorder (FASD)*, The Hospital for Sick Children, Toronto, ON.

Lotz, J. and Macintyre, G. (2003) *Sustainable People: A New Approach to Community Development*, Cape Breton University Press, Sydney Nova Scotia.

McFarlane, A. and Rajani, H. (2007) Rural FASD diagnostic services model: Lakeland Centre for fetal alcohol spectrum disorder. *Can. J. Clin. Pharmacol.*, **14** (3), 301–306.

Malbin, D. (1993) *Fetal Alcohol Syndrome/Fetal Alcohol Effects: Strategies for Professionals*, Hazelden, Center City, MN.

Simon Fraser University Community Economic Development Centre (1997) *Principles of Community Economic Development*. Simon Fraser University, Burnaby, BC Canada. Available at: www.ccednet.ca (accessed 17 September 2008).

Stonehocker, D. (2004) Final Evaluation Report Lakeland Centre for Fetal Alcohol Spectrum Disorder, Cold Lake, AB Project #242, www.lcfasd.com (accessed 4 October 2009).

Stonehocker, D. (2007) Adult Support Coordinator Program Evaluation, Lakeland Centre for Fetal Alcohol Spectrum Disorder, www.lcfasd.com (accessed 4 October 2009).

Waage, M. and Debolt, D. (not dated) Lethbridge Community Justice Project: Case Management for Fetal Alcohol Spectrum Disorder, http://www.cvfasd.org/pdf/commjustice.pdf (accessed 15 January 2009).

20
A Social Work Perspective on Policies to Prevent Alcohol Consumption during Pregnancy
Mary Diana (Vandenbrink) Berube

I am the mother of two adopted children (now grown) with Fetal Alcohol Spectrum Disorder (FASD). I am trained as a Social Worker, and have held various positions with ties to the Ministries of Health and Children's Services for 17 years. My experiences and challenges in raising my sons inform much of the following presentation. In tribute to that experience, I have coined the phrase "industrial-strength traffic on domestic-strength bridges," in describing the road that families travel in caring for these youngsters.

In North America, the societal conundrum created by prenatal drug and alcohol exposure has received significant publicity – largely due to the efforts of those concerned about the lifelong impacts of living with FASD. The spectrum has life-altering consequences for affected individuals, birth mothers, family, and society, and its prevention and intervention strategies are, unfortunately, inherently complicated and influenced by competing policies, politics, and resources allocated to the field.

This chapter explores maternal substance use and subsequent mother-blaming from racial, socioeconomic and gender perspectives, and the challenges that this presents for communities, social work interventions, and advocacy.

Relational and women-centered programming is currently viewed as best-practice intervention modalities for maternal substance use. Without addressing this complex issue through eco-systems lenses and feminist-informed practice, there is little hope of reducing the numbers of substance-using pregnant mothers (Alston and McKinnon, 2005; Germain and Gitterman, 1996; Boyle *et al.*, 2006; Healy, 2005; Marcellus, 2004; Sands and Nuccio, 1992; Taylor, 1999; Turner, 2005; Wakefield, 1996; Wharf and Mckenzie, 2004). In the past, Turner (2005) has defined eco-system intervention as:

> "... an art in which knowledge of the science of human relations and skill in relationship are used to mobilize the capacities in the individual and resources in the community appropriate for better adjustment between the client and all or any part of his total environment" (p. 111).

It is the conclusion of many of those who work with persons with FASD and their birth mothers, that they may be a catalyst for meaningful change, mostly within health promotion and prevention initiatives, as the challenge is to address this issue from a health rather than moral perspective (Wilson and Martell, 2003; Deville and Kopelman, 1998). Most of the programming as well as awareness-raising materials produced in the field of FASD have a common theme; namely, that the community must take on this challenge in a way that is nonstigmatizing and represents a respect for the values and traditions of people involved in FASD. This "whole community concept" also characterizes any of the current models of intervention.

As we sought to create awareness and improved intervention regarding FASD in Alberta, we began by exploring our biases and preconceived notions about who might be affected by FASD, followed by discussions about how to send out a message about healthy, alcohol-free pregnancies that was neither ambivalent, nor condemning. We have not been completely successful. It remains very challenging to present the message in a unified and respectful manner. We have had to compromise by reaching out to select groups with prevention information tailored to their needs. There are huge complications inherent in bridging "... the gap between existing medical knowledge and the situation of everyday society" (Atchison and Bujak, 2001).

There remain fundamental questions to tackle in order to address the complexity of FASD:

1) What will it take to make it possible for every woman in every community to make it through a pregnancy without using drugs or alcohol? Is that the same for everyone? Whose responsibility is this?

2) What is the most respectful way to intervene in the life of an individual with FASD (and her/his family)? How can we ensure that the diagnostic and assessment services they need are available? Can we address prevention of the secondary disabilities with which they so often live?

3) What do the caregivers of affected individuals need in order to remain connected with these challenging persons? How can we prevent the family breakdown that accompanies this disability?

4) When will the determinants of health be taken seriously also for people with FASD?

Warnings regarding the risks of prenatal drug and alcohol exposure were largely unheeded in North America until Drs Smith and Jones confirmed that the health and well-being of babies could conceivably be compromised by alcohol consumption during gestation. However, not until the first infants diagnosed with Fetal Alcohol Syndrome (FAS) were further along in their development was the problem framed more persuasively (Chudley *et al.*, 2005; Jones and Smith, 1973; Streissguth, 1997; Sokol and Clarren, 1989). The subsequent discovery that the challenges experienced by alcohol-affected individuals were more significant than merely compromised growth patterns and physical anomalies (often including a distinguishing facial gestalt), was sobering. It was recognized that FASD lasts a lifetime.

When FASD was identified as an outcome of maternal alcohol use during gestation, public opprobrium focused on addicted females as "bad" and immoral, as opposed to living under poor and stressing circumstances, or having a disability and being in need of a range of supports (Armstrong, 1998; Rutman et al., 2000). The context within which female alcohol/substance use occurs is generally ignored in media stories, and most prevention efforts begin with a legalistic approach (Lofaro, 2006; see also Appendix 20.1). Hence, the response to the tragic problem of gestational alcohol use is all too often articulated using comments such as "... these women should be thrown in jail, or, if they would just stop drinking while pregnant there would never be another baby born with FASD" (Murphy and Rosenbaum, 1999; Poole, 2003; Rutman et al., 2000). Oversimplifications such as these deny the inconceivable challenge and complexity that attends female problem substance use, especially as alcohol and drugs are frequently employed to cope with and manage a difficult social situation, depression, or significant emotional pain (Mate, 2008; Armstrong, 1998; DeVille and Kopelman, 1998; Poole and Isaac, 2001; Poole and Dell, 2005; Tait, 2000; Whiteford and Vitucci, 1997).

There are estimates about substance abuse in Canada demonstrating that approximately one in seven women engage in some form of it during pregnancy, and that the rate of incidence of FASD in Canada is currently at 9.1 per 1000 births (Health Canada, 2005). The majority of pregnant women seem to use substances "moderately," and they generally are influenced by messages about healthy pregnancies, and by widespread public service advertising regarding gestational substance use (AADAC, 2004). However, of notable concern is a cohort of substance using women who are the least responsive to broad informational campaigns; they are not accessing medical services during pregnancy, are poorly connected to social support networks and/or community services and, therefore, are most at risk for becoming birth mothers of children with FASD (Weiner and Morse, 1989). According to the 2005 National framework for action to reduce the harms associated with alcohol and others drugs and substances in Canada, programs aimed at preventing gestational alcohol misuse cost far less than the cost of raising one child with FASD.

Substance abuse during gestation is part of a much broader social situation for the individual as well as part of societal structures and values, which makes it complicated to develop truly effective interventions for prevention.

Substance-using birth mothers were not a priority in the early years of FASD diagnosis in Canada. That they became the focus of compassionate intervention at all in North America is due to the work of the University of Washington, US, pediatric diagnostic assessment team lead by Dr Sterling Clarren who, during the late 1980s, after completing many FASD diagnoses, began to inquire about the mothers of their small patients. Their subsequent research revealed that 80% of the children who had been diagnosed were no longer in the care of their birth mothers, and that 28% of 286 birth mothers of children with FASD had died;

"... being the mother of a child with FASD was in fact fatal" for some of these mothers (Astley et al., 2000). The study unearthed grim statistics regarding the lives of the 80 birth mothers who were later interviewed: 80% lived with partners who did not want them to stop drinking, 80% had major mental health problems, and 100% of them had at one point in their lives experienced some form of abuse. Additionally, many of the women had a parent who suffered from addictions and/or had themselves been alcohol-exposed during gestation; the majority had inadequate housing and education, were unemployed, and their average age of first alcohol and tobacco use was 13 years. These women had made several attempts to stop their substance use; they stated that the reasons for not succeeding were: not wanting to quit because they needed alcohol to cope (87%); being too discouraged/depressed to quit (79%); fear of losing their children to the child protection system (47%); and not having anyone to care for their children (46%) (Astley et al., 2000). Not surprisingly, the study concluded that the diagnosis of FASD was in fact a diagnosis for two: an affected child and a vulnerable birth mother.

An alcohol- and/or drug-exposed fetus is a marker for compromised parenting, just as a woman's substance use is a marker for many deeper troubles in her life, and for society at large, since problem substance use to some extent is a social problem, and not simply an indicator of individual failure (Armstrong, 1998).

In general, public messages, policies and interventions appear to be directed equitably towards all women who use substances during gestation; however, a closer look belies that view (Ladd-Taylor, 2004; Murphy and Rosenbaum, 1999; Paltrow, 1998). Janet Golden (2000) traces media messaging as it evolved from generalized concern over the drinking patterns of middle-class women during the early 1970s towards blaming marginalized women for consuming substances presumably without concern for the outcomes, and from addictions as a public health concern to "danger to society" viewpoints. Both, Deville and Kopelman (1998) and Salmon (2003) observed that fetal protection policies focus predominantly on poor women, and/or those who are part of a visible minority, that these women are ten-fold as likely to be reported to child protection agencies for gestational substance use, and that selective enforcement of reporting and urine testing laws favor privileged middle- and upper-class women. Courtwright (2004) reported the details of a study in which it was suggested that middle-class white women tended to stop substance using during pregnancy, whereas poor uneducated, unmarried and minority women, who exhibited lower drinking rates prior to pregnancy, did not stop. This may speak to the increased stress and lack of social support that marginalized women experience; the multiple environmental risk factors for them and their offspring include a particularly disturbing finding concerning the "... common presence of violence and abuse reported by ... low-income substance using women" (Klien, Crim, and Zahnd, 1997).

Courtwright (2004) postulated that these and other environmental factors may be as influential in predicting poor gestational outcomes as substance use; he further contended that the notion of FAS is a social/moral rather than medical construct, and that FAS only "'began to occur" once it was diagnosed, implying that prior to 1972 no such cases existed. Courtwright's controversial point may challenge us to rethink the blaming and shaming that pregnant substance-using mothers experience at the hands of society, especially those women in minority groups, those living in poverty, and those living under duress (Bouchard and Kemenende, 2005; Courtwright, 2004; Murphy and Rosenbaum, 1999; Poole, 2003; Salmon, 2003).

Previously, Paltrow (1998) had reiterated that stark racial and socioeconomic differences/inequalities existed in FASD prevention policies in North America (Deville and Kopelman, 1998). Both, Salmon (2003) and Poole and Isaac (2001) have suggested that the scrutiny directed at Aboriginal women with regards to the issue of gestational substance use is out of proportion to the numbers engaging in this practice, and that the majority of punitive and legal actions have focused on this group, to the exclusion of most other women in Canadian society. These authors posited that all women need to be the focus of prevention efforts equally, and those who are in need of assistance need nondiscriminatory interventions, not punishment. One infamous example of such discrimination was the case of the young Aboriginal Manitoba woman, Ms G, who was court-ordered to remain in treatment while she was pregnant and substance using (McCormack, 1999; Murphy and Rosenbaum, 1999; Poole and Isaac, 2001).

Besides the racially and socioeconomically biased material, there is gender bias in the messaging regarding both substance use and mothering. Weitz (1998) and Ladd-Taylor (2004) contended that our current preoccupation with blame has nowhere been more damaging than when it has been directed at mothers, and that painting them as societal enemies when they do not appear to prioritize the needs of their offspring is damaging. According to Ladd-Taylor (2004), disadvantaged mothers deserve neither the entire blame for a generation of troubled youth, nor the inadvertent or "deliberate" fetal harm stance currently being promulgated.

Rutman et al. (2000) and Greaves et al. (2004) pointed out that mothering occurs in various circumstances and contexts, and that while expectations for all mothers are the same and mothers share the same willingness to lay aside their own needs for their children, both born and unborn, there exists no level playing field from which to actualize this. DeVille and Kopelman (1998) examined how society has taken an increasingly harsh stance towards the behaviors of pregnant women, and increasingly holds women accountable for gestational outcomes. Whenever parents – and mothers in particular – are singled out as the ones responsible for societal well-being in their role as child bearers and rearers, the blaming and shaming of those who "fail" is a logical outcome (Greaves et al., 2004).

Unfortunately, society rarely acknowledges any aspect of the role of the male partner as well as the social and physical milieu in pregnancy outcomes. Paltrow (1998) and Rutman et al. (2000) have queried the reasons that society and the medical community remain focused on the female in the equation, with "punishments" for poor prenatal care-giving. Dineen (1994) concurred that:

> "... maternal behavior is only a small part of the total picture of the problem of prenatal substance abuse yet it is the primary focus of fetal rights proponents ... this is a medical problem and it is frightening that society is so ready to punish this behavior."

The promotion of child-centered interventions, the use of child protection legislation to "punish" mothers, and the pitting the rights of parents against those of children has resulted in a splitting of the family wellness model into competing parts (Greaves et al., 2004).

Despite evidence of the limited ability of behaviorally focused strategies to prevent or ameliorate pregnant substance use, efforts to utilize approaches addressing the societal determinants of health are rarely attended to within the public health system in Canada (Raphael, 2003). That women often cope with violence through substance abuse is either ignored or condemned (Poole and Isaac, 2001; Jos, Perlmutter, and Marshall, 2003).

Most successful interventions that address social problems such as pregnant substance misuse are both environment and relationship focused (Germain and Gitterman, 1996; Corse, McHugh, and Gordon, 1995: Stepney and Ford, 2000). For example, Poole and Isaac (2001) have proposed that any interventions for an addicted woman be framed and /or addressed almost exclusively within the context of her entire set of relationships, in order to prevent post-treatment drift and/or alienation from her support group. These authors emphasized the detrimental effects that shame and blame have on women as they contemplate addictions treatment, as well as the need for treatment that increases a woman's access to various health services while allowing her to remain with family, and especially her children, without fear of child protection intervention. In studies of barriers to treatment, Murphy-Brennan and Oei (1999), Tait (2000), and Klien, Crim, and Zahnd (1997) reiterated that women frequently self-reported that they were not able to reduce their substance use during pregnancy on their own, largely because their fears of child welfare intrusion kept them from seeking help; that their social situations were severely complicated; and that they faced significant emotional challenges. They further identified the need for comprehensive services to address a broad range of unmet needs in the lives of these women: financial aid, safe housing, adequate nutritious food, medical care, education and or vocational training, support from community services for childrearing.

This underscores the need for women-centered, non-punitive care for mothers, focused on keeping the mother–child bond intact while addressing the multiple problems that underlie problematic female substance use. Given the isolation and lack of mobility that this population so frequently experiences, outreach and home visitor intervention may be the most viable prospect, especially as program outcome reporting demonstrates gains for the mothers and reduced rates of subsequent drug and alcohol-affected births within this population when they have been connected to such services (Pepler et al., 2002; T. Grant, personal communication, June 2005). Current best-practice FASD prevention models are characterized by an approach that includes mentoring (coaching) combined with skill building and connection to community services; known in ecosystems as "casework" (Boyle et al., 2006).

In Washington, one model of intervention, formerly known as the "Seattle Birth to Three Program" but now the Parent Child Assistance Program (PCAP), has been widely replicated. The model was developed by Drs Ann Streissguth and

Therese Grant, subsequent to Dr Clarren's birth mother study mentioned earlier. In Canada, the most well-known programs are *Breaking the Cycle*, Toronto; *Sheway*, Vancouver's East side; *First Steps*, Alberta, and *Stop FAS*, Manitoba (Kyskan and Moore, 2005). In these programs, women-centered casework is accomplished by well-supervised staff persons identified as coaches/ mentors who provide comprehensive, specialized assistance and support, who recognize that change is not a trait or personality characteristic, but rather an outcome of the style of the helper and supportive interactions within the context of relationship building (Greaves, Poole, and Cormier, 2002). Operating within the context of women's lives means that those women's day-to-day considerations are built into the program, and concepts such as flexibility, providing a range of service approaches, and linking women to existing community services are the standard.

Given the horrendous experiences and memories that many of these women are attempting to obliterate through substance use, researcher Nancy Poole (2003) has identified essential components/principles of intervention consistent with ecosystems approaches: a strong emphasis on harm reduction stance towards recovery from addiction; trauma-informed practice; integrated service provision; assisting women to seek safety; being present-focused; teaching coping skills; addressing stigma; and providing information in a nonjudgmental atmosphere. Optimum environments for leaving addiction behind can be compared to intensive care units–that is, spaces where women are assisted with integrating their past experiences, present realities and future coping into complex new identities (N. Poole, personal communication, 2006).

Children born with FASD are at risk for being "hard to care for," commonly manifested through irritability and endless crying, and coupled with problems such as irregular feeding and sleeping rhythms. These, combined with being born to a mother who may still be struggling with emotional pain and/or addiction, make the early months very challenging for the mother–child dyad. The most likely outcome of this scenario is that the infant will land in the home of a substitute caregiver, and that bonding may be compromised. Bonding both requires and produces neurological development within the context of a stable relationship; this in turn leads to adaptability and the ability of the infant to process information–that is, an ability to make sense out of the world. Even in an optimum environment, infants with FASD are challenged to adequately complete the tasks of adaptation and information processing. Further, the caregiving challenges posed by these infants may lead to a series of placements, each shorter than the last, and this too does not bode well for neurological development and attachment. In this context, either the dis-maturity and delays in development that are part of FASD go unnoticed, or the blame for these problems is placed on the frequent moves and/or caregiver shortcomings.

A further complicating factor for children with FASD is their inability to understand boundaries and familiar versus unfamiliar persons and situations. Caregivers report that these children are prone to be indiscriminate in bestowing their affections, and will often "go with anyone"– a characteristic usually associated with lack of attachment. Hence, they have the unfortunate malady of appearing unattached and unconnected whether or not this is actually the case. Caregivers are often challenged to accurately "read" the cues that these children give out, while the children are not good at deciphering the cues that caregivers are sending—an unfortunate combination when being parented by a birth mother still in the grips of emotional chaos/addiction, or an anxious substitute caregiver. The challenges of caring for a child who is likely to have many odd and difficult behaviors related to FASD, in combination with a shaky pre- and postnatal start, often leads to placement breakdown, especially in cases where the diagnosis has not been made. Even under the most ideal circumstances, parenting a child with FASD presents families with enough challenges that might be compared to placing industrial-strength traffic on a domestic strength bridge. Consequently, the normal trajectory of childhood and adolescent development will be compromised, and when this child reaches childbearing age, she or he is highly at risk of becoming another substance-using parent.

The primary disabilities related to an FASD diagnosis, along with the attendant secondary negative outcomes (such as non-completion of school, problems with employment, breaking the law, or being victimized, inability to participate in meaningful relationships, addictions, homelessness, etc.), pose tremendous challenges to the communities in which these persons find themselves. Given that the risk of burn-out for their families and caregivers is high, a harm-reduction approach is as essential for management of this disability as it is for prevention efforts. When the same principles that lead to respectful community-based intervention with birth mothers are applied to interventions for the affected persons, the risk of catastrophic outcomes for persons living with FASD decreases. Best practices for attending to the needs of affected individuals and their caregivers (which may include birth families) have been articulated by Ann Streissguth (1997) and Nanson, Roberts, and Gary (2000) as follows:

For families and caregivers:

- Family-centered care and advocacy to address needs of child(ren) and parents/caregivers.

- Support (including grief and loss counseling, respite, financial aid) for families/caregivers to prevent placement breakdown.

- Regular screening for substance use, mental health problems, reproductive health issues, and provision of assistance to cope with relationships, parenting, family life for both affected persons and their caregivers.

- Parent/caregiver training for every stage of development.

For affected individuals:

- (Researched) early intervention strategies to maximize cognitive, social, emotional development.

- Allowances at every developmental stage for dis-maturity; anticipating and preparing for the corresponding challenges.

- Recurring developmental assessments at every stage to inform case planning.
- Lifelong, continuous advocacy and case-management.
- Increased supervision/one-on-one supports to navigate the years from age 12 to 21 (prevent disruption from school/home placements, trouble with law, etc.).
- Pre-planned transitions from children's to adult services.
- Provision of ongoing and explicit training for the affected person with respect to socialization, employment, sexual activity, life skills, accepting help.
- Recognizing and planning for various degrees of lifelong dependency/daily living supports.
- Specialized intake and interventions for persons with FASD when they interact with the substance use/mental health/justice system.
- Professional, multidisciplinary teams that attend to medical issues related to FASD.
- Periodic screening for hearing, vision speech/language problems.
- Educational placement and supports that sustain both students and teachers.
- Planning/preparing early for future capacity for meaningful work and community engagement.
- Supported employment (from search through structured day/work duties).
- Provision of a range of living/housing options not contingent on their life-management skills.

Programming aimed at reducing gestational substance use began to appear in its present form in North America during the early 1990s, and has been evaluated with some rigor. Perhaps the best example in North America is the PCAP program (as mentioned above), which utilizes intensive questionnaires that are completed at intake and at every six months thereafter. Information in five domains is collected and analyzed for funding/program evaluation purposes: substance use patterns; family and social relationships; education and employment; housing, nutrition, health and reproductive health status, including subsequent pregnancies, and connections to community services. Program outcomes demonstrate that, in all areas of functioning, women who have intensive outreach services, relationship-based support, and who stay connected to the program for several years (three) show significant improvements in all five domains (Grant et al., 2003).

Change efforts at the macro level are much harder to evaluate and measure. There remains significant room for improvement with respect to increasing public awareness of the antecedents of problematic pregnant substance use; reducing bias toward poor, minority women; building support for problematic substance-use programming; addressing policies that pit women against their unborn children; increasing knowledge of health providers regarding the issue; and attending to the scarce funding for prevention projects.

Experiences of trauma and victimization, limited choices, an uneven playing field, a lack of genuine autonomy, powerlessness, and public opprobrium characterize the lives of many substance-using pregnant women. Those women – and, by extension, their offspring with FASD – caught in the cycle of poverty, trauma and problematic substance use pose tremendous challenges for community intervention. Gestational substance misuse with attendant private conflicts and public stigma requires respectful, compassionate, and non-blaming practice methods at the micro level in addition to organizational, social and policy change to address inequalities at the macro level.

Policy changes are needed to reduce barriers to safe and welcoming programming for addictions treatment for women; to reduce the institutionalized class and racial bias that are currently part of addressing gestational substance use; and to provide sustainable funding to continue operating healthy pregnancy outreach programs. Areas for improved intervention include commitments to address addictions issues from a health versus criminal perspective, exploring the effectiveness of family-based addictions treatment, addressing the role of male partners in reducing substance use in pregnancy, and addressing the needs of women living in remote and rural areas.

The enactment of harm-reduction strategies is subject to the tides of public opinion, government ideology, bureaucratic will, and available resources; however, once interventions are entrenched in policy they are less likely to be subject to change.

References

Alberta Alcohol and Drug Abuse Commission (AADAC) (2004) *Windows of Opportunity: A Statistical Profile of Substance Use among Women in Their Childbearing Years in Alberta. Executive Summary*, AADAC, Edmonton, AB, Canada.

Alston, M. and McKinnon, J. (2005) *Social Work; Fields of Practice*, Oxford University Press, Australia.

Armstrong, E. (1998) Diagnosing moral disorder: the discovery and evolution of foetal alcohol syndrome. *Soc. Sci. Med.*, **47**, 2025–2042.

Astley, S.J., Bailey, D., Talbot, C., and Clarren, S.K. (2000) Fetal Alcohol Syndrome (FAS) Primary Prevention through FASD Diagnosis II: a comprehensive profile of 80 birth mothers of children with FAS. *Alcohol Alcohol.*, **35**, 509–519.

Atchison, T.A. and Bujak, J.S. (2001) *Leading transformational change: The physician-executive partnership*. Health Administration Press, Chicago, IL.

Bouchard, L., Roy, J.-F., and van Kemenade, S. (2005) What Impact Does Social Capital Have on the Health of Canadians? Conclusions Drawn from the 2003 General Social Survey, Cycle 17. Working paper series 010, Government of Canada.

Boyle, S., Grafton, H., Mather, J., Smith, L., and Farley, O. (2006) *Direct Practice in Social Work*, University of Utah, US.

Chudley, A., Conry, J., Cook, J., Loock, T., Rosales, T., and LeBlanc, N. (2005) Fetal alcohol spectrum disorder: Canadian guidelines for diagnosis. *Can. Med. Assoc. J.*, **172** (Suppl. 5), s1–s21.

Corse, S.J., McHugh, M.K., and Gordon, S.M. (1995) Enhancing provider effectiveness in treating pregnant women with addictions. *J. Subst. Abuse Treat.*, **12**, 3–12.

Courtwright, D. (2004) How real is foetal alcohol syndrome? *Perspect. Biol. Med.*, **47**, 608–616.

DeVille, K.A. and Kopelman, L.M. (1998) Substance abuse in pregnancy: moral and social issues regarding pregnant women who use and abuse drugs. *Obst. Gynaecol. Clin.*, **25**, 238–254.

Dineen, C.E. (1994) Fetal Alcohol Syndrome: The Legal and Social Responses to its Impact on Native Americans. *North Dakota Law Rev.*, **1**, 40.

Germain, C. and Gitterman, A. (1996) *The Life Model of Social Work Practice*, Columbia University Press, NY, USA.

Golden, J. (2000) A tempest in a cocktail glass: mothers, alcohol and television 1977–1996. *J. Health Polit. Policy Law*, **25** (3), 473–500.

Grant, T., Ernst, C.C., Pagalilauan, G., and Streissguth, A.P. (2003) Post-program follow-up effects of paraprofessional intervention with high-risk women who abused alcohol and drugs during pregnancy. *J. Community Psychol.*, **31**, 211–222.

Greaves, L., Poole, N., and Cormier, R. (2002) *Fetal Alcohol Syndrome and Women's Health: Setting A Women-Centred Research Agenda*, British Columbia Centre of Excellence for Women's Health.

Greaves, L., Pederson, A., Varcoe, C., Poole, N., Morrow, M., Johnson, J., et al. (2004) Mothering under duress: Women caught in a web of discourses. *Assoc. Res. Mothering J.*, **6** (1), 16–27.

Health Canada (2005) *National Framework for Action to Reduce the Harms Associated with Alcohol and Other Drugs and Substances in Canada*, 1st edn, Health Canada, Ottawa, Ontario.

Healy, K. (2005) *Social Work Theories in Context*, Paulgrave McMillan, USA.

Jones, K.L. and Smith, D.W. (1973) Recognition of the fetal alcohol syndrome in early infancy. *Lancet*, **2**, 999–1001.

Jos, P., Perlmutter, M., and Marshall, M. (2003) Substance abuse during pregnancy: clinical and public health approaches. *J. Law Med. Ethics*, **31**, 340–350.

Klien, D., Crim, D., and Zahnd, E. (1997) Perspectives of pregnant substance-using women: findings from the California needs assessment. *J. Psychoactive Drugs*, **29**, 55–66.

Kyskan, C. and Moore, T. (2005) Global perspectives on Fetal Alcohol Syndrome: assessing practices, policies, and campaigns in four English-speaking countries. *Can. Psychol.*, **46**, 153–165.

Ladd-Taylor, M. (2004) Mother-worship/mother blame: politics and welfare in an uncertain age. *Polit. Public Policy*, **6**, 7–15.

Lofaro, T. (2006) Judge Rejects Lawyers' Advice, Sends Pregnant Woman to Jail. The Ottawa Citizen, April 5, www.canada.com (accessed 5 April 2006).

McCormack, T. (1999) *Fetal Syndrome and the Charter: the Winnipeg Glue-Sniffing Case*, Institute for Social Research, York University, pp. 77–99.

Marcellus, L. (2004) Feminist ethics must inform practise: interventions with perinatal substance users. *Health Care Women Int.*, **25**, 730–742.

Mate, G. (2008) *In the Realm of Hungry Ghosts: Close Encounters with Addicition*, Knopf, Canada.

Murphy, S. and Rosenbaum, M. (1999) *Pregnant Women on Drugs: Combating Stereotypes and Stigma*, Rutgers University Press, USA.

Murphy-Brennan, M.G. and Oei, T.P.S. (1999) Is there evidence to show that foetal alcohol syndrome can be prevented? *J. Drug. Educ.*, **29**, 5–24.

Nanson, J., Leslie, M., and Roberts, G. (2000) *Best Practices: Fetal Alcohol Syndrome/Fetal Alcohol Effects and the Effects of Other Substance Use During Pregnancy*, Health Canada CAMH.

Paltrow, L. (1998) Punishing women for their behavior during pregnancy: An approach that undermines the health of women and children, in *Drug Addiction Research and the Health of Women* (eds C.L. Wetherington and A.B. Roman), National Institute on Drug Abuse, , Rockville, MD.

Pepler, D., Moore, J., Motz, M., and Leslie, M. (2002) *Breaking the Cycle: the Evaluation Report (1995–2000)*, Mothercraft, Toronto, Canada.

Poole, N. (2003) *Mother and Child Reunion: Preventing Fetal Alcohol Spectrum Disorder* by *Promoting Women's Health*, British Columbia Centre of Excellence for Women's Health.

Poole, N. and Dell, C. (2005) *Girls, Women, and Substance Use*, British Columbia Centre of Excellence for Women's Health, and Canadian Centre on Substance Abuse.

Poole, N. and Isaac, B. (2001) Apprehensions: barriers to treatment for substance-using mothers. Report for British Columbia Centre of Excellence for Women's Health.

Raphael, D. (2003) Barriers to addressing the societal determinants of health: health units and poverty in Ontario, Canada. *Health Promot. Int.*, **18**, 397–404.

Rutman, D., Callahan, M., Lundquist, A., Jackson, S., and Field, B. (2000) Substance and pregnancy: conceiving women in the policy-making process. Status of Women Canada Report, Canada.

Salmon, A. (2003) "It takes a community" Constructing aboriginal mothers and children with FAS/FAE as objects of moral panic in/through FAS/FAE prevention policy. *J. Assoc. Res. Mothering*, **6**, 113–123.

Sands, R. and Nuccio, K. (1992) Postmodern feminist theory and social work. *Soc. Work*, **37**, 489–494.

Sokol, R.J. and Clarren, S.K. (1989) Guidelines for the use of terminology describing the impact of prenatal alcohol on the offspring. *Alcohol. Clin. Exp. Res.*, **84**, 1421–1428.

Stepney, P. and Ford, D. (2000) *Social Work Models, Methods and Theories*, Russell House Publishing, Dorset, England.

Streissguth, A. (1997) *Fetal Alcohol Syndrome: A Guide for Families and Communities*, Paul H. Brookes, Baltimore.

Tait, C. (2000) *A Study of the Service Needs of Pregnant Addicted Women in Manitoba*, Policy research project, Prairie Women's Health Centre of Excellence.

Taylor, Z. (1999) Values theories and methods in social work education. *Int. Soc. Work*, **42**, 311–317.

Turner, F.J. (ed.) (2005) *Canadian Encyclopaedia of Social Work*, Wilfrid Laurier University Press, Canada.

Wakefield, J. (1996) Does social work need the eco-systems perspective? Reply to Alex Gittermann. *Soc. Serv. Rev.*, **70**, 476–481.

Weiner, L. and Morse, B. (1989) FAS/FAE: focusing prevention on women as risk. *Int. J. Addict.*, **24**, 385–395.

Weitz, R. (ed.) (1998) *The Politics of Women's Bodies: Sexuality Appearance and Behaviour*, Oxford University Press, New York.

Wharf, B. and Mckenzie, B. (2004) *Connecting Policy to Practice in the Human Services*, Oxford University Press, Canada.

Whiteford, L.M. and Vitucci, J. (1997) Pregnancy and addiction: translating research into practice. *Soc. Sci. Med.*, **44**, 1371–1380.

Wilson, S.A. and Martell, R. (2003) The Story of Fetal Alcohol Syndrome. *Women Environ. Int.*, **60/61**, 35–37.

Appendix to Chapter 20

Headline: Judge rejects lawyers' advice, sends pregnant woman to jail

A 22-year-old pregnant woman is appealing a three-month jail sentence given to her by an Ottawa judge who decided she needed a "wake-up call" and overruled lawyers' recommendations she be given six months of house arrest. Angel Fergus, who is expecting her baby at the end of June, had pleaded guilty to theft and assault charges stemming from incidents in 2005 and this year. In sentencing her last week, Ontario Court Justice Richard Lajoie decided that the conditional sentence agreed upon by the Crown and Ms Fergus' lawyer was not in her best interest. Judge Lajoie pointed out that Ms Fergus had smoked marijuana in the first three months of her pregnancy, and that she had reoffended while still on probation. "Ms Fergus is pregnant and she has not found the strength to stop using drugs," he said. "She has cut back, cut back to one joint day a day," the judge said in his decision. He continued: "You are a young lady with a multitude of problems which will not go away by themselves, but it is clear to me that it is time for a wake-up call."

Yesterday, Ms Fergus's appeal lawyer, James Foord, filed an application to have her released on bail prior to the outcome of her appeal. If successful, she could be released by tomorrow, but if not, she may have to serve out her sentence and give birth in jail, he said. James Harbic, the woman's defense lawyer, said he and assistant Crown attorney Mark Moors were surprised by the judge's ruling, saying they didn't have a chance to respond to the decision and address some of the judge's concerns. Yesterday, Ms Fergus's mother was at the Elgin Street courthouse waiting for news about her daughter's appeal. "I'm really worried about my daughter and her health. She's lost 10 pounds and she's 100 pounds soaking wet when she's not pregnant," said Denise Daigle. She said the judge should have been more considerate of her daughter's condition, saying she requires proper medical attention – more than the basic care she is getting in jail. In addition to her pregnancy, Ms Fergus suffers from anxiety and depression.

Ms Fergus led a troubled life prior to appearing before Judge Lajoie on March 24 on three different offences. She was charged with assault on April 30, 2005, after she grabbed an emergency nurse's throat at the General campus of the Ottawa Hospital. On Sept. 30, 2005, Ms Fergus tried to cash a $200 check from the National Capital Marathon at the Money Mart on Montreal Road. A store employee found her signature didn't match their files for the organization and it was later discovered the check had been taken after a break-in at the downtown offices of the National Capital Marathon. The woman was arrested and when police went to her apartment to retrieve her medication, they found a number of items taken from the break-in. She was also faced another theft charge related to a Jan. 21, 2006, incident. Ms Fergus was on a probation order at the time and she had breached her conditions.

Lofaro, T. (2006, April 5) Judge rejects lawyers' advice, sends pregnant woman to jail. *The Ottawa Citizen*, retrieved April 5/06 from www.canada.com

21
A Cross-Ministry Approach to FASD Across the Lifespan in Alberta

Denise Milne, Tim Moorhouse, Kesa Shikaze, and Cross-Ministry Members

On behalf of the Government of Alberta FASD Cross-Ministry Committee (FASD-CMC) include:

- Alberta Aboriginal Relations
- Alberta Advanced Education and Technology
- Alberta Children and Youth Services (co-chair)
- Alberta Education
- Alberta Employment and Immigration
- Alberta Health and Wellness (co-chair)
- Alberta Justice and Attorney General
- Alberta Seniors and Community Supports
- Housing and Urban Affairs
- Alberta Solicitor General and Public Security
- Alberta Gaming and Liquor Commission

Ad hoc partners:

- Public Health Agency of Canada
- First Nations Inuit Health Branch
- Alberta Health Services

21.1
Introduction

In 2002, the Government of Alberta made a commitment to work toward the prevention of Fetal Alcohol Spectrum Disorder (FASD), and to support those affected by FASD in the province (www.fasd-cmc.alberta.ca). While great strides have been made to address FASD in Alberta, the reality is that an estimated 23 000 Albertans are living with FASD and, each year, approximately 350 more babies are born with FASD in the province (www.fasd-cmc.alberta.ca). Although successes are being seen at the individual, family and community levels, it is important to acknowledge that there is still much more work to be done.

Fetal Alcohol Spectrum Disorder–Management and Policy Perspectives of FASD. Edited by Edward P. Riley, Sterling Clarren, Joanne Weinberg, and Egon Jonsson
Copyright © 2011 WILEY-VCH Verlag GmbH & Co. KGaA, Weinheim
ISBN: 978-3-527-32839-0

FASD is a complex issue. The FASD-CMC employs a cross-jurisdictional approach, including representation from 10 provincial government ministries, including Alberta Health Services. The Public Health Agency of Canada and Health Canada (First Nations and Inuit Health), Alberta Region, also have a role in providing FASD-related services, and are ad-hoc members of the Alberta FASD-CMC. Through collaborative planning and delivery of government FASD programs and services, the FASD-CMC works to ensure consistency, alignment and synergy in government goals and priorities for FASD as per the 10-Year FASD Strategic Plan.

The FASD 10-Year Strategic Plan (2007–2017) (www.fasd-cmc.alberta.ca) was created through a cooperation of the ministries and other community-based agencies. The Strategic Plan identifies seven key strategies, along with initiatives to support and realize those strategies. A multilevel, made-in-Alberta, *community-based* approach ensures that supports and services are developed and delivered to the FASD client population through a comprehensive range of strategies, and that the distinct needs of individual communities drive the Strategic Plan's activities. The government supports these solutions by outlining shared goals, developing policy, and providing sustainable funding and other supports that engage and motivate cross-sectoral collaboration.

Other provincial initiatives that support the FASD Strategic Plan include the Safe Communities Secretariat, the Children's Mental Health Plan for Alberta, and a Plan for Alberta Ending Homelessness in 10 Years. Alberta's approach is also coordinated with other provincial and territorial jurisdictions through its partnership with the Canada Northwest Fetal Alcohol Spectrum Disorder Partnership (CNFASDP) (www.cnfasdpartnership.ca).

Through the FASD 10-Year Strategic Plan, the Government of Alberta coordinates support and services to a diverse group of stakeholders who are impacted by FASD:

- children, youth, and adults diagnosed with FASD;
- children, youth and adults with suspected FASD;
- families and caregivers of individuals with FASD;
- at-risk populations (individuals who participate in high-risk activities or have other circumstances that place them at higher risk); and
- Alberta communities (individuals and groups who may not be generally considered at-risk, but would benefit from prevention and awareness-based services and activities).

21.2
The Impact of FASD

FASD is a serious social and health problem for the child welfare, health, and education systems in North America (Gough and Fuchs, 2006). The term describes a wide range of disorders caused by women drinking alcohol during pregnancy. These include Fetal Alcohol Syndrome (FAS), partial FAS, Fetal Alcohol Effect (FAE), Alcohol-Related Neurodevelopmental Disorder (ARND), and Alcohol-Related Birth Defects (ARBD). (It should be noted that the terms FAS and FAE are often used interchangeably with FASD.)

FASD is an umbrella term used to describe the range of disabilities and diagnoses that result from drinking alcohol during pregnancy. FASD is a permanent birth effect caused by the maternal use of alcohol during pregnancy, and is characterized by pre- and/or postnatal growth deficiency, central nervous system (CNS) dysfunction, and a unique cluster of minor facial anomalies (Clarren and Smith, 1978). The presentation of each individual feature of FASD may be expressed with age. Estimates of the incidence of FASD range broadly from 1 to 3 per 1000 live births as documented in epidemiological studies, to 1 per 10 000 live births as documented in birth defect registries (Stratton, Howe, and Battaglia, 1996). Currently, FASD is under-recognized and often goes undiagnosed; it is, therefore, difficult to be certain how many individuals have FASD.

In Canada, it is estimated that nine per 1000 live births are impacted with FASD, and this value is considered conservative (Public Health Agency of Canada, 2006). The birth defects associated with FASD are caused by the consumption of alcohol during pregnancy, and represent a nationwide health concern in Canada. FASD does not discriminate on the basis of race, socio-economic status, or gender (Public Health Agency of Canada, 2006); moreover, because of a lack of recognition and diagnosis, it is difficult to ascertain exactly how many individuals are directly affected. The incidence of FASD is greater than that of either Down syndrome or spina bifida (Public Health Agency of Canada, 2006); typically, the incidence of FAE is five- to 10-fold higher than that of FAS. This means that, each year in Canada, between 123 and 740 babies are born with FAS, and around 1000 with FAE (based on 370 000 births per year). The prevalence of FAS/FAE in high-risk populations, including First Nations and Inuit communities, may be as high as one in five. Fuchs *et al.* (2005), when investigating children in the care of child welfare agencies in Manitoba, estimated that 17% had FASD. Unfortunately, there is a degree of variance among prevalence rates in Canada, as there is at present no coordinated approach to gathering this information.

The FASD-CMC recognizes that individuals with FASD, across their lifespan, may require extensive support and services related to health, social services, education and training, justice, addictions and family supports. Without these supports, an individual with FASD may experience a number of secondary disabilities and negative outcomes, including homelessness, unemployment, involvement in the criminal justice system, mental health problems, school drop-out, inappropriate sexual behavior, and family and placement breakdown.

According to estimates, for each child with FASD, up to $CAN 2 million may currently be required to provide special care, supports and/or supervision during his/her lifetime (Public Health Agency of Canada, 2006). Indeed, during their lifetimes, individuals with FASD currently alive in Canada will cost taxpayers approximately $600 million. The cost of FASD goes far beyond the financial implications, however, with other consequences including: the loss of human potential and employability; services that do not build on an individual's strengths due to a lack of understanding of FASD; and the burden that FASD puts on families, caregivers, and society in general. It has been shown that a significant number of individuals in the criminal justice, child protection, health and disability systems have FASD. Nonetheless, the FASD-CMC believes that, with sufficient and appropriate supports, people affected by FASD can be supported to reach their full potential and lead very successful lives.

21.3
Overview of Strategies

Currently, seven priority strategy areas have been identified in the FASD 10-Year Strategic Plan that are cooperatively supported by the 10 ministries that make up the FASD-CMC. These strategies include:

- Awareness and prevention
- Assessment and diagnosis
- Supports for individuals and caregivers
- Research and evaluation
- Strategic planning
- Training and education
- Stakeholder engagement

Although each of these strategies is inter-related with the others, they are identified as individual strategies for planning, measuring and reporting purposes. Activities related to each strategy work together, provide leverage, and support each other to achieve the desired outcomes.

The first three strategies – Awareness and prevention, Assessment and diagnosis, and Supports for individuals and caregivers – are supported primarily by the FASD Service Network Program.

21.4
FASD Service Network Program

Alberta's 12-system FASD Service Network program consists of groups of government and community agencies that coordinate the delivery of FASD services across Alberta. They form an important part of the FASD community, as they represent a broad range of stakeholders who have provided meaningful input into the development of the 10-Year Strategic Plan. The FASD Service Network Program is also supported by the other identified strategies: Training and Education; Strategic Planning; Stakeholder Engagement; and Research and Evaluation.

The Networks provide community-based solutions and a region-based point of access to FASD programs and services. Moreover, they are recognized for their comprehensive range of access to programs, services and information, and are presently at varying stages of development. All of the networks engage community partners to provide a continuum of coordinated services, which aim to meet the needs of those living with FASD and their caregivers. Services vary in each network and may include:

- Enhanced support for at-risk women;
- In-home and outreach support for children and youth whose parents have FASD;
- Support for caregivers of children, youth, and adults with FASD;
- Life skills programs for youth and adults living with FASD; and
- Assessment and diagnostic services.

With funding directed through the FASD-CMC, the networks have helped to increase broad community awareness for both the disorder and the regionally available supports and services. Through the networks, individuals and families have an organized, central resource that can assist them in accessing services and programs, and provide them with new hope for the future.

21.5
Ministry Initiatives Based on the Strategic Plan

Each of the strategies supports the programs, services and supports that are under development or are being set into action by the FASD-CMC to improve the quality of life for Albertans affected by FASD.

21.5.1
Awareness and Prevention

Awareness and prevention services are those that educate and inform Albertans about the risks of drinking alcohol while pregnant, the effects of FASD, and services that increase overall awareness about healthy pregnancy. They also include activities that increase the Albertans' awareness of services and supports available to individuals impacted by FASD. These services range from broad information concerning the risks of substance abuse to the general public, to very specific strategies and services targeted towards at-risk individuals. These services are categorized into the following areas of: (i) information; (ii) universal prevention; (ii) targeted prevention; and (iv) indicated prevention. These are developed to be culturally appropriate for Aboriginal and other communities.

Examples of current initiatives supporting awareness and prevention under the Strategic Plan are the Parent–Child Assistance Program (PCAP) and generating awareness and skill development in Justice.

21.5.1.1 Parent–Child Assistance Program
A home-visitation program for mothers at high-risk of abusing alcohol and drugs, the PCAP aims to prevent future births of alcohol- and drug-exposed children, to

help mothers maintain healthy family lives, and to ensure that children are living in safe and stable homes. Trained and supervised case managers provide home visits and intervention services including: advocacy for families in need of intervention; assistance for clients in obtaining birth control, housing and employment; and assistance for clients in obtaining FASD assessment and diagnosis.

21.5.1.2   Generating Awareness/Skills Development in Justice

The Solicitor General and Public Security (SGPS) has undertaken several initiatives to increase the assessment capacity, build awareness of FASD, and incorporate behavior management training for supervisors, sheriffs, probation officers and front-line staff in correctional, young offender, and attendance centers. Two FASD Program Coordinators coordinate FASD initiatives and distribute resource materials for the Correctional Services Division and the Sheriffs Branch of SGPS. These initiatives include consulting with experts on FASD in case management for high-risk young offenders, seminars incorporating practical skills for offender management, and the sharing of first-hand accounts of the challenges faced by FASD-impacted individuals for SGPS staff, partners, and stakeholders.

21.5.2
Assessment and Diagnosis

Assessment and diagnostic services include medical, cognitive, behavioral, communication, adaptive and executive functioning information, provided by a multidisciplinary ream trained in the current best-practice model. "Referrals to the team require the history of Prenatal Alcohol Exposure (PAE) and areas of suspected dysfunction" (Dr Gail Andrew, personal communication, 2009). The assessment and diagnostic services may or may not lead to a confirmed diagnosis of full FAS, pFAS, ARND, and ARBD. Examples of initiatives related to assessment and diagnosis under the 10-Year Strategic Plan include the following.

21.5.2.1   Development of an Assessment and Diagnosis Model for Aboriginal and Remote Communities

The Central Alberta FASD Service Network is currently exploring the feasibility of an alternative model of FASD assessment and diagnosis that is aimed at populations which experience multiple barriers with the traditional model. Services for the assessment and diagnosis of FASD in major cities do not always meet the needs of those located in remote communities and Aboriginal reserve communities. Consequently, many barriers – some of which are often insurmountable – result in fewer children and adults accessing the diagnosis and assessment services. As better outcomes are clearly linked to early intervention for individuals affected by PAE, an alternative model is being designed to reduce or remove barriers, such as transportation, child care needs and finances, while offering components that would not form part of the traditional model, such as healing circles with elders, sibling workshops, and health and nutrition support, in an effort to increase the likelihood of engaging families to access assessment and diagnosis services.

21.5.2.2 Adult Assessment and Diagnosis Demonstration Project

The assessment and diagnosis of adults with FASD can be challenging as it requires the confirmation of PAE beyond evidence of organic brain damage. In addition to reviewing birth and family records, reports from reliable family historians may need to be considered in the diagnosis of FASD. The Adult Assessment and Diagnosis Demonstration Project includes the development and evaluation of a diagnostic model that is appropriate for adults with suspected FASD. In addition, individuals will be linked to appropriate community services through the FASD Service Networks, and a training module developed to implement the diagnostic model in both rural and urban FASD networks. In addition, the project aims to provide recommendations for a provincial policy for adult FASD assessment.

21.5.2.3 Development of FASD Clinical Capacity

The number of active FASD assessment and diagnostic clinics in Alberta has recently doubled in a one-year period, from 10 in 2008 to 20 in 2009, with at least three additional clinics under development. Models for the diagnostic clinics are diverse, and have been very creative in providing this service to individuals across the province. The models range from hospital-based clinics to private and traveling clinics, as well as clinics specializing in forensic clients.

A leading practices workshop held in 2008 resulted in a *FASD Adult Assessment and Diagnosis Discussion Paper* (www.fasd-cmc.alberta.ca) that provided seven recommendations which the FASD–CMC has agreed to pursue:

- To assist in the identification and or development of screening tools for adults.
- To enhance the assessment and diagnostic capacity for FASD across Alberta.
- To institute a translator function into Alberta's system of FASD supports and services.
- To build up the system capacity by focusing on youth transitioning to adulthood.
- To work towards the development of clinical standards across the lifespan for adult assessment and diagnosis.
- To develop data management standards and practices.
- To enhance knowledge brokering and continuous quality improvement efforts with regards to adult assessment and diagnosis.

The current leading practice initiative is to pilot a data collection tool in all Alberta clinics. This will support the consistency of practice, regardless of the clinic model used, and ensure that the assessment process fully informs any subsequent service delivery tailored to an individual's needs. This pilot will also form the basis for further quality improvement in clinics, and provide key facts to FASD-CMC and networks in planning a better service delivery (Figure 21.1). The data contained in this figures (as self-reported by the clinics) show the number of active clinics as 11 in 2007, 14 in 2008, and 20 in 2009. It has been estimated that, in 2010, there would be 20 active FASD clinics in the Province of Alberta alone. The estimated provincial range provided by all clinics is 554 to 614 assessments (the lower estimate has been portrayed in the graph). In total, eight additional clinics are

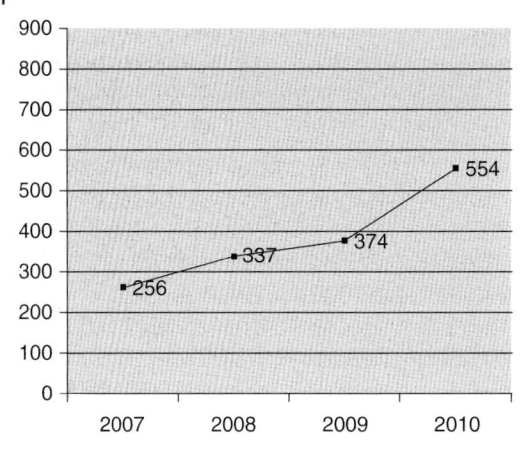

Figure 21.1 Numbers of multidisciplinary assessments completed by Alberta FASD clinics (actual for 2007, 2008 and 2009; estimated for 2010). For details, see the text.

currently under development, and are expected to enter service during 2010. No estimates for 2010 were included from this group. It should be noted that one clinic provided data based on fiscal years rather than calendar years; this information was still included in the calendar year chart, by applying the 2008–2009 numbers on the 2008 calendar, and the same for 2009.

21.5.3
Supports for Individuals and Caregivers

The category of Supports for individuals and caregivers addresses the needs of individuals with FASD and their informal, unpaid support networks. It does not include formal, paid caregivers. Programs and services in the community aimed at enabling individuals affected by FASD to reach their potential, and supports and assistance to families and caregivers of individuals affected by FASD, are included. Consideration is required to ensure that supports for individuals and caregivers are inclusive of the needs of Aboriginal, cultural and rural groups. Examples of such initiatives include the following.

21.5.3.1 Employment Supports for People Affected by FASD

This pilot project (conducted in Medicine Hat, Southern Alberta) was designed to assist individuals with FASD to obtain and maintain employment by developing employability, life management and coping skills, and employment maintenance skills through job coaching and other supports. The purpose was to re-examine the traditional model of vocational support to meet the employability needs of individuals affected by FASD, and develop an approach that would lead to future employment. The follow-up component of the project will be used to evaluate the need for ongoing employment-specific support to individuals living with FASD.

21.5 Ministry Initiatives Based on the Strategic Plan

This pilot study, which will provide a range of employment-related supports to individuals aged 18 years or more, and also demonstrate the characteristics of FASD, already has life supports in place.

21.5.3.2 Employment Supports and Services

This pilot project (conducted in Cold lake, Eastern Alberta) seeks to develop employment supports and services for individuals affected by FASD or who are suspected of living with FASD. The pilot focuses on the following three areas:

- Helping youth transition to adult supports through case management, by supporting families to develop a transition plan, identifying possibilities for individuals, and providing access to services available to meet the needs of that youth.
- Providing awareness and training support to the area by working with community, government, social development departments and agencies, four Métis Settlements and seven First Nations communities.
- Developing and producing free resources and tools to assist employment counselors and support agencies in their work with individuals with FASD.

21.5.3.3 FASD: Supporting Adults Gain and Maintain Employment

This training program has been developed to enable Employment and Immigration professional staff and their delivery partners to work more effectively with clients affected by FASD. The areas covered in this two-day workshop included:

- Characteristics associated with FASD.
- How the characteristics impact the functioning and behavior of adults.
- How to develop helpful and supportive relationships with adults affected by FASD.
- Accommodations and supports needed by those affected by FASD to find success in employment.

21.5.3.4 AVENTA Addiction Treatment for Women Demonstration Project

The aim of the AVENTA project is to provide treatment services for women addictions to alcohol, drugs, nicotine, and gambling. Pregnant women are identified as a priority, and receive support care relative to their pregnancy, after-care addiction treatment, and parenting resources.

21.5.3.5 Kaleidoscope Demonstration Project

The aim of the Kaleidoscope project is to provide transitional supports and pre-employment training to youth and young adults affected by FASD.

21.5.3.6 FASD Community Outreach Program Demonstration Project

The FASD Community Outreach Program demonstration project aims to provide services and supports for families and individuals affected by FASD, from a lifespan approach, in order to live in and be a valued member of their community.

The supports include service linkages, advocacy, ensuring that children and youth have the opportunity to participate in age-appropriate community activities, and the development of enhancing coping strategies for parents and caregivers.

21.5.3.7 Step-by-Step Demonstration Project

The Step-by-Step Project provides coaching, mentoring, and advocacy support for women who have, or are suspected of having, FASD, and who are parenting children who may or may not have FASD.

21.5.3.8 Well Communities–Well Families Demonstration Project

The Well Communities–Well Families multilevel community development program is aimed at preventing and managing FASD by working with the Central Edmonton community to build capacity to support healthy pregnancies and families.

21.5.3.9 Service Coordination and Mentorship

Demands on costly supports such as the court systems, correctional facilities, hospitals, addiction and mental health services are reduced when the social problems and secondary disabilities commonly experienced by adults with FASD are managed. The Ministry of Seniors and Community Supports (SCS) has contracted 12 community agencies to provide supports to adults with FASD, including those who were involved with the courts, had been released from correctional services, or who could be served through the FASD Service Networks. Through co-coordinated supports and mentorship, participants have been helped to make positive life choices in addressing legal, financial, medical, housing and employment concerns. Support staff provide practical teaching and develop structure to help adults with planning and problem solving; managing their time and money; and anticipating outcomes or consequences of their behaviors. In addition, SCS has initiated a project focused on defining common and best practices for supporting adults with FASD. Over the next two years, four agencies will consult with the 12 FASD Service Networks and share their experiences; this should not only increase understanding but also build expertise on which support strategies provide the best results for adults with FASD.

21.5.3.10 FASD Videoconference Learning Series

The FASD Videoconference Learning Series is delivered digitally to communities across the province, to increase both community and individual capacity to support people living with FASD and their caregivers across the lifespan. The series is designed for specific target audiences, including: parents and caregivers; Government of Alberta staff; community agencies; justice; educators; researchers; and allied health professionals, including physicians. In the period 2008–2009, 21 sessions were held, with well over 150 participants at some sessions. The series is due to finish its second year, with 17 sessions held in 2009–2010. Sessions are professionally filmed and posted online to enhance accessibility for those people unable to attend the scheduled sessions. The FASD-CMC website, on which the sessions are posted, attracted almost 5500 visits between March and September 2009 (www.fasd-cmc.alberta.ca).

21.5.3.11 Supports through Justice

Correctional Services Division staff members of the Ministry of Solicitor General and Public Security participate at FASD Service Network meetings in many Alberta communities, and a Justice Support Project for Youth provides assessment and case-conferencing in Edmonton. In correctional centers, two pilot programs – one for aboriginal female offenders impacted by FASD and the other for male offenders impacted by FASD – address needs through psychological support, intensive workshops, and transition worker follow up. A life skills program for adult male offenders builds on existing programming by taking into account the special learning needs often associated with individuals affected by FASD.

21.5.3.12 First Nations and Inuit Supports

First Nations and Inuit Health, Alberta Region focuses its FASD-related activities on four primary objectives which include: early identification, assessment and diagnosis; FASD-related education and training; supports for parents and families of children affected by FASD; and targeted programming for populations at risk. Activities related to these objectives are implemented in communities by using a proposal-driven process, a *per capita* allocation, or a blended model. Activities include an on-reserve mentorship model based on best practice evidence offered in seven Alberta sites: Enoch Cree Nation; Ermineskin Cree Nation; O'Chiese First Nation; Saddle Lake First Nation; Samson Cree Nation; Blood Tribe; and Tsuu T'ina Nation.

21.5.3.13 The WRaP (Wellness, Resiliency and Partnerships) Coaching Demonstration Project

The goal of this project is to identify innovative and effective supports for increasing school success for junior and senior high students with FASD in 10 selected school sites through services provided by five on-site WRaP coaches.

Coaching activities at each school include:

- Goal-setting and information coaching with individuals and small groups of students in the areas of wellness, personal resiliency and fostering community partnerships.

- Providing support, training and innovative strategies to support these students during less-structured parts of the school day (e.g., getting to school on time, managing breaks and lunch hours).

- Supporting students participating in school activities such as field trips, homework clubs, noon-hour and afterschool clubs and activities, and special events.

- Sharing information, research and strategies with school staff.

- Supporting families and caregivers in their efforts to get their children to school and to support their children's academic and social success.

- Collaborating with school staff, regional FASD networks and other community partners to build a circle of support for participating students and school staff.

21.5.4
Training and Education

Training and Education initiatives consist of formal education offered through post-secondary institutions, and pre-service and in-service training programs, typically targeted at program and service providers (including health, medical, and human services professionals) and/or community groups. It is particularly important that training and education services be culturally appropriate for Aboriginal and other populations. Initiatives underway in support of this strategy include the following.

21.5.4.1 Development of e-Learning Modules
A number of e-Learning modules to support school staff working with students with FASD have been developed, specifically in the area of enhancing relationships and supporting positive behavior. These anytime, anywhere learning opportunities will include:

- short video clips;
- case examples;
- visual diagrams;
- interactive activities; and
- reflective questions.

This project represents collaboration with the Family Support for Children with Disabilities program of the Ministry of Children and Youth Services, and is being coordinated by the Social Work and Disabilities Studies Department of Mount Royal College. Located in Calgary, Alberta. These modules are part of a larger e-Learning website *Positive Behavior Supports for Children*, that is freely available to communities, schools, and families across Alberta. The site will be expanded with additional modules, including those related specifically to FASD, through March 2010. The first phase of the site is targeted for community aides and school staff, while a second phase will include learning modules for parents and caregivers.

21.5.4.2 Promising Practices, Promising Futures: Alberta FASD Conference 2009 and 2010
The 2009 Alberta FASD Conference, which was held during two days in February 2009 in Edmonton, Alberta, focused on specific FASD topics based on the strategic pillars of Alberta's FASD 10-Year Strategic Plan. The well-attended, successful conference brought together a full spectrum of FASD stakeholders, and provided presentations and workshops on topics that included: strategies for language competency for children and adolescents with FASD; the effectiveness of FASD programs on outcomes for those affected by FASD; and bridging the communication gap in FASD service delivery.

In 2010, the Alberta FASDD Conference was held for two days in February in Calgary, Alberta, with a focus on supports for the caregivers such as the biological mother, adoptive and foster mothers of children of all ages diagnosed with FASD. With over 600 participants, this conference addressed the strategic pillars of education and training, research and evaluation and supports for the individual living with FASD and their caregivers.

21.5.4.3 IHE Consensus Development Conference on FASD: Across the Lifespan
This conference, which was hosted in partnership with the Institute of Health Economics, was held during three days in 2009 in Edmonton to develop, in the form of a consensus statement, practical recommendations on how to improve prevention, diagnosis and treatment of FASD: Across the Lifespan. The consensus statement will be widely distributed in the Canadian healthcare system, and also to those working in fields related to FASD.

21.5.4.4 FASD Education and Training
Advanced Education and Technology (AET), which is one of the 10 partnering ministries involved in the Strategic Plan, provides funding to post-secondary institutions to offer programming in areas such as early childhood development, teacher education, special needs educational services and community health that include coursework aimed at educating and training students on various aspects of FASD. This includes caregivers and professionals who may interact on a daily basis with individuals affected by FASD.

While AET supports the inclusion of FASD courses or modules within courses in such programming areas, the development of curriculum is an institutional responsibility. Alberta's post-secondary institutions are also involved in valuable research associated with FASD. AET also supports Community Adult Learning Councils and community literacy organizations to provide adult literacy, family literacy, English language learning, and other adult learning programs and services that are accessed by individuals affected by FASD.

21.5.4.5 Leading Practices Workshops
A Leading Practices Workshop on Supports for Individuals and Caregivers for staff from Alberta's 12 FASD Service Networks was held in March, 2009. This workshop promoted the importance of collaboration among care providers through the sharing of goals, objectives, and other key information. The goal was that agencies do not make conflicting demands on individuals or undermine each other's efforts, and that resources be used more efficiently. Future training will develop case management skills to support collaboration and enhance an understanding of how to best provide supports to individuals and caregivers.

21.5.4.6 Building an Educated Workforce
Employment and Immigration, a partnering ministry in the Strategic Plan, is working with the Lakeland FASD Service Network to encourage the local college to develop and deliver a post-diploma program on FASD in order to build a strong

work force that will be better equipped to serve individuals with FASD. The FASD Videoconference Learning Series included several presentations on employing individuals living with FASD and transitional employment supports for individuals moving from young adulthood to adulthood in employment.

21.5.5
Strategic Planning

Strategic planning is the foundation for all FASD initiatives, programs, service and supports offered within the community. Strategic planning is the wide range of activities performed by the government that are aimed at recommending priority areas for government policy and action. Although some ministries are not directly responsible for FASD initiatives, all ministries represented on the FASD-CMC contribute to the overall FASD-CMC outcomes through strategic planning. In addition, the general work of some ministries may serve Albertans who are affected by FASD, whether the programs are targeted specifically to this audience, or not.

Strategic planning initiatives include activities performed by government that are aimed at recommending priority areas for government policy and action. Examples of strategic planning include the CMC-FASD Operational Plan, FASD Service Network Plan, FASD Service Network Program Guidelines, and many other documents located on the FASD-CMC website (www.fasd-cmc.alberta.ca) (FASD-CMC, 2010).

21.5.6
Research and Evaluation

The strategy related to research refers to basic scientific and applied research leading to an increased understanding of FASD, its epidemiology (i.e., incidence and prevalence), leading practices in the prevention and treatment of FASD, and to the development of standards to guide the delivery of FASD clinical services and/or FASD programming. Initiatives underway in support of this strategy include the following.

21.5.6.1 Corrections and Connections to Community
The goal of this demonstration project, Corrections and Connections to Community, coordinated through the University of Alberta, is to determine if recidivism can be reduced. Men at the Fort Saskatchewan Correctional Center, north of the Edmonton area, suspected of having FASD are assessed and, if diagnosed as being affected, are provided with supports during their time at the correctional facility and after their release to the community.

A key project component here is to link with the 12 FASD Service Networks so as to ensure a coordinated and collaborative system of services and supports during transition to community. The demonstration model also captures prevalence and incidence data in Alberta for adults with FASD currently incarcerated and being transitioned back into community, that will allow the institution and FASD Service Networks to adjust service delivery based on these data.

21.5.6.2 FASD Community of Practice Research

A FASD Community of Practice (CoP) Research Project is evaluating promising practices for working with families affected by FASD; this was previously developed by the Southwest Child and Family Service Authority, one of 10 of the Child and Family Services Authorities in Alberta. The purpose and mission of the FASD CoP is to bring together stakeholders who share a passion for improving outcomes for children and youth diagnosed or suspected of having FASD. The CoP collaboratively implements promising practices in FASD and evaluates their impact towards improving service delivery, diagnostic opportunities, and intervention for children, adults and families affected by FASD. By interacting on an ongoing basis, the community is able to deepen its understanding and build expertise in the area of FASD by sharing experiences, learning best practices, and testing new ideas.

FASD and its related effects are disproportionately over-represented in children and families receiving supports and services from Children and Youth Services. Traditional case management practices and beliefs are often unsuccessful, as the parents of a child affected by FASD may also be affected.

Currently, all FASD Service Networks are actively involved in implementing the FASD Promising Practices Framework, and have already experienced positive impacts to practice and collaborative efforts. In respecting the diversity of the regions, the Framework has been revised to include all types of caregiver options including foster care, kinship care and residential care, and incorporates their support and training requirements. Training for caregivers began in three regions in May 2009, with the goal of establishing the same competency in each participating region.

21.5.6.3 Research Project on School Experiences of Children with FASD

Alberta Education, which is one of the 10 partnering ministries in the Strategic Plan, is currently funding a research project to explore the school experience of children with FASD in Alberta. This is part of a larger research project on Life Course Trajectories and Service Utilization Patterns of Children with FASD conducted by the Alberta Network Action Team of the Western/Northern FASD Research Network. This research is being conducted by Dr John D. McLennan, Department of Community Health Sciences, University of Alberta.

21.5.7
Stakeholder Engagement

Stakeholder engagement refers to the sharing of information among the government, practitioners, and the FASD client population in order to facilitate informed and balanced decisions regarding government priorities and action. Engagement occurs among stakeholders at different levels of the system (e.g., between ministries and regional organizations) and also among stakeholders at the same level of the system (e.g., among regional organizations). Stakeholders include provincial, Federal and local governments, community organizations, research organizations, and advocacy groups. The FASD Service Network Program is one example of how stakeholder engagement is incorporated in all initiatives related to the FASD 10-Year Strategic Plan.

The Strategic Plan will be evaluated at year five or in 2011. As part of the mandate of the Cabinet Policy Committee of the Alberta Government, all aspects of the Strategic Plan will be involved in addressing whether the performance targets have been achieved.

In summary, this is a very exciting initiative that the Alberta Government has undertaken to better address the needs of Albertans with FASD, or suspected of living with FASD. Alberta continues to focus on a coordinated response to FASD, and has been working in cooperation with the Canada Northwest FASD partnership and Federal departments to achieve the outcome of ensuring better outcomes for individuals living with FASD, across their lifespan.

References

Clarren, S.K. and Smith, D.W. (1978) Fetal alcohol syndrome. *N. Engl. J. Med.*, **298**, 1063–1067.

FASD-CMC (2010) Fetal Alcohol Spectrum Disorder across the Lifespan, www.fasd-cmc.alberta.ca (accessed 10 March 2010).

Fuchs, D., Burnside, L., Marchenski, S., and Mudry, A. (2005) *Children with Disabilities Receiving Services from Child Welfare Agencies in Manitoba*, Centre of Excellence for Child Welfare, Winnipeg, MB, www.cecw-cepb.ca (accessed 10 January 2010).

Gough, P. and Fuchs, D. (2006) *Children with disabilities receiving services from child welfare agencies in Manitoba. CECW Information*. Centre of Excellence for Child Welfare, Winnipeg, MB, www.cecw-cepb.ca (accessed 1 March 2010).

Public Health Agency of Canada (2006) It's Your Health, http://www.hc-sc.gc.ca/hl-vs/iyh-vsv/diseases-maladies/fasd-etcaf-eng.php (accessed 10 January 2010).

Stratton, K., Howe, C., and Battaglia, F. (1996) *Fetal Alcohol Syndrome: Diagnosis, Epidemiology, Prevention and Treatment*, National Academy Press, Washington, DC.

22
Critical Considerations for Intervention Planning for Children with FASD
John D. McLennan

22.1
Introduction

From a health and human service delivery perspective, there is currently no compelling evidence to warrant the development of unique or separate treatment and service programs for children with Fetal Alcohol Spectrum Disorders (FASD). The service and treatment needs of children with FASD and their families should be driven by the specific needs and problems of the individual child and family, and not the FASD diagnosis. This could be attained by linking children with FASD and their families to evidence-based interventions for the specific area(s) of need or difficulty. The broader service system for children at risk should be improved to better serve children with FASD and their families, along with other at-risk children. This chapter outlines factors behind these recommendations by considering a rational approach to service planning for at risk children.

22.2
The Development of a Rational Service System for At-Risk Children

Healthy child development may be at risk as a consequence of many different factors, including environmental exposures (both social and biological), from the point of conception to the attainment of adulthood, genetic inheritance from both parent lineages, and the complex interaction between genes and environment. Societies frequently commit considerable resources to services, programs and interventions aimed at improving the life outcomes of at-risk children. Although they are often well-intentioned, societal efforts to improve the outcomes for at-risk children may frequently be illogically conceptualized, poorly designed, and inadequately implemented, resulting in inefficient and ineffective (or under-effective) services, programs, and interventions. The end result is missed opportunities to help at-risk children reach their potential.

Many factors must be considered in the conceptualization, development, and implementation of efficient and effective services, programs and interventions for

Fetal Alcohol Spectrum Disorder–Management and Policy Perspectives of FASD. Edited by Edward P. Riley,
Sterling Clarren, Joanne Weinberg, and Egon Jonsson
Copyright © 2011 WILEY-VCH Verlag GmbH & Co. KGaA, Weinheim
ISBN: 978-3-527-32839-0

at-risk children. Critical information is needed from rigorous studies of child development, epidemiology, intervention evaluations, and health and human service delivery. Often, information from these fields may be inadequately or inaccurately used to inform actual service delivery and related policy formation, resulting in significant research-practice gaps.

Service delivery to at-risk children is further undermined by inadequate resources. A consequence of inadequate resources is a need to triage and prioritize, and key within these processes is the decision as to who receives preferential access to what. Prioritization may be driven by a combination of the extent of need and anticipated benefit from given interventions. Within some jurisdictions, including Alberta, Canada, there has been a call to prioritize interventions for children with FASD. Several variables should be considered to rationally inform calls for system change for particular subpopulations of children in need. Some of the important factors to consider when contemplating service reform for subpopulations of at-risk children will be outlined in the following section. These factors will then be applied to the case of FASD, to determine whether the creation of separate service tracks is appropriate at this time.

22.3
Factors Supporting the Development of Separate Specialized Services for Subgroups of At-Risk Children

Key issues for decision-makers to consider in informing policy around *prevention* interventions for at-risk children have been proposed previously (McLennan and Offord, 2002), and have relevance to the prevention domain in the FASD field. Historically, the area of FASD prevention has been fraught with problems, including a likely over-reliance on education campaigns with little evidence of effectiveness. An additional concern has been the extent of neglect of factors that appear to contribute to the risk of FASD beyond the role of alcohol exposure *in utero*, such as the impact of co-occurring poverty and nutritional compromise during pregnancy (e.g., Bingol, 1987; May et al., 2008). However, as prevention interventions are the focus of other chapters, the following discussion will address only the issue of treatment interventions for children identified as having FASD.

Adressing the following six questions will aid in determining the appropriateness of developing and prioritizing specialized services for subgroups of at-risk children (Table 22.1). Although this is not an exhaustive list of considerations, a number of areas are proposed that are often not adequately considered when organizing services for at-risk children.

22.3.1
What is the Prevalence of the Special Subpopulation?

Knowing the prevalence of a given problem or disorder is critical to informing the size of the population that may need and benefit from the specialized services

Table 22.1 Key questions to address when considering the development of separate specialized services for subpopulations of at-risk children.

1) What is the prevalence of the special subpopulation?
2) What is the prevalence of specific difficulties/needs within the subpopulations?
3) What is the attributable risk of the particular disorder for these specific difficulties?
4) What is the effectiveness of interventions for the subpopulation?
5) Is there evidence supporting separating out services for the subpopulation?
6) Are there risks in delivering services separately for the subpopulation?

under consideration. The prevalence can also provide an indicator of prioritization when considered against the larger context of the at-risk population of children – that is, contrasting with the size of other at-risk or otherwise special-needs populations of children.

22.3.2
What is the Prevalence of Specific Difficulties/Needs within the Special Subpopulation?

Depending on the subpopulation in question, what may be equally important is the prevalence of particular problems within the subpopulation in need of intervention or service support. For some fixed conditions, the intent is not to change the fixed condition, but rather to address the challenges that can be seen within the condition. For example, if a child has Down syndrome, the aim is not to "cure" the syndrome but rather to address challenges seen within such children. Now, while information about the prevalence of Down syndrome is important (it is estimated to occur in about 0.1% of the population; Shin et al., 2009), information on specific needs within this population are required for intervention and service planning. Thus, children with Down syndrome are known to be at-risk for such difficulties as hearing loss, congenital heart abnormalities and thyroid disease, which are estimated to occur in approximately 75%, 50%, and 15% of cases, respectively (American Academy of Pediatrics, 2001). This combination of information is informative when planning for the service needs of this population.

22.3.3
What Is the Attributable Risk of the Particular Disorder for These Specific Difficulties?

Related to the above is the importance of determining the attributable risk of the particular disorder (e.g., FASD) or risk exposure (e.g., prenatal alcohol exposure) for specific difficulties/needs (e.g., behavioral or learning problems). This helps to clarify whether the difficulties are appropriately attributed to the disorder/risk exposure in question, or might be better accounted for by other risk exposures. The extent of attribution may be an important factor to rationally inform how to structure services.

22.3.4
What Is the Effectiveness of Interventions for the Subpopulation?

In order to properly justify the need for specialized services, it is not sufficient to demonstrate the extent of need within a given subpopulation. It is critical to have interventions with demonstrated effectiveness for addressing the needs within the given population.

Although there is a growing call to implement interventions that are either evidence-based or evidence-informed, this is not often realized in practice. Services and interventions continue to be heavily driven by tradition, anecdotes, and influential stakeholders—a process which often fails to effectively draw from the growing empirical database which could inform the development of a more effective and efficient services system.

An additional consideration is whether the intervention is specific to a particular disorder, or addresses a given problem across disorders.

22.3.5
Is There Evidence for Unique Benefits to Support Separating Out Services for the Subpopulation?

Demonstrating that services or interventions are effective for a subpopulation is a necessary, though not sufficient, reason to carve out a specialized treatment track. What needs to be demonstrated to support a specialized track is that the intervention or service is *uniquely* beneficial for the subpopulation, and/or is *better delivered separately*.

22.3.6
Are There Risks in Delivering Services Separately for the Subpopulation?

Although carving out services for a particular subpopulation may lead to benefits for the targeted subpopulations, this process is not without risk. Two particularly important considerations are whether a unique service track can lead to: (i) service inefficiencies, and hence may disadvantage other subpopulations; and/or (ii) any disadvantages to the targeted subpopulation.

22.4
Should Separate Specialized Services Be Developed for Children with FASD?

22.4.1
What Is the Prevalence of FASD?

Given variations in different populations, different case ascertainment approaches, and a lack of clarity or consensus as to the edges of the spectrum of FASD, obtaining overall precise prevalence values is not likely. However, estimates can be used to inform service planning. Based on an earlier review of data from the United States, May and Gossage (2001) estimated the full fetal alcohol syndrome prevalence to be between 0.05% to 0.2% among the population, while the broad spectrum may occur in about 1% of the population (May and Gossage, 2001). Findings from more recent in-school studies have estimated levels of 0.7% for FASD from US and Italian studies, and about 2.5% for FASD (May et al., 2009) (this topic is addressed in much greater detail elsewhere in this book).

It is informative to consider the prevalence of other subpopulations of children at-risk, so as to place the values for children with FASD in context. Child mental disorders are estimated to affect 10–20% of the population (Waddell et al., 2002), with about 5% considered to have severe mental illness (Costello et al., 1998). Some of the most common disorders include anxiety, attention deficit/hyperactivity, and conduct, at prevalence rates of approximately 6%, 5%, and 4%, respectively (Waddell et al., 2002).

From a developmental disability perspective, it is estimated that approximately 17% of children in the US fall in this category, including about 6.5% with learning disabilities (Boyle, Decouflé, and Yeargin-Allsopp, 1994). Pervasive developmental disorders are estimated to occur in about 0.6% of children (Fombonne, 2003). While these values capture children with diagnostic disorders, there is even a larger population who are at-risk secondary to important and common risk factors such as poverty, nutritional compromise, and violence exposure.

22.4.2
What Is the Prevalence of Specific Difficulties/Needs within a Population of Children with FASD?

The core defining characteristics of FASD, such as growth impairment or facial dysmorphology, are not normally targets of interventions. Rather, the target of interventions are the apparent sequelae of the brain injury or abnormal brain development, which may manifest as neurocognitive deficits or behavioral problems.

Unfortunately, this is an extremely problematic area in the case of FASD, given the substantial gaps in the research literature and misinterpretation or misrepresentation of some study findings. The most important sources of data to answer this question of prevalence of specific difficulties or needs are derived from studies of children recruited from community samples rather than clinical samples. However, the vast majority of the present understanding of FASD is derived from children recruited from clinics, and who are at particularly high risk for referral bias. or what is known as "Berksonian bias." Indeed, if this bias is not taken into consideration, *there is a substantial risk for drawing inaccurate conclusions as to the prevalence rates of specific difficulties within this population as well as misattributions.*

A referral bias or Berksonian bias can occur when factors of interest in the clinic population also influence whether a patient is seen in the clinic (Schwartzbaum, Ahlbom, and Feychting, 2003). Children who come to clinics with FASD are not representative of all children with FASD in the general population; rather, they represent a particular subpopulation of children with FASD. The growth impairment and facial dysmorphology in children who might meet criteria for FASD are not typically main factors that would precipitate entry into services; instead, factors such as entry into foster care, significant behavioral struggles, or substantial developmental delays would prompt clinic referrals. Hence, clinics will typically receive a subgroup of children with FASD who are more likely to come from more severely impaired biological family life (and hence foster care placement), and/or have more severe behavioral difficulties and/or more pronounced neurocognitive deficits than is reflective of the full population of children meeting the criteria for FASD. This in and of itself is not a problem, as children with more severe problems should preferentially be seen in clinics. The problem arises when those treating or studying these more severely impaired subgroups make assumptions that they are representative or typical of FASD when they are not. This pattern is shown schematically in Figure 22.1.

In order to obtain a more accurate picture of the typical needs and problems of children with FASD for service planning, there is a need to examine results from studies of such children from samples drawn from representative community populations. Today, there is now a small–but growing–database upon which to draw, with some of the newest and most rigorous data being obtained from a series of in-school studies from Italy, South Africa, and the United States (May et al., 2009).

These studies provide an unbiased sample which can provide more information as to the typical prevalence of problems such as emotional, behavioral, social, and cognitive problems. Although, to date, the ranges of these types of variables considered in these population-based studies are limited, they are generating some important initial findings. At this point, for illustrative purposes, one dimension

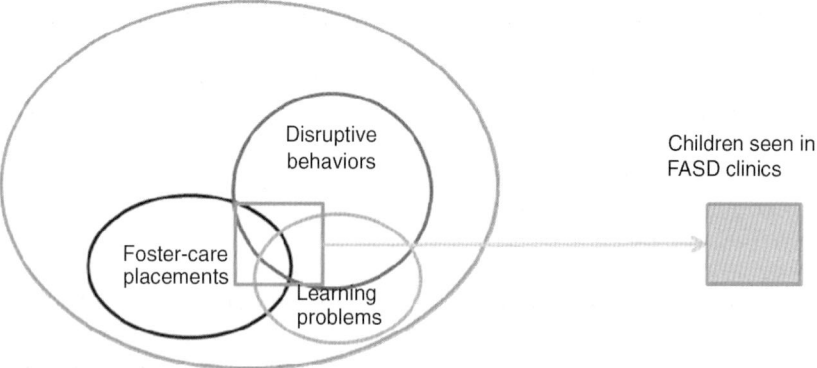

Figure 22.1 Illustration of factors that can contribute to a referral or Berksonian bias for children seen with FASD.

Table 22.2 ADHD disorders within children with and without FASD based on a school-based population in Italy (Aragón et al., 2008).

ADHD type	Teacher		Parent	
	FASD	Comparison	FASD	Comparison
Inattentive	17%	0%	4%	0%
Hyperactive-impulsive	0%	4%	0%	0%
Combined	9%	5%	0%	0%

of problems examined in one of these population studies will be considered, namely problems of attention and hyperactivity within one of the Italian studies (Aragón et al., 2008).

Problems of attention and hyperactivity are a particularly important focus, as they very frequently reported in children with FASD, either as a core manifestation of the brain involvement or a very frequent comorbidity. Burd, Carlson, and Kerbeshian (2007) summarized findings from a group of FASD clinic studies and estimated the prevalence of Attention-Deficit/Hyperactivity Disorder (ADHD) at 48%, although some studies have reported much higher levels (Burd et al., 2003; Iosub et al., 1981; Steinhausen, Gobel, and Nestler, 1984). The prevalence of ADHD problems for children with and without FASD from one of the rigorous population-based studies in Italy are summarized in Table 22.2 (Aragón et al., 2008). With the teacher as the informant, higher rates of problems of attention were found within the FASD group. However, the same pattern was not observed with the parent as the informant, and nor was there an excess of hyperactivity problems. In addition, the prevalence of inattention problems was dramatically lower than that reported in the clinical review by Burd, Carlson, and Kerbeshian (2007). This difference was most likely due, in part, to a substantial referral bias within the clinic-based studies.

Attention and hyperactivity problems are but one of the difficulties reported in children with FASD. This same approach could be used to determine the prevalence of other difficulties such as social and cognitive problems, and it is expected that there would be similarly lower rates of impairment, in contrast to the clinic-based studies. The Italian and South African studies have examined some cognitive domains in their population-based studies (Adnams et al., 2001; Aragón et al., 2008; Kodituwakku et al., 2006).

22.4.3
What Is the Attributable risk of FASD for these Specific Difficulties?

One important related factor that is not actually well addressed by the community/ population-based studies detailed above is the extent to which given problems are a function of prenatal alcohol exposure and/or FASD versus other co-occurring risk factors. Children who have been exposed to alcohol prenatally, often (though not always) may be exposed to other risk factors that influence the manifestation of problems such as neurocognitive deficits and behavioral problems. Yet, it is striking that this basic issue receives inadequate attention in much of the clinical and research work in FASD.

There are many other risk factors that may co-occur with prenatal alcohol exposure. These may include other prenatal period exposures including nicotine and illicit drugs, as well as other factors such as inadequate prenatal nutrition. Similarly, there can be multiple postnatal exposures including neglect and abuse, and multiple changes in caregivers, as well as more subtle parent–child difficulties which can influence the outcomes of children, both with and without FASD. These other risk factors appear to be extremely common within clinic-referred populations of children with FASD (e.g., Streissguth et al., 2004). Although such environmental risk factors have been acknowledged by clinicians and researchers alike, there is infrequent grappling with the relative influence of these different factors versus prenatal alcohol exposure for the frequently reported poor outcomes of clinical populations of children with FASD. This is not to minimize the teratogenic effects of prenatal alcohol exposure, for which there is clear evidence; rather, it serves as a reminder that many other factors may also impinge on healthy child development, for both FASD and non-FASD children alike.

Currently, an additional major factor receiving inadequate attention in the FASD field is the extent to which genetic endowment from the child's biological parents influence child development. Genetic factors are clearly at play in many developmental and mental health problems, and this is equally true of children with or without prenatal alcohol exposure or FASD. There appears to be a striking failure to factor in parental genetic contributions within parts of the FASD field, despite decades of studies investigating genetic, environmental and genetic–environmental interactions for a host of child outcomes. More recently, this expertise has been extended to studies of prenatal alcohol exposure (e.g., Knopik et al., 2006; D'Onofrio et al., 2007).

To further illustrate this issue, the example of attentional problems within children with FASD is considered again. Although attention problems seen in children with FASD may (in part) be a function of prenatal alcohol exposure, at least some of the attentional problems are likely a function of genetic predisposition if the child's mother or father had ADHD. In fact, there is very strong evidence of a heritability factor for ADHD (Albayrak et al., 2008). Furthermore, persons with ADHD are at risk for alcohol and drug abuse. This pattern is reflected in a case example, in which a woman with a child with FASD was found (on more detailed clinical review) to have had ADHD as a child and subsequently to have developed problematic alcohol use during her teenage years, which then extended to alcohol use prenatally. The resultant child with FASD may have attentional problems as a function of a heritable ADHD risk, rather than from prenatal exposure to alcohol, or FASD; this scenario is depicted schematically in Figure 22.2. This proposal, again, is not intended to downplay the teratogenic impact of alcohol in general, nor the potential for attentional problems in particular. That said, it is wrong to

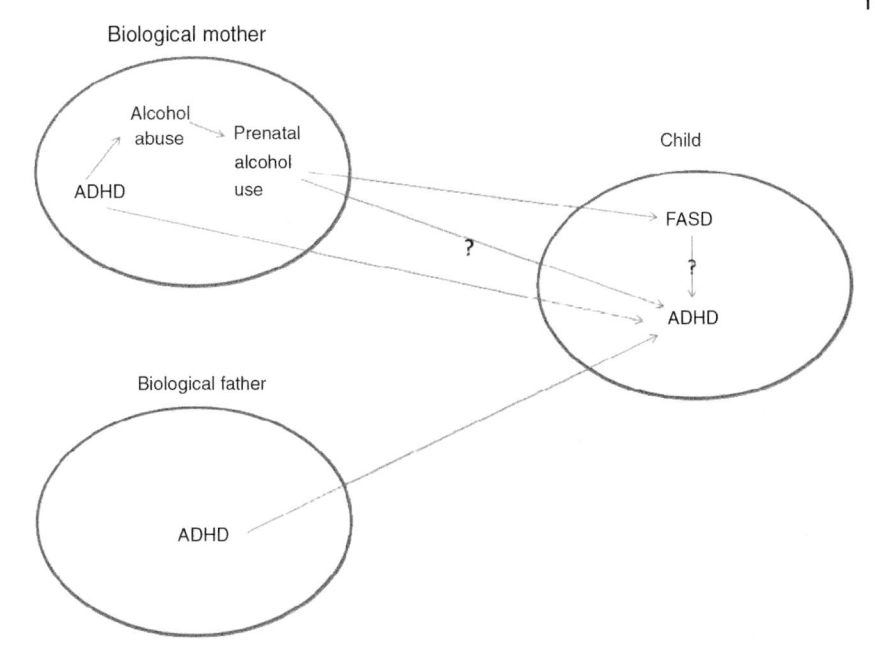

22.4 Should Separate Specialized Services Be Developed for Children with FASD? | 377

Figure 22.2 Different potential contributors to ADHD problems in a child with FASD.

assume (and hence attribute) *all* attentional difficulties of children with FASD to prenatal alcohol exposure or to a manifestation of FASD.

It is hypothesized that a significant proportion of the ADHD seen within the clinical population of children with FASD is *not* secondary to the alcohol exposure *in utero*, but is more likely to be a function of the typical etiology of ADHD, considered primarily due to genetic factors (Barkley, 2006). This is not to say that *some* of the ADHD pattern in FASD children is not a function of alcohol exposure, nor that *some* of the attentional problems in FASD may differ from typical ADHD (Coles et al., 1997). However, at least some – if not most – of the ADHD seen in clinical populations of FASD children is not likely a direct alcohol effect. This may account for the lack of any robust relationship found between prenatal alcohol exposure and ADHD (Linnet et al., 2003; Rodriguez et al., 2009).

Although attentional difficulties and a role for genetic predisposition have been used to illustrate this point, the same reasoning can be applied for other difficulties (e.g., behavioral problems) with other risk exposures (e.g., prenatal poverty). For example, Fryer et al. (2007) examined psychiatric problems in children with alcohol exposure and FASD and found that, within the alcohol-exposed group, family placement had a stronger association with having a psychiatric disorder than did having a diagnosis of FAS. Another study examining mental disorder outcomes in a birth cohort that had been followed-up, calculated the effect sizes for a number of risk factors (Barr et al., 2006). While alcohol binge-drinking mid-pregnancy and FASD were related to mental problems at follow-up, measures of change in home placement, prenatal socioeconomic status, and mother–infant interactions all demonstrated even stronger associations with the child mental health outcomes. A third study examining behavior problems in children also found prenatal alcohol exposure to be a significant risk factor; however, the maternal psychopathology proved to be a much stronger predictor of child behavioral problems (Sood et al., 2001).

Finally, *delinquency* is another important concern that has been flagged for youth with FASD. Often, an assumption is made that the delinquency is likely or even predominantly a function of the prenatal alcohol exposure and/or the FASD. However, the results of a sophisticated study conducted by Lynch et al. (2003) showed that delinquency was best explained by adolescent life stress, low parental supervision, and adolescent drug use, and *not* to prenatal alcohol exposure. Yet, these findings were not particularly surprising, as extensive research has been conducted on delinquency, including the complex contributions of genes, environment, and gene–environment interplay (e.g., Beaver et al., 2009). However, it is striking that, despite the extensive studies of delinquency, important stakeholders in the FASD field seemingly ignore this work in favor of simplistic linkages between a single risk exposure (prenatal alcohol) and an outcome as complex as delinquency. Once again, this is not to minimize the detrimental role of prenatal alcohol exposure, but rather to emphasize the need to consider a range of risk factors and their relative strengths when planning interventions to improve child outcomes.

22.4.4
What Is the Effectiveness of Interventions for FASD?

As noted above, the intent of interventions is not to change the FASD diagnosis but rather to address the common struggles that occur within this population. Unfortunately, very few rigorous studies of interventions with children with FASD have been conducted to date. A commissioned systematic review carried out in 2004 identified very few studies of interventions used with FASD populations; moreover, those that were identified had significant methodological limitations (Premji, Benzies, and Hayden, 2004). Since then, several studies have employed an improved methodological rigor (for a review, see Peadon et al., 2009) that have targeted typical difficulties identified in children with FASD, including social skills deficits, mathematics difficulties, and attentional problems (O'Connor et al., 2006; Coles, Kable, and Taddeo, 2009; Doig, McLennan, and Gibbard, 2008).

These studies have provided useful information for demonstrating that certain interventions can have a positive impact for children with FASD. However, it is critical to note that none of these interventions was particularly unique to the FASD populations. In fact, they were at least (in part) all drawn from interventions which have been successful with non-FASD populations, where they were used to target shared difficulties such as social skill deficits, struggling with mathematics, and attentional weaknesses. Although it is important to demonstrate that children with FASD and these particular weaknesses can improve with certain interventions, this finding does not, in and of itself, support the creation of separate treatment services. In fact, these same findings could be used to support the inclusion (integration) of children with FASD with particular difficulties (e.g., attentional problems) with other children that share the same difficulties, by using the same interventions. For example, in the case of the study conducted by Doig, McLennan, and Gibbard (2008), the ADHD problems seen in FASD were managed using standardized best-practice medication management procedures recommended for children with ADHD for whom medications have been deemed necessary (Pliszka et al., 2006).

22.4.5
Is There Evidence for Unique Benefits to Support Separating-Out Services for Children with FASD?

As noted above, interventions which have demonstrated at least some evidence of effectiveness with FASD children are not unique interventions specifically for FASD. In some cases, it has been noted by the authors that the interventions were modified or "tailored" for children with FASD, although there are no clear data available indicating that these modifications were differentially effective for this population. Furthermore, given the heterogeneity in the FASD population, intervention modifications may be more suitable at the individual child and family level, rather than at a group level. Again, the tailoring of more general interventions is not unique to FASD populations. For example, a social skills intervention for a child with autism with moderate mental retardation is likely to be modified for a similar child without co-morbid mental retardation.

The findings that the above-reported interventions have demonstrated effectiveness with non-FASD populations actually provides more support for the delivery of integrated interventions compared to via separate service tracks. For example, difficulties with mathematics has been identified as a particular problem area for children with FASD. However, there are obviously children without FASD who have similar difficulties and who may benefit from supplemental mathematics support. Yet, until evidence is provided that children with and without FASD who have co-occurring math problems require substantially different mathematics support programs, presumably all of these children might be supported by a similar support program. The corollary is that children with FASD who do *not* have mathematics problems should not automatically be assigned to supplemental math programs. In other words, it should not be assumed that *all* FASD children require a particular intervention, without considering their *individual* needs. This same argument could be extended to question the use of specialized classrooms specifically for children with FASD.

As of yet, there is no unique behavioral phenotype specific to FASD. Many of the challenges and problems described for children with FASD have clear overlap with the challenges and problems faced by children with other developmental disorders and/or mental health disorders. Even if some proposed characteristics were eventually to be confirmed (e.g., specific neuropsychiatric deficits such as impairments in complex number processing and explicit memory; Kodituwakku, 2007), the majority of challenges and problems would still overlap substantially with other populations. Given this, it would seem a reasonable strategy to meet these challenges within an integrated system along with populations of other children with developmental and mental disorders within a broader service system. Evidence that would challenge this recommendation would be findings identifying treatment approaches that are uniquely or preferentially beneficial to FASD children and their families, and not other children and their families with overlapping needs. Furthermore, there would have to be an advantage to delivering these separately. No evidence yet exists of this type.

A challenge to the above argument may be that the presumed unique maternal component that differentiates the FASD population from others, namely the mothers' *inappropriate* alcohol use during pregnancy, which may signal an addiction problem and hence the need for an extra dimension to service delivery that might not otherwise be provided. However, inappropriate alcohol use by a caregiver is not only seen in families with a child with FASD. Adult alcohol abuse and alcohol dependence are very common problems in society, and have a much broader manifestation than FASD (e.g., Kessler *et al.*, 1994; Osborne and Berger, 2009). Providing support and interventions for parents with alcohol addictions (and other addictions) is an important service, but it is certainly not unique to families with a child with FASD. It should also be noted that not all parents of children with FASD have had, or have, an alcohol addiction. The need for substance-abuse interventions for caregivers should be based on need, and not whether or not the child has FASD.

Another challenge to the proposed emphasis on integrated services is based on a perception of the greater severity or complexity of the difficulties seen in children with FASD versus other populations of children with special needs. This perception is partly driven by advocates responding to the often great severity and complexity of children seen within clinically referred populations; that is, a subset of children with FASD as described above (see Figure 22.1). If the full population of children with FASD is considered, however, there is a varying severity and complexity of difficulties similar to most other developmental and mental health disorders. For example, children with mood disorders referred to secondary and tertiary clinical facilities can have extremely complex and severe problems, while there is a large range of severity if the broad range of manifestation of mood problems is considered from a population perspective.

Service provision can be inadequate for children with severe and complex difficulties; however, this, again, is not unique to children with FASD. Service reforms that are aimed at improving help to children with severe and complex difficulties could also be harnessed for children with FASD who have severe and complex difficulties. This, again, may support an integrated service rather than a separate approach particularly for children with severe and complex presentation.

A more credible challenge to the proposed emphasis on integration is the severe gaps that exist in the broader service system. In most jurisdictions, there is clearly an inadequate access to a broad array of effective interventions and services for children with developmental and/or mental health problems who might benefit from such services. However, the creation of separate treatment tracks for one subpopulation in response to a broader service system weakness is inefficient, and penalizes those children with similar needs who do not belong to the subpopulation. Hence, a child with complex neuropsychiatric problems secondary to a mother's polysubstance use, which did not include alcohol, would not be eligible for special services that are reserved for a subpopulation who have been exposed to alcohol.

22.4.6
Are There Risks in Delivering Services Separately for Children with FASD?

As outlined above, there are two important risks to consider when contemplating the development of separate services for a particular subpopulation. One is the potential inefficiencies which may disadvantage other subpopulations as suggested above. In cases where similar or same treatment interventions may benefit both children with FASD and children with another developmental and/or mental disorder, the creation of a separate parallel service is likely to be inefficient.

In addition to inefficiencies and the potential disadvantages to other groups, are there any disadvantages to children with FASD if separate services are developed for them? It is proposed that there may be two concerns here: (i) potential distortions that limit access to effective treatments; and (ii) the potential impact of stigma.

Once again, the example of ADHD can be used in the context of an FASD diagnosis. Although the different potential linkages between ADHD and FASD have been outlined previously (e.g., Oesterheld and Wilson, 1997), there continues to be significant confusion around the issue. Some commentators have set up a false dichotomy whereby a child either has FASD or ADHD; this is reflected in statements about children with FASD being inaccurately diagnosed with ADHD, as if one excluded the other. This represents a lack of understanding of diagnostic approaches for mental and developmental disorders in general, and ADHD in particular. The most common diagnostic system used in North America for mental disorders, which includes ADHD, is the Fourth Edition of the Diagnostic and Statistical Manual of Mental Disorders (American Psychiatric Association, 1994). A phenomenological approach is used within this and other diagnostic systems; that is, a disorder is diagnosed if a person demonstrates a specific cluster of signs and symptoms with various qualifiers. This is contrasted with an etiological system, which makes assumptions about causality. In the phenomenological approach, if a child has a cluster of symptoms and signs consistent with ADHD with associated functional impairment, they have ADHD, whether or not there was prenatal alcohol exposure or FASD. It is not an "either–or" question.

Beyond the diagnostic issue, the critical treatment issue is whether ADHD in the context of prenatal alcohol exposure or FASD should be managed differently than ADHD without this context. There is no evidence at this time to support a separate approach. Given the previous argument, that at least some ADHD seen in FASD likely has a similar basis to those of other children with ADHD, the offer of best-practice ADHD treatments should presumably be the first-line approach for dealing with the ADHD within FASD.

Currently, the most evidenced-based treatment approaches for ADHD are select behavioral modification strategies and certain medications (Chronis, Jones, and Raggi, 2006; Pliszka et al., 2007). There is currently a belief in some sectors of the FASD practice community that behavioral modification strategies are ineffective for the FASD population, although this is not based on any existing rigorous scientific evidence. Given the extensive scientific evidence for the effectiveness of behavioral modification strategies for a wide variety of problems in many different populations of special needs children, it is unlikely that at least some of these benefits could not be extended to children with FASD. However, in a unique FASD-specific service, if this belief in the noneffectiveness of behavioral modification approaches were to prevail, there would be a risk that these important intervention strategies would be withheld, and hence limit a potentially effective treatment for children with FASD. Similarly, medication for the treatment of ADHD may also be withheld, despite growing evidence of a role for medication for ADHD within this population (e.g., Doig, McLennan, and Gibbard, 2008; Oesterheld et al., 1998).

An additional concern is the potential stigma related to carved-out services, with FASD populations perhaps being at particular risk for associated stigma. Unlike most other mental health and developmental disorders, this diagnosis highlights a single morally laden exposure – that is, prenatal alcohol exposure. Given that the abuse of alcohol by women in particular has traditionally been targeted with a pronounced moral condemnation and blame (Finkelstein, 1994), it would be expected that condemnation might easily be extended to pregnant women whose drinking could potentially harm a developing fetus. One recent study of the experiences of caregivers of children with FASD from specialty clinics found that a number of primary caregivers had experiences with stigmatization and judgment about FASD towards their child or themselves, in their attempts to obtain needed services or supports (Huculak and McLennan, 2009). This suggests that certain negative prevailing attitudes surrounding alcohol use may continue to promote a degree of social marginalization for women who consume alcohol during pregnancy, which would presumably impose barriers to help-seeking for them and/or their children, who could otherwise benefit from timely access to supports or treatment, including (at times), substance-abuse treatments. There is evidence, for example, that one important barrier to help seeking for treatment for alcohol dependency may be the stigma associated with drinking problems, in particular directed against women (Schober and Annis, 1996; Smith, 1992). The use of FASD unique services could, then, serve to heighten this marginalization and associated stigma, by virtue of inappropriately or even falsely highlighting the ostensive unique impacts of a single-risk exposure (alcohol *in utero*) among consumers of the service.

22.5
Policy Considerations: Strengthening the Service System for a Broader Range of Children At-Risk

There is no doubt that there are significant gaps in the service system for children at-risk. It is estimated that only about 20% of children with mental disorders ever see a specialist (Waddell et al., 2002), let alone obtain access to evidence-based interventions. Hence, to better serve all children with special needs – including those with FASD – the existing service system must be strengthened. Given limited resources, the system must be efficient and, where possible, services should be available to a broad range of children who may benefit from them. However, the service system should have sufficient flexibility to create specialized tracks for certain subgroups or certain needs, as evidence demonstrates that this subgroup preferentially or uniquely responds to this service above other subpopulations. As yet, there is no evidence to support carving out specialized treatment tracks for children with FASD. Indeed, FASD children and their families have experienced significant gaps in service delivery, as have many other children and families with other mental and developmental disorders. These experiences need to inform system reform so that all special needs children, and their families, could benefit.

Acknowledgments

The author thanks Susan Huculak for critical feedback and editing support on earlier versions of this chapter. The author also thanks the Alberta Heritage Foundation for Medical Research and the Canadian Institutes of Health Research for personnel research awards which provided time for research work in this field.

References

Adnams, C., Kodituwakku, P., Hay, A., Molteno, C., Viljoen, D., and May, P. (2001) Patterns of cognitive-motor development in children with fetal alcohol syndrome from a community in South Africa. *Alcohol. Clin. Exp. Res.*, **25**, 557–562.

Albayrak, O., Friedel, S., Schimmelmann, B.G., Hinney, A., and Hebebrand, J. (2008) Genetic aspects in attention-deficit/ hyperactivity disorder. *J. Neural Transm.*, **115** (2), 305–315.

American Academy of Pediatrics (2001) Health supervision of children with Down syndrome. *Pediatrics*, **107** (2), 442–449.

American Psychiatric Association (1994) *Diagnostic and Statistical Manual of Mental Disorders (DSM-IV)*, American Psychiatric Association, Washington, DC.

Aragón, A., Conriale, G., Diorentino, D., Kalberg, W., Buckley, D., Gossage, J.P., Ceccanti, M., Mitchell, E., and May, P. (2008) Neuropsychological characteristics of Italian children with fetal alcohol spectrum disorders. *Alcohol. Clin. Exp. Res.*, **32** (11), 1909–1919.

Barkley, R. (2006) Etiologies in Attention-Deficit Hyperactivity Disorder, in *A Handbook for Diagnosis and Treatment*, 3rd edn (ed. R. Barkley), Guilford Press, New York, pp. 219–247.

Barr, H., Bookstein, F., O'Malley, K., Connor, P., Huggins, J., and Streissguth, A. (2006) Binge drinking during pregnancy as a predictor of psychiatric disorders on the structured clinical interview for DSM-IV in young adult offspring. *Am. J. Psychiatry*, **163**, 1061–1065.

Beaver, K., DeLisi, M., Wright, J.P., and Vaughn, M.G. (2009) Gene environment interplay and delinquent involvement: evidence of direct, indirect, and interactive effects. *J. Adolesc. Res.*, **24**, 147–168.

Bingol, N. (1987) The influence of socioeconomic factors on the occurrence of fetal alcohol syndrome. *Adv. Alcohol Subst. Abuse*, **6** (4), 105–118.

Boyle, C., Decouflé, P., and Yeargin-Allsopp, M. (1994) Prevalence and health impact of developmental disabilities in US children. *Pediatrics*, **94** (93), 399–403.

Burd, L., Klug, M., Martsolf, J., and Kerbeshian, J. (2003) Fetal alcohol syndrome: neuropsychiatric phenomics. *Neurotoxicol. Teratol.*, **25**, 697–705.

Burd, L., Carlson, C., and Kerbeshian, J. (2007) Fetal alcohol spectrum disorders and mental illness. *Int. J. Disabil. Hum. Dev.*, **6** (4), 383–396.

Chronis, A., Jones, H., and Raggi, V. (2006) Evidence-based psychosocial treatments for children and adolescents with attention-deficit/hyperactivity disorder. *Clin. Psychol. Rev.*, **26**, 486–502.

Coles, C., Kable, J., and Taddeo, E. (2009) Math performance and behavior problems in children affected by prenatal alcohol exposure: intervention and follow-up. *J. Dev. Behav. Pediatr.*, **30**, 7–15.

Coles, C., Platzman, K., Raskind-Hood, C., Brown, R., Falek, A., and Smith, I. (1997) A comparison of children affected by prenatal alcohol exposure and attention deficit, hyperactivity disorder. *Alcohol. Clin. Exp. Res.*, **21** (1), 150–161.

Costello, E.J., Messer, S.C., Bird, H.R., Cohen, P., and Reinherz, H.Z. (1998) The prevalence of serious emotional disturbance: a re-analysis of community studies. *J. Child Fam. Stud.*, **7** (4), 411–432.

D'Onofrio, B., Van Hulle, C., Waldman, I., Rogers, J.L., Rathouz, P., and Lahey, B. (2007) Causal inferences regarding prenatal alcohol exposure and childhood externalizing problems. *Arch. Gen. Psychiatry*, **64** (11), 1296–1304.

Doig, J., McLennan, J.D., and Gibbard, B. (2008) Medication effects on symptoms of attention-deficit/hyperactivity disorder in children with fetal alcohol spectrum disorder. *J. Child Adolesc. Psychopharmacol.*, **18** (4), 365–371.

Finkelstein, N. (1994) Treatment issues for alcohol- and drug-dependent pregnant and parenting women. *Health Soc. Work*, **19** (1), 7–15.

Fombonne, E. (2003) Epidemiology of autism and other pervasive developmental disorders: an update. *J. Autism Dev. Disord.*, **33**, 365–381.

Fryer, S., McGee, C., Matt, G., Riley, E., and Mattson, S. (2007) Evaluation of psychopathological conditions in children with heavy prenatal alcohol exposure. *Pediatrics*, **119**, e733–e741.

Huculak, S. and McLennan, J. (2009) "Finding the Right Kind of Help:" The Service Needs of Families Seen through FASD Specialty Clinics in Alberta. 22nd Annual Sebastian K. Littman Research Day, University of Calgary, March 6, 2009.

Iosub, S., Fuchs, M., Bingol, N., and Gromisch, D. (1981) Fetal alcohol syndrome revisited. *Pediatrics*, **68** (4), 475–479.

Kessler, R., McGonagle, K., Shao, S., Nelson, C., Hughes, M., Eschleman, S., *et al.* (1994) Lifetime and 12-month prevalence of DSM-III-R psychiatric disorders in the United States: results from the National Comorbidity Survey. *Arch. Gen. Psychiatry*, **51**, 8–19.

Knopik, V., Health, A., Jacob, T., Slutske, W., Bucholz, K., Madden Pm Waldron, M., and Martin, N. (2006) Maternal alcohol use disorder and offspring ADHD: disentangling genetic and environmental effects using a children-or-twins design. *Psychol. Med.*, **36**, 1461–1471.

Kodituwakku, P. (2007) Defining the behavioral phenotype in children with fetal alcohol. *Neurosci. Biobehav. Rev.*, **31**, 192–201.

Kodituwakku, P., Coriale, G., Fiorentino, D., Aragon, A., Kalberg, W.O., Buckley, D., Gossage, J.P., Ceccanti, M., and May, P.A. (2006) Neurobehavioral characteristics of children with fetal alcohol spectrum disorders in communities from Italy: preliminary results. *Alcohol. Clin. Exp. Res.*, **30** (9), 1551–1156.

Linnet, K., Dalsgaard, S., Obel, C., Wisborg, K., Henriksen, T., Rodriguez, A., Kotimaa, A., Moilanen, I., Thomsen, P., Olsen, J., and Jarvelin, M.R. (2003) Maternal lifestyle factors in pregnancy risk of attention deficit hyperactivity disorder and associated behaviors: review of current evidence. *Am. J. Psychiatry*, **160**, 1028–1040.

Lynch, M., Coles, C., Corley, T., and Falek, A. (2003) Examining delinquency in adolescents differentially prenatally exposed to alcohol: the role of proximal and distal factors. *J. Stud. Alcohol.*, **64**, 678–686.

McLennan, J. and Offord, D. (2002) Should postpartum depression be targeted to improve child mental health? *J. Am. Acad. Child Adolesc. Psychiatry*, **41** (1), 28–35.

May, P.A. and Gossage, J.P. (2001) Estimating the prevalence of Fetal Alcohol Syndrome. A summary. *Alcohol Res. Health*, **25** (3), 159–167.

May, P.A., Gossage, J.P., Marais, A.S., Hendricks, L.S., Snell, C.L., Tabachnick, B.G., *et al.* (2008) Maternal risk factors for fetal alcohol syndrome and partial fetal alcohol syndrome in South Africa: a third study. *Alcohol. Clin. Exp. Res.*, **32** (5), 738–753.

May, P., Gossage, P., Kalberg, W., Robinson, L., Buckley, D., Manning, M., and Iiloyme, H.E. (2009) Prevalence and epidemiologic characteristics of FASD from various research methods with an emphasis on recent in-school studies. *Dev. Disabil. Res. Rev.*, **15**, 176–192.

O'Connor, M., Frankel, F., Paley, B., Schonfeld, A., Carpenter, E., Laugeson, E., and Marquardt, R. (2006) A controlled social skills training for children with fetal alcohol spectrum disorders. *J. Consult. Clin. Psychol.*, **74** (4), 639–648.

Oesterheld, J. and Wilson, A. (1997) ADHD and FAS. *J. Am. Acad. Child Adolesc. Psychiatry*, **36**, 1163.

Oesterheld, J., Kofoed, L., Tervo, R., Fogas, B., Wilson, A., and Fiechtner, H. (1998) Effectiveness of methylphenidate in Native American children with fetal alcohol syndrome and attention deficit/hyperactivity disorder: a controlled pilot study. *J. Child Adolesc. Psychopharmacol.*, **8** (1), 39–48.

Osborne, C. and Berger, L. (2009) Parental substance abuse and child well-being. *J. Fam. Issues*, **30** (3), 341–370.

Peadon, E., Rhys-Jones, B., Bower, C., and Elliott, E. (2009) Systematic review of interventions for children with Fetal Alcohol Spectrum Disorders. *BMC Pediatr.*, **9** (35), 1–9.

Pliszka, S., and AACAP Work Group on Quality Issues (2007) Practice parameters for the assessment and treatment of children and adolescents with attention-deficit/hyperactivity disorder. *J. Am. Acad. Child Adolesc. Psychiatry*, **46** (7), 894–921.

Pliszka, S.R., Crismon, M.L., Hughes, C.W., Corners, C.K., Emslie, G.J., Jensen, P.S., McCracken, J.T., Swanson, J.M., and Lopex, M. (2006) The Texas Children's Medication Algorithm Project: revision of the algorithm for pharmacotherapy of attention-deficit/hyperactivity disorder. *J. Am. Acad. Child Adolesc. Psychiatry*, **45**, 642–657.

Premji, S., Serrett, K., Benzies, K., Hayden, K., Dmytryshyn, A., and Williams, A. (2004) The State of the Evidence Review: Interventions for Children and Youth with a Fetal Alcohol Spectrum Disorder (FASD). Report prepared for the Alberta Centre for Child, Family, and Community Research. Available at: http://www.research4children.com/public/data/documents/SOTEIntpdf.pdf

Rodriguez, A., Olsen, J., Kotimaa, A., Kaakinen, M., Moilanen, I., Henriksen, T., Linnet, K., Miettunen, J., Obel, C., Tannila, A., Ebeling, H., and Järvelin, M. (2009) Is prenatal alcohol exposure related to inattention and hyperactivity symptoms in children? Disentangling the effects of social adversity. *J. Child Psychol. Psychiatry*, **50** (9), 1073–1083.

Schober, R. and Annis, H.M. (1996) Barriers to help-seeking for change in drinking: a gender-focused review of the literature. *Addict. Behav.*, **21** (1), 81–92.

Schwartzbaum, J., Ahlbom, A., and Feychting, M. (2003) Berkson's bias reviewed. *Eur. J. Epidemiol.*, **18**, 1109–1112.

Shin, M., Besser, L.M., Kucik, J.E., Lu, C., Siffel, C., and Correa, A. (2009) Prevalence of Down syndrome among children and adolescents in 10 regions of the United States. *Pediatrics*, **124**, 1565–1571.

Smith, L. (1992) Help-seeking in alcohol-dependent females. *Alcohol Alcohol.*, **27** (1), 3–9.

Sood, B., Delaney-Black, V., Covington, C., Nordstrom-Klee, B., Ager, J., Templin, T., Janisse, J., Martier, S., and Sokol, R. (2001) Prenatal alcohol exposure and childhood behavior at age 6 to 7 years: I. Dose-response effect. *Pediatrics*, **108** (2), e34.

Steinhausen, H.C., Gobel, D., and Nestler, V. (1984) Psychopathology in the offspring of alcoholic parents. *J. Am. Acad. Child Psychiatry*, **23** (4), 465–471.

Streissguth, A., Bookstein, F., Barr, H., Sampson, P., O'Malley, K., and Young, J.K. (2004) Risk factors for adverse life outcomes in fetal alcohol syndrome and fetal alcohol effects. *Dev. Behav. Pediatr.*, **25**, 228–238.

Waddell, C., Offord, D.R., Shepherd, C.A., Hua, J.M., and McEwan, K. (2002) Child psychiatric epidemiology and Canadian public policy-making: the state of the science and the art of the possible. *Can. J. Psychiatry*, **47** (9), 825–832.

Part Five
Research Needed on FASD

23
FASD Research in Primary, Secondary, and Tertiary Prevention: Building the Next Generation of Health and Social Policy Responses

Amy Salmon and Sterling Clarren

23.1
Introduction

For many of those associated with Fetal Alcohol Spectrum Disorder (FASD) diagnosis, intervention, and/or prevention, there remains a mysterious aspect to the condition. When FASD was first identified during the 1970s as the leading known cause of developmental delay, and the only disability of its type that might be termed "100% preventable," it was thought that a road map to prevention had been well established. This road map was considered to be clear, uncomplicated, and achievable: namely, that FASD – and the range of physical, cognitive, and daily living challenges linked to Prenatal Alcohol Exposure (PAE) – could be prevented in their entirety if women planning pregnancies would stop drinking alcohol. Likewise, once the range of impairments and functional difficulties of people diagnosed with FASD had been codified and elaborated, then a road map for providing appropriate supports to individuals and families, and preventing what have come to be known as "secondary disabilities," would also be achievable. Yet, although many promising practices have been developed and implemented, efforts to achieve and sustain primary, secondary, and tertiary prevention at a population level have been – and remain – illusive.

This chapter is guided by a persistent belief that the diagnosis of FASD and its treatments and prevention can be both demystified and achieved through evidence-based research that promote integrated approaches to prevention that effectively link: (i) a compassionate and timely care for pregnant women and mothers with substance-use problems; (ii) accurate and meaningful FASD diagnosis; and (iii) supportive care for people living with this disability in their communities. These steps need to be documented, as they are implemented for quality assurance and quality improvement. It is believed that answers to such questions will provide the evidence needed to drive the next stages of policy development in this field.

Re-drawing the road map for prevention has been informed by experiences with the Canada Northwest FASD Research Network (CanFASD Northwest), which was formed when Ministers from the Governments of British Columbia, Alberta,

Fetal Alcohol Spectrum Disorder–Management and Policy Perspectives of FASD. Edited by Edward P. Riley, Sterling Clarren, Joanne Weinberg, and Egon Jonsson
Copyright © 2011 WILEY-VCH Verlag GmbH & Co. KGaA, Weinheim
ISBN: 978-3-527-32839-0

Saskatchewan, Manitoba, Nunavut, the Northwest Territories, and the Yukon all realized that they needed meaningful data to be collected in a way capable of informing public policy needs related to FASD. CanFASD Northwest was the organizational response to this need (Clarren and Lutke, 2008), and the structure of the organization evolved through forums that were held with those actively engaged in all aspects of FASD studies. It became apparent that, whilst over 170 demonstration, implementation and research projects on various aspects of FASD were underway in western Canada, virtually none of these was being assessed, and few of the groups were even talking to or learning from one another. On analyzing the findings from these forums, it was clear that FASD could be advanced only if all aspects of the condition–diagnosis, intervention, prevention and population health surveillance–were considered together.

One reason why so many basic investigations remain to be performed appears to be related to a failure of FASD to be incorporated into the general cloth of medical work, nor easily addressed and accommodated in systems such as education, social services, mental health, and justice. Why is this the case? As Don Berwick (2004) frequently teaches, "Every system is perfectly constructed to give the result that is achieved." But, if we do not like the result achieved from our own systems, we must try to understand what produced the less-than-expected result. Indeed, this is what is needed in order to move this field forward; there is a need to evaluate what is being done, how it might be done better, and then to implement and disseminate change. The aim of this chapter is to highlight some of the mysteries that are preventing the identification and implementation of the effective, integrated policy responses required not only to prevent FASD but also to improve the outcomes for those affected.

23.2
Mapping Prevention: What Research is Needed Now, and Why?

23.2.1
Primary Prevention: Social Support and Determinants of Women's Health

One of the biggest mysteries confronting this field is the question of why pregnant women continue to drink when most know that doing so could harm their fetus and lead to lifelong disabilities in their child.

Mothers consistently describe social support as critical to having a healthy pregnancy (Creamer et al., 1998; Grant et al., 2005; Parkes et al., 2008; Sweeney et al., 2000). It is now abundantly clear that women who give birth to children with FASD are most often women whose own health and well-being are compromised in many ways. The lives of the birth mothers of children diagnosed with FASD are frequently imbued with violence, isolation, poverty, mental ill health (including diagnosed psychiatric conditions, very high stress levels, and trauma), addictions, and lack of supportive health and social care before, during, and after their pregnancy (Astley et al., 2000; Badry 2007; Salmon, 2007a). Undoubtedly, the complexity of these issues demands a timely and coordinated approach to care. Despite increasing acknowledgment that isolation and lack of social support are common among pregnant women and mothers with substance-use problems, the challenges accessing timely and supportive services remain for many women most vulnerable to having a child with FASD, including addictions treatment (UNODC, 2004), parenting support (Badry, 2007; Salmon, 2007b), and support for experiences of violence and trauma (Moses et al., 2003). Studies have also shown that a lack of collaboration between child welfare and addictions treatment services will result in further barriers to care for both mothers and children, as does a lack of support for including fathers in service delivery (Weaver, 2009).

While public health messaging campaigns exhort pregnant women to identify their drinking as problematic, and to seek professional help if they cannot stop drinking on their own, health systems and services are often unprepared to provide help when women seek it. More often than not, pregnant women facing concurrent problems with violence, mental health, and substance use are shuffled between uncoordinated systems of care with competing and contradictory service mandates and access criteria. To illustrate this, women presenting to addictions services are often told that they need to be treated for their mental health issues before they can enter addictions treatment, while mental health services often require abstinence from alcohol and (nonprescribed) drugs before women can be admitted into their care. Likewise, transition houses and other services for women experiencing violence have often been unprepared to provide service to women with untreated mental health and substance-use problems (Godard, Cory, and Abi-Jaoude, 2008; Morrissey et al., 2005). Thus, women who seek help without the economic resources to purchase private services (where they are available), or the social capital to advocate for access to much-needed assistance, often find themselves bounced around and between systems of care, until they are bounced out of them entirely.

Regardless of the circumstances in which they become pregnant, give birth to, and raise their children, contemporary western societies invest mothers with the primary responsibility of ensuring that their children achieve an optimal level of health and well-being. Yet, when mothers are unable to do so, they become objects of derision, particularly if these outcomes are understood to be the result of "poor choices" rather than circumstances beyond their control (Bell, McNaughton, and Salmon, 2009). Although common, "shame and blame" approaches to FASD prevention have been shown repeatedly to be ineffective at reducing drinking among women at highest risk for having a child with FASD, and result in many missed opportunities for providing supportive care (Armstrong, 2003; Badry, 2007; Salmon, 2007a).

However, it is often (and ironically) the case that isolation and marginalization also occur (or are reinforced) when women reach out to healthcare and social services. In the words of one mother:

> It's like I have to beg, and if it's not begging for what I need, for the support for what my kids need, I'll have to fall on my face before they recognise that I have some issues or problems, and that's the frustrating thing ... it makes life a lot harder. (Salmon and Ham, 2008)

While multiple barriers to care exist, evidence is also accumulating that interventions to increase social support for pregnant women and new mothers by addressing social determinants of women's health can improve outcomes for mothers and children, and also reduce the likelihood of future substance-exposed pregnancies (Grant et al., 2005, Motz et al., 2006; Parkes et al., 2008; Poole, 2000).When asked for her views on why this approach was so important, one mother stated:

> They still took me for who I was, they didn't care that I used, they didn't care that I was using when I was pregnant, they just wanted to make sure that I was fed and had somewhere to go ... I honestly think that if it wasn't for this place, my children probably wouldn't have survived my whole pregnancy. (Salmon and Ham, 2008)

While evidence accumulates that preventing FASD requires policy responses to improve social determinants of women's health, such determinants have been unevenly supported by governments. Specifically, policy development (and the funding for programs and services which follow such policies) in Canada has tended to prioritize initiatives that focus on prompting individual behavior change, and which privilege child health, particularly in the 0–6 years age group. For example, a 2007 study performed by Frankish and colleagues surveyed health regions across Canada for an assessment of their allocation of resources directed toward each social determinant of health (Frankish et al., 2007). Their findings showed that child development and personal health practices were self-reported by the majority of health regions to receive greatest attention, both in terms of internal policy and program attention and through inter-sectoral activities with other health regions and levels of government. Conversely, culture, gender, and employment/working conditions (all of which are linked closely to both income, social status, and access to health services) received least attention in most regions.

The uneven support for these specific social determinants of health has important implications for efforts to advance public policy support for the primary prevention of FASD. To date, the FASD prevention activities most commonly undertaken by governments have been those aimed at changing personal health practices. For the most part, this has taken the form of primary prevention campaigns aimed at increasing public awareness of the risk posed by prenatal alcohol exposure and urging pregnant women to abstain from drinking (Deshpande et al., 2005). In tandem, child health and development services have emerged that respond to concerns about FASD through diagnosis and intervention with pediatric populations. Multidisciplinary diagnostic clinics have been, and continue to be, developed across Canada, and integrated teams of professionals including physicians, psychologists, occupational therapists, speech language pathologists, social workers, and educators have sought to identify new and better ways of improving developmental outcomes for children and youth living with these disabilities. Together, these efforts have yielded substantive increases in public awareness of the importance of avoiding alcohol use in pregnancy, and of the challenges faced by children with FASD, both of which are clearly important and much needed. At the same time, these efforts have also constructed FASD in the public (and political) imagination as a *children's* health issue, and have inspired reductionist approaches to FASD prevention as primarily a problem of maternal ignorance and/or malfeasance. In other words, the task of preventing FASD has come to be understood by many as an effort to improve children's health by intervening with their mothers, to ensure that women who drink are aware that they are "hurting their babies," and to "protect" those babies whose mothers continue to knowingly put them at risk through action which they are believed to have complete control to stop. In so doing, opportunities have been missed to understand that FASD and its prevention are also *women's* health issues, and to act to reduce FASD prevalence at a population level by improving social, economic, and political contexts which give rise to problematic substance use among women and compromise maternal and child health (Clarren and Salmon, 2010).

23.2.2
Accurate Diagnosis of FASD: Preventing Secondary Disabilities and Reaching out to Mothers (and Potential Mothers)

Historically, FASD prevention activities (particularly primary prevention efforts) have been separated from initiatives to develop diagnostic services and supports for people living with FASD.

An accurate diagnosis of FASD is critical for prevention and intervention. Indeed, the problems that can accrue for individuals and communities when a diagnosis of FASD is made inaccurately or haphazardly – including experiences of stigma and stereotyping of FASD as being a problem for some groups (such as Aboriginal peoples) and but not others – have been well described (Tait, 2003; Van Bibber, 1997). A proper diagnosis – which requires specialist multidisciplinary teams of professionals – depends on a diagnostic capacity which is, at present, significantly lacking. To illustrate this, assume (as a simple estimate) that the prevalence of FASD in Canada is one in 100 people, and that of Fetal Alcohol Syndrome (FAS) itself is one in 1000 (Center for Disease Control and Prevention, 2002). Given the current population of Canada, the approximate number of cases of FAS is likely to be about 32 000, and the number of cases of FASD about 340 000. At present, no firm figures are available on the incidence of new births with FASD, the death rate of people with FASD, or how may people with FASD enter Canada from other countries, but even so, the number of unrecognized cases is huge. The number of evaluations that can be made annually in all of Canada today now is less than 2000. Such a discrepancy between the numbers of those who might require diagnosis and those who can actually be diagnosed is beyond reason.

Diagnosis is needed to prove to governments that there is a need for services. If efforts are not made to systematically collect information demonstrating the scope of the need for support services, then effective and responsive systems cannot be constructed. Systems will not be built for hypothetical clients. Diagnosis also represents the starting point for specific treatment planning in either intervention or prevention. Put simply, no diagnosis – no problem; and no problem – no need for a solution.

How then can this gap between diagnostic capacity and diagnostic need be approached? The answer may lay in the degree of mystery that individual patients display. Normally, adaptive people are those who have a normal temperament and mood that directs their affect, normal cognition and information processing, normal environments, and normal ability to learn. In contrast, those with FASD have maladaptation that may be related to any or all of these internal or external aspects of brain function. Moreover, not only may maladaptation have many causes, but society has various expectations for resolving maladaptation, depending on its cause. If it is thought that a maladaptation is due to a health or environmental issue, the societal judgment is that it is a disease and the intervention goal is to treat it or cure it. If it is believed that a maladaptation is due to temperament or mood – that it is psychiatric – then it may still be thought of as a disease and an intervention effected with treatment or cure. However, that same maladaptation of temperament might be termed disobedience in the Court, and the intervention might be punishment or separation. If the issue is abnormal cognition or performance deficits, societal judgment then – and only then – is that it is a disability that must be accepted or treated. What is mysterious about the diverse population of people diagnosed with FASD is that they may have problems that stem from any or all of these factors, at the same time or at different times.

If it is done well, efforts to link up policies and programs to enhance social supports and determinants of health for birth mothers with diagnostic and support services for their children with FASD could also help to dismantle the shame and blame that underscore much of contemporary secondary and tertiary prevention efforts. It has been (and in many instances it remains) the case that the perspectives of birth mothers have been rendered invisible in diagnostic and support services for children with FASD. This has created conditions in which shame and mother-blaming have flourished, frustrating opportunities for secondary and tertiary prevention within a broad range of systems. Because FASD has been constructed as a consequence of maternal behavior that often viewed as reckless, hedonistic, or self-indulgent, this diagnosis generates different responses than other developmental disabilities that result from random chance or previously undetected genetic variation. The fact that many children with FASD come into contact with diagnostic and support services through the foster-care system has also created conditions in which it has been seen acceptable to attribute difficulties with learning, self-regulation, and life skills experienced by many people living with FASD as a product of poor parenting rather than disability. In resource-constrained environments, where health, social service, and educational systems must constantly seek ways of reducing costs and cutting back on the delivery of expensive and time-consuming services, mother-blaming discourses may be more readily invoked to offer a rationale for shifting responsibility away from governments and systems and onto families. As a result, birth mothers raising children with FASD who struggle in school and their communities, or who come into conflict with authorities, have become accustomed to hearing that "the problem is at home" when they attempt to advocate for additional supports for their child. Data collected systematically to demonstrate the range of impairments and support needs common to people living with FASD, their birth mothers, and their families, could also demonstrate gaps in service provision that could subsequently render governments accountable for ensuring that these needs are met.

Envision a system of screening for FASD based on alcohol exposure, facial features and, most importantly, on the evidence of maladaptation that appears mysterious and confusing. Those individuals screened positive would be assessed by primary triage specialists, who would evaluate the history and determine how much evaluation of temperament and mood, cognition and performance, and global environment has already been done, and how much success has already been achieved. Referral for many would be to general medicine, a specialist physician, or to various mental health services, for specific types of intervention. Those who continue to struggle would then be referred on to FASD multidisciplinary programs (that would need to be built at a pace in keeping with demand). These multidisciplinary groups could, at the same time, access the more challenging needs of individuals and, preferably over time, determine what percentage of the problems were the direct results of brain damage, and which were social, environmental, or something else. These final clinics should be organized to collect data that allow for consistent, cross-clinic and cross-provincial and territorial studies.

These data would also suggest the specific functional diagnoses of the patients and their treatment needs. The data might also suggest problems in the ways that the patients were managed in their interaction (or lack of interaction) with services prior to diagnosis that could help set targets for the prevention of secondary disabilities. Specifically, when a diagnosis is not linked to supportive services in a timely and ongoing fashion, children with FASD are placed at further risk for developing problems often described as "secondary disabilities," including poor performance in school, disrupted school experience, mental health problems, problems with employment and maintaining independence as an adult, and coming into conflict with the law (Streissguth and Kanter, 1997). Data-tracking outcomes for individuals seen at these clinics could help to identify what types of support reduce the likelihood of developing secondary disabilities, and also identify those areas in which governments need to invest more in order to meet secondary and tertiary prevention needs. Finally, the data would help with the identification of birth mothers who might have FASD themselves, who might be at risk for subsequent pregnancies exposed to alcohol, and who may themselves be in need of support.

If the achievement of primary, secondary, and tertiary prevention of FASD requires policies that help to link more effectively the supports for at-risk women with supports for their children, then data are required from both women-serving and children-serving programs, and from clinics that can show how best this care should be delivered. These data would be key to building models that would allow for an appropriate planning to expand the capacity for diagnostic clinics and for services for "at-risk" women. The type of questions that should perhaps be asked right now are as follows:

- How much do we need to increase capacity?
- Where should new programs be located?
- Who should staff them?
- How do we monitor program processes and outcomes?

Answers to these questions are obtained by analyzing the composition and work of programs themselves. Who is on the staff? How much time do they spend with each client? Do they need more time with each woman/family, or with specific women/families? Could they use professional time more efficiently and, if so, how? Are experiences or needs different in different places? Do urban services need a different composition from rural services? Some of these data can be gathered at this time through the use of a common data set and common data analysis at diagnostic clinics. Even today, clinics serving children are routinely collecting almost all of this information. However, the data are collected in varying ways, and often not kept after the evaluation has been completed. Similar processes for data collection appropriate to the range of programs currently offered to women are urgently needed, and efforts to develop mechanisms for collecting comparable data across women-serving programs are still in their very early stages. A common form for pediatric diagnostic clinics is now available from the CanFASD Northwest, and many clinics would be ready to work collaboratively to create this data set. CanFASD Northwest is also supporting a similar initiative to create common evaluation and outcome measures for women-serving programs. Unfortunately, however, political will and adequate long-term funding for both initiatives remain elusive.

23.3
Conclusions: Drawing a Road-Map for Integrated, Supportive, and Effective Care

Re-drawing a road map for FASD prevention requires policy responses that are meaningful, effective, and compassionate. Such approaches must demonstrate an understanding of the contexts in which women drink during pregnancy, including the recognition that alcohol use is often an attempt to cope with exceptionally difficult life circumstances in the absence of other supports. Such responses also must be compassionate, based on understandings of the needs and strengths of mothers, families, and children who are dealing with the consequences of alcohol use in pregnancy. This requires an increased system capacity for inter-agency cooperation wrapped around the mother, child, and family. In order to achieve this understanding, there is a need for reliable data demonstrating the scope of the challenges experienced by individuals with FASD at various developmental stages, the supports needed by people with FASD to achieve positive outcomes, and the most effective means of delivering these supports. Screening materials are available that could be used for a demonstration system within the confines of one or more provinces that already have reasonable FASD diagnostic capacity, so as to increase their load two- or threefold over a period of a few years. During this time, the work of the clinics should be analyzed so that the procedures and processes of the programs can be standardized and itemized for both personnel and cost. Similar data are also needed from programs serving birth mothers, including reliable outcome measures appropriate for the range of supports needed by and delivered to substance-using pregnant women and mothers, which can identify the aspects of the programs that are most effective in fostering positive outcomes for women. Together, these data will provide a strong evidence base which can be used by governments and health planners to identify models for organizing and delivering integrated, supportive care for women and their children.

References

Armstrong, E. (2003) *Conceiving Risk. Bearing Responsibility. Fetal Alcohol Syndrome and Diagnosis of Moral Disorder*, The John Hopkins University Press, Baltimore.

Astley, S.J., Bailey, D., Talbot, C., and Clarren, S.K. (2000) Fetal Alcohol Syndrome (FAS) primary prevention through FASD Diagnosis II: A comprehensive profile of 80 birth mothers of children with FAS. *Alcohol Alcohol.*, **35** (5), 509–519.

Badry, D. (2007) Birth mothers of fetal alcohol syndrome children, in *Social Justice in Context, Vol. 3, 2007–2008* (eds M.J. Jackson, M.J. Pickard, and E.W. Brawner), Carolyn Freeze Baynes Institute for Social Justice, East Carolina University, Greenville, pp. 88–103.

Bell, K., McNaughton, D., and Salmon, A. (2009) Medicine, morality, and mothering: public health discourses on fetal alcohol exposure, smoking around children, and childhood overnutrition. *Crit. Public Health*, **19** (2), 155–170.

Berwick, D. (2004) Acceptance Speech for the 2004 TRUST Award, www.ihi.org (accessed 13 April 2010).

Center for Disease Control and Prevention (2002) Fetal alcohol syndrome – Alaska, Arizona, Colorado, and New York: 1995–1997. *Morb. Mort. Wkly Rep.*, **514**, 433–435.

Clarren, S.K. and Lutke, J.M. (2008) Building clinical capacity for fetal alcohol spectrum disorder diagnoses in Western & Northern Canada. *Can. J. Clin. Pharmacol.*, **15** (2), e223–e237.

Clarren, S.K. and Salmon, A. (2010) Prevention of fetal alcohol spectrum disorder proposal for a comprehensive approach. *Expert Rev. Obstet. Gynecol.*, **5** (1), 23–30.

Creamer, S. and McMurtrie, C. (1998) Special needs of pregnant and parenting women in recovery: A move toward a more woman-centered approach. *Women's Health Issues*, **8** (4), 239–245.

Deshpande, S., Basil, M., Basford, L., Thorpe, K., Piquette-Tomei, N., Droessler, J., Cardwell, K., Williams, R.J., and Bureau, A. (2005) Promoting alcohol abstinence among pregnant women: potential social change strategies. *Health Mark. Q.*, **23** (2), 45–67.

Frankish, C.J., Moulton, G.E., Quantz, D., Carson, A.J., Casebeer, A.L., Eyles, J.D., Labonte, R., and Evoy, B.E. (2007) Addressing the non-medical determinants of health: a survey of Canada's health regions. *Can. J. Public Health*, **98** (1), 41–47.

Godard, L., Cory, J., and Abi-Jaoude, A. (2008) *Summary Report of Building Bridges: Linking Woman Abuse, Substance Use and Mental Ill Health*. BC Women's Hospital and Health Centre, Vancouver.

Grant, T.M., Ernst, C.C., Streissguth, A., and Stark, K. (2005) Preventing alcohol and drug exposed births in Washington State: intervention findings from three parent-child assistance program sites. *Am. J. Drug Alcohol Abuse*, **31** (3), 471–490.

Morrissey, J.P., Ellis, A.R., Gatz, M., Amaro, H., Reed, B.G., Savage, A., Finkelstein, N., Mazelis, R., Brown, V., Jackson, E.W. and Banks, S. (2005) Outcomes for women with co-occurring disorders and trauma: program and person-level effects. *J. Subst. Abuse Treat.*, **28** (2), 121–133.

Moses, D.J., Reed, B.G., Mazelis, R., and D'Ambrosio, B. (2003) *Creating Services for Women with Co-Occurring Disorders*, Substance Abuse and Mental Health Services Administration and Centre for Mental Health Services, Alexandria.

Motz, M., Leslie, M., Pepler, D.J., Moore, T.E., and Freeman, P.A. (2006) Breaking the cycle: measures of progress 1995–2005. *J. FAS Int., Spec. Suppl.*, **4**, e22, http://www.motherisk.org/JFAS_documents/BTC_JFAS_ReportFINAL.pdf

Parkes, T., Poole, N., Salmon, A., Greaves, L., and Urquhart, C. (2008) *Double Exposure: A Better Practices Review on Alcohol Interventions during Pregnancy*, British Columbia Centre of Excellence for Women's Health, Vancouver.

Poole, N. (2000) *Evaluation Report of the Sheway Project for High Risk Pregnant and Parenting Women*, British Columbia Centre of Excellence for Women's Health, Vancouver.

Salmon, A. (2007a) Dis/abling states, dis/abling citizenship: young Aboriginal mothers, substantive citizenship, and the medicalization of FAS/FAE. *J. Crit. Educ. Policy*, **5** (2). Available at: http://www.jceps.com/?pageID=article&articleID=103.

Salmon, A. (2007b) Beyond shame and blame: aboriginal mothers, fetal alcohol spectrum disorders, and barriers to care, in *Highs and Lows: Women and Substance in Canada* (eds L. Greaves, N. Poole, and J. Greenbaum), Centre for Addiction and Mental Health, Toronto, pp. 227–236.

Salmon, A., and Ham, J. (2008) *Evaluation of Client Outcomes Associated with MEIA-Funded Supports at Sheway*, BC Centre of Excellence for Women's Health, Vancouver.

Streissguth, A.P. and Kanter, J. (eds) (1997) *The Challenge of Fetal Alcohol Syndrome: Overcoming Secondary Disabilities*, University of Washington Press, Seattle.

Sweeney, P.J., Schwartz, R.M., Mattis, N.G. et al. (2000) The effect of integrating substance abuse treatment with prenatal care on birth outcome. *J. Perinatol.*, **4**, 19–24.

Tait, C.L. (2003) *Fetal Alcohol Syndrome Among Canadian Aboriginal People in Canada: Review and Analysis of the Intergenerational Links to Residential Schools*, Aboriginal Healing Foundation, Ottawa.

UNODC (2004) *Substance Abuse Treatment and Care for Women: Case Studies and Lessons Learned*, The United Nations Office on Drugs and Crime, Vienna.

Van Bibber, M. (1997) *It Takes A Community: A Resource Manual for Community-Based Prevention of FAS/FAE*, Health Canada First Nations and Inuit Health Branch, Ottawa.

Weaver, S. (2009) *"Left out:" Father Exclusion in Addictions Health Services*. University of British Columbia, Vancouver.

24
Focusing Research Efforts: What Further Research into FASD is Needed?

Sara Jo Nixon, Robert A. Prather, and Rebecca J. Gilbertson

24.1
Introduction

Over the past three decades, significant advances have been made in describing the dysmorphology and neurobehavioral deficits associated with fetal alcohol exposure. Careful, systematic studies have characterized the derailing of neurodevelopmental trajectories and social-behavioral maladaptation accompanying prenatal alcohol exposure (Coles, Kable, and Taddeo, 2009; Lebel *et al.*, 2008; Greenbaum *et al.*, 2009; Paley and O'Connor, 2009). The results of these studies, which were derived from prospective animal studies and observational human investigations, have led to significant alterations in medical practice, educational considerations, and, increasingly, judicial decisions (Grant *et al.*, 2009). Despite these exceptional advances, there is still much to be done; consequently, in the following sections an outline has been provided of three general areas of inquiry that demand systematic consideration. Initially, attention is focused on the complexity of observed heterogeneity in outcomes, and the potential clinical and scientific benefits of disentangling risk and resiliency factors. The benefits which might be derived by the application of conceptual models grounded in behavioral and cognitive neuroscience are then explored, by moving from descriptive overviews of neurobehavioral deficiencies to place a greater emphasis on process and component processes that, in turn, might direct interventional efforts. Finally, a preliminary discussion is provided of the research opportunities and clinical interventions that might arise if the strengths of the neurosciences were to be applied more directly to the question, which obviously is bounded by the parameters of brain and behavior plasticity.

24.2
FASD and Heterogeneity: An Encouraging Outcome

If the association between exposure and negative outcome is absolute, linear, and clearly observable, then clinical recommendations are generally direct and

Fetal Alcohol Spectrum Disorder–Management and Policy Perspectives of FASD. Edited by Edward P. Riley, Sterling Clarren, Joanne Weinberg, and Egon Jonsson
Copyright © 2011 WILEY-VCH Verlag GmbH & Co. KGaA, Weinheim
ISBN: 978-3-527-32839-0

anecdotal evidence supportive. Unfortunately, however, for the majority of human maladies these parameters cannot be approached, and Prenatal Alcohol Exposure (PAE) is similarly uncooperative. Predictive models based on animal studies, and observational data from human studies, have demonstrated the generally recognized association between quantity/frequency and timing as key factors in determining the locus and severity of a neurobehavioral impact. These findings, when spanning the gestational period, have guided the generally offered recommendation that "there is no safe time to drink during pregnancy." For those with a responsibility to future generations, this seems appropriate advice, but it must be realized that approximately 14.6% (estimated) of women drink at some point during their pregnancy, while the combined rate of Fetal Alcohol Spectrum Disorder (FASD) and Alcohol-Related Neurodevelopmental Disorder (ARND) is estimated to be 9.1 per 1000, or about 1% of live births (Ebrahim et al., 1998; Sampson et al., 1997). Therefore, while PAE is necessary for the development of FASD (and of Fetal Alcohol Syndrome; FAS), it is not sufficient. Even if the less obvious neurobehavioral complications, which may not be evident until entry to school (Streissguth et al., 1986) are disregarded, there is a clear dissociation between exposure levels, dysmorphology, and neurobehavioral compromise. Finally, there is significant incongruity among measures within the same individuals. For example, a host of studies has demonstrated an absence or minimal congruence between structural abnormalities and performance, and/or between intelligence scores and functional adaptation (Lebel et al., 2008).

Although such heterogeneity creates challenges in public health messaging, it also serves to force movement. These inconsistencies moved the field from reliance on a diagnostic dichotomy to a more complex ordinal scale engaging a spectrum of potential outcomes. Yet, the questions remain:

- Why are some fetuses relatively spared from alcohol's toxic effects?

- When exposure dose and timing are accounted for, what are the genetic or environmental factors which impact outcome?

- Does this protection arise from maternal and/or fetal characteristics?

- What is the nature of the maternal/fetal interaction and how might addressing this question inform a larger field concerned with fetal development and maternal and child health?

- Finally, what role might paternal alcoholism play in the neurobehavioral dysregulation observed in PAE children?

But, underlying such questions is a more fundamental problem. Given the heterogeneity in the association between dose exposure and negative outcome, what other factors might account for variability in outcome? That is, what factors may increase fetal vulnerability to developmental stressors or toxins, of which alcohol is one? One such frequently discussed factor is that of psychiatric comorbidity among pregnant, drinking women. It can be assumed that, among women who are unable to quit drinking during pregnancy, a significant number also smoke

(Burns, Mattick, and Wallace, 2008). The negative impact of smoking, absent of alcohol consumption, on pregnancy is well documented (for a review, see Cornelius and Day, 2009), unfortunately the effects of smoking and drinking are synergistic rather than additive. Thus, the potential negative impact on the fetus is significantly heightened (Aliyu et al., 2009).

Obviously, smoking does not "cause" FASD; rather, smoking increases the biologic vulnerability, heightening the threats to central nervous system (CNS) integrity, such as those associated with alcohol exposure. Although responsibility for ensuring prenatal health is typically assigned to the mother, the contributing influences are not that restricted, and such a distributed responsibility is especially evident in the consideration of environmental toxins, such as smoking. In short, the detrimental effects of smoking are not always directly administered. For example, many mothers who successfully quit smoking during pregnancy have partners and/or close friends who continue to smoke in their presence, providing fetal exposure to these toxins through "secondhand" smoke (El-Mohandes et al., 2010). While general recommendations to avoid such situations might be offered, there are insufficient data available regarding the exposure and vulnerability factors to inform more specific direction.

Fetal vulnerability may also be increased by sociocultural risks associated with the environments in which many of these women live, including hypoxic episodes accompanying domestic violence and/or common pregnancy risks, such as poor glycemic or blood pressure control. Additionally, higher gravidity and parity are also risk factors (May et al., 2008; Burd and Christensen, 2009). Importantly, nutritional factors are increasingly considered; recently acquired provocative data have suggested that the timely administration of supplements such as zinc and choline may minimize alcohol-related dysmorphology and/or mitigate behavioral compromise (Summers, Rofe, and Coyle, 2009; Thomas, Abou, and Dominguez, 2009; Ryan, Williams, and Thomas, 2008). Yet, further studies must be conducted to determine the efficacy and effectiveness of these interventions for at-risk babies.

In this discussion, it would be remiss if paternal alcoholism were not explored as a source of fetal vulnerability. Indeed, many reports have demonstrated that the children of male alcoholics are at increased risk for poor behavioral control, problems in abstracting/problem-solving, poor school and social achievements, and an early age of the first drink (Tarter et al., 2003; Giancola et al., 1996; Aytaclar et al., 1999). Neurobehavioral dysregulation in brain activation and electrophysiological responses have also been noted (Begleiter et al., 1984; Carlson, Iacono, and McGue, 2004). Whilst these findings are not dependent on the mother drinking during pregnancy, the behavioral compromise is remarkably similar to those exhibited by PAE children. Although these data do not suggest that paternal alcoholism causes FASD, they do show that the neurobehavioral deficits that often are attributed to maternal drinking may arise from multiple genetic and environmental sources. Systematic, prospective, longitudinal research directed at disentangling these interactive sources is essential to the development of more effective intervention efforts.

In summary, the hallmark of FASD is heterogeneity in apparent vulnerability and in outcome. The etiological and neurobehavioral complexity dictates the engagement of multimodal research strategies and the systematic consideration of individual differences in risk and resiliency attributes. Whilst heterogeneity creates challenges in creating credible public messages, it also reflects the reality of a dynamic neurobehavioral system which includes the capacity for recovery and compensatory action.

24.3
Models: Moving Beyond Description

In this section, attention is focused on a more effective modeling of the neurobehavioral deficits associated with PAE. Many reports have described the breadth of these deficits, and have illustrated their impact on independent living, psychosocial adaptation, peer-group and social effectiveness, and vulnerability to high-risk behaviors (Baer et al., 1998, 2003; Barr et al., 2006; Connor et al., 2000). As the first stage of scientific discovery, these descriptive data have provided an exceptional foundation for conceptual development. In order to progress FASD clinical and scientific studies to a more advanced stage of scientific thought, it will be necessary to encourage current efforts, shifting from the cataloging of deficits to that of developing explanatory and predictive accounts.

As demonstrated recently by Joanne Weinberg and colleagues, multimodal strategies can facilitate these efforts. In noting the high comorbidity of negative affect among children and adults with FASD, Weinberg et al. proposed that this vulnerability might be associated with hypothalamic–pituitary–adrenal (HPA) axis dysregulation, consequent to PAE. Using a mouse model, Weinberg's group conducted a series of studies in which PAE and later-life stress were systematically manipulated (Hellemans et al., 2008; Oberlander et al., 2008; Hellemans et al., 2009; Hellemans et al., 2010; Verma et al., 2010). Consistent with a working hypothesis, when PAE adult animals were subjected to mild chronic stress they displayed an altered stress-related behavior when compared to control animals. Furthermore, the results were sexually dimorphic; PAE males showed a greater hyperactivity and a disrupted social behavior, whereas PAE females showed a behavior consistent with behavioral despair (Hellemans et al., 2010). Thus, the vulnerability for affective disorders commonly observed in FASD persons may arise from an alcohol-related disruption in HPA tone and regulation, rather than (or in addition to) ineffective coping mechanisms and impaired problem-solving skills that might arise from cortical compromise. These types of research efforts offer explanation and direct future hypothesis-driven research.

Going beyond endpoint performance to examine component processes is gaining appreciation. For example, deficits in episodic memory might be clarified by examining attention, working memory, and/or retrieval processes (Kaplan, 1988; Nixon, 1993; Lezak, 1995; Sullivan and Pfefferbaum, 2005). A specific advantage of this strategy is that it not only permits, but rather encourages, comparisons across ostensibly distinct tasks, and thus facilitates the development of systems-based analyses of neurobehavioral outcomes. Further, process-oriented studies lend themselves to more effective education and intervention efforts by providing insight with regard to compromised and spared processes, and directing the consideration of both rehabilitation and compensatory processes.

Several groups in this field of research have employed this approach, with exciting findings. As noted, PAE children frequently demonstrate deficits over a wide range of skills essential to academic achievement (Kable, Coles, and Taddeo, 2007; Jacobson et al., 2003); among those most frequently compromised are basic math skills, for which many parents/caregivers and educators assume that little progress can be achieved. Yet, as part of a multisite, collaborative project with the Center for Disease Control and Prevention, Claire Coles and her colleagues confronted this assumption when they developed and tested a math intervention for children [e.g., Math Interactive Learning Experience (MILE)] with FASD (Kable, Coles, and Taddeo, 2007; Coles, Kable, and Taddeo, 2009). The program engaged families and caregivers as well as the students, by focusing on the development of compensatory skills to support any impaired function. The intervention improved math skills not only immediately after the intervention (Kable, Coles, and Taddeo, 2007), but also at the six-month follow-up (Coles, Kable, and Taddeo, 2009). Although, others have also suggested that this process approach showed much promise (Adnams et al., 2007; Kalberg and Buckley, 2007; O'Connor et al., 2006; Padgett, Strickland, and Coles, 2006), systematic longitudinal studies are required to assess the impact over time, and to provide insight regarding continuing interventions, and their cost and benefit.

24.4
Applying Neuroscience: Beyond the Mother?

The study of ARNDs suffers because current developments in the neurosciences are not fully applied. Consistent with the preventable nature of ARNDs, existing messages regarding abstinence during pregnancy are both appropriate and essential although, as discussed by Medina and Krahe (2008), these admonitions are not always followed. Furthermore, in some cases, they are not convincingly presented, and/or fail to be supported by significant others within the larger community. Finally, it is not unusual for women to have consumed alcohol prior to recognizing their pregnancy. Thus, while prevention via abstinence is the preferred approach, realities demand that a broader perspective be adopted regarding PAE. Traditional classroom educational efforts at maximizing individual strengths would no doubt enhance psychosocial adaptation, but are not purposed with directly altering the neurobehavioral systems underlying the impairment.

Assumptions regarding the timeframes and rigidity in brain development have been significantly challenged in recent years. Both animal and human studies have demonstrated an enormous plasticity in neural development throughout infancy, childhood and adolescence, and into young adulthood. (Ballantyne et al., 2008;

Blake et al., 2006; Giedd et al., 2009; McCutcheon and Marinelli, 2009; Shaw et al., 2008) Importantly, some studies have suggested the presence of neurotrophic processes, even among adults (Nixon et al., 2010; Crews and Nixon, 2009; Griesbach, Hovda, and Gomez-Pinilla, 2009; Liu et al., 2009). Although periods of plasticity may reflect critical vulnerable periods (Dobbing and Sands, 1979; Gilbertson and Barron, 2005; Goodlett and Johnson, 1997; 1999; West et al., 1989), they also denote the dynamic nature of neural systems and the potential for continuing intervention. As major efforts are made to understand and intervene in FASD, it is important to apply systematically what has been learned regarding "normal" brain development, the on-set/off-set of systems, the process of neuronal growth and pruning, and the role of environmental factors in mitigating and perhaps even reversing the negative consequences of PAE.

Environmental enrichment has been examined as a means of compensating for a variety of developmental disorders (e.g., Dong and Greenough, 2004). Their success, even when limited, speaks of the potency of environmental context in postnatal development. Furthermore, neurobehavioral enhancement has been observed with both mental and physical challenges, thus reinforcing the importance of integrated neural systems and the relevance of comprehensive interventions (Sim et al., 2008; Hannigan, O'Leary-Moore, and Berman, 2007). While these data continue to show promise, very little is known regarding the appropriate "dosing" of such exposures, or how to most effectively combine physical and mental challenges without over-stressing the compromised systems. Given the relative low costs associated with such interventions, even if the absolute improvement were to be minimal, these interventions merit prospective longitudinal evaluation.

Pharmacologic interventions have also offered promise. For example, maternal treatment with thyroid hormone has been shown to block some behavioral abnormalities among rodents (Wilcoxon et al., 2005). Similarly, zinc supplementation throughout pregnancy has been shown to prevent fetal dysmorphology and to increase postnatal survival in a rodent model (Summers, Rofe, and Coyle, 2009). By contrast, zinc supplementation with chronic ethanol exposure failed to demonstrate any beneficial effects; rather, the excess zinc was associated with a number of negative developmental outcomes, including problems of short-term memory (Moazedi Ahmad et al., 2007). It would appear that these interventions may be beneficial only under specific and limited conditions.

Of particular promise are interventions derived from the findings that PAE-related neurobehavioral deficits result from an interference in the neuronal plasticity essential for normal development (Medina and Krahe, 2008). Although the specific mechanisms underlying this interference are not yet known, certain pharmacologic interventions may block or mitigate these deficits. Although a complete review of this developing subject is beyond the scope of this chapter, promising data obtained by Jennifer Thomas and colleagues provides optimism and direction for additional research. Specifically, Thomas, Abou, Dominguez (2009) observed in rats that maternal choline supplementation accompanying alcohol exposure caused an attenuation of fetal alcohol-related effects in the offspring. Furthermore, in an earlier study Ryan, Williams, and Thomas (2008)), again using rats, showed the administration of choline after alcohol exposure during the third trimester (equivalent period) to mitigate deficits in spatial learning. That this positive effect was not limited by the coadministration of alcohol and choline suggested that choline, if administered after a period of maternal alcohol consumption, might be efficacious.

Phosphodiesterase (PDE) inhibitors are widely used as enhancers of neuronal plasticity, and the PDE4 and PDE1 families have each been linked to learning and memory, despite acting via different mechanisms, additionally may provide the opportunity for intervention. Medina, Krahe, and Ramoa (2006), showed that the PDE1 inhibitor, vinpocetine, could restore plasticity in ocular dominance compromised by alcohol exposure. Given the prevalence of visual deficits in FASD, this outcome demands further study.

In summary, although behavioral and pharmacologic interventions offer significant promise in FASD, the potential benefits will be significantly enhanced if, and when, the neurodevelopmental processes that are compromised in PAE are better understood. If compromise in neuronal plasticity lays at the core of the neurobehavioral deficits, then the supplementation/intervention of relevant biochemical, cellular and molecular processes should produce the greatest benefit. There is, however, much to be revealed concerning these processes, most notably their timeframe and flexibility.

As the success of these interventions relies on maternal compliance, one of the most important areas of research and education centers on the process of engaging drinking mothers and securing their continued participation. Clearly, even the most complete neurobiological models will not produce an effective change if they cannot be applied!

This reality demands that interventions be considered that are not solely dependent on maternal cooperation. An increasing recognition regarding the plasticity of the brain throughout adolescence and into young adulthood (and indeed across the lifespan) offers significant direction for research and intervention. Systematic studies using a combination of pharmacologic and behavioral approaches throughout these developmental periods could identify effective interventions, with clinically relevant outcomes. Whilst these studies will require an initial investment, the potential benefits in the quality of life, independence and productivity will surely surpass these initial costs. Finally, these studies will not only impact the lives of those affected by PAE; rather, they will also improve the present understanding of the etiology and intervention with other developmental disorders.

24.5
Summary

It has been suggested, based on a literature review, that the greatest gains in impacting FASD may, arguably, be achieved with research focusing on three broad areas of study:

- First, it is important to embrace the heterogeneity of FASD by systematically examining contributing factors. Maternal–fetal interactions during critical developmental periods must be more closely examined, and clinical and basic research engaged in these efforts. It is also necessary to disentangle the effects of alcohol from those associated with other overlapping disorders, by using appropriate control groups and accounting for biological and environmental confounds.

- A second area of research is directed towards building and then testing conceptual models that can provide differential hypotheses; in this way, the field can be advanced from a description of FASD to an explanation of the condition, and a more effective intervention.

- The third area of research incorporates the increasing emphasis that is being placed on neurobiological and behavioral processes that will accelerate such effort and enhance the opportunity to benefit from advances in other sciences.

This comprehensive research initiative has significant relevance to clinical development, scientific explanation, and policy and resource allocation.

References

Adnams, C.M., Sorour, P., Kalberg, W.O., Kodituwakku, P., Perold, M.D., Kotze, A., et al. (2007) Language and literacy outcomes from a pilot intervention study for children with fetal alcohol spectrum disorders in South Africa. *Alcohol*, **41** (6), 403–414.

Aliyu, M.H., Wilson, R.E., Zoorob, R., Brown, K., Alio, A.P., Clayton, H., et al. (2009) Prenatal alcohol consumption and fetal growth restriction: potentiation effect by concomitant smoking. *Nicotine Tob. Res.*, **11** (1), 36–43.

Aytaclar, S., Tarter, R.E., Kirisci, L., and Lu, S. (1999) Association between hyperactivity and executive cognitive functioning in childhood and substance use in early adolescence. *J. Am. Acad. Child Adolesc. Psychiatry*, **38** (2), 172–178.

Baer, J.S., Barr, H.M., Bookstein, F.L., Sampson, P.D., and Streissguth, A.P. (1998) Prenatal alcohol exposure and family history of alcoholism in the etiology of adolescent alcohol problems. *J. Stud. Alcohol*, **59** (5), 533–543.

Baer, J.S., Sampson, P.D., Barr, H.M., Connor, P.D., and Streissguth, A.P. (2003) A 21-year longitudinal analysis of the effects of prenatal alcohol exposure on young adult drinking. *Arch. Gen. Psychiatry*, **60** (4), 377–385.

Ballantyne, A.O., Spilkin, A.M., Hesselink, J., and Trauner, D.A. (2008) Plasticity in the developing brain: intellectual, language and academic functions in children with ischaemic perinatal stroke. *Brain*, **131** (Pt 11), 2975–2985.

Barr, H.M., Bookstein, F.L., O'Malley, K.D., Connor, P.D., Huggins, J.E., and Streissguth, A.P. (2006) Binge drinking during pregnancy as a predictor of psychiatric disorders on the Structured Clinical Interview for DSM-IV in young adult offspring. *Am. J. Psychiatry*, **163** (6), 1061–1065.

Begleiter, H., Porjesz, B., Bihari, B., and Kissin, B. (1984) Event-related brain potentials in boys at risk for alcoholism. *Science*, **225** (4669), 1493–1496.

Blake, D.T., Heiser, M.A., Caywood, M., and Merzenich, M.M. (2006) Experience-dependent adult cortical plasticity requires cognitive association between sensation and reward. *Neuron*, **52** (2), 371–381.

Burd, L. and Christensen, T. (2009) Treatment of fetal alcohol spectrum disorders: are we ready yet? *J. Clin. Psychopharmacol.*, **29** (1), 1–4.

Burns, L., Mattick, R.P., and Wallace, C. (2008) Smoking patterns and outcomes in a population of pregnant women with other substance use disorders. *Nicotine Tob. Res.*, **10** (6), 969–974.

Carlson, S.R., Iacono, W.G., and McGue, M. (2004) P300 amplitude in nonalcoholic adolescent twin pairs who become discordant for alcoholism as adults. *Psychophysiology*, **41** (6), 841–844.

Coles, C.D., Kable, J.A., and Taddeo, E. (2009) Math performance and behavior problems in children affected by prenatal alcohol exposure: intervention and follow-up. *J. Dev. Behav. Pediatr.*, **30** (1), 7–15.

Connor, P.D., Sampson, P.D., Bookstein, F.L., Barr, H.M., and Streissguth, A.P. (2000) Direct and indirect effects of prenatal alcohol damage on executive function. *Dev. Neuropsychol.*, **18** (3), 331–354.

Cornelius, M.D. and Day, N.L. (2009) Developmental consequences of prenatal tobacco exposure. *Curr. Opin. Neurol.*, **22** (2), 121–125.

Crews, F.T. and Nixon, K. (2009) Mechanisms of neurodegeneration and regeneration in alcoholism. *Alcohol Alcohol.*, **44** (2), 115–127.

Dobbing, J. and Sands, J. (1979) Comparative aspects of the brain growth spurt. *Early Hum. Dev.*, **3** (1), 79–83.

Dong, W.K. and Greenough, W.T. (2004) Plasticity of nonneuronal brain tissue: roles in developmental disorders. *Ment. Retard. Dev. Disabil. Res. Rev.*, **10** (2), 85–90.

Ebrahim, S.H., Luman, E.T., Floyd, R.L., Murphy, C.C., Bennett, E.M., and Boyle, C.A. (1998) Alcohol consumption by pregnant women in the United States during 1988–1995. *Obstet. Gynecol.*, **92** (2), 187–192.

El-Mohandes, A.A., Kiely, M., Blake, S.M., Gantz, M.G., and El-Khorazaty, M.N. (2010) An intervention to reduce environmental tobacco smoke exposure improves pregnancy outcomes. *Pediatrics*, **125** (4), 721–728.

Giancola, P.R., Martin, C.S., Tarter, R.E., Pelham, W.E., and Moss, H.B. (1996) Executive cognitive functioning and aggressive behavior in preadolescent boys at high risk for substance abuse/dependence. *J. Stud. Alcohol*, **57** (4), 352–359.

Giedd, J.N., Lalonde, F.M., Celano, M.J., White, S.L., Wallace, G.L., Lee, N.R., et al. (2009) Anatomical brain magnetic resonance imaging of typically developing children and adolescents. *J. Am. Acad. Child Adolesc. Psychiatry*, **48** (5), 465–470.

Gilbertson, R.J. and Barron, S. (2005) Neonatal ethanol and nicotine exposure causes locomotor activity changes in preweanling animals. *Pharmacol. Biochem. Behav.*, **81** (1), 54–64.

Goodlett, C.R. and Johnson, T.B. (1997) Neonatal binge ethanol exposure using intubation: timing and dose effects on place learning. *Neurotoxicol. Teratol.*, **19** (6), 435–446.

Goodlett, C.R. and Johnson, T.B. (1999) Temporal windows of vulnerability within the third trimester equivalent: why knowing when matters, in *Alcohol and Alcoholism: Effects on Brain and Development* (eds J.H. Hannigan, L.P. Spear, N.E. Spear, and C.R. Goodlett), Lawrence Erlbaum Associates, Mahwah, pp. 59–84.

Grant, T.M., Huggins, J.E., Sampson, P.D., Ernst, C.C., Barr, H.M., and Streissguth, A.P. (2009) Alcohol use before and during pregnancy in western Washington, 1989–2004: implications for the prevention of fetal alcohol spectrum disorders. *Am. J. Obstet. Gynecol.*, **200** (3), e271–e278.

Greenbaum, R.L., Stevens, S.A., Nash, K., Koren, G., and Rovet, J. (2009) Social cognitive and emotion processing abilities of children with fetal alcohol spectrum disorders: a comparison with attention deficit hyperactivity disorder. *Alcohol. Clin. Exp. Res.*, **33** (10), 1656–1670.

Griesbach, G.S., Hovda, D.A., and Gomez-Pinilla, F. (2009) Exercise-induced improvement in cognitive performance after traumatic brain injury in rats is dependent on BDNF activation. *Brain Res.*, **1288**, 105–115.

Hannigan, J.H., O'Leary-Moore, S.K., and Berman, R.F. (2007) Postnatal environmental or experiential amelioration of neurobehavioral effects of perinatal alcohol exposure in rats. *Neurosci. Biobehav. Rev.*, **31** (2), 202–211.

Hellemans, K.G., Verma, P., Yoon, E., Yu, W., and Weinberg, J. (2008) Prenatal alcohol exposure increases vulnerability to stress and anxiety-like disorders in adulthood. *Ann. N. Y. Acad. Sci.*, **1144**, 154–175.

Hellemans, K.G., Sliwowska, J.H., Verma, P., and Weinberg, J. (2009) Prenatal alcohol exposure: fetal programming and later life vulnerability to stress, depression and anxiety disorders. *Neurosci. Biobehav. Rev.*, **34** (6), 791–807.

Hellemans, K.G., Verma, P., Yoon, E., Yu, W.K., Young, A.H., and Weinberg, J. (2010) Prenatal alcohol exposure and chronic mild stress differentially alter depressive- and anxiety-like behaviors in male and female offspring. *Alcohol. Clin. Exp. Res.*, **34** (4), 633–645.

Jacobson, S.W., Dodge, N., Dehaene, S., Chiodo, L.M., Sokol, R.J., and Jacobson, J.L. (2003) Evidence for a specific effect of prenatal alcohol exposure on 'number sense. *Alcohol. Clin. Exp. Res.*, **27**, A121.

Kable, J.A., Coles, C.D., and Taddeo, E. (2007) Socio-cognitive habilitation using the math interactive learning experience program for alcohol-affected children. *Alcohol. Clin. Exp. Res.*, **31** (8), 1425–1434.

Kalberg, W.O. and Buckley, D. (2007) FASD: what types of intervention and rehabilitation are useful? *Neurosci. Biobehav. Rev.*, **31** (2), 278–285.

Kaplan, E. (1988) A process approach to neuropsychological assessment, in *Clinical Neuropsychology and Brain Function: Research, Measurement, and Practice* (eds T. Boll and B.K. Bryant), American Psychological Association, Washington, D.C., pp. 125–167.

Lebel, C., Rasmussen, C., Wyper, K., Walker, L., Andrew, G., Yager, J., et al. (2008) Brain diffusion abnormalities in children with fetal alcohol spectrum disorder. *Alcohol. Clin. Exp. Res.*, **32** (10), 1732–1740.

Lezak, M.D. (1995) *Neuropsychological Assessment*, 3rd edn, Oxford University Press, New York.

Liu, F., You, Y., Li, X., Ma, T., Nie, Y., Wei, B., et al. (2009) Brain injury does not alter the intrinsic differentiation potential of adult neuroblasts. *J. Neurosci.*, **29** (16), 5075–5087.

McCutcheon, J.E. and Marinelli, M. (2009) Age matters. *Eur. J. Neurosci.*, **29** (5), 997–1014.

May, P.A., Gossage, J.P., Marais, A.S., Hendricks, L.S., Snell, C.L., Tabachnick, B.G., et al. (2008) Maternal risk factors for fetal alcohol syndrome and partial fetal alcohol syndrome in South Africa: a third study. *Alcohol. Clin. Exp. Res.*, **32** (5), 738–753.

Medina, A.E. and Krahe, T.E. (2008) Neocortical plasticity deficits in fetal alcohol spectrum disorders: lessons from barrel and visual cortex. *J. Neurosci. Res.*, **86** (2), 256–263.

Medina, A.E., Krahe, T.E., and Ramoa, A.S. (2006) Restoration of neuronal plasticity by a phosphodiesterase type 1 inhibitor in a model of fetal alcohol exposure. *J. Neurosci.*, **26** (3), 1057–1060.

Moazedi Ahmad, A., Ghotbeddin, Z., and Parham, G.H. (2007) Effect of zinc supplementation of pregnant rats on short-term and long-term memory of their offspring. *Pak. J. Med. Sci.*, **23**, 405–409.

Nixon, S.J. (1993) Application of theoretical models to the study of alcohol-induced brain damage, in *Alcohol-Induced Brain Damage*, vol. **22** (eds W.A. Hunt and S.J. Nixon), National Institutes of Health, Rockville, MD, pp. 213–230.

Nixon, K., Morris, S.A., Liput, D.J., and Kelso, M.L. (2010) Roles of neural stem cells and adult neurogenesis in adolescent alcohol use disorders. *Alcohol*, **44** (1), 39–56.

Oberlander, T.F., Weinberg, J., Papsdorf, M., Grunau, R., Misri, S., and Devlin, A.M. (2008) Prenatal exposure to maternal depression, neonatal methylation of human glucocorticoid receptor gene (NR3C1) and infant cortisol stress responses. *Epigenetics*, **3** (2), 97–106.

O'Connor, M.J., Frankel, F., Paley, B., Schonfeld, A.M., Carpenter, E., Laugeson, E.A., et al. (2006) A controlled social skills training for children with fetal alcohol spectrum disorders. *J. Consult. Clin. Psychol.*, **74** (4), 639–648.

Padgett, L.S., Strickland, D., and Coles, C.D. (2006) Case study: using a virtual reality computer game to teach fire safety skills to children diagnosed with fetal alcohol syndrome. *J. Pediatr. Psychol.*, **31** (1), 65–70.

Paley, B. and O'Connor, M.J. (2009) Intervention for individuals with fetal alcohol spectrum disorders: treatment approaches and case management. *Dev. Disabil. Res. Rev.*, **15** (3), 258–267.

Ryan, S.H., Williams, J.K., and Thomas, J.D. (2008) Choline supplementation attenuates learning deficits associated with neonatal alcohol exposure in the rat: effects of varying the timing of choline administration. *Brain Res.*, **1237**, 91–100.

Sampson, P.D., Streissguth, A.P., Bookstein, F.L., Little, R.E., Clarren, S.K., Dehaene, P., *et al.* (1997) Incidence of fetal alcohol syndrome and prevalence of alcohol-related neurodevelopmental disorder. *Teratology*, **56** (5), 317–326.

Shaw, P., Kabani, N.J., Lerch, J.P., Eckstrand, K., Lenroot, R., Gogtay, N., *et al.* (2008) Neurodevelopmental trajectories of the human cerebral cortex. *J. Neurosci.*, **28** (14), 3586–3594.

Sim, Y.J., Kim, H., Shin, M.S., Chang, H.K., Shin, M.C., Ko, I.G., *et al.* (2008) Effect of postnatal treadmill exercise on c-Fos expression in the hippocampus of rat pups born from the alcohol-intoxicated mothers. *Brain Dev.*, **30** (2), 118–125.

Streissguth, A.P., Barr, H.M., Sampson, P.D., Parrish-Johnson, J.C., Kirchner, G.L., and Martin, D.C. (1986) Attention, distraction and reaction time at age 7 years and prenatal alcohol exposure. *Neurobehav. Toxicol. Teratol.*, **8** (6), 717–725.

Sullivan, E.V. and Pfefferbaum, A. (2005) Neurocircuitry in alcoholism: a substrate of disruption and repair. *Psychopharmacology (Berl.)*, **180** (4), 583–594.

Summers, B.L., Rofe, A.M., and Coyle, P. (2009) Dietary zinc supplementation throughout pregnancy protects against fetal dysmorphology and improves postnatal survival after prenatal ethanol exposure in mice. *Alcohol. Clin. Exp. Res.*, **33** (4), 591–600.

Tarter, R.E., Kirisci, L., Mezzich, A., Cornelius, J.R., Pajer, K., Vanyukov, M., *et al.* (2003) Neurobehavioral disinhibition in childhood predicts early age at onset of substance use disorder. *Am. J. Psychiatry*, **160** (6), 1078–1085.

Thomas, J.D., Abou, E.J., and Dominguez, H.D. (2009) Prenatal choline supplementation mitigates the adverse effects of prenatal alcohol exposure on development in rats. *Neurotoxicol. Teratol.*, **31** (5), 303–311.

Verma, P., Hellemans, K.G., Choi, F.Y., Yu, W., and Weinberg, J. (2010) Circadian phase and sex effects on depressive/anxiety-like behaviors and HPA axis responses to acute stress. *Physiol. Behav.*, **99** (3), 276–285.

West, J.R., Goodlett, C.R., Bonthius, D.J., and Pierce, D.R. (1989) Manipulating peak blood alcohol concentrations in neonatal rats: review of an animal model for alcohol-related developmental effects. *Neurotoxicology*, **10** (3), 347–365.

Wilcoxon, J.S., Kuo, A.G., Disterhoft, J.F., and Redei, E.E. (2005) Behavioral deficits associated with fetal alcohol exposure are reversed by prenatal thyroid hormone treatment: a role for maternal thyroid hormone deficiency in FAE. *Mol. Psychiatry*, **10** (10), 961–971.

Part Six
Personal Views from People Living with FASD

Fetal Alcohol Spectrum Disorder–Management and Policy Perspectives of FASD. Edited by Edward P. Riley,
Sterling Clarren, Joanne Weinberg, and Egon Jonsson
Copyright © 2011 WILEY-VCH Verlag GmbH & Co. KGaA, Weinheim
ISBN: 978-3-527-32839-0

25
Living with FASD

Myles Himmelreich

I want a new brain! This is a statement I have thought about, felt, and even said throughout my life. They say if you work at something it becomes easier to do. Why does this not seem true for me, an individual such as myself who has been living with Fetal Alcohol Spectrum Disorder (FASD) for over thirty years? Five years ago, in 2004 I started going to conferences and sharing what it was like for me growing up with a disability. As I traveled around attending conferences on FASD, I couldn't help but notice that much of the information shared seemed to be negative. I would hear the numbers on the extremely high percentage of individuals with FASD who could not graduate high school, were unable to maintain full time employment, and often ended up in jail. While growing up, one of the nicknames I had was *Smiley Myley*, because I smiled a lot and liked to make others happy. I decided that when I would do my presentations I would share not only the loss and hardships but also all the triumphs and successes I have had in my life. I wanted people to leave my presentations smiling.

I have had the opportunity to work with some amazing and highly educated people in this field. These people have gone to school for many years, done tons of research, and some have even published articles and books on the topic of FASD. I have been asked many times how I got into this field and what kind of education I had. I can proudly say that I was born with FASD and have come to learn techniques to live successfully. I have not always understood why I had certain struggles in my life. Going through school there were very few things that made any sense to me. I would go to my classes and try to do the work, day after day, but I always felt I had learned little, and retained even less. I struggled to concentrate in class, and the teachers would get upset that I wasn't paying attention to them. In junior high school I was in my early teen years, and that can be a tough time for anyone. Although I was actually 14–15 years old, in terms of maturity I was only a child (I know this now, but I didn't know it then). I would quite often do silly things and, because the other kids would laugh, I thought they liked me. But I soon realized they were laughing *at* me, not *with* me. So there I was, unable to do the schoolwork and having very few true friends.

At home I felt I didn't fit in either. Though I was adopted, raised and loved by an amazing family, I still felt like I didn't belong. There was anywhere from five

Fetal Alcohol Spectrum Disorder–Management and Policy Perspectives of FASD. Edited by Edward P. Riley, Sterling Clarren, Joanne Weinberg, and Egon Jonsson
Copyright © 2011 WILEY-VCH Verlag GmbH & Co. KGaA, Weinheim
ISBN: 978-3-527-32839-0

to ten people living at home at any given time, yet I felt alone. It was hard to explain what I didn't understand at the time. This caused problems because people in my family started to make their own conclusions on why I acted the way I did. I started to be labeled as lazy, weird, silly, and unmotivated. As school continued on, so did the labels, and I started to not only accept but also to buy into these labels. I figured it was easier to act like I didn't care about the schoolwork, than face the fact that I couldn't do it. I felt stuck, like I was trapped between four walls most of the time. It was also frustrating not being able to do certain things. (I didn't know why I couldn't do them). Things such as homework, getting up on time or understanding why I acted the way I did were all challenges for me. There was many times where I would lie in my room feeling angry, upset and confused, wondering why I was not like the other kids, and why I acted in the way I do. Why doesn't my brain work? Looking back now I can understand that it had a lot to do with having affects of alcohol exposure. I simply didn't realize that my brain wasn't working the same as other kids.

Unfortunately, at the time I couldn't see any answers, just more and more questions. I do have some good memories of junior high school: I remember being one of the top athletes in the school; I enjoyed making my teachers and classmates laugh; and I did really well in drama. I totally understand that it is not easy being a teacher – one teacher to thirty-plus students is unbelievable. I do wish though, that more training on FASD was given in the university for teachers. I feel that if my teachers had known more than they had about FASD, they may have been able to understand, and perhaps could have offered me better support.

Instead of thinking I wasn't doing my homework or showing up late on purpose, if they (the teachers) had understood that I didn't get proper sleep and would forget how to do the work and feel overwhelmed, they would have worked with me, rather than just punishing me. As I said, one of the things I enjoyed a lot was sport. I remember the sports coach, who was also my special education teacher, wouldn't let me participate one day because of something that had happened earlier in class. It was hard for me to understand the consequences of something I did hours ago. But, the hardest part was that they never knew how much it hurt me. There was something I was good at naturally (sports), something I enjoyed, and one of the few things that made me feel normal and here they were, taking it away from me. I continued on through school because I figured that's what you do till I failed a grade again, or somehow made it through to grade 12.

My life has never seemed to go on the same path as many other people I have known. Due to this, there's been a lot of frustration and depression, but there have also been a lot of surprises and rewards. School was a place where my path decided to veer to the side instead of going straight through. I went into high school with a plan – I was going to be more reserved, do my schoolwork, and let my athletic skills shine. The high school I went to concentrated more on vocational subjects rather than just academic subjects – there was hairdressing, mechanics, woodworking, and a day care. Many of the students that attended that school were in the same boat as me. Many had either learning or behavioral problems, and it was a great school for those who learned better with hands-on (support), or needed a bit more time to finish their work. It was a good school for me during my first year because my plan seemed to be working – my grades weren't great, but I was getting by. Kids in the school near me said that I was good at basketball instead of a funny, nerdy guy. And I had a few girlfriends, so overall things were okay.

The year after grade 10 my life started to go down a path that would be dark for what seemed to be forever. Though some of my thoughts and actions made me seem younger than I was, there were still times when I acted just like the other teenagers around me. I thought I knew it all! I felt my parents were too strict, and that I never got to do anything. The neighborhood elementary school I went to as a child had just got a new outdoor basketball court, and I was spending most of the days that summer there. While playing there I met these two brothers who had just moved into the neighborhood. I introduced myself, we soon became friends, and they invited me over to hang out. Up until that point in my life the friends I had were from the church I grew up in, or other neighborhood kids who I had played video games with. It was a new experience for me to be in a house where the parents smoked and so did the kids. So, it didn't take long for me at age 17 to end up getting involved in these new experiences of smoking, drinking, and doing drugs. As the partying become more of a regular thing, so did the fighting with my parents. I figured I had two out of three things in my life that were going well, and that good enough for me. But it didn't take long for the late nights out with friends to get my parents fed up, and state they were okay with me leaving home. I didn't need my parents, I had friends, and that was enough – or at least, that's what I thought. Though I was doing okay in school I ended up having to leave because I hadn't paid my school fees. So, there I was with no school, no money, and no real place to stay. When you are 18 years old with no job experience or education there are not a lot of jobs open to you. I tried a few different jobs but nothing ever lasted. I was having the same frustrations and problems I did at school. I couldn't remember the steps and, just like the teachers, my bosses would get frustrated at having to repeat things to me. Also, like school I didn't want to ask for help because I didn't want them to think I was stupid. Eventually, I ended up getting government assistance so I could have some financial income.

My days consisted of drinking and playing basketball. I continued doing odd jobs and going to different life skills programs. I was feeling really lost and alone. The people I thought were my friends ended up being the same ones that stole from me and even beat me up, leaving me in a pool of blood because I didn't buy them cigarettes. You would think this would make me stop hanging out with them, but this was not the case. I was so afraid of living alone I took the abuse they dished out. I was 19 at the time, and it was at this point that I decided to try and find my biological family. I knew I had two older brothers and two younger sisters. I wrote an article in the local newspaper that included my name and the people I was looking for. At the same time, I contacted the adoption agency for information. One of my brothers, Harrison (Harry), saw the article and contacted me from Lethbridge, Alberta. I was very excited, as I thought that if I found my family that would change everything in my life for the better. I was excited to meet my (biological) Mom and Dad – I had so many questions. I spent the next few days talking with my brother on the phone and making plans to meet. It was only a few hours on the bus from Calgary to Lethbridge, so I decided to go there to meet my brother. The day before I was due to go I was talking to Harry on the phone and I asked him about our parents. He explained to me that I had a different father than him and our brother Mike. He told me how he and Mike grew up with Mom, and that there was a lot of partying. He then went on to tell that our Mom died from drinking and my father was killed in jail. I didn't know what to do or what to say, but I told him that I'd see him tomorrow and acted like I was okay. I hung up the phone and went to a friend's house and drank; the feelings of excitement about meeting my family quickly became feelings of anger, loss, and regret.

Later, I was contacted by the adoption agency telling me of a few other brothers and sisters I that I didn't know about, and over the next year I met my other siblings. We had all grown up in different homes, and had such different backgrounds that unfortunately it wasn't the happy family reunion I had hoped for. I felt such a sense of loss because of this. I was done with it all – school, work, friends, and family. I thought my whole life was nothing but a huge failure. By this point the only thing that seemed to a steady constant for me was alcohol, and I drank with family, friends, strangers at the bars, and at home alone. I thought that someone or something would make everything all right. It took me being so low to realize that nothing and no one could better my life until I started to care enough about myself to pick myself up and move forward. I did a life skills program that helped me to not only write a resume, but also to learn how to try to live a healthy lifestyle. My life was starting to go down a smoother path.

I still had my struggles, but I was beginning to see a glimpse of hope. Though there were people around me who cared, I knew they also questioned how well I would succeed in society. I had my doubts too, but I figured I had nothing more to lose, so I got a full time job, moved to my own place and got back into sports. Sports became a huge part of my life – it was a place I could go to have fun, make new, healthy friends and stay busy. I had to learn what it was like to live independently. I found some strengths and strategies that I still try to use from this life skills course. I pay bills on the same day of the month and now use the conference feature of online banking; my grocery list is usually the same. and I use Saturday mornings as my house-cleaning day. These routines, and the repetition of them, are what helped me to start living a more balanced life. A lady I was working with on my budget told me of an FASD conference she was helping to host in 2004, and she asked if I would be interested in speaking at it. This is the conference I wrote about at the start of my chapter. As I stood up there, sharing my life story, I had an overwhelming feeling come over me – I was feeling nervous and excited, but as the talk went on I experienced the strangest feeling, that I now know was a sense of success. I couldn't explain it at the time but I knew this was it; this is what I am meant to do. As the years have gone on I have had great opportunities to speak at conferences, to audiences of teachers, doctor, lawyers, psychologists and many others. I have come to realize that I have been blessed with an amazing talent to do public speaking.

After that first conference (2004) I was offered a job as a mentor. I have now been mentoring for six years, and have been lucky to meet so many great individuals living with FASD. Through the mentoring process I have come to see many of the same struggles we share, but also, and more importantly, I have come to see the great strength and self-determination that we all have. I have been able to accept the struggles and losses in my life. I feel that, without these struggles, I wouldn't have truly appreciated my success. I've been blessed with an adoptive family who I've reconnected with, a biological family, though it has been tough. It helps keep things interesting. I have a great group of friends that are understanding and supportive. I would love to say that everything is now perfect, but I am realistic and understand that I have, and will always have, FASD. This is a reminder to me as I continue working. I do amazing things with my presentations and my one-on-one mentoring, but the paperwork and keeping timelines is something I still struggle with. It has taken this past year for me to summon up the strength to get (let) my coworkers to understand that, although I present well and have a lot of strengths, there are still things I need help and support with. Fortunately, I've been able to build a great support team around me – I felt alone for so many years, but now I know I'm not alone. Today, I know I'm not just another statistic living with a disability – I'm Myles Himmelreich, a successful young man who has found his ability.

26
Charlene's Journey
Charlene Organ

> "The first step to getting the things you want out of life is this: Decide what you want." Ben Stein.

I'm a young girl, 20 years old, and I feel I'm just like any other person trying to get by on life, even though I was born with Fetal Alcohol Syndrome (FAS) and have problems – just like any other person.

I have a mother who loves and cares for me, a grandmother who worries about me, an aunt who spoils me, and friends who I can talk to and hang out with.

My knowledge about FAS increased significantly during my second year in High School. At that time we had to do a final project about something dealing with any kind of topic that involves physiology. I wanted to do something about FAS, although that was not one of the topics my teacher suggested. He also pointed out that FAS was a hard topic to deal with, but I wanted to research it so much.

My main reason for researching FAS was because I was told by my Mom that I was born with it, and I felt this was a good opportunity to find out more about what I really had gotten from FAS.

I asked my Mom for help with my project. I always ask for help when I don't know how to start, but that's part of life – asking for help when you need it. During the time it took to do the research I learnt a lot about FAS that I didn't know about before. The proper term about the implications of FAS seemed to be called "challenges" – so I learned that I have challenges!

I also learned that FAS is a permanent defect that happens to a fetus when a woman is pregnant and drinks alcohol. I learned that FAS can actually happen to anyone – no matter if rich, poor, healthy, sick, or what race you are. FAS doesn't discriminate. Among the many other things I found out about FAS were that if women drink while pregnant, alcohol can stunt the growth or weight of the baby, cause facial deformities, and damage the brain, as well as causing mental, physical and behavioral problems for the child once born. The second thing I learned was that FAS was identified and named by Dr David Smith and Dr Kenneth Lyons Jones, in 1973. There were many more interesting things I learned about FAS during that time of research for my school work in physiology.

I wanted to invite a speaker for my project, and my Mom volunteered to talk about FAS from a parents' point of view and how it is when dealing with a child who has FAS. Both our presentations went fine, but I was inspired by my performance and just wanted to tell more people about myself and how I became a success story; namely, through the fact that not many people with FAS can actually go to school. Although most parents quit trying to teach their FAS children, my Mom tried and tried and tried, and finally I actually learned to do the things I can do today. I went to school, I learned, and I did many other things that most people with FAS could not do.

I was born on a Monday, May 1, 1989 in Edmonton, Alberta. I weighed 6 pounds (about 3 kg), was 14–15 inches (about 38 cm) tall, and went into surgery two weeks later for the philtrum (the groove between the nose and upper lip) to be sewn together because it was open at birth. I was also born with a cleft lip and palate.

The main challenges that I face include learning disabilities, Attention Deficit/Hyperactivity Disorder (ADHD), confusion under pressure, seizures, poor generalization ability, poor handwriting, lack of inconsistent memory retention, asthma, delayed motor skill development, tactile defensiveness, and social communication.

I have learned that, during the early teens – at age 12 years and older – the following challenges are often experienced: mental depressions, 60% are expelled from school, 60% have trouble with the law, 50% are in jail, 50% display inappropriate sexual behavior, and 35% have alcohol or drug problems. When people with FAS reach 21 years or older, they often live in group homes (80%) with family, friends or assisted living, and have a hard time getting a job and keeping it. Luckily for me, I didn't endure that – I have had a job at Zellers Northgate mall since July 2008, and am doing fine at work.

I did have one detention in school, although it wasn't for anything serious, I am very proud of myself for not being a part of the statistics that I've mentioned. I have other challenges too – I have trouble handling money, I sometimes have problems telling the time, and interacting socially. I've lived in a good home for 95% of my life. I was diagnosed with FAS at the age of 3 years, but in all of my life I've never experienced violence, which I'm happy about. During the first two years of my life I was in a foster home, but then I was adopted by my current mother Debra, who is a single parent and works full time.

Mama and Papa (my grandparents) and my Auntie Glenda were a very important part of my ability to succeed. Soon after I came home I had my second birthday with a cake that was my very own – my new family was all there. But just after that I got really sick and ended up in the hospital – in fact, I was sick most of the time for the first couple of years that I came home.

My Mom got some time off work when she adopted me, but she had to go back to work soon after. Papa often took me in his car to see horses, to visit his friends, and to see our neighbor's German Shepherd dog called Sasha, that I became very close to. Sasha would often bark like crazy, but not for me, and that made my Mom jealous because she barked at her. I used to spend hours swinging on the swing set in my grandparent's back yard, and taking walks with Mama. Before I was adopted, my Mom had been given medical documentation on all of my challenges – FAS wasn't included in the list, but it was suspected as one of my challenges. The first time I was in the hospital with my new family, the doctor on call happened to be Dr Berhmann, who delivered me. He was a neat doctor who was very nice to me and my Mom – he told Mom that he wasn't taking on new patients, because he had delivered me into the world, so I was an "old" patient and he would like to be our doctor. He never told Mom until I was about three years old that I had FAS, but when he did my Mom was devastated and cried for a whole day. But she decided it wouldn't stand in the way of me being successful. Because I had been adopted, I was already involved with social services for additional help. Early intervention came to my house to help Mom with different methods to help me speak and learn to focus, but I got really sick again and ended up in hospital for a third time that year. This time it was over Christmas, and Mom was sad because this was our first Christmas together, but she said she was just glad that we had each other. That was the first time I saw Santa Claus – one of the ladies where my Mom worked had given me a toy Santa, but when I saw the real one I guess the surprised look on my face made Santa, the nurse and my Mom all start to cry! I don't really remember that part, but Mom talks about it a lot. That Christmas I also got a stuffed Dalmatian puppet that we called Booboo, so that I could practice making the sound of "B"s, which I found difficult.

I attended Mayfield School, which had a program called "early intervention" for children with challenges who needed extra help to in order to overcome some of their problems. I went to Mayfield between the ages of three and six. Every morning, Mom would put me on the bus and then sign to me "I love you" and "Charlene is beautiful" from the curb. I would sign back to her through the window; it turned into our own little "trade mark." Every day I went on the bus, Mom was outside and we would do the same sign, so I wouldn't miss Mom too much when I was away. She made a picture book of my family and my favorite things, and put it in my back pack. That way, when I got lonely I could look at the pictures and feel better. At home, Mom put pictures, alphabets and books all over our house so that I would learn. We lived in a townhouse with children all around us, but unfortunately I didn't make any friends as I couldn't talk well enough for anyone, except Mom, to understand me. I didn't understand what it meant to interact with someone who wasn't challenged, or someone who didn't love me and understand my gestures. The other children didn't always understand – some of them in the complex started to make comments about my thumb. That was when my Mom decided that the extra thumb on my left hand should be removed before I went to regular public school, so I wouldn't be made fun of. Mom found the best plastic surgeon in Canada – Dr Wilkes; he removed my little thumb and later on helped with my facial reconstruction, even though it meant I had to go through many, many sessions of surgery.

While I was going to Mayfield I had a speech therapist, an occupational therapist, a teacher's aide, a physiologist and handicap children services to help me succeed. I also went to a speech therapist at the University hospital, who showed

Mom how to get me to use words and to help make them understandable. There was a specialist there who examined my cleft palate so that we could try to get it repaired; they also suggested that one day I would get my teeth done, so I would look just like everyone else.

Some of Mom's and Mama's first memories were of me sitting in my crib running my fingers over all the printed words on the white pages of a telephone book, and slowly turning each page, giving the impression that I was reading each line. From that point on, Mom made sure that books would be a major part of my life. I had hundreds of books in my play room growing up. I loved reading books from the library, and to this day I often read books more than once. When I was younger, before prayers, we read a story every night. Mom also would buy me movies to watch on the television, listen to the music, and then read the book, every single day. Mom encouraged me to make up my own stories, which was very helpful when I had to write a story for grade four. I wrote about a unicorn, and Mom drew the pictures which I then colored. I was even chosen to read my story as a special presentation for the school.

We had help from Handicap Children Services which helped me get into Mayfield. Being diagnosed with all of my disabilities, this helped Mom ask the schools for the special help I needed, so I was able to learn.

When Dr Berhmann first told Mom that I had FAS she was extremely upset, because for most children diagnosed with FAS their future is not very bright. Mom decided rather than seeing this as a "door closed," she would use whatever resources were available so that I would succeed. She never said I couldn't do something – she always said I would do it at my own time whenever my brain was ready. Mom always used the word "patience" – if I was getting frustrated she would say "Have patience, try again." If I was standing in line, wanting it to go faster, she would say "patience," and always let someone behind us go first. Mom knew I could do whatever I wanted; as a team we just had to find the special way to get there.

My Mom took me to daycare every day before she went to work. The daycare had an aide who was assigned to me to help with communication and social skills, and this allowed for a basic normal childhood interacting with other children who were able to do so naturally. Without this help I wouldn't have been able to have fit in. A school bus would come to pick me up and take me to school. I went to many different elementary schools, because Mom had to find a school that was able to teach me in the way that I could learn. This meant that the teacher needed to understand what my limitations and challenges were, as well as wanting and being able to communicate on a daily basis with Mom, the principal, and me. Mom said "we all had to be on the same page". Mom spent many hours talking to teachers and ensuring that any homework they sent home was done, and that the lessons learned at school were understood when I got home. This would mean that on many nights Mom would have to re-teach me what had been taught at school, because I would forget things easily. But she went over things again and again until I remembered them – sometimes she would even have to help me again the next day, because I had forgotten it again.

Often, questions asked at school would be too complicated for me to understand, so Mom would break the questions down into smaller questions, so that I could understand what was being asked. She would write notes to the teachers telling them if I had a hard time remembering something, or if I didn't understand something, asking them to give me extra help.

My grade one teacher, Mrs Fontaine taught me that I needed to pay attention in class, to listen to others' words, and show respect. She also taught me to make sure that my homework was done and all my assignments were complete, even if I was away from school because I was sick or in the hospital. Mrs Fontaine is the teacher that my Mom and I have kept in contact after all these years, and she has always had faith in me. She said that someday I would become someone special, and it looks like she was right. I had to do homework every night and read every night as well. Mom helped me, even when she was tired.

Even though I was very shy, I made some very good friends in grade one because my teacher made sure that everyone in the class showed respect to all classmates. It was not a class for only challenged students, and some of the friendships I made were with kids who had no challenges. A favorite nickname they gave me was Little Pocahontas – I had long hair that was down past my bum (see Figure 26.1), and I am treaty status – so it kind of fitted! My Papa used to call me his little Indian Princess, he thought I was beautiful.

Figure 26.1 Charlene's long hair.

Even though we had very little money, Mom always made sure my clothes were clean, mended and pressed. I had a shower or bath every day, and my hair was always clean and brushed. Mom said appearance was half the battle; she also insisted that I said "please" and "thank you," and to always respect others – because you can never have too many manners.

My Mom was unemployed for a period while I was in grade two, so she decided to help organize and build a school playground for the school that I went to. I really liked this because I got to see my Mom at lunch time, and if I needed her she could take me right to school. I even helped Mom plant some of the flowers for the grand opening of the playground. Mom always seemed to be there, no matter where I was.

Grade two showed a different challenge, as the teacher didn't understand my limitations or accommodations required to make me a success. The same happened in grade three, at a different school. But Mom was there through it all, trying to explain to everyone what it was that I needed, and what they needed to do to help me – she never gave up. Finally, Mom found me a different school, so in grade four in a segregated class room with a teacher and teacher's aide, we finally found a place where I could learn and fit in. The teacher took the time to focus on my abilities, not my challenges, and I started to learn. If someone in my class was having a hard time I was encouraged to help them out, and this made me feel good. I felt safe with my teachers and thought I could tell them anything.

Unfortunately, even though I was in a great class room, I still found it hard to make friends. I was bullied by other students and relied on Mom and the teachers for intervention. Mom had to actually come to the school to address a physical attack on me several months after it happened – I didn't tell her about it right away. I didn't understand that it was important for me to make sure that I tell Mom everything when it happened, so she could make sure I was safe and tell me how I could avoid it from happening again. I am still trying to conquer this challenge to this day – some decisions are hard for me to make on my own.

My Mom said that one way to know if I was making a wrong or right decision, was to think about what would the answer would be if she had made it. Then, if I wasn't sure, I should ask myself, "do I feel happy, scared, or sad?" If I did whatever it was that I was feeling when I made the decision, it would help me to make the decision, but if I was not sure then to call her and ask. When my Mom came to the school, and made it known that this was not acceptable and that she was the type of Mom that would stand up for me, the school knew that they needed to make sure this never happened to me again. It never did happen again in elementary school.

I changed to afterschool care that was in the neighborhood, and made some friends there who are still my friends today. It's nice when I can visit with friends like everybody else. Some of the other kids in the neighborhood would come over and want to go to the playground – they would ask if I could go because they knew that my Mom would then take us all and play with everyone. Mom did this for years until she knew I was safe on my own, and that I was with people that she knew too.

Mom was able to have someone come once or twice a week (Handicap Children Services helped with the funding). She interviewed a lot of people until she found somebody that she could trust and I could relate to. The girl from services would come once a week, and take me to the library or swimming, or just go for walks. She was someone else that I could talk to and trust, other than my Mom. About three different girls came over the years, and I learned to really like each of them. The last girl who used to come and take me out had to move because her fiancé was moving to a different city. I really miss her, but am writing to her so we can stay in touch. Mom really liked her too.

While I looked at all of the other kids being able to ride their bikes, I was sad because I wasn't able to master riding a two-wheeler. But Mom, Mama and Papa always said "be patient and it will come with time." It took many years and many hours of practice, but I kept trying and trying so that, finally, I could ride a two-wheeler. Mom's famous saying is "Not our time, but in Charlene's time."

At home I had a play room with hundreds of stuffed animals, hundreds of books, toys, and costumes that I would spend hours making sure things were organized, and everything had its place. When it was not in its place, I would have to ask for help, and Mom would show me how to put things back in order. My Papa built me a blackboard that was 8 feet wide and 5 feet high (2.5 × 1.5 m) that covered a whole wall. I used to do homework on it, writing and pretending to be a teacher. I would teach my Mom what I learned from school; that always helped me remember what I had learned in class that day. Mom said being able to see a problem on a blackboard made it easier for me to find a solution.

Papa also built me a sand box so that I could play outside and have friends from the neighborhood complex come into my yard. In that way, Mom could keep an eye on me, and make sure I was safe and not being taken advantage of. Mom would also put the tepee outside, and we would play there for hours using our imagination. During the winter I used to play in the snow, with Mom right there with me. Papa also built me a kitchen cupboard and stove so that I could cook in my kitchen while Mom was making dinner for the two of us. When Mom baked something or measured something she would always get me involved, which helped me to learn my numbers or alphabet.

While I was in elementary school (I was often away from school because of surgeries and respiratory issues), Mom would make sure to get all of the homework that I missed and we would work on it when I got out of the hospital. She would spend hours teaching me everything I missed so I wouldn't fall behind. When I was in the hospital, my Mama would stay with me during the day while my Mom was at work; Mom would then stay with me at night so my Mama could get some sleep. Papa would always be there just in case a ride was needed, or we needed to get to a doctor's appointment. He would meet Mom at the appointment, so that Mom wouldn't miss too much work.

Every summer Papa and Mama would take Mom and me to the mountains or the lake for vacation. These were very special times. Papa taught me how to build a fire and to use a hammer and screwdriver. At the age of three I had my own toolbox with real tools, which came in real handy in woodworking class in Junior

High. It took a little more than an hour by bus to get to Junior High and back. Unfortunately, I didn't know anyone there, so it was very hard to make friends – the teachers seemed friendly, but the kids were not.

One of my teachers decided that she was only going to get me to draw pictures in class because of something called my IQ. She didn't think I was able to do homework or understand regular classroom work, but Mom was furious and had a meeting with the principal and assistant principal. She demanded that they challenge my limitations and ignore the IQ, so that I had a chance to succeed. Mom gave them a binder with all kinds of teaching tools that have worked for me in the past, and all kinds of information about FAS so that this wouldn't happen to anybody else who might have the same challenges I have. Mom always attended every parent–teacher interview, and this didn't change when I went to Junior High. She said that she needed to make sure that we were all on the same page. Mom got involved with the counselors at the school because I wasn't able to make friends. I would come home very unhappy on many nights.

All I wanted was a friend, someone who didn't take advantage of me. Many of the kids who pretended to be my friend were only using me for one reason or another. I was never able to understand how they were using me, or why. Mom would talk to me about how I was being taken advantage of, and tried to get me to understand – sometimes I would, sometimes I wouldn't. She always seemed to know ahead of time who was going to take advantage of me and who was not. So she stood by to give me a hug when they would hurt me, and then tell me what it was they did to make me sad.

Because my Mom went to the school and told the principal and teachers how unhappy she was about the way I was being taught, the school decided to put me in a high-achieving class with a teacher who expected success, not failure, from her students. All the way through school Mom never expected grade As or 100% marks from me – only for me always to do my very best. If I did my very best but didn't pass I wasn't to worry because we would work at it together, to make sure that we did achieve a passing grade. But the passing grade was not the most important thing.

The next year, I got a teacher who made learning fun. I was able to meet other teachers who saw within me many talents. I had an art teacher who encouraged me to draw and always told me to do my best; the picture I drew was chosen by the Edmonton Public School Board to hang in their office building for one year. I can remember how proud my Papa, Mama and Mom were – they even took me to the head office building to see my picture on the wall. I got an award from L.Y. Cairns called the Stevenson Trophy for outstanding achievement in Art. (*Note:* L.. Y. Cairns is a congregated public school for children aged 7 to 12 years with all levels of disabilities, be they physical or mental challenges. The school specializes in student instruction to provide them with the tools to succeed and transition into adulthood.) I also got the Al Jamieson Award for outstanding service to others for being the most helpful student in the school.

Although I started to make friends a little easier this year, I was still being taken advantage of. The teachers who insisted on success pushed daily for me to achieve this, and because of the consistent positive reinforcement I did succeed and wanted to learn more.

Between grade eight and nine my Mom and I moved from our home down the road to take care of my Mama and Papa, who were getting old and needed to have someone look after them. When we moved in with them, I am proud to announce that I got the master bedroom – mainly because I had so much stuff I needed some place to hold it all. Mom had to put all of her stuff in storage (it's still there). It was great moving in with Mama and Papa, because I got to see them every single day, anytime I wanted. My Papa was very proud of me and I was very proud of him.

In grade nine I received an award for being a good person and for leading by example in wanting everyone to show respect to each other. I took different classes, such as food, child care, and physical education. I had my very first boyfriend in grade nine. He was a year younger than me, and we met on the bus going to school. We went to the school dances together, and had lots of fun, but it lasted only for a short time. I guess it was not true love.

When I finally moved on to Senior High, which was in the same school, I felt more important. Even though I was in Senior High, there were many words and things that people said which I didn't understand. Sometimes I would just say "yes" if that's what I thought they wanted to hear, but later on I would ask my Mom at home what the words meant. I still do to this today, even though I am not in school.

I was able to take work experience and other optional courses. It was kind of scary doing work experience in the real world. I went to Lauderdale Elementary School and helped them because I had learned how to use the computer (sometimes better than my Mom). I also did work experience at Lauderdale After school care, looking after little kids. I found out this isn't what I want to do for a living, but I still want to get married and have kids – just not a whole load of them.

I also had many surgeries while in high school. Because we managed to get my allergies under control, we got an adorable extra small, white, fluffy fur ball puppy. Although I wanted a big dog, Mom convinced me to get a small dog. She is a purebred Maltese, and was so tiny that she could almost fit into both palms of my hands. I named her Alu, meaning "White Bear" from my native language, Cree.

My Mom works for Finning Canada, which sells Allu products, and when she told one of the Canadian salesmen about me (and my jaw surgery in December 2006) and my new puppy Alu, he contacted the Allu representative in Norway. She sent me an e-mail and a package with lots of things with the Allu logo. It was neat having someone from a different country sending me an e-mail. Alu became Papa's little buddy, and they would take naps together every day in Papa's rocking chair. I was very excited to have my very own puppy – just like other kids had. Everyone loved my dog because she is such a princess and a total Diva. I liked holding her when we went out in public, because people would stop me and ask me all about her. We even got her a doggy buggy because she had leg surgery and could not walk for a couple of months. All of my friends fell in love with her too.

In grade 11, I had a chance to work at Future Shop Northgate; one of my duties was to organize the DVD section, and this is where I found I had a real talent. I like things in their place – I have all of my movies and CDs at home (over 1000 of them) organized by alphabet and type. I even have a computer listing of them, so I know exactly what I have and where they belong. I took a special class in grade 11 that was so much fun. I found I had more talent in computers – I even did a PowerPoint display about relationships with a class mate. I liked doing the animation in the PowerPoint presentation and the different sound effects that we used. I enjoy working on the computer and doing different things in Word, Excel, PowerPoint, and especially Publisher, so I can build special wallpapers from the various pictures I find on the internet. It seems like I am redoing my wallpaper every month. I am downloading and organizing my songs on my i-pod classic that I have on my computer at home.

Finding a boyfriend is something that I have always wanted, somebody who will treat me the way that Mom says a boy should treat a girl. Whenever any of my friends are having a fight or a hard time with their partner, they would come to me for help and I would give them advice. Mom bought me many books about relationships and boyfriends, and I have read them all. I just want to be able to find a nice guy to fall in love with. Although I have dated a couple of guys, we both decided to be just friends because there was no "chemistry."

One day I found what I thought was true love. We went out for more than ten months – we went swimming, dancing, movies, restaurants, shows, to the park and hanging at home. My Mom picked him up and drove us everywhere – she was always there to make sure we were both safe. Unfortunately, or fortunately, this relationship did not last. It broke my heart, but Mom was there to listen to my tears. So was a close friend of mine by the name of Krissy who was there to listen to me and help me heal. She was the one friend that I can always count on, and I still keep in touch with her.

One of my teachers at L.Y. Cairns was someone that I was able to trust, and I felt that she was a friend as well. Mom and she had a connection and they would talk all the time. She told me that no matter what, I could come to her and talk about anything, boy trouble, friend trouble or even if there was something that I did not understand, she would explain to me what it meant.

I found it very hard to talk in front of people because I am shy. One of the teachers, Mr Reese, asked if I would like to recite a poem for Remembrance Day. I wanted to do this for my Papa. I wore my Papa's Air Force uniform to school – the whole sha-bang, from the blue shirt and tie, and I even did my hair the way that ladies did in the Air Force during World War II. I invited my Papa as an honorary veteran to hear my poem, and he was so proud to be invited that he even wore the medals that he had received. When I got home, my Papa told me that he was so proud of me, and started to cry – I'll never forget that day. I've kept in contact with Mr and Mrs Reese since I have left L.Y. Cairns, and have seen movies with them, and even invited them to my eighteenth birthday dinner. Mr Reese always made me feel special.

There were many things about grade 12 that were exciting, and many that were scary. I was disappointed about many friendships that I had made and thought were true but in the end found out I was being used. I found out that I could accomplish lots if people believed in me, and I continue to do my very best. Even though this was my last year of high school, Mom didn't allow me to stop doing my best – she was always checking on my homework making sure that things were getting done when they should have, and helping me study and study and study for final exams. I was very worried about passing, but Mom kept saying to me that it didn't matter if I didn't pass, as long as I did my best. If I didn't pass, we would just take it over again, not to worry. Sometimes it was hard to concentrate as I wanted to make friends and just hang out. Most of my friends lived on the other side of town, which made it hard to meet them because Mom would have to drive me there and back. I still was unable to take the bus to and from school yet – Mom said that I wasn't ready.

In December 2006, and over Christmas, I went into hospital and had facial surgery. It was extremely painful; they built me a chin and moved my top jaw out and my bottom jaw back, and I couldn't eat regular food for weeks. I had never experienced pain like that before – it was just awful.

I was going to graduate that year from L.Y. Cairns. Mom and I went looking for a dress and found a pink one that we bought at the beginning of grade 12. Later in the year I decided that I was not going to go because I didn't have a date, but Mom told me that, no matter what, I was going to go to my grad. So a couple of my girlfriends who didn't have dates either decided that we would all go to grad together. One day Mom and I were shopping and found another dress that was unreal, so we bought it. It was white with soft pink flowers, and looked like a wedding dress. When I wore it I felt like a princess, and everyone thought that I looked like one too. Mom bought tickets for the five of us, Mama, Papa, Auntie Glenda, Mom, and me. Then, just before graduation I met a boy who wanted to take me to grad. I was treated like a princess with flowers and everything. This was a very special night because my Mom gave a speech addressing the grade 12s that were graduating. My Papa was there (he had been very sick and I wasn't sure if he was going to be able to make it), my Mama was there, my auntie was there, and I had a date who was proud of me. I received a certificate – one of 14 that were presented outside of Toronto – that was called "Employability Skills and Achievement Certificate." I would be able to use this certificate in placement of a regular high school diploma. I danced all night and had a dream come true. This was a wonderful night, just like a fairy tale. My date and I parted shortly after as friends, but I'm still looking for that special one. Hopefully someday it will happen.

The next summer (2007) was an awful time. My Papa died, and I really, really miss him. I didn't understand why he had to leave us – things just didn't make sense any more. I did all sorts of things that I didn't normally do, not understanding the consequences of my actions; I just wanted my Papa back home. My Mom was going through a really rough time as well. I missed not being able to talk to her, so I guess this is why I was acting out the way I was. I went on to the internet, but thank goodness my Mom found out. She tried to explain to me that the internet was not a safe place to be without supervision. I still was very confused, so Mom found a counselor for me to talk to until I was able to feel normal again. Because I never really showed my emotions during my Papa's funeral my Mom was worried. She had decided that she would get me another puppy, and maybe this would somehow make everything all better. It didn't replace my Papa, but I sure do love her. We think she came from a puppy mill because she was so ill, and thank goodness for our vet who helped her get better so we could bring her home and love her. She is not a princess at all, she plays in the water with me and brings her toys to me all the time. I even signed the papers for her, so she is "really" my dog. I named her Aniu, which means "White Wolf"; again it came from my language; Cree. Aniu is a 5-pound (2.5 kg) pure-bred Yorkshire Terrier. Alu and Aniu are not really buddies because they like different things; one is quiet and the other is very busy, but Aniu sleeps with me all the time.

Mom told me that I needed to get my high school diploma so I needed to go to upgrading school. We decided that I would go to Center High; this meant that I would have to learn how to take the transit system to get to there. I now feel much more independent and I tell Mom whenever I am taking it. I always take my cell phone with me, so I'll be safe and if I need to get a hold of Mom, I can.

I went to Center High for two years and finished my grade 12. I passed all of my courses and really enjoyed meeting new people. I am friends with many of them still and go to see them occasionally. While I was at Center High, I won a bursary as the teachers thought I "had what it takes" to succeed in post secondary school. I also competed against many students, not all of whom were challenged, and won a scholarship—the Centennial Scholarship Award, 2009, and Mom was very proud of me. While I was at Center High, sometime in September, I finally got a bridge for the roof of my mouth, which meant that I would have a smile like everyone else. Since I got my bridge, I have not stopped smiling!

In my second year at Center High, I finally had my last surgery in December 2008, performed by Dr Wilkes. I got a new nose and lip, and just before the doctor took off my bandages he asked me if there was anything I wanted to ask him. I asked him if I would be beautiful, but he said that I was already beautiful before he started the surgery. He then took the bandage off and stepped outside of the examining room. I stood in front of the three-way mirror and just stared for about a minute at the person in the mirror looking back at me. Then I looked at my Mom and said, "I look normal." This was a great day—I just wish that my Papa could have seen me (see Figure 26.2).

Another thing that I learned from my research paper is that, people born with FAS find that there is a possibility of emphases in the areas of talent that others may not be able to do. Some examples include music, playing instruments, singing, art, reading, computers, writing poetry, mechanics, and welding.

The talent and interest I am most interested in is reading. I enjoy reading all kinds of books; the most recent books I have enjoyed reading is Manga, comic books that are from Japan, Vampires and Werewolves, and novels. With my computer I can type 65 words per minute, and I have talent in Publisher, Microsoft, PowerPoint, Access, Excel, and many more. I like to sing, but I have to admit I am not very good at it—it's not a career for me, that's for sure. I enjoy music—any kind of music, soundtracks, classical, specific artists. My favorite artist of all time is Billy Idol, Pink, Fall Out Boy, Panic At The Disco, and a few more. I met a

Figure 26.2 Charlene all complete!

friend, Christine, who lives just down the alley from me; she is a little older than me, but is a great friend, and we have so much in common. We like the same things and she is my best friend – I'm so glad we found each other. Now, if I can just find that someone special?

I love to go shopping on my own or with my girlfriends. Mom has given me a good tip, so I don't spend my money on silly things. I go shopping in all the stores – everything that I like and think I want to buy, I put on hold and then I walk away. This gives me time to think out everything I have on hold and what it is that I really want. I have my own bank card, but it has a limit on it which is good, so I don't spend too much money. I get to make my own decisions, and if I am really not sure about what I am supposed to buy, or if it's a good idea, I can always call Mom.

Life with FAS doesn't happen without medication, and I'm taking some to help me with my challenges. One medication is to help me with my allergies, one is for my ADHD, to stay focused, and one is to help me sleep at night. I don't really enjoy taking these pills every day, but I have to.

I would like to make sure that I want to get across is that, even though I have FAS, I am still a person and need to be treated with respect. I have dreams, I have hopes, and with the help from my family, special friends and teachers, I am hoping to have a great life.

I am going to apply to Northern Alberta Institute of Technology and hope to get in next term, so I can learn more things and meet more people.

I miss my Papa very much, and I hope that I can make him very proud of me, because I loved him lots. This is not the end of my story – it's only the beginning of a brand new chapter. My Journey is still going on, and I'm looking forward to many more successes and a good life, thanks to all the chances I have been given rather than the challenges I have.

Mom wanted to add that she is very proud of me, and knows that I am here for a special reason. She wants the rest of the world to know that I am special and will show the world that, with love and understanding, anything can be accomplished.

Appendix: FASD Consensus Statement of the Jury

Acknowledgments

The Honourable Anne McLellan (LL.M, King's College, University of London; Alberta Institute for American Studies, University of Alberta; Academic Director and Distinguished Scholar in Residence, Institute for United States Policy Studies) led a distinguished jury of citizens and experts to develop practical recommendations on how to improve prevention, diagnosis, and treatment of Fetal Alcohol Spectrum Disorder (FASD).

Expert Chair, **Dr Gail Andrew** (MDCM FRCP(C); Member, Board of Directors, Canada Northwest FASD Research Network; Medical Site Lead–Pediatrics, Medical Director–FASD Clinical Services and Pediatric Consultant, Pediatric Programs, Glenrose Rehabilitation Hospital) led a panel of experts in presenting available scientific evidence on FASD.

Process

This consensus statement was prepared by an independent jury of health professionals, academics, and public representatives based on: (1) Relevant published studies assembled by the scientific committee of the consensus development conference; (2) Presentations by experts working in areas relevant to the conference questions; (3) Information by people who have been touched by FASD; (4) Questions and comments from conference attendees during open discussion periods; and (5) Closed deliberations by the jury.

The conference was held in the province of Alberta, Canada. The consensus statement therefore often refers to the situation in Alberta, although data were not only drawn from that area, but also from other parts of Canada, the US, and internationally.

This statement is an independent report of the jury, and is not a policy statement of the conference partners, conference sponsors, or the Government of Alberta.

Fetal Alcohol Spectrum Disorder–Management and Policy Perspectives of FASD. Edited by Edward P. Riley, Sterling Clarren, Joanne Weinberg, and Egon Jonsson
Copyright © 2011 WILEY-VCH Verlag GmbH & Co. KGaA, Weinheim
ISBN: 978-3-527-32839-0

Conference Questions

The jury used the evidence presented to them at the conference to determine answers – in the form of a consensus statement – to the following questions:

1) What is FASD, and how is it diagnosed?
2) Do we know the prevalence and incidence of FASD in different populations, and can the reporting be improved?
3) What are the consequences of FASD for individuals, their families and society?
4) How can FASD be prevented?
5) What policy options could more effectively support individuals with FASD and their families across the lifespan?
6) What further research into FASD is needed?

Introduction

Fetal Alcohol Spectrum Disorder (FASD) is an umbrella term used to describe the range of disabilities caused by prenatal exposure to alcohol. It is a significant Canadian health concern, and concerted action is required from all levels of government, researchers, communities, families and individuals if we are to deal with it effectively.

Compared to many other areas of study, FASD is relatively new. It was first identified in 1973, when a pattern of malformations among infants born to alcohol abusing women was noted. Since then research has been conducted and knowledge gained, and it is time to move forward, building on the good work that has been done by researchers, clinicians, and communities across the country.

It would be a simple but short-sighted strategy to say to all women of child-bearing years – *"just don't drink alcohol."* That would ignore the complexity of the lives of women and their families and the communities in which they live. This is not only a "women's" issue; it is one for which all of us, women and men, mothers and fathers, families and communities, need to take responsibility.

The difficulties for Canadian families living with FASD cannot be overestimated, and actions to support them should be comprehensive, integrated and timely. Multidisciplinary assessment and multisectoral responses are necessary.

It is time for a National Agenda integrating research done and lessons learned. The Agenda must increase awareness of FASD and promote the development of effective prevention and treatment programs, as well as family support systems. The time for further action is now.

Question 1

What is FASD?
FASD refers to a complex range of brain injuries that can result from Prenatal Alcohol Exposure (PAE). It is an umbrella term that has evolved over time, and

is used to denote an array of developmental, physical, learning and behavioral conditions.

The bottom line is that PAE, in combination with other risk factors, may cause brain injuries, which are expressed in unique and individual ways.

FASD can occur in all segments of society. Poverty, genetics, maternal stress, poor nutrition and other prenatal exposures can influence the severity of FASD. PAE, while not the sole component contributing to FASD, is a necessary one, and therefore FASD is preventable.

How is FASD diagnosed?

There are no definitive biological markers for FASD, such as a blood test or the use of imaging technology. While there is promising research in a number of areas, there is as yet no definitive or cost-effective test. National guidelines for the diagnosis of FASD were accepted across Canada in 2005, and involve a comprehensive multidisciplinary assessment of brain function. The challenge with diagnosis is not simply to identify brain injury but to assess a person's ability in the exercise of judgment, planning, memory and the ability to cope independently in day-to-day life.

The National Guidelines for FASD Diagnosis include demonstrated maternal alcohol consumption during pregnancy, physical examination for growth and physical features, and neurodevelopmental assessment. Early identification and diagnosis can support better interventions and can affect long-term outcomes. However, there are shortcomings with respect to the current system and ensuring consistent implementation of the guidelines. There are also administrative challenges, including:

- Limited human and financial resources for neurodevelopmental assessment across the country.
- Cost in both time and resources.
- Lack of training of personnel in conducting assessments.

In addition, stigma can create barriers to active participation and accuracy of diagnosis (shame and blame).

Recommendations:

1) There is a need for national funding for research to develop accurate and cost-effective neurobiological and/or functional markers of FASD.
2) Comprehensive diagnostic capability needs to be available across the lifespan.

Question 2

Do we know the prevalence and incidence of FASD in different populations, and can the reporting be improved?

There is clearly a need for major improvement in the reporting of FASD. The current provincial/territorial and national estimate for FASD in Canada is nine cases per 1000 infants born. This is based on an extrapolation of US data. Some recent international data have been gathered from a variety of in-school screening and diagnosis studies, which suggest that the overall incidence may be higher. There is Canadian data that indicate a greater prevalence in rural communities, foster care systems, juvenile justice systems and Aboriginal populations. The high prevalence of FASD in Aboriginal populations is symptomatic of a historical and multigenerational trauma, associated with events such as the residential school system.

Obtaining accurate information is extremely important, as the details of regional and local prevalence and incidence are important to target and determine the effectiveness of prevention and intervention efforts.

Surveillance and screening tools need to be simple, cost-effective, and accurate before they can be effectively implemented across the country. A starting point is accurate data regarding prenatal exposure to alcohol. There are impediments to the collection of these data, including stigma, reluctance of care providers, and limited availability of support services once prenatal exposure has been identified.

Recommendations:

1) A national surveillance strategy needs to be implemented to assess progress in the prevention and treatment of FASD in Canada. Questions on FASD should be included in the regular Canadian Community Health Survey.

2) Registries of nonpersonalized data for FASD surveillance should be established in each province to increase the capacity for screening, diagnosis, and reporting of FASD nationally. This should build on projects already under way regionally and provincially to increase data collection.

3) Reliable methods of early detection of developmental delays and disorders, including FASD, should be introduced into early school years, and be available throughout the lifespan.

4) A strategy to reduce the stigma associated with a diagnosis of FASD is needed in order to ensure maximum participation in screening, prevention and diagnostic processes.

Question 3

What are the consequences of FASD for individuals, their families, and society?

The consequences of FASD are widespread, affecting individuals, their families, communities, and society as a whole. FASD is a highly heterogeneous disorder.

Individuals with FASD are most directly affected. FASD, as a brain disorder, is associated with a high incidence of cognitive and behavioral problems. People affected by FASD may have significant difficulties with memory, attention, self-care, decision making and social skills, as well as mental health disorders including depression and addiction. They may have problems with organization and planning their activities, difficulty controlling their emotions, and completing tasks that would allow them to lead productive lives. FASD is often further complicated by medical issues, including a higher rate of heart disease, hearing, and vision problems.

FASD has a dramatic impact on families, whether it is the biological, adoptive, or foster family. Families must be aware that there will be additional costs in raising a child affected by FASD, and this may cause additional family stress. The biological mother is dramatically impacted, regardless of whether she is raising the child or not. Guilt can be considerable. When a woman is under stress or depressed, she may continue to drink; indeed, a major risk factor for having an alcohol affected child is having a previous child affected with FASD.

FASD also affects all other members of the immediate family, including siblings and the extended family. Emotional, financial, and social burdens can be considerable. Indeed, the stress of living with a child affected with FASD may result in family discord or breakup. Adoptive and foster families confront similar issues in dealing with the needs of affected children. Again, proper supports are essential.

One cost that is more difficult to measure is that of lost human potential. The needs of individuals affected by FASD currently generate considerable costs for the social welfare, educational, medical, judicial and correctional systems and significant challenges for communities.

In Alberta, the annual economic cost of FASD is estimated to be between $130–400 million per year. Of this total, educational and medical costs take up 60% (including addictions and drug treatments), additional costs to families account for 20%, and the remaining 20% is for social services, supportive housing, lost productivity costs and other services, such as costs to the justice system. Clearly, addressing this issue is crucial not only from the perspective of social justice but also from the economic perspective.

Recommendations:

1) As FASD is a lifelong disability, there should be a commitment by governments to provide seamless and equitable services across the lifespan.

2) Important transitions from child to adult services need to be pre-planned and allow for effective wraparound services which will support individuals and families and communities at each stage of life.

3) People affected with FASD will require lifelong intersectoral services. Consistent standards between provinces should be established to reduce variations in the funding and provision of these services.

4) Adults with FASD will require ongoing life skills and socialization assistance and support.

5) For those children who enter the child welfare system, there should be improvements to ensure the ability to provide stable foster care. Multiple placements should be avoided wherever possible.

Question 4

How can FASD be prevented?

FASD prevention requires complex, culturally sensitive, multilevel initiatives that address very specific barriers and opportunities for learning, engagement and supportive change. To successfully prevent FASD, it is critical to involve women, men, their support systems, community advocates, health promotion experts, researchers, health/social system planners, and service providers in designing these initiatives.

The Canadian Prevention Framework describes four levels of FASD prevention: (1) Raising awareness for the whole population; (2) Discussing alcohol use with all girls and women of childbearing age; (3) Reaching and providing specialized care and support to girls and women who use alcohol during pregnancy; and (4) Supporting new mothers with alcohol problems.

Coordination and integration of prevention strategies must occur at all levels. We must learn from experience and build on and use existing umbrella strategies where available.

"Shame and blame" approaches to FASD prevention result in many missed opportunities to provide women with the timely, appropriate, and respectful supports needed to reduce the negative impacts of their alcohol use on their health and that of their children. Systems must be meaningful, effective, and compassionate in responding to the challenge of FASD and its prevention.

Because of the negative human and economic impacts of FASD, prevention is a good public investment.

Recommendations:

1) A national primary prevention strategy must include a clear message consistent with Canadian values. This should include education about the effects and risks of alcohol beginning in elementary school and continuing through post-secondary education. It should also include education about birth control.

2) Prevention programs should target the Social Determinants of Health.

3) Prevention programs should be designed with built-in evaluation frameworks.

4) Prevention efforts should be community driven, culturally appropriate, and should honor traditional knowledge. This is especially true in Aboriginal and immigrant communities.

5) A high priority should be placed on ensuring that prevention services are provided to women and families at highest risk of having a child with FASD. The Parent–Child Assistance Program (PCAP) has shown great success. Canadian programs based on the PCAP model should be encouraged.

6) Governments should require messaging about FASD in pregnancy testing kits and in contraceptive packages.

7) The reforms being made in primary care have the potential to improve the relationships required for effective prevention and support. Physicians and other primary health providers should take full advantage of "teachable moments" to discuss pregnancy prevention and the risks of alcohol consumption with their patients/clients of child-bearing age.

8) Increase the number of women-centered alcohol treatment programs and beds. Keeping mothers and children together during interventions should be a priority.

9) National, provincial and territorial alcohol strategies must address FASD.

Question 5

What policy options could more effectively support individuals with FASD and their families across the lifespan?

Ideally, policies and programs should reflect evidence-based best practices. Unfortunately, evidence is not yet available on how to best support individuals affected by FASD and their families across the lifespan. However, there are many examples of promising practices which may well be helpful and cost-effective. These programs should be nurtured and shared, within the context of evaluation, so that the findings can inform future service delivery. Such evaluations should ensure the outcomes assessed are linked to functional improvements in the lives of those living with FASD and their families.

The heterogeneity of FASD requires the ability to tailor services to the needs of the individuals and their families, and recognize that these needs may change over time. Arbitrary eligibility criteria such as IQ, chronological age and place of residence (e.g. rural/urban, on/off reserve) are counter-productive and can be unjust. FASD is a lifelong condition and special attention needs to be paid to key points of transition.

Since FASD involves so many sectors, an inter-disciplinary approach is critical. Currently, different approaches may be taken by social services, education, health, the courts and the corrections system. The resulting fragmentation can be frustrating to people affected by FASD and expensive for tax payers. There should be '*no wrong doors*' for people affected by FASD who need support; mechanisms need to be in place to ensure such support is seamless. Services should address cumulative risk, both environmental and biological, and not be based on silos of care. Policy and services also need to be culturally sensitive.

Students affected by FASD continue to show low rates of school completion, high rates of suspension, poor academic achievement, and limited positive social involvement.

Numerous studies have identified the presence of adult offenders with FASD in Canada's correctional systems. The range and complexity of community re-entry needs of people affected by FASD require interdisciplinary and multisectoral approaches to connect them with services and supports that match their functional capacity. Introducing services while in jail has the ability to increase the effectiveness of connecting released offenders to community resources.

While services and treatments for FASD have unique characteristics, they should be part of a larger system of delivering supports for people with disabilities. The services to individuals and their families should be needs-based. Lessons from effective evidence-based approaches dealing with other developmental disorders should be adopted where appropriate to FASD, and *vice versa*. How best to place FASD programming within the broader framework of services for individuals with developmental and behavioral challenges is a matter of debate.

Recommendations:

Lifelong Services

1) Services should be: (a) based on functional need rather than arbitrary eligibility criteria; (b) lifelong; (c) seamless; (d) individualized; (e) culturally sensitive; and (f) sustainably funded.

2) There should be funding for systematic evaluations of programs and sharing of findings to develop best practices.

3) Build communities of support for individuals affected by FASD and their families. In particular, encourage mentorship and activity-based programs.

4) There should be special attention and support for First Nations, Inuit and Metis peoples affected by FASD who have experienced societal breakdown due to historical and multigenerational trauma.

Diagnosis/Assessment

1) There must be equitable and timely access to diagnosis for individuals with suspected FASD, including appropriate communication of findings with the individual, family, and other service providers.

2) Functional reassessments should be undertaken as needed.

3) Amend the Criminal Code to allow for Court-ordered assessments, including FASD.

Education and Training

1) Educational instruction and materials should be provided to promote awareness, understanding and knowledge of best practices for those who are or will be working with people affected by FASD.

2) An individualized educational plan needs to be developed focusing on skill development, inclusion, participation and recognition of existing strengths, to facilitate becoming a contributing member of society.

Legal

1) Improve outcomes and reduce costs to the legal system by utilizing an alternative measures program for adults affected by FASD charged with nonviolent

first offences. Take into account a FASD diagnosis on subsequent nonviolent offences when sentencing.

2) Improve outcomes and reduce cost to the youth justice system by utilizing an alternative measures program for all young offenders affected by FASD, charged with nonviolent offences.

3) Pre-release and post-release programs for individuals need to be established.

4) Enhance the correctional environments to respond to the special needs of persons affected by FASD to protect them from exploitation and abuse.

Question 6

What further research into FASD is needed?

We still do not understand all of the basic mechanisms that create the spectrum of severity within FASD. Effects of exposure are highly variable. For example, the spectrum lies on a continuum from still birth to children with subtle learning and behavioral problems. Nor do we know enough about the factors that may magnify or reduce the risk from prenatal exposure.

What we do know is that research more than pays for itself. It reduces costs by more effective prevention, intervention, and treatment of FASD.

Current FASD research activities across Canada remain fragmented and underfunded, and this leads to a risk of omission or duplication of effort. There remain significant gaps in our knowledge, including outcomes across the lifespan and for special populations. Research is still in its infancy for corrections, justice and social services, and is not comprehensive with respect to rates, outcomes, costs and co-occurring conditions. We continue to incarcerate the disabled!

We need new approaches to research to reflect current realities. We must embed research into service delivery. Research must include all levels of evidence including traditional knowledge. Rigorous and culturally appropriate research should include both quantitative and qualitative methodologies. Involvement and participation of the community in research is essential.

We need to support the ongoing development of interdisciplinary, integrated research networks that include health, education, social services, corrections, and communities. Active population-based monitoring must be in keeping with ethical and privacy standards.

Research must include basic as well as translational studies which are relevant to the lived experiences of persons affected by FASD. It needs to encompass issues in diagnosis, interventions, and all levels of prevention. Such research must be culturally appropriate, and address the needs of the individual, family, community, and the nation.

Recommendations

1) More translational research from basic science to the human experience, such as the beneficial effects of nutrition.

2) More research on prenatal alcohol exposure on brain structure and function, with the aim of improving interventions and outcomes.

3) More reliable biological and/or behavioral indicators of maternal alcohol consumption during pregnancy are needed.

4) Ongoing research for the development of better screening and surveillance tools that are specific and sensitive to prenatal alcohol exposure. These should be adaptable, culturally appropriate and lead to accurate referrals for diagnosis and supports.

5) Encourage uniform approaches to recording clinical findings found during FASD assessments by using standardized forms and definitions.

6) Support intersectoral research with education, health, and social services.

7) More research between corrections, justice and social services to identify rates, outcomes, costs, and co-occurring conditions.

8) Initiate research into the role of parents, including fathers.

9) Embed research and evaluation into programs and services to allow for self-correction and continuous improvement.

10) Promote research on interventions based on social determinants of health that could modify the incidence and severity of FASD.

Conclusion

We, the jury, believe that comprehensive and lifelong services for people affected by FASD can, and must, be improved. Ongoing prevention efforts must be expanded.

FASD is a complex issue. It has profound short- and long-term consequences for individuals, families, and communities. A multidisciplinary and multisectoral approach is needed if we are to improve the lives of those living with and affected by FASD.

To be successful, we need to be informed as much as possible by research and evidence but, at the end of the day, we must remember we are dealing with individual persons and families. We must respond with compassion for the challenges that they face and respect for the unique capabilities that they bring.

Government ministries and health systems owe it to everyone to fund and develop programs and explore new ways to help families, researchers and service providers to address this important issue. It is just. It makes sense. It is an investment in our future.

We, the jury, believe a national agenda to address FASD and its prevention is necessary, and are pleased that Alberta is a leader in those efforts.

Jury Members

Chair: The Honourable Anne McLellan, Alberta Institute for American Studies, University of Alberta; Academic Director and Distinguished Scholar in Residence, Institute for United States Policy Studies; LL.M, King's College, University of London

Judith Bossé, Associate Assistant Deputy Minister, Public Health Agency of Canada

Jennifer Coppens, Medical Student, University of Alberta

Raisa Deber, Professor, Department of Health Policy, Management and Evaluation, Faculty of Medicine, University of Toronto; Director, CIHR Team in Community Care and Health Human Resources

David Elton, President, Norlien Foundation and Max Bell Foundation

Mark Hattori, Acting Assistant Deputy Minister, Program Quality and Standards, Alberta Children and Youth Services

James Hees, Reporter, CBC Radio Edmonton

Malcolm King, Professor, Department of Medicine, University of Alberta; Scientific Director, CIHR Institute of Aboriginal Peoples' Health

Christine Loock, Professor, Department of Pediatrics, Faculty of Medicine, University of British Columbia; Developmental Pediatrician, Children's and Women's Centre of British Columbia

Rebecca Martell, Clinical Associate, Occupational Performance Analysis Unit (OPAU), Department of Occupational Therapy, University of Alberta

Edward Riley, Distinguished Professor, Psychology; Director, Center for Behavioral Teratology, San Diego State University

Marguerite Trussler, Chairperson, Alberta Liquor and Gaming Commission

Lee Ann (Weaver) Tyrrell, Project Manager, (Initial) Alberta/Prairie Province FASD Strategy; (First) Director, Yellowhead Tribal Services Agency; Retired

Conference Speakers and Topics

What is Fetal Alcohol Spectrum Disorder (FASD) and how is it diagnosed? An overview of FASD
Gail Andrew, *Member, Board of Directors, Canada Northwest FASD Research Network; Medical Site Lead–Pediatrics, Medical Director–FASD Clinical Services and Pediatric Consultant, Pediatric Programs, Glenrose Rehabilitation Hospital*

A personal perspective
Myles Himmelreich, *Director of Programming, Canadian FASD Foundation*

Do we know the prevalence and incidence of FASD in difference populations, and can the reporting be improved? Prevalence and incidence in Alberta and Canada
Suzanne Tough, *Scientific Director, Alberta Centre for Child, Family and Community Research*

Prevalence and incidence internationally
Philip May, *Professor of Sociology and Family and Community Medicine, University of New Mexico; Senior Research Scientist, Center on Alcoholism Substance Abuse, and Addictions (CASAA)*

Extent and impact on child development
Ben Gibbard, *Developmental Pediatrician, Alberta Children's Hospital; Assistant Professor, Department of Pediatrics, Faculty of Medicine, University of Calgary*

Prevalence of FAS in Foster Care
Susan Astley, *Professor of Epidemiology/Pediatrics, University of Washington; Director, Washington State Fetal Alcohol Syndrome Diagnostic and Prevention Network*

Genetic predisposing factors
Albert Chudley, *Medical Director, Winnipeg Regional Health Authority Program in Genetics and Metabolism; Professor, Department of Pediatrics, University of Manitoba*

Direct and indirect mechanisms for alcohol damage to the brain
Joanne Weinberg, *Professor and Distinguished University Scholar and Acting Department Head, Cellular and Physiological Sciences, University of British Columbia*

What are the consequences of FASD for individuals, their families, and society? Economic implications for individuals and families
Philip Jacobs, *Professor, Gastroenterology Division, Department of Medicine, University of Alberta; Director of Research Collaborations, Institute of Health Economics*

Consequences on the community
Mary Berube, *Director, Intergovernmental Initiatives, Ministry Support Services Division, Alberta Children and Youth Services*

Impact on system usage within foster care
Linda Burnside, *Executive Director, Disability Programs, Manitoba Family Services and Housing*

Co-morbidities with mental health for an individual with FASD
Dan Dubovsky, *FASD Specialist for the Substance Abuse and Mental Health Services Administration (SAMHSA), FASD Center for Excellence*

Efficacy of a neurobehavioral construct: interventions for children and adolescents with FASD
Diane Malbin, *Executive Director, Fetal Alcohol Syndrome Consultation, Education and Training Services Inc. (FASCET)*

How can FASD be prevented? Preconception initiatives
Lola Baydala, *Associate Professor of Pediatrics, University of Alberta; Misericordia Community Hospital*

Inventory of primary prevention campaigns
Robin Thurmeier, *FASD Resources Researcher, Saskatchewan Prevention Institute*

Primary care physician perspective
June Bergman, *Associate Professor, Department of Family Medicine, Faculty of Medicine and Dentistry, University of Calgary*

Mentoring programs for at-risk mothers
Nancy Whitney, *Clinical Director, King County Parent-Child Assistance Program, University of Washington*

Addressing FASD as a women's health issue
Amy Salmon, *Managing Director, Canada Northwest FASD Research Network; Clinical Assistant Professor, School of Population and Public Health, Faculty of Medicine, University of British Columbia*

Prevention of FASD: a broader strategy in women's health
Nancy Poole, *Research Associate, British Columbia Centre of Excellence for Women's Health; Research Consultant, Women and Substance Use Issues, British Columbia Women's Hospital*

What policy options could more effectively support individuals with FASD and their families across the lifespan? Educational system, parental, and community support
Frank Oberklaid, *Director, Centre for Community Child Health, Royal Children's Hospital and Professor Pediatrics, University of Melbourne*

Shifting responsibility from the individual to communities of care
Audrey McFarlane, *Executive Director, Lakeland Centre for Fetal Alcohol Spectrum Disorder*

Education policy directions for supporting children and youth with FASD, and their families
Elizabeth Bredberg, *Director, Society of the Advancement of Excellence in Education*

Development of life skills: education, parenting, and family mentoring
Claire Coles, *Professor, Department of Psychiatry and Behavioral Sciences and Pediatrics, Emory University School of Medicine; Director, Fetal Alcohol and Drug Exposure Clinic, Marcus Autism Center, Children's Health Care of Atlanta*

Life stages and transitions
Brenda Bennett, *Executive Director, FASD Life's Journey Inc.*

Social services and corrections
Sharon Brintnell, *Professor, Department of Occupational Therapy, and Director, Occupational Performance Analysis Unity, Faculty of Rehabilitation Medicine, University of Alberta*

Treatment for FASD
John McLennan, *Assistant Professor, Departments of Community Health Sciences, Psychiatry, and Paediatrics, University of Calgary*

Justice issues
Mary Kate Harvie, *Associate Chief Judge, Provincial Court of Manitoba*

Policy development and FASD
Dorothy Badry, *Assistant Professor, Faculty of Social Work, University of Calgary*

What further research into FASD is needed? Health and social policy
Sterling Clarren, *CEO and Scientific Director, Canada Northwest FASD Research Network; Clinical Professor of Pediatrics, School of Medicine, University of Washington;*
    Clinical Professor of Pediatrics, Faculty of Medicine, University of British Columbia;

Focusing research efforts ... Where?
Sara Jo Nixon, *President, Research Society on Alcoholism. Fellow, Division 28 and 50, American Psychological Society; Professor and Chief Division of Addictions Research; Director, Neurocognitive Laboratory, Department of Psychiatry, College of Medicine, University of Florida*

Child health and well-being
Bruce Perry, *Senior Fellow, The Child Trauma Academy*

Planning Committee

Egon Jonsson, Executive Director and CEO, Institute of Health Economics

Amanda Amyotte, Project Officer, Alberta Children and Youth Services

Mary Berube, Director, Intergovernmental Initiatives, Ministry Support Services Division, Alberta Children and Youth Services

Laurie Beverley, Executive Director, Community Treatments and Supports, Alberta Health Services

Jewel Buksa, President, BUKSA Conference Management and Program Development

Corine Frick, Program Director, Alberta Perinatal Health Program

Tara Hanson, Director of Operations, Alberta Centre for Child, Family and Community Research

Braden Hirsch, Acting Director, Community Partnerships, Alberta Seniors and Community Supports

Marty Landrie, Interim Executive Director, Poundmaker's Lodge

Rhonda Lothammer, Communications Manager, Institute of Health Economics

Thanh Nguyen, Health Economist, Institute of Health Economics

Julie Peacock, Director, Primary Care, Children and Youth Interventions, Alberta Health Services–Addiction and Mental Health

Nancy Reynolds, President and Chief Executive Officer, Alberta Centre for Child, Family and Community Research

Kesa Shikaze, Project Manager, Healthy Living, Alberta Health and Wellness

Rob Skrypnek, Sumera Management Consulting

John Sproule, Senior Policy Director, Institute of Health Economics

Melissa Waltner, Executive Assistant, Institute of Health Economics

Scientific Committee

Gail Andrew, Member, Board of Directors, Canada Northwest FASD Research Network; Medical Site Lead–Pediatrics, Medical Director–FASD Clinical Services, and Pediatric Consultant, Pediatric Programs, Glenrose Rehabilitation Hospital

June Bergman, Associate Professor, Department of Family Medicine, Faculty of Medicine and Dentistry, University of Calgary

Sterling Clarren, Chief Executive Officer and Scientific Director, Canada Northwest FASD Research Network; Clinical Professor, Pediatrics, School of Medicine, University of Washington; Clinical Professor, Pediatrics, Faculty of Medicine, University of British Columbia

Corine Frick, Program Director, Alberta Perinatal Health Program

Denise Milne, Senior Manager, FASD Initiatives/ Children's Mental Health, Alberta Children and Youth Services

Hannah Pazderka, Director of Research, CASA Child, Adolescent and Family Mental Health

Nancy Reynolds, President and Chief Executive Officer, Alberta Centre for Child, Family and Community Research

John Sproule, Senior Policy Director, Institute of Health Economics

Bonnie Stonehouse, Coordinator, Program Development for Persons with Disabilities, Alberta Seniors and Community Supports

Melissa Waltner, Executive Assistant, Institute of Health Economics

Communications Committee

Roxanne Dubé Coelho, Public Affairs Officer, Alberta Children and Youth Services

Rhonda Lothammer, Communications Manager, Institute of Health Economics

Jewel Buksa, President, BUKSA Conference Management and Program Development

Disclosure Statement

All of the jury members who participated in this conference and contributed to the writing of this statement were identified as having no financial or scientific conflict of interest, and all signed forms attesting to this fact. Unlike the expert speakers who present scientific data at the conference, the individuals invited to participate on the consensus panel are reviewed prior to selection to ensure they are not proponents of an advocacy position with regard to the topic.

Questions or comments can be directed to:

Institute of Health Economics
1200, 10405 Jasper Avenue
Edmonton Alberta, Canada T5J 3N4
Tel: 780 448 4881; Fax: 780 448 0018
info@ihe.ca
www.ihe.ca

Institute of Health Economics

The Institute of Health Economics (IHE) is an independent, not-for-profit organization that performs research in health economics and synthesizes evidence in health technology assessment to assist health policy making and best medical practices.

IHE Board of Directors

(as of October 2009)

Chair

Dr Lorne Tyrrell – Professor and CIHR/GSK Virology Chair, University of Alberta

Government

Ms. Linda Miller – Deputy Minister, Alberta Health and Wellness

Ms Annette Trimbee – Deputy Minister, Advanced Education and Technology

Dr Jacques Magnan – Acting President and CEO, Alberta Heritage Foundation for Medical Research

Academia

Dr Andy Greenshaw – Associate Vice-President Research, University of Alberta

Dr Tom Feasby – Dean, Faculty of Medicine, University of Calgary

Dr Philip Baker – Dean, Faculty of Medicine and Dentistry, University of Alberta

Dr James Kehrer – Dean, Faculty of Pharmacy and Pharmaceutical Sciences, University of Alberta

Dr Tom Noseworthy – Professor and Head, Community Health Sciences, University of Calgary

Dr Herb Emery – Professor, Department of Economics, University of Calgary

Dr Doug West – Chair, Department of Economics, University of Alberta

Industry

Mr Terry McCool – Vice President, Corporate Affairs, Eli Lilly Canada Inc.

Mr Gregg Szabo – Vice President, Corporate Affairs, Merck Frosst Canada Inc.

Dr Bernard Prigent – Vice President and Medical Director, Pfizer Canada Inc.

Mr Grant Perry – Vice President, Public Affairs, GlaxoSmithKline Inc.

Mr William Charnetski – Vice President, Corporate Affairs and General Counsel, AstraZeneca Canada Inc.

Other

Mr Doug Gilpin – Chair, Audit and Finance Committee

CEO

Dr Egon Jonsson – Executive Director and CEO, Institute of Health Economics, Professor, University of Alberta, University of Calgary

FASD Research and Resources

Ongoing research and evaluation of programs will help to determine best practices for preventing FASD and supporting those already affected.

The Institute of Health Economics is currently working on a series of project related to FASD. This conference also is part of a series of consensus development conference produced by the Institute. Visit www.ihe.ca for more information.

The Government of Alberta's Fetal Alcohol Spectrum Disorder Cross-Ministry Committee has a comprehensive website with extensive resources on FASD. Visit their site at www.fasd-cmc.alberta.ca.

The Public Health Agency of Canada has also developed a site to provide basic information on FASD and to report what they are doing in the area. Visit their site at www.phac-aspc.gc.ca/fasd-etca.

Index

a

aboriginal communities 34, 268
– assessment and diagnosis 358
accountability 320
acetaldehyde dehydrogenase (ALDH) 112
active treatment hospital 262
activity-dependent neuroprotective protein 76
activity-dependent neurotrophic factor (ADNF) 76
Adaptations Committee 154
addiction
– neurobiology 91
Addiction Severity Index (ASI) 200
ADHD, see attention deficit hyperactivity disorder
adolescence 264
adrenocorticotropic hormone (ACTH) 79ff.
adult 354
– assessment and diagnosis 359
– correctional environment in Canada 238
– intervention after release 244
– policy consideration 251
– supporting gain and maintain employment 361
adult offender
– disability and social support 233ff.
adulthood 265
agoraphobia 164
Alberta
– community 354
– cross-ministry approach to FASD across the lifespan 353ff.
– FASD Conference 364
alcohol
– cell adhesion 76
– cell–cell interaction 76
– direct effect on fetus 74
– indirect effect on fetus 75

– oxidative stress 76
– prostaglandin 76
alcohol dehydrogenase 5, 112
– *ADH1B* 112ff
– *ADH1C* 112
– *ALDH2* 112f.
– human isoenzyme 113
– polymorphism 18, 113f.
alcohol metabolism
– genetic factor 111
alcohol teratogenesis
– direct and indirect mechanism 73ff.
– implication for understanding alterations in brain and behavior in FASD 73ff.
– maternally mediated 73
alcohol-exposed birth
– prevention 199
alcohol-exposed pregnancy
– prevention 202
Alcohol-Related Birth Defect (ARBD) 4, 129, 355
Alcohol-Related Neurodevelopmental Disorder (ARND) 1ff., 20, 128, 355
aldehyde dehydrogenase (ALDH)
– polymorphism 18
Alexis Nakota Sioux Nation 153f.
Alexis Working Committee 153f.
appraisal process 186
ARBD, see Alcohol-Related Birth Defect
arginine vasopressin (AVP) 79
ARND, see Alcohol-Related Neurodevelopmental Disorder
assessment 358
– strengths and challenges 229
assessment team 134
attention deficit hyperactivity disorder (ADHD) 138, 278, 375ff.
AUDIT (Alcohol Use Disorders Identification Test) 203

Fetal Alcohol Spectrum Disorder–Management and Policy Perspectives of FASD. Edited by Edward P. Riley, Sterling Clarren, Joanne Weinberg, and Egon Jonsson
Copyright © 2011 WILEY-VCH Verlag GmbH & Co. KGaA, Weinheim
ISBN: 978-3-527-32839-0

AVENTA Addiction Treatment for Women Demonstration Project   361
awareness   181, 357
– generating   358

b
behavior
– alteration in FASD   73
behavioral change   189
behavioral despair   89
Berksonian bias   373
birth
– policy development   261
blood alcohol concentration (BAC)   17
Born Free Campaign   178
brain
– alteration in FASD   73
British Columbia's Key Worker and Parent Support Program   297ff.
– challenge   308
– model and components   298
– perception   307

c
campaign evaluation   182
campaign implementation   189
Canada
– correctional environment for adults with FASD   238
– cost   47ff.
– current state of FASD primary prevention in north-western Canada   176
– FASD   27ff.
– The Young Offenders Act (YOA)   216
Canadian Institute of Health Research (CIHR)   153ff.
capacity building   156
care
– collaborative   166ff.
– effective   396
– family-centered   346
– integrated   396
– supportive   396
– trauma-informed   167
– women-centered   165ff.
– youth with FASD leaving care   280
caregiver   346, 354
– access to services and resources   311
– perceptions of the program   307
– support   310, 360
case manager   197
cell adhesion
– alcohol-induced disruption   76
cell adhesion molecule (CAM)   76

cell–cell interaction
– alcohol-induced disruption   76
central nervous system (CNS)   3
– dysfunction   115
child welfare system   31, 276ff.
– cost for children with FASD   282
childhood
– policy development   261
children
– care   277f.
– disability   277
– FASD   275, 345, 354, 367, 369ff.
– intervention planning   369ff.
– positive behavior support   364
– prevalence of specific difficulties/needs   373
– research project on school experience   367
– risk for being "hard to care for"   345
– separate specialized service for subgroups   370
children in care (CIC)   284ff.
chronic mild stress (CMS)   89
CIHR guidelines for research involving aboriginal people   156
client-centered lifelong multisectoral support   245
Clinic for Alcohol and Drug Exposed Children (CADEC)   220
collaborative approach   312
collaborative care   166
– research   169
community
– corrections and connections   366
community care
– model   330
community development strategy   176
community economic development (CED)   335
community member participation   155
community needs   189
community partner
– perceptions of the program   307
community service provider   198
community support
– student   322
conference
– recommendation   224
connection
– community   366
convergent functional genomics   115
coping appraisal   187
correction
– community   366

correction system 33
correctional system 243
corticosterone (CORT) 80ff.
corticotropin-releasing hormone (CRH) 79
cortisol 83
cortisone 83
cost 45ff.
– adjustment 47
– annual 51
– annual cost per case 55f.
– child welfare care 282
– direct 46
– FASD 45ff., 249
– indirect 46
– lifetime cost per case 56
– method 49f.
– perspective 46
– target population 47
cost effectiveness
– PCAP 201
criminal justice issue 215
cross-ministry approach
– Alberta 353ff.
– lifespan 353ff.
cultural fairness 268
cytochrome P450 (CYP2E1) 112
– polymorphism 113

d

de-medicalization 237
11-dehydrocorticosterone 83
delinquency 378
demographic factor 18
depression 87ff.
– FASD 89f.
– prevention and treatment 90
developmental disabilities (DD) 250f
– assistance 250
diagnosis 330, 358
– accurate 393
– adolescence 140
– adulthood 140
– early infancy 137
– FASD 127ff.
– history 128
– implication 140
– neonatal period 137
– policy consideration 142
– school age 138
– toddlerhood 138
diagnostic assessment 323
disability
– children in care 277
– FASD 234

– preventing secondary disabilities 393
– primary 234
– secondary 235, 393ff.
disability paradigm
– FASD 267
disciplinary measure 323
DNA methylation 85, 117
dopamine (DA) 88
drinks per drinking day (DDD) 19
drug-exposed birth
– prevention 199

e

e-learning module 364
eco-system intervention 339
ecological approach 312
economic impact
– children in care 284
educated workforce 365
education 262, 324, 364f.
– policy 317ff.
– system 318
education professional 324
– FASD 321
educational goal 323
educational programming 325
– student with FASD 325
educational strategy 176
eligibility 319
employment 246
– support 360f.
endocrine balance
– disruption 77
epigenetic regulation
– altered gene expression 84
epigenetic reprogramming of the HPA axis 85
epigenetics 116
ethnicity 18
evaluation 189, 366
– formative findings 301
– method 300
– summative findings 309
external executive function support 248

f

familial factor 18
family 197, 346, 354
family planning 199
family-centered care 346
family-centered model of practice 312
FASD clinical capacity 359
FASD Community of Practice (CoP)
   Research Project 367

FASD Community Outreach Program Demonstration Project   361
FASD cross-ministry committee (FASD-CMC)   353ff.
– strategy   356
FASD prevention   161ff.
– women's health perspective   161
FASD Service Network Program   356
FASD Videoconference Learning Series   362
FASD Youth Justice Program
– recommendation   230
fatty acid ethyl ester (FAEE)   132
Fetal Alcohol Effects (FAE)   128, 176, 355
Fetal Alcohol Spectrum Disorder (FASD)   1ff., 89ff., 355
– alteration in brain and behavior   73
– annual cost per case   55f.
– assessment   229
– awareness   181, 357
– barriers to screening   210
– baseline rate   37
– care for children   277f.
– characteristics   17
– cost   45ff., 249, 282
– cross-ministry approach   353
– depression   89
– diagnosis   127ff., 393
– diagnostic process   133
– disability paradigm   267
– economic impact   284
– education policy   317ff.
– education professional   321
– frequency in Canada   27
– genetic factor   109ff.
– heterogeneity   399
– history   2
– history of diagnosing   128
– human right   237
– impact   5, 275ff., 354
– implication of a diagnosis   140
– incarceration   237
– incidence   27ff.
– intervention   8, 38f.
– knowledge   181
– legal system   212ff.
– lifespan   136, 353
– lifetime cost per case   56
– maternal risk factor   17
– national rate in Canada   29
– policy   317ff.
– policy development for individuals and families across the lifespan   259ff.
– preconception prevention initiative   151
– prevalence   5, 17ff., 27ff., 372
– prevalence from in-school study   21
– prevalence rate   27
– prevention   193ff.
– prevention and treatment of depression   90
– prevention effort   27ff.
– primary healthcare   207ff.
– primary prevention   175
– program   247
– provincial rate in Canada   30
– rate   22
– rate of exposure   35
– research   387ff., 399ff.
– risk   35
– risk factor   4
– risk for specific difficulties   375
– screening   131, 210, 221
– social determinant of health   236
– stress responsiveness   81
– training   247
Fetal Alcohol Syndrome (FAS)   1ff., 17, 176, 355
– annual cost per case   55f.
– baseline rate   37
– diagnosing the effect   3
– history   2
– impact   5
– intervention   8, 38f.
– lifetime cost per case   56
– prevalence   5, 20
– risk   35
– risk factor   4
fetal hypothalamic–pituitary–adrenal (HPA) axis   130
fetal/early programming   82
– HPA axis   84
fetus
– direct effect of alcohol   74
– indirect effect of alcohol   75
First Nations and Inuit support   363
folic acid (FA)   117
formative evaluation findings   301
functional assessment   323
functional behavior analysis   324

g

gender-based analysis   162
gene expression
– altered epigenetic regulation   84
genetic factor   18, 109ff.
– alcohol metabolism   111
– evidence   110

– FASD   109ff.
genomic imprinting   116
glial cell
– peptide secretion   76
glucocorticoid   83
glucocorticoid hormone   80
glucocorticoid receptor (GR)   85
glutathione S-transferase (GST)   114f.

h
harm-reduction   197
– orientation   165
– practice   168
health   389
– social determinant   236, 390
health promoter   188
healthcare
– primary   207ff.
– system   324
healthcare reform
– impact   211
Hedgehog signaling pathway   115f.
heterogeneity
– FASD   399
high-risk mother   193
hippocampus   80
histone modification   85
holistic approach   312
housing   246
HPA dysregulation   81
HPG–HPA interaction   82
human right
– FASD   237
11β-hydroxysteroid dehydrogenase type 2   83
hydroxytryptamine (5-HT) system   84ff.
hypothalamic–pituitary–adrenal (HPA) axis   78ff.
– altered neuroendocrine-immune interaction   84
– dysregulation   402
– epigenetic reprogramming   85
– fetal   130
– fetal programming   84
– monoaminergic neurotransmitter   88
– programming by PAE   82
hypothalamic–pituitary–gonadal (HPG) axis   82

i
icons project   227f.
IHE Consensus Development Conference on FASD   365
imprinting   116

incarceration   237
incidence   27ff.
– FASD   29
incidence rate in Canada   32
individual with FASD   259ff., 354ff.
– need   379
– policy development   259ff.
– support   360
individual education plan (IEP)   324
information gap   217
inter-agency   322
intervention   8, 38f., 333
– adult with FASD after release   244
– effectiveness   371ff.
– special subpopulation   372

j
justice
– skill development   358
– support   363

k
Kaleidoscope Demonstration Project   361
key worker   297ff.
– family- and community-focused activities   303
– program   299
– regional and provincial support   305

l
Lakeland Center for Fetal Alcohol Spectrum Disorder (LCFASD)   327ff.
– future   335
– history   334
– policy consideration   336
learning disability (LD)   318
learning outcome   320
legal strategy   176
legal system   212ff.
legislative context   216
Life Book   227
Life Trajectory Policy Model   269
LifeSkills Training (LST) program   152
lifespan   136, 353
– IHE Consensus Development Conference on FASD   365

m
Manitoba Adolescent Treatment Center (MATC)   220
Manitoba Child Welfare System   275ff.
Manitoba Clinic for Alcohol and Drug Exposed Children   220

Manitoba Fetal Alcohol Spectrum Disorder (FASD) Youth Justice Program   215ff.
– statistical outcome   225
maternal alcohol consumption
– pregnancy   36
maternal risk factor
– FASD   17
maternal substance use   339
maternal–fetal interaction   406
meconium   132
medial PFC (mPFC)   80f.
medical assessment   222
mesocorticolimbic pathway   91
message strategy   179
microRNA (miRNA)   117f.
ministry initiative
– strategic plan   357
mitogen-activated protein kinase (MAPK) signaling pathway   115f.
monoaminergic neurotransmitter   88
– HPA axis   88
Mother Kangaroo Campaign   178ff.
Mothers-to-be Mentorship program   333
motivational interviewing (MI) strategy   196

n

National Registry of Evidence-Based Programs and Practices (NREPP)   152
Neonatal Abstinence Syndrome (NAS)   137
neuroendocrine-immune interaction
– HPA programming   84
neuroscience   403
neurotransmitter   88
– monoaminergic   88
– HPA axis   88
noradrenaline (NA)   88
nutrition   18

o

OAR(S) (own, act and reflect) tool   170
occupational therapist   135
offender   233ff.
– adult   233ff.
– disability and social support   233ff.
– rehabilitation   240
– treatment program   240
outcome   333
ownership, control, access and possession (OCAP)   157
oxidative stress
– alcohol   76

p

P4502E1   112
paraventricular nucleus (PVN)   79f.
parent
– access to services and resources   311
– support   310
Parent Support Program   297ff.
– perception   307
Parent–Child Assistance Program (PCAP)   194, 332, 357
– cost effectiveness   201
– intervention   193ff.
– outcome   200
parent–child interaction therapy (PCIT)   9
parenting support and management (PSM)   9
parenting strategy   310
partial FAS (pFAS)   19f., 128
peptide
– secreted from glial cell   76
perceived response efficacy   186
perceived self-efficacy   187
perceived severity   186
perceived vulnerability   186
phosphodiesterase (PDE) inhibitor   405
physician   135
policy   317ff.
– evaluation of Key Worker and Parent Support Program   297ff.
– social work perspective   339ff.
policy consideration   313, 383
– adult with FASD   251
– LCFSAD   336
policy development
– for individuals and families across the lifespan   259ff.
policy indication   322
policy maker   188
policy recommendation   202
polymorphism
– alcohol dehydrogenase (ADH)   18
– aldehyde dehydrogenase (ALDH)   18
Porsolt forced swim test   89
practice workshop   365
preassessment period   222
prefrontal cortex (PFC)   80f.
pregnancy
– maternal alcohol consumption   36
– prevention of alcohol-exposed pregnancy   202, 339
– social work perspective   339ff.
Prenatal Alcohol Exposure (PAE)   1ff., 87, 127ff.

– depression   87
– history   2
– HPA dysregulation   81
– immune system   86
– intervention   8
– prevention   7
– programming of the HPA axis   82
– stress responsiveness   87
prevalence   27ff.
– FASD   17ff., 31, 372
– in-school study   21
– method-depending rate   20
– special subpopulation   370
prevalence rate in Canada   32
prevention   193, 332, 357
– alcohol-exposed birth   199
– current state of FASD primary prevention in north-western Canada   176
– drug-exposed birth   199
– indicated   194
– mapping   390
– preconception prevention initiative   151
– primary   208, 389f.
– primary prevention measure   184
– secondary   208, 245, 389
– selective   8, 193
– strategy   151ff.
– tertiary   208, 389
– universal   7, 193
– women-centered practice   163
primary care
– FASD   208
primary healthcare
– FASD   207ff.
programming   323
– fetal/early   82
– HPA axis by PAE   82
prostaglandin
– alcohol effect   76
Protection Motivation Theory (PMT)   184f.
psychological assessment   222
psychologist   135

q
QFT (quantity, frequency, timing) factors   17

r
race and ethnicity   18
rational service system
– at-risk children   369
recidivism   240
rehabilitation   240

relational approach   312
relational theory   195
release
– adult with FASD   244
– planning   242
remote communities
– assessment and diagnosis   358
research   366, 387ff., 399ff.
– relationship   155
responsibility
– community   327
– individual   327
role-modeling   198

s
screening   131, 221
– barrier   210
Seattle Birth to Three Program, see Parent–Child Assistance Program
self-efficacy   196
sentencing
– alternative   240
sentencing conference   224
sentencing process   225
Separating Out Services   379
– subpopulation   372
serotonergic system   84
serotonin   88
Service Coordination and Mentorship   362
sex, gender and diversity based analysis (SGDBA)   162
skill development
– justice   358
social marketing strategy   176ff.
social policy response   389
social support
– adult with FASD after release   244
– women's health   390
social work perspective
– policy   339ff.
social worker   134
socioeconomic status   18
Solicitor General and Public Security (SGPS)   358
sonic hedgehog (Shh)   116
special subpopulation   370
– effectiveness of intervention   372
– prevalence of specific difficulties/needs   371
– Separating Out Services   372
speech language pathologist   135
spirituality   18
stages-of-change   196

stakeholder engagement 367
Step-by-Step Demonstration Project 362
strategic planning 366
stress 87ff.
– neurobiology 78
stress responsiveness
– FASD 81
– PAE 87
student
– community support 322
– curriculum 320
– education system 318
– educational programming 325
– eligibility 319
– exclusion 319
– learning disability 318
– learning outcome 320
– special education system 320
substance abuse 91
– treatment 199
substance use
– maternal 339
sudden infant death syndrome (SIDS) 4

t

teenage years 264
teratogenesis 109
"This is Me" – A Tool for Learning About and Working with People Affected by FASD 226
"This is Me Life Book" 227
threat appraisal 187
– process 186

training 247, 364f.
transforming growth factor β (TGF-β) signaling pathway 115
treatment 245
– program 240
trust 197

u

USA
– FAS/FASD cost 51

w

Well Communities–Well Families Demonstration Project 362
With Child/Without Alcohol Campaign 178ff.
women-centered care 165
women's health
– determinant 390
– social support 390
women's health perspective
– FASD prevention 161ff.
WRaP (Wellness, Resiliency and Partnerships) Coaching Demonstration Project 363

y

YCJA (Youth Criminal Justice Act) 216f.
YOA (The Young Offenders Act) 216
youth
– FASD 280, 354
Youth Accommodation Counsel 229